The Effect Of Sterilization Methods

On Plastics And Elastomers

Plastics Design Library

ISBN 1-884207-10-3
Library of Congress Card Number 94-66484

First Printing, May 1994
Second Printing, May 1996
Third Printing, October 1996

Published in the United States of America, Norwich, NY by Plastics Design Library a division of William Andrew Inc.

Comments, criticisms and suggestions are invited, and should be forwarded to Plastics Design Library.

Please Note: Although the information in this volume has been obtained from sources believed to be reliable, no warranty, expressed or implied, can be made as to its completeness or accuracy. Design processing methods and equipment, environment and others variables effect actual part and mechanical performance. Inasmuch as the manufacturers, suppliers and Plastics Design Library have no control over those variables or the use to which others may put the material and, therefore, cannot assume responsibility for loss or damages suffered through reliance on any information contained in this volume. No warranty is given or implied as to application and to whether there is an infringement of patents is the sole responsibility of the user. The information provided should assist in material selection and not serve as a substitute for careful testing of prototype parts in typical operating environments before beginning commercial production.

Manufactured in the United States of America.

Plastics Design Library, 13 Eaton Avenue, Norwich, NY 13815 Tel: 607/337-5080 Fax: 607/337-5090

Table of Contents

Thermoplastic Elastomer

Silicone Rubber

Indices and Appendices

Introduction

This reference publication presents an extensive compilation of how sterilization methods, and their assistant media affect the properties and characteristics of plastics and elastomers. The primary focus of sterilization is killing microorganisms. But the impact of sterilization methods extends beyond the cleaning of medical devices. For instance, irradiation, a widely used sterilization method, is also useful in modifying the physical properties of polymer materials to improve performance. The information contained in this book is therefore not only valuable to those involved in the design, manufacture, and sterilization of medical devices but to many other industries, including cosmetics, food packaging, and agriculture.

The basic physical characteristics of polymers are generally well defined by manufacturers. The effects of sterilization methods, however, are not well compiled; nor is raw data easily accessed. This volume serves to turn the vast amount of disparate information from wide ranging sources (i.e., conference proceedings, materials suppliers, test laboratories, monographs, patents, trade and technical journals) into useful engineering knowledge.

The data provided ranges from a general overview of the compatibility of various plastics and elastomers to sterilization methods to detailed discussions and test results. For users to whom sterilization is a relatively new field, the detailed glossary of terms, including descriptions of test methods, will prove useful. For those who wish to delve beyond the data presented, source documentation is presented in detail. This book also presents data on the resistance of plastics and elastomers to chemicals used in sterilization and to other fluids of medical significance.

In compiling data, the philosophy of Plastics Design Library is to provide as much information as is available. This means that complete information for each test is provided. At the same time, an effort is made to provide information for as many sterilization methods and material combinations as possible. Therefore, even if detailed test results are not available (i.e., the only information available is that a material is resistant or degrades), information is still provided. The belief is that some limited information serves as a reference point and is better than no information. Flexibility and ease of use were also carefully considered in designing the layout of this book. We trust you will greet this reference publication with the same enthusiasm as previous Plastics Design Library titles and that it will be a useful tool in your work.

How a material performs in its end use environment is a critical consideration and the information presented here gives useful guidelines. However, this or any other information resource should not serve as a substitute for actual testing in determining the applicability of a particular part or material in a given end use environment.

How To Use This Book

The first section of this book presents the results of exposure of families of plastics and elastomers to various sterilization methods. Each chapter represents a single generic family. Data appears in textual, tabular, and graphical forms. Textual information is useful as it is often the only information available or the only way to provide an expansive discussion of test results.

Tables and graphs provide detailed test results in a clear, concise manner. Careful study of a table will show how variations in exposure conditions influence a material's physical characteristics. The effect of radiation dose, number of sterilization cycles, post exposure time, post exposure temperature, type of carrier gas and other variables can be garnered from close examination of the tables. Endnotes associated with data in the tables are presented as an appendix and appropriately referenced in the tables.

Graphs are another way of viewing trends in property variable relationships and are used in this book to present additional information. Each table or graph is designed to stand alone, be easy to interpret and provides all relevant and avaiable details of test conditions and results. The information's source is referenced to provide an opportunity for the user to find additional information. The source information might also help to indicate any bias which might be associated with the data.

Resistance of plastics and elastomers to chemicals used in sterilization and to other fluids of medical significance is included in the Chemical Resistance Tables. Test results on a range of chemicals from cold sterilants and disinfectants to carrier gases and boiler water additives are given. Each row in the Chemical Resistance Tables represents one specific reagent and associated test conditions for the given material. This data is a subset of information contained in other Plastics Design Library products which are detailed on the inside back cover of this book. More detailed information on specific tests and exposure to many more reagents can be found in these publications.

The following table explains different fields from the Chemical Resistance Tables found in this book:

FIELD HEADING	DESCRIPTION
REAGENT	environment to which the plastic was exposed
REAGENT NOTE	provides additional information about the reagent and conditions of exposure
CONC.	% concentration of the given reagent
TEMP.	exposure temperature in °C
TIME	exposure time in days
LOAD	exposure stress on sample in Mpa; exposure strain on sample in % or other note on any stress or strain to which the sample may have been subject during exposure
PDL RATING	based on a scale of 0 to 9 (with 9 as the highest resistance) developed by PDL; details of how rating is calculated are given later in this section
RESISTANCE NOTE	gives additional information about the resistance of the plastic to the reagent (i.e. observed changes, specimen details, cautionary notes, etc.)
MATERIAL NOTE	gives details on the specific material tested; this includes, if available, supplier, trade name, grade, filler

PDL Resistance Rating

The PDL Resistance Rating is determined using a weighted value scale developed by PDL and reviewed by experts. Each of the ratings is calculated from test results provided for a material after exposure to a specific environment. It gives a general indication of a materials resistance to a specific environment. In addition, it allows the user to search for materials most likely to be resistant to a specific exposure medium.

After assigning the weighted value to each field for which information is available, the PDL Resistance Rating is determined by adding together all weighted values and dividing this number by the number of values added together. All numbers to the right of the decimal are truncated to give the final result. If the result is equal to 10, a resistance rating of 9 is assigned. Each reported field is given equal importance in assigning the resistance rating since, depending on the end use, different factors play a role in the suitability for use of a material in a specific environment. Statistically, it is necessary to consider all available information in assigning the rating. Supplier resistance ratings are also figured into the calculation of the PDL Resistance Rating. Weighted values assigned depend on the scale used by the supplier.

The following tables give the values and guidelines used in assigning the PDL Resistance Rating. The guidelines - especially in the case of visual observations - are sometimes subject to an educated judgement. An effort is made to maintain consistency and accuracy..

Weighted Value	Weight Change	Diameter Length change	Thickness Change	Volume* Change	Mechanical** Property Retained	Visual*** Observed Change	BTT (min.)	Permeation Rate (μg/cm^2/min.)	Hardness Change (units)
10	0-0.25	0-0.1	0-0.25	0-2.5	≥97	no change	≤1	≤0.9	0-2
9	>.25-0.5	>0.1-0.2	>0.25-0.5	>2.5-5.0	94-<97		>1 to ≤2		>2-4
8	>0.5-0.75	>0.2-0.3	>0.5-0.75	>5.0-10.0	90-<94		>2 to ≤5	>0.9-9	>4-6
7	>0.75-1.0	>0.3-0.4	>0.75-1.0	>10.0-20.0	85-<90	slightly discolored slightly bleached	>5 to ≤10		>6-9
6	>1.0-1.5	>0.4-0.5	>1.0-1.5	>20.0-30.0	80-<85	discolored yellows slightly flexible	>10 to ≤30	>9-90	>9-12
5	>1.5-2.0	>0.5-0.75	>1.5-2.0	>30.0-40.0	75-<80	possible stress crack agent flexible possible oxidizing agent slightly crazed	>30 to ≤120		>12-15
4	>2.0-3.0	>0.75-1.0	>2.0-3.0	>40.0-50.0	70-<75	dostorted, warped softened slight swelling blistered known stress crack agent	>120 to ≤240	>90-900	>15-18
3	>3.0-4.0	>1.0-1.5	>3.0-4.0	>50.0-70.0	60-<70	cracking, crazing brittle plasticizer oxidizer softened swelling surface hardened	>240 to ≤480		>18-21
2	>4.0-6.0	>1.5-2.0	>4.0-6.0	>60.9-90.0	50-<60	severe distortion oxidizer and plasticizer deteriorated	>480 to ≤960	>900-9000	>21-25
1	>6.0	>2.0	>6.0	>90.0	>0-<50	decomposed	>960		>25
					0	solvent dissolved disintegrated		>9000	

*All values are given as percent change from original.

**Percent mechanical properties retained include tensile strength, elongation, modulus, flexural strength and impact strength. If the % retention is greater than 100%, a value of 200 minus the %property retained is used in the calculation.

***Due to the variety of information of this type reported, this table can be used only as a guideline.

Sterilization Methods

Four common types of sterilization are in use today: gas - ethylene oxide (EtO), steam autoclave, irradiation -gamma, electron beam, and beta radiation, and dry heat. Cold sterilization and disinfectants are also covered.

Ethylene Oxide:

Ethylene Oxide, a dominant sterilization technique is declining in use due to the following:

1) Physical property changes in polymers due to reactivity of the gas.
2) Length of degassing time, product aeration and elimination of gas toxic residues.
3) Absorption and adsorption of the gas, leaving residues and damaging the optical properties of the polymer.
4) The Environmental Protection Agency has found EtO to be mutagenic and has initiated steps to restrict its use.
5) Operator safety

Residual levels of EtO and related gas sterilants should fall within FDA guidelines, as follows:

GAS	BLOOD/CARDIAC DEVICES AND IMPLANTS	TOPICAL MEDICAL DEVICES
Ethylene Oxide	25 ppm	250 ppm
Ethylene Chlorohydrin	25 ppm	250 ppm
Ethylene Glycol	250 ppm	5000 ppm

Nevertheless EtO is the least aggressive form of sterilization for many materials. And the replacement of the most common EtO carrier gas - CFC-12 (Freon) - with non-ozone depleting alternatives, such as carbon dioxide and chloro-tetrafluoroethane - HCFC-124, will cause EtO to remain a viable choice for many users of sterilization services.

Irradiation:

Radiation sterilization can be accomplished using one of three forms of radiation: gamma, electron beam, or beta. Because repeated irradiation is equivalent to on going aging treatments, irradiation techniques have been successful for disposable articles, where only one dose is required. Polymers can also be modified, using irradiation, to improve physical properties and performance. Not all polymers will benefit and, in fact, many will degrade. Two factors predominate the change in physical properties; first, chain scission of the polymer molecule resulting in reduced molecular weight; and second, crosslinking of the polymer molecules resulting in the formation of large three dimensional matrices. These changes occur at the same time, and the predominance of one or the other, dependent on the polymer irradiated and radiation dose, will determine the outcome.

Those requiring radiation sterilization services or irradiation for polymer modification are increasingly looking to contract providers. This is a result of the ongoing cost and environmental pressures associated with EtO, coupled with the substantial capital required to construct an irradiation facility. The trend towards outsourcing should cause manufacturers to be more circumspect in their specification of materials and sterilization procedures since they give up direct control of this important function. For example, those using contract irradiation services should specify the minimum and maximum dose. All irradiation will result in the product getting a range of exposure. For gamma irradiation the range will typically be from 1.1 to 1.4 times the minimum dose; for electron beam the range is greater. If a minimum dose of 2.5 Mrads (25 kGy) is specified, the maximum dose may range from 2.75 Mrads (27.5 kGy) to 3.5 Mrads (37.5 kGy). The range in dosage absorbed results from a combination of packaging, part density and differences at irradiation facilities.

Gamma radiation is the most widely used sterilization method after EtO and the fastest growing of all sterilization methods. Gamma irradiation involves the bombardment of photons from a source such as Cobalt 60. It is characterized by deep penetration and low dose rates. Its high energy renders it capable of breaking any chemical bond present in a molecule, making it a good method for sterilizing devices composed of multiple resins. Gamma can penetrate large bulky objects. Prepackaged articles may therefore be sterilized, as packaging materials like cellophane, cellulose acetate, polyethylene, nylon, and even metal can be penetrated. Gamma is considered to have five times the penetration capability, as compared to electron beam irradiation.

Electron beam irradiation is bombardment of high energy electrons. In this method, sterilization is quick but with limited penetration. Beta radiation (electrons generated from a linear accelerator) has seen increasing use in Europe and test results are presented. Its use is limited in much the same way as electron beam due to limited penetration.

Steam:

Steam autoclave is a method by which a device is exposed to live steam in a pressurized chamber known as an autoclave. For non-critical applications, autoclaving is an acceptable method of sterilization. Types of autoclave sterilization include prevacuum, usually performed at 132°C (270°F), and gravity displacement cycle at 121°C (250°F). As condensation in an autoclave can cause impurities, steam is usually fed from a central boiler system where low concentrations of rust inhibiting feed water additives often are used to act as acid scavengers in steam condensate. Medical devices may be exposed to a variety of potentially aggressive agents such as soaps, lubricants, common disinfectants and rust inhibiting feed water additives. Generally a single exposure to these agents will not cause a change in properties. However, residue from these reagents left on a part during sterilization drying cycles can increase the chemical's concentration, causing the substance to react with the resin over time.

Dry Heat:

Dry heat sterilization is generally accomplished by subjection to high heat (165 °C to 170 °C) for 2 to 3 hours in ovens.

Radiation Resistance

Dow Chemical: Magnum 2620 (features: high gloss)

Samples lost 15% of Izod impact strength after low doses of radiation. Further reduction (24-30%) in Izod impact strength was seen after 10 Mrads of radiation. Instrumented dart impact showed minimal decreases after sterilization. The reduction of impact strength is caused by radiation crosslinking the butadiene rubber phase which reduces ductility. Both E-beam and gamma radiation sterilization cause a decrease in impact strength.

Reference: Hermanson, Nancy J., Steffens, John F., *The Physical and Visual Property Changes in Thermoplastic Resins After Exposure to High Energy Sterilization - Gamma Versus Electron Beam,* ANTEC 1993, conference proceedings - Society of Plastics Engineers, 1993.

Dow Chemical: Magnum 2642 (features: low gloss)

After sterilization samples showed some decrease in physical properties. The Izod impact strength decreased by 20% when exposed to 2.5 Mrads and by 29% when exposed to 10 Mrads of radiation. Exposing low gloss ABS to 10 Mrads of gamma or E-beam radiation caused a 30% decrease in instrumented dart impact as seen in total energy.

Reference: Hermanson, Nancy J., Steffens, John F., *The Physical and Visual Property Changes in Thermoplastic Resins After Exposure to High Energy Sterilization - Gamma Versus Electron Beam,* ANTEC 1993, conference proceedings - Society of Plastics Engineers, 1993.

Gamma Radiation Resistance

GE Plastics: Cycolac GTM5300 (features: transparent, medical grade)

Cycolac GTM5300 retains its properties through 5 megarads of gamma radiation, but yellowing occurs.

Reference: *Sterilization Comparison,* supplier marketing literature (JRW/3/91) - General Electric Plastics, 1991.

GE Plastics: Cycolac HP20 (features: medical grade)

Cycolac HP20 retains its properties through 5 megarads of gamma radiation, but yellowing occurs.

Reference: *Sterilization Comparison,* supplier marketing literature (ACB/3/93) - General Electric Plastics, 1993.

Dow Chemical: Magnum CLR95 (features: transparent)

The physical properties of transparent ABS are not significantly affected by up to 10 Mrads of gamma radiation. Appearance is improved by the photo-bleaching phenomenon. Data collected two weeks after exposure show no significant loss in tensile yield strength. The yellowness index and ΔE values roughly doubled between exposure to 2.5 Mrads and 10.0 Mrads. Exposure to 10.0 Mrads does cause permanent discoloration. No difference is seen when comparing the physical properties of the irradiated samples stored in fluorescent light versus those stored in complete darkness. Storage in light does not affect the bleach-back or optical properties.

Reference: Sturdevant, Marianne F., *Sterilization Compatibility of Rigid Thermoplastic Materials,* supplier technical report (301-1548) - Dow Chemical Company, 1988.

Dow Chemical: Magnum 9020 (features: high gloss)

High-gloss ABS loses impact strength upon exposure to gamma radiation. There is a slight linear increase in tensile strength with the increase in gamma dosage. The optical properties are not affected by the photo-bleaching phenomenon. Samples lost 28% of its Izod impact strength after exposure to 2.5 Mrads and 55% after exposure to 10.0 Mrads. The losses in impact properties and increase in tensile strength are thought to be attributed to breakdown and/or crosslinking occurring in the rubber phase. The yellowness index and ΔE values roughly doubled between exposure to 2.5 and 10.0 Mrads. Exposure to 10.0 Mrads induces a permanent color change. No difference is seen when comparing the physical properties of the irradiated samples stored in fluorescent light versus those stored in complete darkness. Storage in light does not affect the bleach-back and optical properties when exposed to 2.5 and 10.0 Mrads.

Reference: Sturdevant, Marianne F., *Sterilization Compatibility of Rigid Thermoplastic Materials,* supplier technical report (301-1548) - Dow Chemical Company, 1988.

Dow Chemical: Magnum 343 (features: low gloss)

Low-gloss ABS phyical properties are not significantly affected by exposure to 10.0 Mrads of gamma radiation. There is a slight linear increase in tensile strength with the increase in gamma dosage. The increase in tensile strength are thought to be attributed to breakdown and/or crosslinking in the rubber phase. No difference is seen when comparing the physical properties of the irradiated samples stored in fluorescent light versus those stored in complete darkness. Appearance is improved by the photo-bleaching phenomenon.

Reference: Sturdevant, Marianne F., *Sterilization Compatibility of Rigid Thermoplastic Materials,* supplier technical report (301-1548) - Dow Chemical Company, 1988.

Dow Chemical: (features: low gloss, natural resin)

The retention of impact strength of gamma sterilized rubber-modified styrenic polymers is dependent upon the degree of crosslinking that occurs in the butadiene rubber phase. The higher the radiation dosage, the greater the crosslinking, and the lower the ultimate impact strength. Low gloss ABS, because of its lower proportion and different type of rubber, does not show any significant change in its notched Izod impact properties at both 2.5 and 10 Mrads exposure.

Reference: Sturdevant, Marianne F., *The Long-term Effects of Ethylene Oxide and Gamma Radiation Sterilization on the Properties of Rigid Thermoplastic Materials,* ANTEC 1990, conference proceedings - Society of Plastics Engineers, 1990.

Dow Chemical: (features: high gloss, natural resin)

The retention of impact strength of gamma sterilized rubber-modified styrenic polymers is dependent upon the degree of crosslinking that occurs in the butadiene rubber phase. The higher the radiation dosage, the greater the crosslinking, and the lower the ultimate impact strength. At the sterilization exposure level of 10 Mrads, high gloss ABS showed losses in impact strength accompanied by a slight increase in tensile strength and a decrease in tensile elongation at break. This change is attributed to the crosslinking of the butadiene rubber matrix.

If the rubber content is high enough, crosslinking becomes the dominating factor in determining the physical property characteristics of the polymer upon irradiation. Crosslinked butadiene rubber losses its impact strength, thus at dosages sufficient to crosslink all the rubber, the enhanced impact properties originally provided by the rubber modifier are lost. The remaining impact strength of the material will be no better than that of the unmodified polymer. Comparing the notched Izod impact strength at 2.5 Mrads and 10 Mrads, one can see the loss in properties with the increse in radiation dosage.

Reference: Sturdevant, Marianne F., *The Long-term Effects of Ethylene Oxide and Gamma Radiation Sterilization on the Properties of Rigid Thermoplastic Materials,* ANTEC 1990, conference proceedings - Society of Plastics Engineers, 1990.

Monsanto: Lustran ABS 248 (features: medium impact)

Injection molded test specimens were irradiated at doses of 1.5, 2.5, 3.5 and 5.0 Mrads of gamma radiation. The Izod impact (ASTM D-256) value of high gloss, mediumimpact ABS (Lustran 248) showed a linear decrease with increasing dose level. At 5.0 Mrads it lost 5% to 10% of its original Izod impact value. There was no measurable change in tensile modulus (ASTM D-638). An increase in tensile stress (ASTM D-638) at yield of 5.0% was noted. There was no difference in tensile stress (ASTM D-638) at fail between the control and irradiated samples. Tensile elongation at yield exhibited a slight increase. The results of tensile elongation at fail (ASTM D-638) were extremely variable. Virtually no change in flexural modulus (ASTM D-790) was noted. No obvious trends in flexural modulus (ASTM D-790) were observed, with values fluctuating plus or minus 5.0%. At 2.5 Mrads discoloration was minimal.. At 10 Mrads the sample turned green, with increasing discoloration as dosages moved toward this level. During experimentation, a "fading" effect was observed. After time, the discoloration of all samples was barely perceptible.

Reference: *Effects Of Electron Beam And Gamma Radiation Sterilization Of Thermoplastics,* supplier technical report (7126A) - Monsanto Company, 1990.

Monsanto: Lustran ABS 743 (features: high impact)

Injection molded test specimens were irradiated at doses of 1.5, 2.5, 3.5 and 5.0 Mrads of gamma radiation. The Izod impact (ASTM D-256) value of high impact ABS (Lustran 743) showed a linear decrease with increasing dose level. There was no measurable change in tensile modulus (ASTM D-638). An increase in tensile stress (ASTM D-638) at yield of 5.0% was noted. There was no difference in tensile stress (ASTM D-638) at fail between the control and irradiated samples. Tensile elongation at yield exhibited a slight increase. The results of tensile elongation at fail (ASTM D-638) were extremely variable. Virtually no change in flexural modulus (ASTM D-790) was noted. No obvious trends in flexural modulus (ASTM D-790) were observed, with values fluctuating plus or minus 5.0%. At 2.5 Mrads discoloration was minimal.. At 10 Mrads the sample turned green, with increasing discoloration as dosages moved toward this level. During experimentation, a "fading" effect was observed. After time, the discoloration of all samples was barely perceptible.

Reference: *Effects Of Electron Beam And Gamma Radiation Sterilization Of Thermoplastics,* supplier technical report (7126A) - Monsanto Company, 1990.

Monsanto: Lustran ABS Ultra HX (features: high impact)

Injection molded test specimens were irradiated at doses of 1.5, 2.5, 3.5 and 5.0 Mrads of gamma radiation. The Izod impact (ASTM D-256) value of high gloss, high impact ABS (Lustran Ultra HX) showed a linear decrease with increasing dose level. There was no measurable change in tensile modulus (ASTM D-638). An increase in tensile stress (ASTM D-638) at yield of 5.0% was noted. There was no difference in tensile stress (ASTM D-638) at fail between the control and irradiated samples. Tensile elongation at yield exhibited a slight increase. The results of tensile elongation at fail (ASTM D-638) were extremely variable. Virtually no change in flexural modulus (ASTM D-790) was noted. No obvious trends in flexural modulus (ASTM D-790) were observed, with values fluctuating plus or minus 5.0%. At 2.5 Mrads discoloration was minimal.. At 10 Mrads the sample turned green, with increasing discoloration as dosages moved toward this level. During experimentation, a "fading" effect was observed. After time, the discoloration of all samples was barely perceptible.

Reference: *Effects Of Electron Beam And Gamma Radiation Sterilization Of Thermoplastics,* supplier technical report (7126A) - Monsanto Company, 1990.

Electron Beam Radiation

Monsanto: Lustran ABS 248 (features: medium impact)

Injection molded test specimens were irradiated at doses of 1.5, 2.5, 3.5 and 5.0 Mrads of electron beam radiation. The Izod impact (ASTM D-256) value of high gloss, mediumimpact ABS (Lustran 248) exhibited a linear decrease with increasing dosages. At 5.0 Mrads it lost 5% to 10% of its original value. There was no measurable change in tensile modulus (ASTM D-638). An increase in tensile stress (ASTM D-638) at yield of 10% to 18% was noted. There was a 5% to 15% difference in tensile stress (ASTM D-638) at fail between the control and irradiated samples. Tensile elongation at yield (ASTM D-638) showed a slight increase. The results of tensile elongation at fail (ASTM D-638) were extremely variable. Virtually no change in flexural modulus (ASTM D-790) was noted. No obvious trends in flexural modulus (ASTM D-790) were observed, with values fluctuating plus or minus 5.0%. At 2.5 Mrads discoloration was minimal. With increases in dosages up to 10.0 Mrads the samples turned green, color intensity increased with higher doses. During experimentation, a "fading" effect was observed. After time, the discoloration of all samples was barely perceptible.

Reference: *Effects Of Electron Beam And Gamma Radiation Sterilization Of Thermoplastics,* supplier technical report (7126A) - Monsanto Company, 1990.

Monsanto: Lustran ABS 743 (features: high impact)

Injection molded test specimens were irradiated at doses of 1.5, 2.5, 3.5 and 5.0 Mrads of electron beam radiation. The Izod impact (ASTM D-256) value of high impact ABS (Lustran 743) exhibited a linear decrease with increasing dosages. There was no measurable change in tensile modulus (ASTM D-638). An increase in tensile stress (ASTM D-638) at yield of 10% to 18% was noted. There was a 5% to 15% difference in tensile stress (ASTM D-638) at fail between the control and irradiated samples. Tensile elongation at yield (ASTM D-638) showed a slight increase. The results of tensile elongation at fail (ASTM D-638) were extremely variable. Virtually no change in flexural modulus (ASTM D-790) was noted. No obvious trends in flexural modulus (ASTM D-790) were observed, with values fluctuating plus or minus 5.0%. At 2.5 Mrads discoloration was minimal. With increases in dosages up to 10.0 Mrads the samples turned green, color intensity increased with higher doses. During experimentation, a "fading" effect was observed. After time, the discoloration of all samples was barely perceptible.

Reference: *Effects Of Electron Beam And Gamma Radiation Sterilization Of Thermoplastics,* supplier technical report (7126A) - Monsanto Company, 1990.

Monsanto: Lustran ABS Ultra HX (features: high impact)

Injection molded test specimens were irradiated at doses of 1.5, 2.5, 3.5 and 5.0 Mrads of electron beam radiation. The Izod impact (ASTM D-256) value of high gloss, high impact ABS (Lustran Ultra HX) exhibited a linear decrease with increasing dosages. There was no measurable change in tensile modulus (ASTM D-638). An increase in tensile stress (ASTM D-638) at yield of 10% to 18% was noted. There was a 5% to 15% difference in tensile stress (ASTM D-638) at fail between the control and irradiated samples. Tensile elongation at yield (ASTM D-638) showed a slight increase. The results of tensile elongation at fail (ASTM D-638) were extremely variable. Virtually no change in flexural modulus (ASTM D-790) was noted. No obvious trends in flexural modulus (ASTM D-790) were observed, with values fluctuating plus or minus 5.0%. At 2.5 Mrads discoloration was minimal. With increases in dosages up to 10.0 Mrads the samples turned green, color intensity increased with higher doses. During experimentation, a "fading" effect was observed. After time, the discoloration of all samples was barely perceptible.

Reference: *Effects Of Electron Beam And Gamma Radiation Sterilization Of Thermoplastics,* supplier technical report (7126A) - Monsanto Company, 1990.

Beta Radiation Resistance

ABS (features: transparent)

Beta sterilization is increasing in importance in Europe. Samples of clear ABS were tested at beta radiation levels of 2.7 to 10.8 Mrads. Physical properties were retained after 10.8 Mrads. In contrast to gamma sterilization, beta sterilization generates more heat. At 10.8 Mrads the surface of the ABS began to adhere to the packaging. The material became yellow. After two weeks of storage there was a reduction in yellowness. Storage in light or dark did not have a meaningful impact.

Reference: Haines, Jim, Hauser, Debbie, *Zylar Clear Alloys For Medical Applications,* supplier technical report - Novacor Chemicals Inc.

Ethylene Oxide (EtO) Resistance

Dow Chemical: Magnum 9020 (features: high gloss); **Magnum CLR95** (features: transparent)

ABS responds favorably to ethylene oxide exposure without loss in properties even after five repeated sterilization cycles. Five cycles did not effect the notched Izod impact strength. The tensile yield strength was also not affected by multiple EtO exposures. On day one transparent ABS demonstrated low residual EtO levels at less than 300 ppm.

Reference: Sturdevant, Marianne F., *Sterilization Compatibility of Rigid Thermoplastic Materials,* supplier technical report (301-1548) - Dow Chemical Company, 1988.

Dow Chemical: Magnum 343 (features: low gloss)

ABS responds favorably to ethylene oxide exposure without loss in properties even after five repeated sterilization cycles. Five cycles did not effect the notched Izod impact strength. The tensile yield strength was also not affected by multiple EtO exposures. After 20 days the resin achieved a residual level of EtO of less than 200 ppm.

Reference: Sturdevant, Marianne F., *Sterilization Compatibility of Rigid Thermoplastic Materials,* supplier technical report (301-1548) - Dow Chemical Company, 1988.

Dow Chemical: ABS (features: low gloss, natural resin); **ABS** (features: high gloss, natural resin)

Physical property retention shows compatibility with ethylene oxide sterilization. Samples retained their physical integrity after one cycle of EtO sterilization. However, because of the styrenic components in these polymer matrices, over exposure to EtO sterilization should be avoided. Five repeated sterilization cycles caused some embrittlement. The embrittlement is seen as a loss of tensile elongation at break and a decrease in instrumented dart impact energy. After multiple EtO cycles the embrittlement appears to compound with time. The elongation and instrumented impact strengths at six months and one year were significantly less than what was observed at two weeks after sterilization. Because of the styrenic component of the bulk polymer matrix, a slight decrease in the tensile elongation at break is also observed.

Reference: Sturdevant, Marianne F., *The Long-term Effects of Ethylene Oxide and Gamma Radiation Sterilization on the Properties of Rigid Thermoplastic Materials,* ANTEC 1990, conference proceedings - Society of Plastics Engineers, 1990.

Dow Chemical: Magnum 2642

Tensile yield is slightly affected (5-8% decrease) by ethylene oxide sterilization regardless of the carrier gas. The sterilant mixture using HCFC-124 has less affect on the instrumented dart impact than the standard cycle of CFC-12/EtO. A decrease in peak energy is seen after 2 cycles of CFC-12/EtO, however, the standard deviation for this value is high so the true significance of the trend is obscured. Because of the styrenic component and its sensitivity to EtO, repeated cycles of EtO sterilization should be avoided.

Reference: Hermanson, Nancy J., *Effects Of Alternate Carriers Of Ethylene Oxide Sterilant On Thermoplastics,* supplier technical report (301-02018) - Dow Chemical Company.

Steam Resistance

GE Plastics: Cycolac GTM5300 (features: transparent, medical grade)

Cycolac GTM 5300 ABS is not recommended for steam autoclave sterilization.

Sterilization Comparison, supplier marketing literature (JRW/3/91) - General Electric Plastics, 1991.

GE Plastics: Cycolac HP20 (features: medical grade)

Cycolac HP20 ABS is not recommended for steam autoclave sterilization.

Sterilization Comparison, supplier marketing literature (ACB/3/93) - General Electric Plastics, 1993.

BASF: Terlux (features: transparent)

Steam sterilization is not recommended due to the material's low resistance to heat deformation.

Reference: Johnson, James A., supplier written correspondence - BASF Corporation, 1994.

TABLE 01: Effect of Gamma Radiation Sterilization on ABS

Material Family	ABS					
Material Supplier/Name	DOW					
Material Note	low gloss, natural resin	low gloss, natural resin	low gloss, natural resin	low gloss, natural resin	low gloss, natural resin	low gloss, natural resin
Reference No.	5	5	5	5	5	5

EXPOSURE CONDITIONS

Type	Gamma Radiation	Gamma Radiation	Gamma Radiation	Gamma Radiation	Gamma Radiation	Gamma Radiation
Details	source: Cobalt 60	source: Cobalt 60	source: Cobalt 60	source: Cobalt 60	source: Cobalt 60	source: Cobalt 60
Radiation Dose (Mrads)	2.5	2.5	2.5	10	10	10
Note	test lab: Radiations Sterilizers Inc.	test lab: Radiations Sterilizers Inc.	test lab: Radiations Sterilizers Inc.	test lab: Radiations Sterilizers Inc.	test lab: Radiations Sterilizers Inc.	test lab: Radiations Sterilizers Inc.

POST EXPOSURE CONDITIONING

Note	type: storage in dark	type: storage in dark	type: storage in dark	type: storage in dark	type: storage in dark	type: storage in dark
Temperature (°C)	21	21	21	21	21	21
Time (hours)	336	4368	8760	336	4368	8760

PROPERTIES RETAINED (%)

Tensile Strength	103.1 (ht)	106.6 (ht)	108.7 (ht)	108.6 (ht)	111.9 (ht)	111.6 (ht)
Tensile Strength @ Yield	103.1 (ik)	102.5 (ik)	102 (ik)	110.1 (ik)	110.6 (ik)	107.5 (ik)
Elongation	97.3 (au)	86.7 (au)	86.7 (au)	61.3 (au)	74.7 (au)	73.3 (au)
Dart Impact (total energy)	162.7 (fc)	188.2 (fc)	175.5 (fc)	117.2 (fi)	124.5 (fc)	106.4 (fc)
Notched Izod Impact	96 (fw)	92 (fw)	104 (fw)	80 (fw)	96 (fw)	88 (fw)

TABLE 02: Effect of Gamma Radiation Sterilization on ABS

Material Family	ABS					
Material Supplier/Name	DOW					
Material Note	high gloss, natural resin	high gloss, natural resin	high gloss, natural resin	high gloss, natural resin	high gloss, natural resin	high gloss, natural resin
Reference No.	5	5	5	5	5	5

EXPOSURE CONDITIONS

Type	Gamma Radiation	Gamma Radiation	Gamma Radiation	Gamma Radiation	Gamma Radiation	Gamma Radiation
Details	source: Cobalt 60	source: Cobalt 60	source: Cobalt 60	source: Cobalt 60	source: Cobalt 60	source: Cobalt 60
Radiation Dose (Mrads)	2.5	2.5	2.5	10	10	10
Note	test lab: Radiations Sterilizers Inc.	test lab: Radiations Sterilizers Inc.	test lab: Radiations Sterilizers Inc.	test lab: Radiations Sterilizers Inc.	test lab: Radiations Sterilizers Inc.	test lab: Radiations Sterilizers Inc.

POST EXPOSURE CONDITIONING

Note	type: storage in dark	type: storage in dark	type: storage in dark	type: storage in dark	type: storage in dark	type: storage in dark
Temperature (°C)	21	21	21	21	21	21
Time (hours)	336	4368	8760	336	4368	8760

PROPERTIES RETAINED (%)

Tensile Strength	99.9 (ht)	109.5 (ht)	103.9 (ht)	104.5 (ht)	105.3 (ht)	105.3 (ht)
Tensile Strength @ Yield	104.1 (ik)	121.6 (ik)	107.5 (ik)	107.7 (ik)	110.7 (ik)	111 (ik)
Elongation	84.2 (au)	60.5 (au)	60.5 (au)	60.5 (au)	50 (au)	60.5 (au)
Dart Impact (total energy)	216.1 (fc)	183.4 (fc)	202.4 (fc)	159.5 (fc)	138 (fc)	105.9 (fi)
Notched Izod Impact	71.9 (fw)	67.2 (fw)	67.2 (fw)	45.3 (fw)	42.2 (fw)	43.8 (fw)

TABLE 03: Effect of Gamma Radiation Sterilization on ABS

Material Family	ABS					
Material Supplier/Name	DOW MAGNUM 9020		DOW MAGNUM 343		DOW MAGNUM CLR95	
Material Note	high gloss	high gloss	low gloss	low gloss	transparent	transparent
Reference No.	3	3	3	3	3	3
EXPOSURE CONDITIONS						
Type	Gamma Radiation	Gamma Radiation	Gamma Radiation	Gamma Radiation	Gamma Radiation	Gamma Radiation
Details	source: Cobalt 60	source: Cobalt 60	source: Cobalt 60	source: Cobalt 60	source: Cobalt 60	source: Cobalt 60
Radiation Dose (Mrads)	2.5	10	2.5	10	2.5	10
POST EXPOSURE CONDITIONING						
Note	type: storage under fluorescent light	type: storage under fluorescent light	type: storage under fluorescent light	type: storage under fluorescent light	type: storage under fluorescent light	type: storage under fluorescent light
Temperature (°C)	21	21	21	21	21	21
Time (hours)	336	336	336	336	336	336
PROPERTIES RETAINED (%)						
Tensile Strength @ Yield	105.4 (ii)	114.3 (ii)	101.8 (ii)	112.7 (ii)	101.3 (ii)	100 (ii)
Notched Izod Impact	71.9 (fp)	45.3 (fp)	92 (fp)	100 (fp)	88.5 (fp)	96.2 (fp)

TABLE 04: Effect of Gamma Radiation Sterilization on ABS

Material Family	ABS					
Material Supplier/Name	DOW MAGNUM 9020		DOW MAGNUM 343		DOW MAGNUM CLR95	
Material Note	high gloss	high gloss	low gloss	low gloss	transparent	transparent
Reference No.	3	3	3	3	3	3
EXPOSURE CONDITIONS						
Type	Gamma Radiation	Gamma Radiation	Gamma Radiation	Gamma Radiation	Gamma Radiation	Gamma Radiation
Details	source: Cobalt 60	source: Cobalt 60	source: Cobalt 60	source: Cobalt 60	source: Cobalt 60	source: Cobalt 60
Radiation Dose (Mrads)	2.5	10	2.5	10	2.5	10
POST EXPOSURE CONDITIONING						
Note	type: storage in dark	type: storage in dark	type: storage in dark	type: storage in dark	type: storage in dark	type: storage in dark
Temperature (°C)	21	21	21	21	21	21
Time (hours)	336	336	336	336	336	336
PROPERTIES RETAINED (%)						
Tensile Strength @ Yield	103.6 (ii)	107.1 (ii)	101.8 (ii)	110.9 (ii)	100 (ii)	100 (ii)
Notched Izod Impact	71.9 (fp)	45.3 (fp)	96 (fp)	80 (fp)	96.2 (fp)	76.9 (fp)

TABLE 05: Effect of Gamma Radiation Sterilization on ABS

Material Family	ABS							
Material Supplier/Name	DOW MAGNUM 2620				DOW MAGNUM 2642			
Material Note	high gloss	high gloss	high gloss	high gloss	low gloss	low gloss	low gloss	low gloss
Reference No.	1	1	1	1	1	1	1	1
EXPOSURE CONDITIONS								
Type	Gamma Radiation	Gamma Radiation	Gamma Radiation	Gamma Radiation	Gamma Radiation	Gamma Radiation	Gamma Radiation	Gamma Radiation
Radiation Dose (Mrads)	2.5	2.5	10	10	2.5	2.5	10	10
Note	test lab: SteriGenics	test lab: SteriGenics	test lab: SteriGenics	test lab: SteriGenics	test lab: SteriGenics	test lab: SteriGenics	test lab: SteriGenics	test lab: SteriGenics
POST EXPOSURE CONDITIONING								
Note	type: aging	type: aging	type: aging	type: aging	type: aging	type: aging	type: aging	type: aging
Time (hours)	168	1344	168	1344	168	1344	168	1344
PROPERTIES RETAINED (%)								
Tensile Strength @ Yield	104.9 (is)	104.9 (is)	100 (is)	107.3 (is)	106.3 (is)	103.1 (is)	106.3 (is)	106.3 (is)
Elongation	106.5 (bp)	80.6 (bp)	77.4 (bp)	77.4 (bp)	107.9 (bv)	96.6 (bv)	103.4 (bv)	94.4 (bv)
Flexural Strength	102.8 (cm)	101.4 (cm)	109.7 (cm)	104.2 (cm)	105.2 (cm)	101.7 (cm)	110.3 (cm)	105.2 (cm)
Modulus	102.2 (hb)	99.6 (hb)	99.6 (hb)	99.1 (hb)	99 (hb)	100.5 (hb)	100.5 (hb)	106.8 (hb)
Flexural Modulus	110.7 (cf)	116.1 (cf)	118.8 (cf)	115.6 (cf)	99 (cf)	100.5 (cf)	100.5 (cf)	106.8 (cf)
Dart Impact (total energy)	102.1 (ei)	104.3 (ei)	104.3 (ei)	85.1 (ei)	91.3 (em)	91.3 (em)	63 (em)	87 (em)
Dart Impact (peak energy)	102.1 (di)	104.3 (di)	104.3 (di)	85.1 (di)	90 (dj)	90 (dj)	73.3 (dj)	86.7 (dj)
Notched Izod Impact	95.6 (gb)	84.9 (gb)	76.5 (gb)	70.1 (gb)	78.7 (gd)	94.9 (gd)	67.5 (gd)	70.6 (gd)
Heat Deflection Temperature	100 (m)	98.8 (m)	97.5 (m)	100 (m)	105.9 (m)	107.4 (m)	105.9 (m)	108.8 (m)
Vicat Softening Point	100 (ku)	98.2 (ku)	98.2 (ku)	98.2 (ku)	99 (ku)	99 (ku)	100 (ku)	99 (ku)
SURFACE and APPEARANCE								
ΔL Color	-7.35 (ac)	-2.11 (ac)	-16.34 (ac)	-3.16 (ac)	-9.3 (ac)	-1.5 (ac)	-14.5 (ac)	-3 (ac)
Δa Color	1.47 (o)	2.34 (o)	2.6 (o)	2.26 (o)	-1.7 (o)	-0.8 (o)	0.4 (o)	-0.6 (o)
Δb Color	-2.39 (t)	1.99 (t)	-3.75 (t)	6.93 (t)	6.3 (t)	5.1 (t)	7.1 (t)	10.5 (t)

TABLE 06: Effect of Gamma Radiation Sterilization on ABS

Material Family	ABS							
Material Supplier/Name	**MONSANTO LUSTRAN ABS 248**				**MONSANTO LUSTRAN ABS 743**			
Material Note	medium impact	medium impact	medium impact	medium impact	high impact	high impact	high impact	high impact
Reference No.	88	88	88	88	88	88	88	88
EXPOSURE CONDITIONS								
Type	Gamma Radiation	Gamma Radiation	Gamma Radiation	Gamma Radiation	Gamma Radiation	Gamma Radiation	Gamma Radiation	Gamma Radiation
Radiation Dose (Mrads)	1.5	2.5	3.5	5	1.5	2.5	3.5	5
PROPERTIES RETAINED (%)								
Notched Izod Impact	90 (fo)	90 (fo)	90 (fo)	90 (fo)	90 (fo)	90 (fo)	90 (fo)	90 (fo)

TABLE 07: Effect of Electron Beam Radiation Sterilization on ABS

Material Family	ABS							
Material Supplier/Name	DOW MAGNUM 2620				DOW MAGNUM 2642			
Material Note	high gloss	high gloss	high gloss	high gloss	low gloss	low gloss	low gloss	low gloss
Reference No.	1	1	1	1	1	1	1	1

EXPOSURE CONDITIONS

Type	Electron Beam Radiation	Electron Beam Radiation	Electron Beam Radiation	Electron Beam Radiation	Electron Beam Radiation	Electron Beam Radiation	Electron Beam Radiation	Electron Beam Radiation
Radiation Dose (Mrads)	2.5	2.5	10	10	2.5	2.5	10	10
Note	test lab: E-Beam Services, Inc.	test lab: E-Beam Services, Inc.	test lab: E-Beam Services, Inc.	test lab: E-Beam Services, Inc.	test lab: E-Beam Services, Inc.	test lab: E-Beam Services, Inc.	test lab: E-Beam Services, Inc.	test lab: E-Beam Services, Inc.

POST EXPOSURE CONDITIONING

Note	type: aging	type: aging	type: aging	type: aging	type: aging	type: aging	type: aging	type: aging
Time (hours)	168	1344	168	1344	168	1344	168	1344

PROPERTIES RETAINED (%)

Tensile Strength @ Yield	104.9 (is)	104.9 (is)	104.9 (is)	107.3 (is)	100 (is)	103.1 (is)	100 (is)	106.3 (is)
Elongation	71 (bp)	67.7 (bp)	71 (bp)	77.4 (bp)	96.6 (bv)	79.8 (bv)	84.3 (bv)	76.4 (bv)
Flexural Strength	104.2 (cm)	102.8 (cm)	106.9 (cm)	102.8 (cm)	105.2 (cm)	101.7 (cm)	108.6 (cm)	103.4 (cm)
Modulus	103.9 (hb)	100.4 (hb)	101.3 (hb)	100.4 (hb)	101.6 (hb)	95.8 (hb)	104.7 (hb)	102.1 (hb)
Flexural Modulus	116.1 (cf)	117.4 (cf)	114.3 (cf)	114.3 (cf)	101.6 (cf)	95.8 (cf)	104.7 (cf)	102.1 (cf)
Dart Impact (total energy)	102.1 (ei)	89.4 (ei)	89.4 (ei)	91.5 (ei)	67.4 (em)	78.3 (em)	63 (em)	69.6 (em)
Dart Impact (peak energy)	102.1 (di)	89.4 (di)	89.4 (di)	91.5 (di)	76.7 (dj)	86.7 (dj)	83.3 (dj)	80 (dj)
Notched Izod Impact	91.2 (gb)	84.9 (gb)	84.9 (gb)	78.5 (gb)	81.2 (gd)	83.8 (gd)	67.5 (gd)	67.5 (gd)
Heat Deflection Temperature	100 (m)	98.8 (m)	103.8 (m)	105 (m)	107.4 (m)	105.9 (m)	113.2 (m)	119.1 (m)
Vicat Softening Point	98.2 (ku)	98.2 (ku)	98.2 (ku)	98.2 (ku)	100 (ku)	99 (ku)	100 (ku)	99 (ku)

SURFACE and APPEARANCE

ΔL Color	-7.86 (ac)	-1.1 (ac)	-14.85 (ac)	-2.26 (ac)	-7.6 (ac)	-1.1 (ac)	-13.6 (ac)	-2.1 (ac)
Δa Color	1.24 (o)	1 (o)	2.69 (o)	1.16 (o)	-1.7 (o)	-0.8 (o)	0.5 (o)	-0.9 (o)
Δb Color	-2.41 (t)	2.6 (t)	-4.91 (t)	5.49 (t)	5.6 (t)	4.2 (t)	6.9 (t)	8.6 (t)

TABLE 08: Effect of Electron Beam Radiation Sterilization on ABS

Material Family	ABS							
Material Supplier/Name	MONSANTO LUSTRAN ABS 248				MONSANTO LUSTRAN ABS 743			
Material Note	medium impact	medium impact	medium impact	medium impact	high impact	high impact	high impact	high impact
Reference No.	88	88	88	88	88	88	88	88
EXPOSURE CONDITIONS								
Type	Electron Beam Radiation	Electron Beam Radiation	Electron Beam Radiation	Electron Beam Radiation	Electron Beam Radiation	Electron Beam Radiation	Electron Beam Radiation	Electron Beam Radiation
Radiation Dose (Mrads)	1.5	2.5	3.5	5	1.5	2.5	3.5	5
PROPERTIES RETAINED (%)								
Notched Izod Impact	100 (fo)	97 (fo)	95 (fo)	90 (fo)	100 (fo)	98 (fo)	96 (fo)	94 (fo)

TABLE 09: Effect of Ethylene Oxide Sterilization on Dow ABS.

Material Family	ABS					
Material Supplier/Name	Dow					
Material Note	low gloss, natural resin	low gloss, natural resin	low gloss, natural resin	low gloss, natural resin	low gloss, natural resin	low gloss, natural resin
Reference No.	5	5	5	5	5	5
EXPOSURE CONDITIONS						
Type	Ethylene Oxide	Ethylene Oxide	Ethylene Oxide	Ethylene Oxide	Ethylene Oxide	Ethylene Oxide
Details	12% EtO and 88% Freon	12% EtO and 88% Freon	12% EtO and 88% Freon	12% EtO and 88% Freon	12% EtO and 88% Freon	12% EtO and 88% Freon
Concentration	660 mg/l	660 mg/l	660 mg/l	660 mg/l	660 mg/l	660 mg/l
Number of Cycles	1	1	1	5	5	5
Note	RH: 60%; test lab: Ethox Corp.	RH: 60%; test lab: Ethox Corp.	RH: 60%; test lab: Ethox Corp.	RH: 60%; test lab: Ethox Corp.	RH: 60%; test lab: Ethox Corp.	RH: 60%; test lab: Ethox Corp.
Temperature (°C)	49	49	49	49	49	49
Time (hours)	≥ 6	≥ 6	≥ 6	≥ 6	≥ 6	≥ 6
PRE EXPOSURE CONDITIONING						
Preconditioning Note	time: 8 hours; temperature: 37.8°C; RH: 60%	time: 8 hours; temperature: 37.8°C; RH: 60%	time: 8 hours; temperature: 37.8°C; RH: 60%	time: 8 hours; temperature: 37.8°C; RH: 60%	time: 8 hours; temperature: 37.8°C; RH: 60%	time: 8 hours; temperature: 37.8°C; RH: 60%
POST EXPOSURE CONDITIONING						
Note	type: evacuation; pressure: 127 mm Hg	type: evacuation; pressure: 127 mm Hg	type: evacuation; pressure: 127 mm Hg	type: evacuation; pressure: 127 mm Hg	type: evacuation; pressure: 127 mm Hg	type: evacuation; pressure: 127 mm Hg
POST EXPOSURE CONDITIONING II						
Note	type: aeration	type: aeration	type: aeration	type: aeration	type: aeration	type: aeration
Temperature (°C)	32.2	32.2	32.2	32.2	32.2	32.2
Time (hours)	≥ 16	≥ 16	≥ 16	≥ 16	≥ 16	≥ 16
POST EXPOSURE CONDITIONING III						
Note	type: storage in dark	type: storage in dark	type: storage in dark	type: storage in dark	type: storage in dark	type: storage in dark
Temperature (°C)	21	21	21	21	21	21
Time (hours)	336	4368	8760	336	4368	8760
PROPERTIES RETAINED (%)						
Tensile Strength	103.9 (hu)	100.4 (hu)	103.1 (hu)	97.8 (hu)	101.2 (hu)	99.9 (hu)
Tensile Strength @ Yield	99.9 (in)	98.3 (in)	101.1 (in)	98.1 (in)	99.2 (in)	98.5 (in)
Elongation	74.7 (aw)	45.3 (aw)	56 (aw)	26.7 (az)	24 (az)	17.3 (aw)
Dart Impact (total energy)	52.9 (fa)	56.9 (fa)	65.7 (fa)	53.9 (fa)	45.1 (fa)	59.8 (fa)
Notched Izod Impact	116 (fx)	120 (fx)	108 (fx)	136 (fx)	128 (fx)	120 (fx)

TABLE 10: Effect of Ethylene Oxide Sterilization on Dow ABS.

Material Family	ABS	ABS	ABS	ABS	ABS	ABS
Material Supplier/Name	Dow	Dow	Dow	Dow	Dow	Dow
Material Note	high gloss, natural resin	high gloss, natural resin	high gloss, natural resin	high gloss, natural resin	high gloss, natural resin	high gloss, natural resin
Reference No.	5	5	5	5	5	5

EXPOSURE CONDITIONS

Type	Ethylene Oxide	Ethylene Oxide	Ethylene Oxide	Ethylene Oxide	Ethylene Oxide	Ethylene Oxide
Details	12% EtO and 88% Freon	12% EtO and 88% Freon	12% EtO and 88% Freon	12% EtO and 88% Freon	12% EtO and 88% Freon	12% EtO and 88% Freon
Concentration	660 mg/l	660 mg/l	660 mg/l	660 mg/l	660 mg/l	660 mg/l
Number of Cycles	1	1	1	5	5	5
Note	RH: 60%; test lab: Ethox Corp.	RH: 60%; test lab: Ethox Corp.	RH: 60%; test lab: Ethox Corp.	RH: 60%; test lab: Ethox Corp.	RH: 60%; test lab: Ethox Corp.	RH: 60%; test lab: Ethox Corp.
Temperature (°C)	49	49	49	49	49	49
Time (hours)	≥ 6	≥ 6	≥ 6	≥ 6	≥ 6	≥ 6

PRE EXPOSURE CONDITIONING

Preconditioning Note	time: 8 hours; temperature: 37.8°C; RH: 60%	time: 8 hours; temperature: 37.8°C; RH: 60%	time: 8 hours; temperature: 37.8°C; RH: 60%	time: 8 hours; temperature: 37.8°C; RH: 60%	time: 8 hours; temperature: 37.8°C; RH: 60%	time: 8 hours; temperature: 37.8°C; RH: 60%

POST EXPOSURE CONDITIONING

Note	type: evacuation; pressure: 127 mm Hg	type: evacuation; pressure: 127 mm Hg	type: evacuation; pressure: 127 mm Hg	type: evacuation; pressure: 127 mm Hg	type: evacuation; pressure: 127 mm Hg	type: evacuation; pressure: 127 mm Hg

POST EXPOSURE CONDITIONING II

Note	type: aeration	type: aeration	type: aeration	type: aeration	type: aeration	type: aeration
Temperature (°C)	32.2	32.2	32.2	32.2	32.2	32.2
Time (hours)	≥ 16	≥ 16	≥ 16	≥ 16	≥ 16	≥ 16

POST EXPOSURE CONDITIONING III

Note	type: storage in dark	type: storage in dark	type: storage in dark	type: storage in dark	type: storage in dark	type: storage in dark
Temperature (°C)	21	21	21	21	21	21
Time (hours)	336	4368	8760	336	4368	8760

PROPERTIES RETAINED (%)

Tensile Strength	103.4 (hu)	105 (hu)	105 (hu)	105.5 (hu)	106.1 (hu)	107.3 (hu)
Tensile Strength @ Yield	103.3 (in)	106 (in)	107 (in)	105.1 (in)	107.5 (in)	108.5 (in)
Elongation	71.1 (aw)	50 (aw)	63.2 (aw)	42.1 (aw)	44.7 (aw)	42.1 (aw)
Dart Impact (total energy)	187.8 (fa)	178 (fa)	169.8 (fa)	148.8 (fa)	116.1 (fa)	107.3 (fa)
Notched Izod Impact	95.3 (fx)	90.6 (fx)	87.5 (fx)	92.2 (fx)	84.4 (fx)	81.3 (fx)

TABLE 11: Effect of Ethylene Oxide Sterilization on ABS

Material Family	ABS							
Material Supplier/Name	DOW MAGNUM 2642							
Reference No.	6	6	6	6	6	6	6	6
EXPOSURE CONDITIONS								
Type	Ethylene Oxide	Ethylene Oxide	Ethylene Oxide	Ethylene Oxide	Ethylene Oxide	Ethylene Oxide	Ethylene Oxide	Ethylene Oxide
Details	12% EtO and 88% Freon	12% EtO and 88% Freon	12% EtO and 88% Freon	12% EtO and 88% Freon	8.6% EtO and 91.4% HCFC-124	8.6% EtO and 91.4% HCFC-124	8.6% EtO and 91.4% HCFC-124	8.6% EtO and 91.4% HCFC-124
Concentration	600 mg/l	600 mg/l	600 mg/l	600 mg/l				
Number of Cycles	1	1	2	2	1	1	2	2
Note	RH: 60%; test lab: Ethox Corp.	RH: 60%; test lab: Ethox Corp.	RH: 60%; test lab: Ethox Corp.	RH: 60%; test lab: Ethox Corp.	RH: 60%; test lab: Ethox Corp.	RH: 60%; test lab: Ethox Corp.	RH: 60%; test lab: Ethox Corp.	RH: 60%; test lab: Ethox Corp.
Temperature (°C)	48.9	48.9	48.9	48.9	48.9	48.9	48.9	48.9
Time (hours)	6	6	6	6	6	6	6	6
PRE EXPOSURE CONDITIONING								
Preconditioning Note	time: 18 hours; temperature: 37.8°C; RH: 60%	time: 18 hours; temperature: 37.8°C; RH: 60%	time: 18 hours; temperature: 37.8°C; RH: 60%	time: 18 hours; temperature: 37.8°C; RH: 60%	time: 18 hours; temperature: 37.8°C; RH: 60%	time: 18 hours; temperature: 37.8°C; RH: 60%	time: 18 hours; temperature: 37.8°C; RH: 60%	time: 18 hours; temperature: 37.8°C; RH: 60%
POST EXPOSURE CONDITIONING								
Note	type: aeration; pressure: 127 mm Hg	type: aeration; pressure: 127 mm Hg	type: aeration; pressure: 127 mm Hg	type: aeration; pressure: 127 mm Hg	type: aeration; pressure: 127 mm Hg	type: aeration; pressure: 127 mm Hg	type: aeration; pressure: 127 mm Hg	type: aeration; pressure: 127 mm Hg
Temperature (°C)	32.2	32.2	32.2	32.2	32.2	32.2	32.2	32.2
POST EXPOSURE CONDITIONING II								
Note	type: ambient conditions	type: ambient conditions	type: ambient conditions	type: ambient conditions	type: ambient conditions	type: ambient conditions	type: ambient conditions	type: ambient conditions
Time (hours)	168	1344	168	1344	168	1344	168	1344
PROPERTIES RETAINED (%)								
Tensile Strength @ Yield	100.7 (il)	100.8 (il)	100.5 (il)	95.2 (il)	97.2 (il)	91.4 (il)	100.3 (il)	95.9 (il)
Elongation	90.9 (bk)	127.3 (bk)	81.8 (bk)	72.7 (bk)	100 (bk)	109.1 (bk)	127.3 (bk)	100 (bk)
Modulus	97.1 (gu)	101.3 (gu)	100.3 (gu)	76.8 (gu)	98.7 (gu)	96.2 (gu)	101.6 (gu)	72.6 (gu)
Dart Impact (total energy)	92.3 (ds)	102.6 (ds)	100 (ds)	53.8 (dw)	87.2 (ds)	82.1 (dx)	89.7 (ds)	94.9 (ds)
Dart Impact (peak energy)	69 (er)	89.7 (er)	72.4 (er)	37.9 (ev)	65.5 (er)	65.5 (et)	69 (er)	79.3 (er)
SURFACE and APPEARANCE								
ΔE Color	0.26	0.41	0.16	0.48	0.24	0.76	0.41	0.61

TABLE 12: Effect of Ethylene Oxide Sterilization on Dow Magnum 2642 ABS

Material Family	ABS			
Material Supplier/Name	DOW MAGNUM 2642			
Reference No.	6	6	6	6
EXPOSURE CONDITIONS				
Type	Ethylene Oxide	Ethylene Oxide	Ethylene Oxide	Ethylene Oxide
Details	12% EtO and 88% Freon	12% EtO and 88% Freon	8.6% EtO and 91.4% HCFC-124	8.6% EtO and 91.4% HCFC-124
Concentration	600 mg/l	600 mg/l		
Number of Cycles	1	1	1	1
Note	RH: 60%; test lab: Ethox Corp.	RH: 60%; test lab: Ethox Corp.	RH: 60%; test lab: Ethox Corp.	RH: 60%; test lab: Ethox Corp.
Temperature (°C)	48.9	48.9	48.9	48.9
Time (hours)	6	6	6	6
PRE EXPOSURE CONDITIONING				
Preconditioning Note	time: 18 hours; temperature: 37.8°C; RH: 60%	time: 18 hours; temperature: 37.8°C; RH: 60%	time: 18 hours; temperature: 37.8°C; RH: 60%	time: 18 hours; temperature: 37.8°C; RH: 60%
POST EXPOSURE CONDITIONING				
Note	type: aeration; note: 10 air changes per hour	type: aeration; note: 30 air changes per hour	type: aeration; note: 10 air changes per hour	type: aeration; note: 30 air changes per hour
Temperature (°C)	32.2	54.4	32.2	54.4
RESIDUALS (ppm)				
Residuals Determined	ethylene oxide	ethylene oxide	ethylene oxide	ethylene oxide
Little or No Aeration	984	984	866	866
24 hour Aeration	465	200	467	383
48 hour Aeration		162		195
72 hour Aeration	152	146	258	169
168 hour Aeration	92		134	

TABLE 13: Effect of Ethylene Oxide Sterilization on ABS

Material Family	ABS					
Material Supplier/Name	DOW MAGNUM 9020		DOW MAGNUM 343		DOW MAGNUM CLR95	
Material Note	high gloss	high gloss	low gloss	low gloss	transparent	transparent
Reference No.	3	3	3	3	3	3
EXPOSURE CONDITIONS						
Type	Ethylene Oxide	Ethylene Oxide	Ethylene Oxide	Ethylene Oxide	Ethylene Oxide	Ethylene Oxide
Details	12% EtO and 88% Freon	12% EtO and 88% Freon	12% EtO and 88% Freon	12% EtO and 88% Freon	12% EtO and 88% Freon	12% EtO and 88% Freon
Concentration	660 mg/l	660 mg/l	660 mg/l	660 mg/l	660 mg/l	660 mg/l
Number of Cycles	1	5	1	5	1	5
Note	RH: 60%	RH: 60%	RH: 60%	RH: 60%	RH: 60%	RH: 60%
Temperature (°C)	49	49	49	49	49	49
Time (hours)	≥ 6	≥ 6	≥ 6	≥ 6	≥ 6	≥ 6
PRE EXPOSURE CONDITIONING						
Preconditioning Note	time: 8 hours; temperature: 37.8°C; RH: 60%	time: 8 hours; temperature: 37.8°C; RH: 60%	time: 8 hours; temperature: 37.8°C; RH: 60%	time: 8 hours; temperature: 37.8°C; RH: 60%	time: 8 hours; temperature: 37.8°C; RH: 60%	time: 8 hours; temperature: 37.8°C; RH: 60%
POST EXPOSURE CONDITIONING						
Note	type: evacuation; pressure: 127 mm Hg	type: evacuation; pressure: 127 mm Hg	type: evacuation; pressure: 127 mm Hg	type: evacuation; pressure: 127 mm Hg	type: evacuation; pressure: 127 mm Hg	type: evacuation; pressure: 127 mm Hg
POST EXPOSURE CONDITIONING II						
Note	type: aeration	type: aeration	type: aeration	type: aeration	type: aeration	type: aeration
Temperature (°C)	32	32	32	32	32	32
Time (hours)	≥ 16	≥ 16	≥ 16	≥ 16	≥ 16	≥ 16
POST EXPOSURE CONDITIONING III						
Note	type: storage in dark; RH: 50%	type: storage in dark; RH: 50%	type: storage in dark; RH: 50%	type: storage in dark; RH: 50%	type: storage in dark; RH: 50%	type: storage in dark; RH: 50%
Temperature (°C)	21	21	21	21	21	21
Time (hours)	336	336	336	336	336	336
PROPERTIES RETAINED (%)						
Tensile Strength @ Yield	103.6 (ii)	105.4 (ii)	100 (ii)	98.2 (ii)	100 (ii)	101.3 (ii)
Notched Izod Impact	95.3 (fp)	92.2 (fp)	116 (fp)	136 (fp)	104.8 (fp)	119 (fp)

TABLE 14: Effect of Ethylene Oxide Sterilization on Dow Magnum ABS

Material Family	ABS		
Material Supplier/Name	**DOW MAGNUM 9020**	**DOW MAGNUM 343**	**DOW MAGNUM CLR95**
Material Note	high gloss	low gloss	transparent
Reference No.	3	3	3
EXPOSURE CONDITIONS			
Type	Ethylene Oxide	Ethylene Oxide	Ethylene Oxide
Details	12% EtO and 88% Freon	12% EtO and 88% Freon	12% EtO and 88% Freon
Concentration	660 mg/l	660 mg/l	660 mg/l
Number of Cycles	1	1	1
Note	RH: 60%	RH: 60%	RH: 60%
Temperature (°C)	49	49	49
Time (hours)	≥ 6	≥ 6	≥ 6
PRE EXPOSURE CONDITIONING			
Preconditioning Note	time: 8 hours; temperature: 37.8°C; RH: 60%	time: 8 hours; temperature: 37.8°C; RH: 60%	time: 8 hours; temperature: 37.8°C; RH: 60%
POST EXPOSURE CONDITIONING			
Note	type: evacuation; pressure: 127 mm Hg	type: evacuation; pressure: 127 mm Hg	type: evacuation; pressure: 127 mm Hg
POST EXPOSURE CONDITIONING II			
Note	type: aeration	type: aeration	type: aeration
Temperature (°C)	32	32	32
RESIDUALS (ppm)			
Residuals Determined	ethylene oxide	ethylene oxide	ethylene oxide
24 hour Aeration	470	470	265
48 hour Aeration			178
72 hour Aeration	291	291	162
168 hour Aeration	169	169	104
720 hour Aeration	163	163	97
768 hour Aeration	141	141	67
840 hour Aeration	75	75	41
888 hour Aeration	60	60	30

GRAPH 01: **Post Gamma Radiation Exposure Time vs. Yellowness Index of ABS**

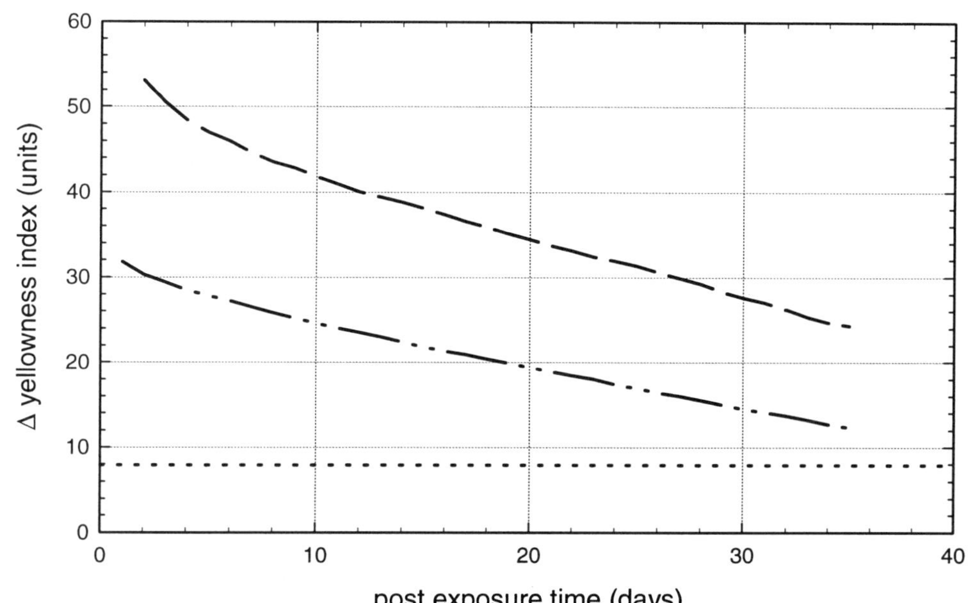

………………	Dow Magnum CLR95 ABS; before exposure
—··—··	Dow Magnum CLR95 ABS; 2.5 Mrads
– – – –	Dow Magnum CLR95 ABS; 10 Mrads
Reference No.	3

GRAPH 02: **Post Gamma Radiation Exposure Time vs. Delta E Color Change of ABS**

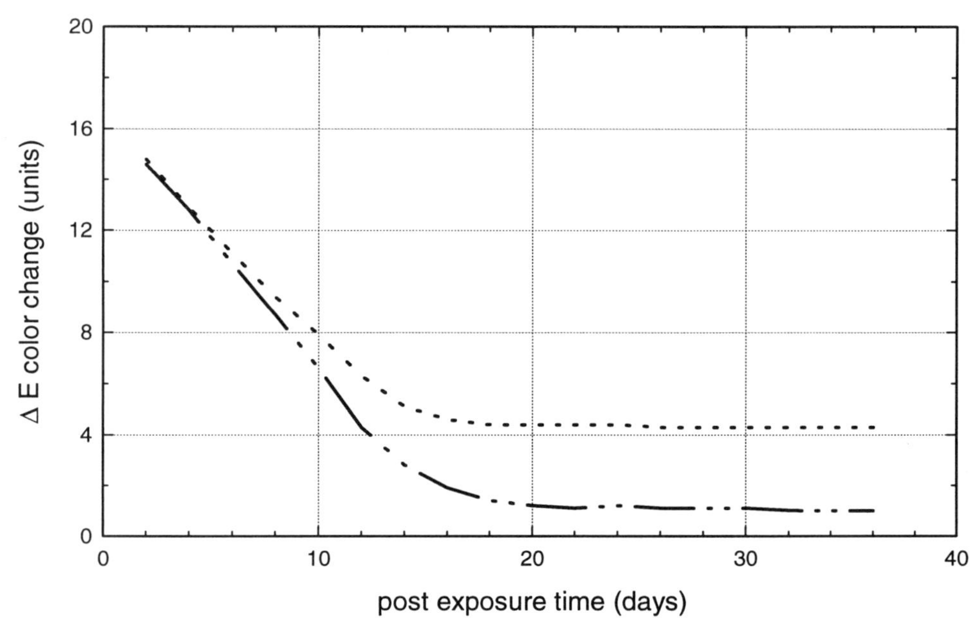

………………	Dow Magnum 343 ABS (low gloss); 2.5 Mrads; dark storage
—··—··	Dow Magnum 343 ABS (low gloss); 2.5 Mrads; light storage
Reference No.	3

GRAPH 03: **Post Gamma Radiation Exposure Time vs. Delta E Color Change of ABS**

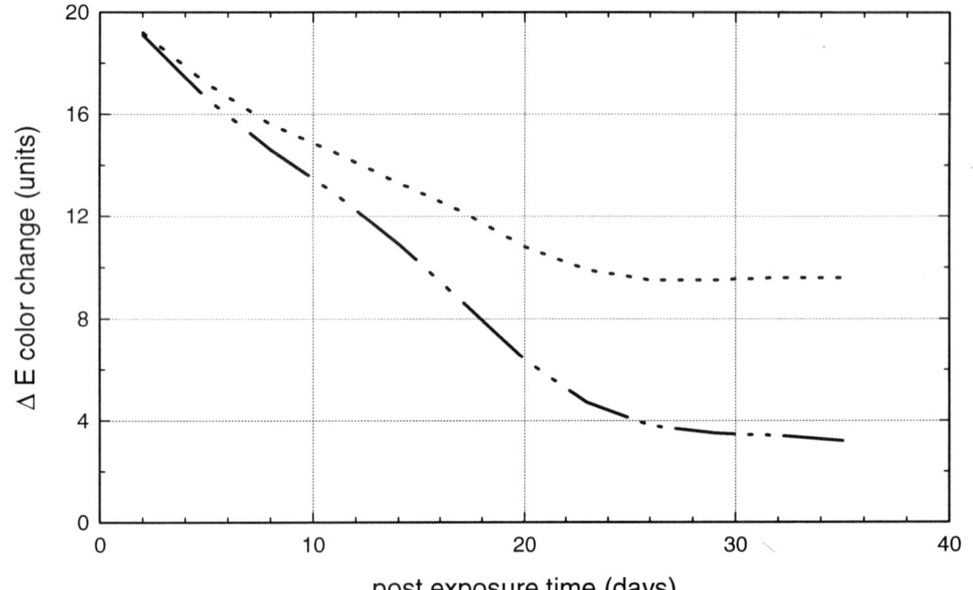

··············	Dow Magnum 343 ABS (low gloss); 10 Mrads; dark storage
—·· —··	Dow Magnum 343 ABS (low gloss); 10 Mrads; light storage
Reference No.	3

GRAPH 04: **Post Gamma Radiation Exposure Time vs. Delta E Color Change of ABS**

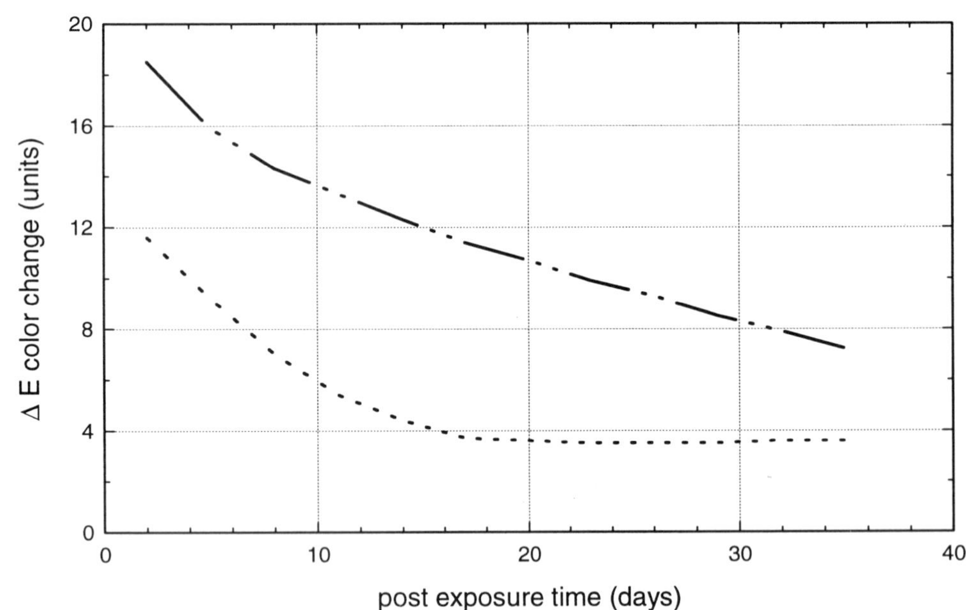

··············	Dow Magnum 9020 ABS (high gloss); 2.5 Mrads
—·· —··	Dow Magnum 9020 ABS (high gloss); 10 Mrads
Reference No.	3

GRAPH 05: **Beta Radiation Dose vs. Tensile Strength of ABS**

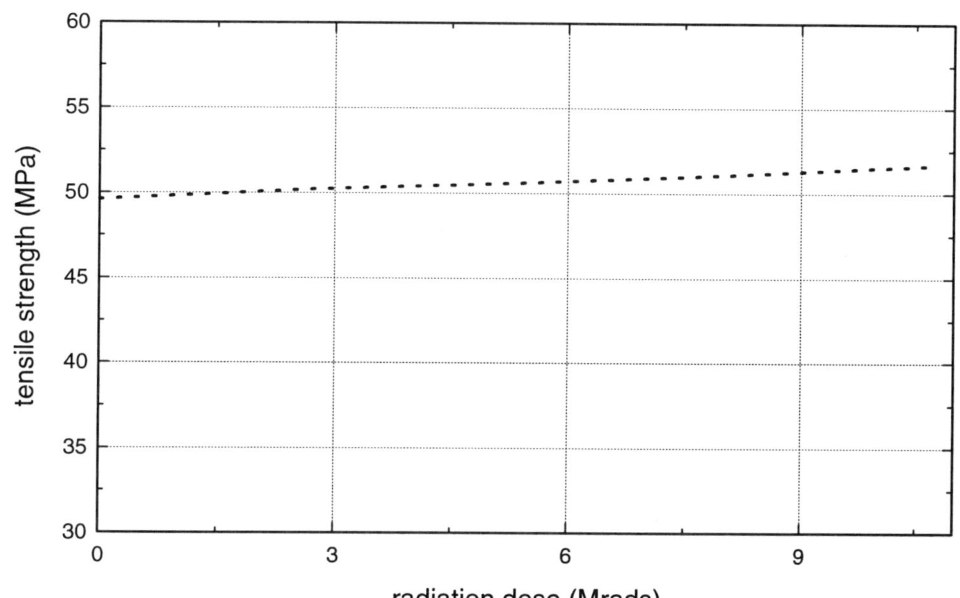

....	ABS (transpar.)
Reference No.	105

GRAPH 06: **Beta Radiation Dose vs. Tensile Modulus of ABS**

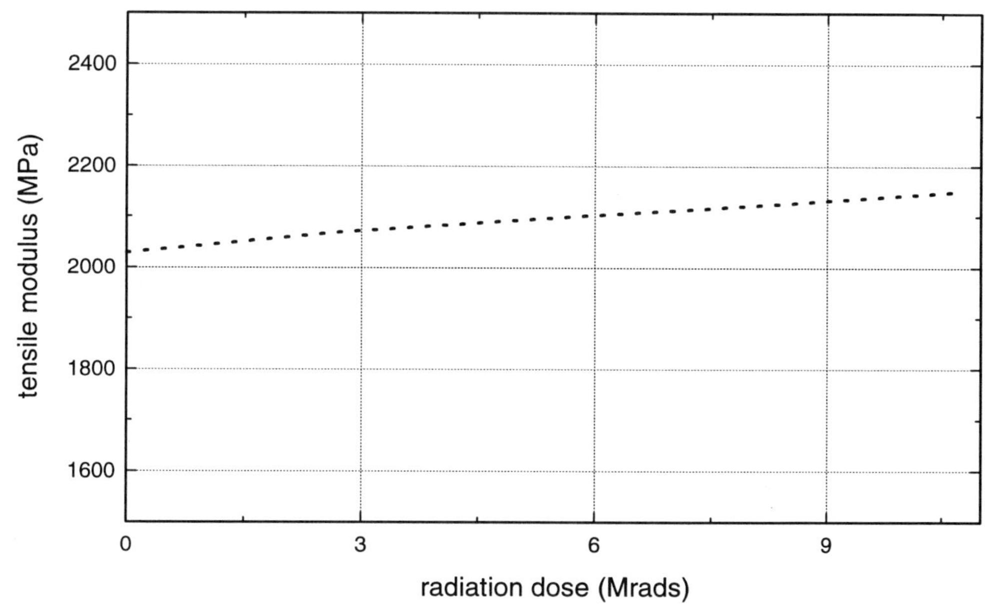

....	ABS (transpar.)
Reference No.	105

GRAPH 07: **Beta Radiation Dose vs. Notched Izod Impact Strength of ABS**

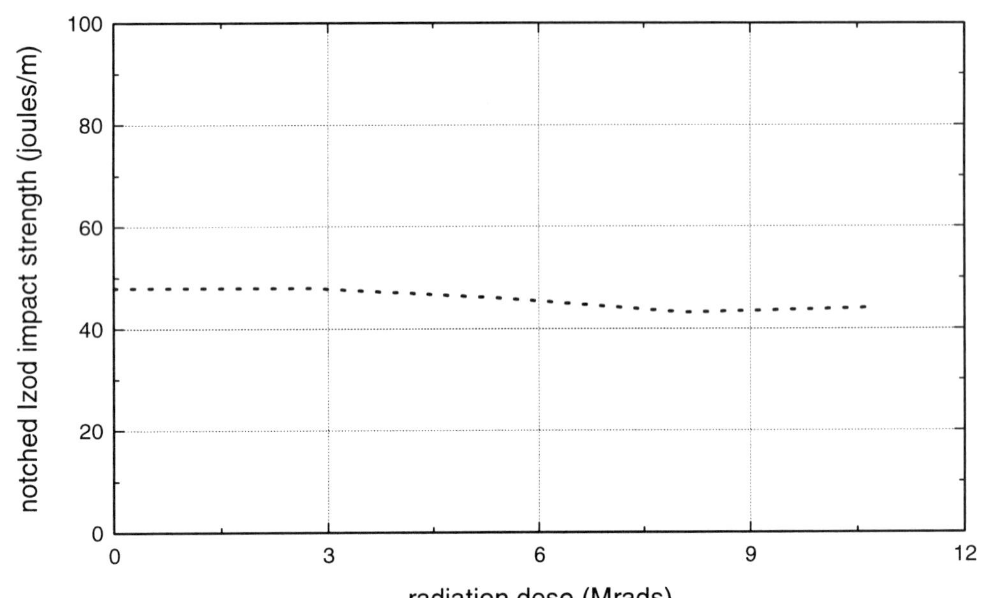

....	ABS (transpar.)
Reference No.	105

GRAPH 08: **Beta Radiation Dose vs. Yellowness Index of ABS**

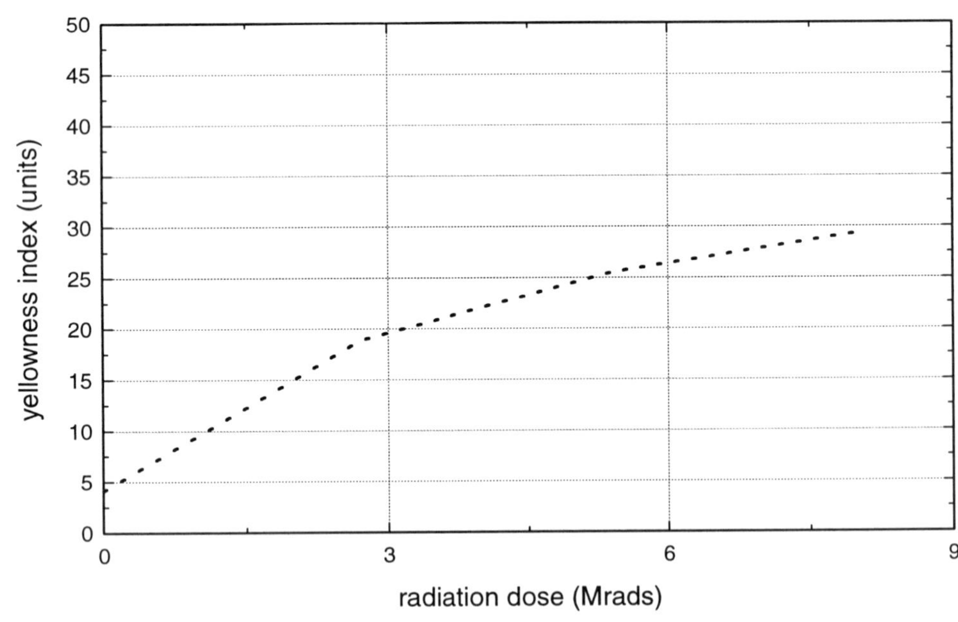

....	ABS (transpar.)
Reference No.	105

GRAPH 09: Post Beta Radiation Exposure Time vs. Yellowness Index of ABS

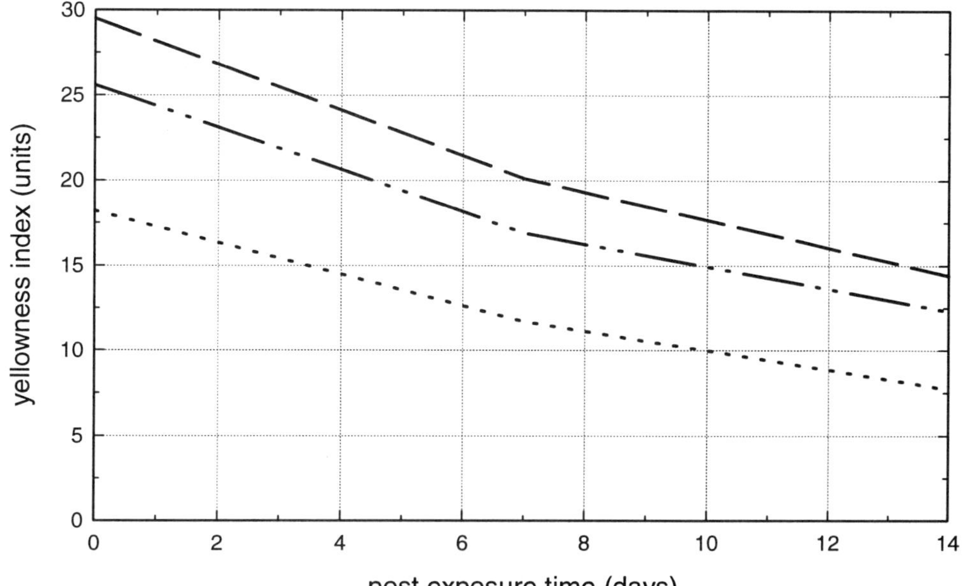

....	ABS (transpar.); 2.7 Mrads
—··—··	ABS (transpar.); 5.4 Mrads
— — — —	ABS (transpar.); 8.1 Mrads
Reference No.	105

Gamma Radiation Resistance

Cyro: Cyrolite G20 HIFLO (features: transparent, impact modified); **Cyrolite G20-001** (features: transparent, impact modified); **XT 250-000** (features: transparent, impact modified); **XT 250-301** (features: transparent, impact modified); **XT 375-000** (features: transparent, impact modified); **XT 375-301** (features: transparent, impact modified); **XT X800-301** (features: transparent, impact modified)

At a level of 2.5 Mrads, Cyrolite polymers acquire a greenish color. The color change was less obvious compared to XT. Cyrolite G20 and G20 Hiflo (G20-001 has a very similar result to G20 Hiflo) develop a light green color while XT 250-301, XT 375-301, and XT X800-301 were dark olive green at medical doses (2.5 Mrads). The yellowness index on the first day of irradiation was around 14.0, while that of XT is in the region of 30.0. To determine if the dyes incorporated were responsible for this green color, polymers containing no dyes were also irradiated. Cyrolite turned yellow and XT polymer (XT 250-000, XT 375-000) gave an amber color. It is preferable to use a dyed system to minimize the noticeable effect of color development. The green color generated was found to have a lower yellowness index value in comparison to an undyed specimen. Both Cyrolite G20 and XT polymer without antioxidant turned yellow with XT darker (close to amber) due to the presence of acrylonitrile in the formulation.

It is believed that gamma radiation causes a chain scission reaction to occur in acrylics by way of radical formation producing lower molecular weight fragments. Interaction between the trapped free radicals or ionic species with oxygen diffusing into the polymer is the reason why the polymer behaves in the above mentioned manner.

It is possible to accelerate the aging process by heating test specimens at 79.5°C for two hours to see what the long term discoloration would look like.

The color that is generated is uniform throughout the whole specimen on the first day of irradiation. After examining sample discs a few days later, we have seen a colorless sharp border which follows the surface contour. It slowly moves away from the surface toward the body of the disc and is less noticeable after a period of eight weeks.

The fact that color disappears starting from the surface has led us to believe that there is an interaction between the trapped free radicals or ionic species and oxygen which is diffusing into the polymer. The presence of the free radicals was also confirmed by ESR spectroscopy which is highly sensitive to free radicals.

The bleaching effect was slowed down considerably when discs were put into a glass jar filled with nitrogen gas. Likely the bleaching did not stop due to a small percentage of oxygen still present in the jar.

A slight increase in hardness and tensile strength may be seen. We believe this arises from the crosslinking taking place in polybutadiene which is used as an impact modifier.

The physical properties after irradiation at medical doses are not reduced to cause lack of performance. Care must be taken to follow all the rules to avoid stressed and oriented products, which is the main reason for product failure. Purple dyes used to mask the yellow color and a few of the antioxidants have yielded promising results.

Cyrolite polymer should be recommended in applications over XT where optical properties and color which is more obvious in thicker sections, can be considered acceptable.

Reference: Atakent, A. I., *Gamma Radiation Study,* supplier technical report - Cyro Industries, 1990.

Cyro: Acrylite H16 (features: transparent)

At level of 2.5 Mrads, Acrylite acquires a yellow color which is easily noticeable and deepens with elevated doses. The maximum dose level attempted is 10.0 Mrads. Acrylite discs irradiated at 2.5 and 5.0 Mrads, which when checked nine months later, were almost identical to unradiated control discs, while those irradiated at 7.5 and 10.0 Mrad levels were only slightly yellow. This yellowness is not apparent when the discs are viewed on a flat surface. The edge colors must be compared to tell the difference

Reference: Atakent, A. I., *Gamma Radiation Study,* supplier technical report - Cyro Industries, 1990.

Rohm & Haas: Plexiglas G (features: transparent)

Plexiglas G is transparent to gamma rays in all ordinary thicknesses. Colorless Plexiglas sheet has about the same gamma ray absorption coefficient as water. However, high dosage and intensity, as is common in sterilizing, may cause discoloration or even the loss of some physical properties. The effect of gamma radiation on methyl methacrylate with a dosage of 1017 nvt is an approximate 50% decrease in tensile and impact strength. The sample became very dark and light transmission decreased.

Reference: *Plastics Technology Center Report,* competitor's technical report (Form 17479) - Atohaas, 1992.

Rohm & Haas: Plexiglas SG-7 (features: transparent, impact modified)

Plexiglas SG-7 exposed to 5.0 Mrads of gamma radiation experiences virtually no yellowing or discoloration. Properties such as impact strength, tensile and flexural strength, modulus of elasticity and percent elongation are constant.

Reference: *Plexiglas Acrylic Resin From AtoHaas Clearly The Best.,* supplier marketing literature (PL-1700b) - Atohaas, 1993.

ICI: Perspex CP1000IG (features: gamma radiation stabilized, transparent, impact modified); **Perspex CP924G** (features: gamma radiation stabilized, transparent, impact modified); **Perspex CP927G** (features: gamma radiation stabilized, transparent, impact modified); **Perspex CP927GHF** (features: gamma radiation stabilized, transparent, FDA grade, impact modified)

Gamma ray radiation generated by Cobalt 60 isotopes has a tendency to discolor (yellow) most acrylics. For Perspex medical grades, this yellowing is temporary and recovery is accomplished via either time or temperature, or both.

Reference: *Perspex Acrylic Polymers,* supplier marketing literature (MG1 1/93) - ICI Acrylics, 1993.

Rohm & Haas: Plexiglas DR-G (features: gamma radiation stabilized, transparent, high impact); **Plexiglas HFI-10G** (features: gamma radiation stabilized, transparent, high impact); **Plexiglas HFI-7G** (features: gamma radiation stabilized, transparent, medium impact); **Plexiglas MI-7G** (features: gamma radiation stabilized, transparent, medium impact); **Plexiglas V-Grade** (features: transparent, general purpose grade)

After exposure to 2.5 Mrads, Plexiglas HFI-7G and Plexiglas HFI-10G acrylics recover quickly, approaching their original color in approximately 56 days. In less than three months, Plexiglas DR-G acrylic recovers to the same endpoint. Recovery time can be dramatically accelerated by storing gamma irradiated parts at elevated temperatures. The temperature selected should be compatible with the specific type of acrylic and the packaging in which the part is stored. The thickness of gamma radiation resistant acrylics has a direct influence on yellowing and recovery time. The thicker the plastic, the greater the initial yellowing and the longer the recovery time. After 84 days, the recovery of thicker samples is less than that of thinner samples.

Reference: *Plexiglas Acrylic Molding Pellets For Medical Applications,* supplier marketing literature (PL-1519a) - Rohm and Haas, 1986.

Novacor Chemicals: (composition: 100% methyl methacrylate; features: transparent)

Acrylic (methyl methacrylate) with various levels of styrene (0 to 80%) were irradiated. Tensile strength, tensile elongation, and Izod were all tested, as well as color, shortly after and five weeks after exposure. The acrylic rapidly lost one third of its tensile strength, and ultimately lost 44%. The three copolymers maintained their original tensile strength, even through 6.0 Mrads of exposure. Tensile elongation followed a similar pattern. All materials start out with 2-3% elongation, but the acrylic resin lost 50% of its elongation, even at a very low radiation dosage, while the copolymers showed good retention over the entire exposure range. The Izod impact strengths of the copolymers were initially stronger than the acrylic resin; there was no appreciable loss of any impact strengths. Generally the yellowing increased with the amount of radiation and the methyl methacrylate content of the polymer. A decrease in color intensity occurred over time in all cases, resulting in the three copolymers exposed at 1.27 and 2.80 Mrads being nearly equal to the unexposed specimens. However the acrylic resin remained light yellow.

Reference: Hauser, Deborah I., Netolicky, William, *The Effects Of Gamma And Ethylene Oxide Sterilization On Styrene Methyl Methacrylate Copolymers And A Toughened Terpolymer,* Medical '91, Ghent, Belgium, conference proceedings - Novacor Chemicals Inc., 1991.

Acrylic (features: transparent, general purpose grade); **Acrylic** (features: transparent, impact modified)

Both the acrylic and the toughened acrylic had no change in flexural modulus at both 3.0 and 5.0 Mrads. The acrylic resin lost 40% tensile strength at 5.0 Mrads exposure; the toughened acrylic showed no meaningful change. The toughened acrylic lost approximately 9% notched Izod impact at 5.0 Mrads. The two acrylic resins yellowed to an index of over 25 at 5.0 Mrads of exposure. After 4 weeks both acrylic resins remained at approximately a 15 yellowness index. These results were obtained with recovery taking place in the light. Results in the dark were not materially different.

Reference: Hauser, Deborah I., Netolicky, William, *The Effects Of Gamma And Ethylene Oxide Sterilization On Styrene Methyl Methacrylate Copolymers And A Toughened Terpolymer,* Medical '91, Ghent, Belgium, conference proceedings - Novacor Chemicals Inc., 1991.

Electron Beam Radiation

ICI: Perspex CP1000IG (features: gamma radiation stabilized, transparent, impact modified); **Perspex CP924G** (features: gamma radiation stabilized, transparent, impact modified); **Perspex CP927G** (features: gamma radiation stabilized, transparent, impact modified); **Perspex CP927GHF** (features: gamma radiation stabilized, transparent, FDA grade, impact modified)

Electron beam radiation that generates electrons via a linear accelerator is attractive due to its speed and the lack of any lingering radioactivity when the E-beam generator is shut down. E-beam radiation will cause yellowing. The higher the dosage, the greater the yellowing and the longer the time of recovery. As the temperature increases the yellowness index is drastically and quickly reduced over a 24 hour time period. Even if left at room temperature, the yellowness index is cut in half after 20 days and almost non-existent after 60 days. After 5 Mrads the flexural strength, tensile strength and impact strength show very little change.

Reference: *Perspex Acrylic Polymers,* supplier marketing literature (MG1 1/93) - ICI Acrylics, 1993.

Beta Radiation Resistance

Acrylic (features: transparent, impact modified)

Beta sterilization is increasing in importance in Europe. Samples of impact modified acrylic were tested at beta radiation levels of 2.7 to 10.8 Mrads. Physical properties were retained after 10.8 Mrads. In contrast to gamma sterilization, beta sterilization generates more heat. At 10.8 Mrads the surface of the acrylic began to adhere to the packaging. The material became quite yellow. After two weeks of storage there was some reduction in yellowness. Storage in light or dark did not have a meaningful impact.

Reference: Haines, Jim, Hauser, Debbie, *Zylar Clear Alloys For Medical Applications,* supplier technical report - Novacor Chemicals Inc.

Ethylene Oxide (EtO) Resistance

Rohm & Haas: Plexiglas DR-G (features: gamma radiation stabilized, transparent, high impact); **Plexiglas HFI-10G** (features: gamma radiation stabilized, transparent, high impact); **Plexiglas HFI-7G** (features: gamma radiation stabilized, transparent, medium impact); **Plexiglas MI-7G** (features: gamma radiation stabilized, transparent, medium impact); **Plexiglas V-Grade** (features: transparent, general purpose grade)

Acrylics and impact modified acrylics are compatible with ethylene oxide gas and can be EtO sterilized without adversely affecting the medical device.

Reference: *Plexiglas Acrylic Molding Pellets For Medical Applications,* supplier marketing literature (PL-1519a) - Rohm and Haas, 1986.

Continental: Acrycal MP CP61 (features: transparent)

Tensile yield properties are not affected with 1 cycle of CFC-12/EtO sterilization. Three other test conditions (2 cycles CFC-12/EtO and 1 and 2 cycles of HCFC-124/EtO) decrease the tensile yield by 14-23%. Acrylic has low instrumented impact values and no significant differences are seen in the data after any sterilization cycle. Appearance tests also show no differences in the sterilization techniques. Any differences in the percent haze values are attributed to the increase in handling of the samples.

Reference: Hermanson, Nancy J., *Effects Of Alternate Carriers Of Ethylene Oxide Sterilant On Thermoplastics,* supplier technical report (301-02018) - Dow Chemical Company.

Rohm & Haas: Plexiglas SG-7 (features: transparent, impact modified)

After EtO sterilization properties such as impact strength, tensile and flexural strength, modulus of elasticity and percent elongation are constant.

Reference: *Plexiglas Acrylic Resin From AtoHaas Clearly The Best.,* supplier marketing literature (PL-1700b) - Atohaas, 1993.

ICI: Perspex CP1000IG (features: gamma radiation stabilized, transparent, impact modified); **Perspex CP924G** (features: gamma radiation stabilized, transparent, impact modified); **Perspex CP927G** (features: gamma radiation stabilized, transparent, impact modified); **Perspex CP927GHF** (features: gamma radiation stabilized, transparent, FDA grade, impact modified)

Acrylics should be sterilized with dry ethylene oxide as wet ethylene oxide will cause part crazing. Relative humidity in the chamber must be kept below 0.5%.

Reference: *Perspex Acrylic Polymers,* supplier marketing literature (MG1 1/93) - ICI Acrylics, 1993.

TABLE 15: Effect of Gamma Radiation Sterilization on Acrylic Resin

Material Family	ACRYLIC RESIN								
Material Supplier/Name	CYRO CYROLITE G			CYRO CYROLITE G10			CYRO CYROLITE G20-001		
Material Note	transparent	transparent	transparent	transparent	transparent	transparent	transparent, impact modified	transparent, impact modified	transparent, impact modified
Reference No.	19	19	19	19	19	19	19	19	19
EXPOSURE CONDITIONS									
Type	Gamma Radiation	Gamma Radiation	Gamma Radiation	Gamma Radiation	Gamma Radiation	Gamma Radiation	Gamma Radiation	Gamma Radiation	Gamma Radiation
Details	source: Cobalt 60	source: Cobalt 60	source: Cobalt 60	source: Cobalt 60	source: Cobalt 60	source: Cobalt 60	source: Cobalt 60	source: Cobalt 60	source: Cobalt 60
Radiation Dose (Mrads)	2.5	5	7.5	2.5	5	7.5	2.5	5	7.5
PROPERTIES RETAINED (%)									
Tensile Strength	84 (he)	88 (he)	74.7 (he)	101.4 (he)	101.4 (he)	101.4 (he)	98.6 (he)	100 (he)	98.6 (he)
Elongation	68.4 (ak)	84.2 (ak)	68.4 (ak)	130.8 (ak)	111 (ak)	90.1 (ak)	93 (ak)	100.9 (ak)	75.7 (ak)
Elongation @ Yield	68.4 (bd)	84.2 (bd)	68.4 (bd)	103.2 (bd)	103.2 (bd)	106.5 (bd)	97.2 (bd)	102.8 (bd)	102.8 (bd)
Flexural Strength	79.8 (ck)	77.4 (ck)	64.3 (ck)	101.9 (ck)	102.9 (ck)	102.9 (ck)	102.9 (ck)	102 (ck)	102.9 (ck)
Modulus	102.1 (gs)	100 (gs)	100 (gs)	100 (gs)	100 (gs)	100 (gs)	106.1 (gs)	100 (gs)	100 (gs)
Flexural Modulus	100 (bz)	100 (bz)	100 (bz)	97.1 (bz)	100 (bz)	100 (bz)	100 (bz)	100 (bz)	100 (bz)
Notched Izod Impact	100 (fm)	95.5 (fm)	100 (fm)	96 (fm)	96 (fm)	84 (fm)	89.5 (fm)	84.2 (fm)	78.9 (fm)
Heat Deflection Temperature	98.9 (k)	98.9 (k)	97.7 (k)	98.9 (k)	100 (k)	97.8 (k)	100 (k)	101.1 (k)	101.1 (k)
Vicat Softening Point	99 (ks)	99 (ks)	98.1 (ks)	97.2 (ks)	100.9 (ks)	97.2 (ks)	100 (ks)	102 (ks)	100 (ks)
SURFACE and APPEARANCE									
Hardness Units Change	M0 (gl)	M-1 (gl)	M0 (gl)	M2 (gl)	M-1 (gl)	M2 (gl)	M1 (gl)	M0 (gl)	M4 (gl)
Δ Yellowness Index	17.3	22	27.6	11.9	17.1	19.9	13.4	20.7	19.7
Haze (x-direction) Retained (%)	120	160	140	189.4	88.8	101.9	107	94.7	96.5
Haze (y-direction) Retained (%)	120	170	140	191.4	89.5	102.5	107	94.7	96.5
Haze (z-direction) Retained (%)	114.3	142.9	128.6	180.7	90.9	103.6	104.5	110.5	107.5
Transmittance Retained (%)	96.2	94.3	91.5	94.1	93.2	91.7	95	93.3	93.4

TABLE 16: Effect of Gamma Radiation Sterilization on Acrylic Resin

Material Family	ACRYLIC RESIN					
Material Supplier/Name	CYRO CYROLITE G20 HIFLO			CYRO XT 250-301		
Material Note	transparent, impact modified	transparent, impact modified	transparent, impact modified	transparent, impact modified	transparent, impact modified	transparent, impact modified
Reference No.	19	19	19	19	19	19

EXPOSURE CONDITIONS

Type	Gamma Radiation	Gamma Radiation	Gamma Radiation	Gamma Radiation	Gamma Radiation	Gamma Radiation
Details	source: Cobalt 60	source: Cobalt 60	source: Cobalt 60	source: Cobalt 60	source: Cobalt 60	source: Cobalt 60
Radiation Dose (Mrads)	2.5	5	7.5	2.5	5	7.5

PROPERTIES RETAINED (%)

Tensile Strength	100 (he)	100 (he)	101.5 (he)	101.3 (he)	101.3 (he)	101.3 (he)
Elongation	93.6 (ak)	72.7 (ak)	150.9 (ak)	163.3 (ak)	153.1 (ak)	115.3 (ak)
Elongation @ Yield	100 (bd)	103.1 (bd)	103.1 (bd)	103.1 (bd)	103.1 (bd)	103.1 (bd)
Flexural Strength	101 (ck)	101 (ck)	102.1 (ck)	92.3 (ck)	100 (ck)	102.6 (ck)
Modulus	100 (gs)	97 (gs)	97 (gs)	102.6 (gs)	100 (gs)	102.6 (gs)
Flexural Modulus	100 (bz)	100 (bz)	100 (bz)	102.6 (bz)	100 (bz)	100 (bz)
Notched Izod Impact	94.4 (fm)	88.9 (fm)	72.2 (fm)	92.9 (fm)	85.7 (fm)	78.6 (fm)
Heat Deflection Temperature	101.2 (k)	98.8 (k)	98.8 (k)	100 (k)	100 (k)	97.8 (k)
Vicat Softening Point	101.9 (ks)	100 (ks)	100 (ks)	101.9 (ks)	101.9 (ks)	104.9 (ks)

SURFACE and APPEARANCE

Hardness Units Change	M0 (gl)	M4 (gl)	M6 (gl)	M1 (gl)	M2 (gl)	M0 (gl)
Δ Yellowness Index	12.2	17.5	21.5	26.2	34.3	39.6
Haze (x-direction) Retained (%)	104.4	111.1	104.4	84.4	71.9	67.2
Haze (y-direction) Retained (%)	105.7	113.6	106.8	84.6	72.3	66.9
Haze (z-direction) Retained (%)	100.8	113	107.3	88.6	75.7	77.1
Transmittance Retained (%)	94.6	92.7	91.2	85.4	79.8	77.1

TABLE 17: Effect of Gamma Radiation Sterilization on Acrylic Resin

Material Family	ACRYLIC RESIN					
Material Supplier/Name	CYRO XT 375-301			CYRO XT X800-301		
Material Note	transparent, impact modified	transparent, impact modified	transparent, impact modified	transparent, impact modified	transparent, impact modified	transparent, impact modified
Reference No.	19	19	19	19	19	19
EXPOSURE CONDITIONS						
Type	Gamma Radiation	Gamma Radiation	Gamma Radiation	Gamma Radiation	Gamma Radiation	Gamma Radiation
Details	source: Cobalt 60	source: Cobalt 60	source: Cobalt 60	source: Cobalt 60	source: Cobalt 60	source: Cobalt 60
Radiation Dose (Mrads)	2.5	5	7.5	2.5	5	7.5
PROPERTIES RETAINED (%)						
Tensile Strength	102.7 (he)	102.7 (he)	102.7 (he)	97 (he)	101.5 (he)	101.5 (he)
Elongation	138.1 (ak)	161.1 (ak)	146.8 (ak)	81.3 (ak)	173.8 (ak)	122.5 (ak)
Elongation @ Yield	103 (bd)	103 (bd)	103 (bd)	100 (bd)	103.4 (bd)	110.3 (bd)
Flexural Strength	92 (ck)	101.8 (ck)	99.1 (ck)	96.1 (ck)	100 (ck)	101 (ck)
Modulus	100 (gs)	102.8 (gs)	100 (gs)	100 (gs)	100 (gs)	100 (gs)
Flexural Modulus	100 (bz)	100 (bz)	97.2 (bz)	97.1 (bz)	100 (bz)	100 (bz)
Notched Izod Impact	90.5 (fm)	81 (fm)	76.2 (fm)	94.7 (fm)	84.2 (fm)	78.9 (fm)
Heat Deflection Temperature	101.1 (k)	100 (k)	97.8 (k)	100 (k)	101.1 (k)	98.9 (k)
Vicat Softening Point	100 (ks)	100 (ks)	100 (ks)	100 (ks)	100 (ks)	99 (ks)
SURFACE and APPEARANCE						
Hardness Units Change	M2 (gl)	M1 (gl)	M3 (gl)	M7 (gl)	M6 (gl)	M6 (gl)
Δ Yellowness Index	26.2	34.4	37.8	24	29.9	35.4
Haze (x-direction) Retained (%)	98.5	93.8	104.7	65.7	68.7	82.8
Haze (y-direction) Retained (%)	100.8	96.1	107.8	65.9	68.9	83.7
Haze (z-direction) Retained (%)	103.5	99.4	110.6	62.9	66.2	84.7
Transmittance Retained (%)	85.9	80.8	79.1	83.8	79.7	77.4

TABLE 18: Effect of Gamma Radiation Sterilization on Acrylic Resin

Material Family	ACRYLIC RESIN							
Material Supplier/Name	NOVACOR							
Material Note	100% methyl methacrylate; transparent	100% methyl methacrylate; transparent	100% methyl methacrylate; transparent	100% methyl methacrylate; transparent	100% methyl methacrylate; transparent	100% methyl methacrylate; transparent	100% methyl methacrylate; transparent	100% methyl methacrylate; transparent
Reference No.	106	106	106	106	106	106	106	106
EXPOSURE CONDITIONS								
Type	Gamma Radiation	Gamma Radiation	Gamma Radiation	Gamma Radiation	Gamma Radiation	Gamma Radiation	Gamma Radiation	Gamma Radiation
Radiation Dose (Mrads)	1.27	2.8	3.55	5.43	1.27	2.8	3.55	5.43
POST EXPOSURE CONDITIONING								
Time (hours)	0	0	0	0	840	840	840	840
SURFACE and APPEARANCE								
Yellowness Index note	yellow	dark yellow	dark yellow	yellow orange	light yellow	light yellow	light yellow	dark yellow

TABLE 19: Effect of Gamma Radiation Sterilization on Acrylic

Material Family	ACRYLIC RESIN			
Material Note	transparent, general purpose grade	transparent, general purpose grade	transparent, impact modified	transparent, impact modified
Reference No.	106	106	106	106
EXPOSURE CONDITIONS				
Type	Gamma Radiation	Gamma Radiation	Gamma Radiation	Gamma Radiation
Radiation Dose (Mrads)	3	5	3	5
PROPERTIES RETAINED (%)				
Tensile Strength	80 (he)	58 (he)	100 (he)	100 (he)
Modulus	100 (bz)	100 (bz)	100 (bz)	100 (bz)
Notched Izod Impact			96 (fm)	89 (fm)
SURFACE and APPEARANCE				
Δ Yellowness Index	20 (kt)	24.5 (kt)	14 (kt)	19 (kt)

TABLE 20: Effect of Ethylene Oxide Sterilization on Acrylic Resin

Material Family	ACRYLIC RESIN							
Material Supplier/Name	CONTINENTAL ACRYCAL MP CP61							
Material Note	transparent	transparent	transparent	transparent	transparent	transparent	transparent	transparent
Reference No.	6	6	6	6	6	6	6	6

EXPOSURE CONDITIONS

Type	Ethylene Oxide	Ethylene Oxide	Ethylene Oxide	Ethylene Oxide	Ethylene Oxide	Ethylene Oxide	Ethylene Oxide	Ethylene Oxide
Details	12% EtO and 88% Freon	12% EtO and 88% Freon	12% EtO and 88% Freon	12% EtO and 88% Freon	8.6% EtO and 91.4% HCFC-124	8.6% EtO and 91.4% HCFC-124	8.6% EtO and 91.4% HCFC-124	8.6% EtO and 91.4% HCFC-124
Concentration	600 mg/l	600 mg/l	600 mg/l	600 mg/l				
Number of Cycles	1	1	2	2	1	1	2	2
Note	RH: 60%; test lab: Ethox Corp.	RH: 60%; test lab: Ethox Corp.	RH: 60%; test lab: Ethox Corp.	RH: 60%; test lab: Ethox Corp.	RH: 60%; test lab: Ethox Corp.	RH: 60%; test lab: Ethox Corp.	RH: 60%; test lab: Ethox Corp.	RH: 60%; test lab: Ethox Corp.
Temperature (°C)	48.9	48.9	48.9	48.9	48.9	48.9	48.9	48.9
Time (hours)	6	6	6	6	6	6	6	6

PRE EXPOSURE CONDITIONING

Preconditioning Note	time: 18 hours; temperature: 37.8°C; RH: 60%	time: 18 hours; temperature: 37.8°C; RH: 60%	time: 18 hours; temperature: 37.8°C; RH: 60%	time: 18 hours; temperature: 37.8°C; RH: 60%	time: 18 hours; temperature: 37.8°C; RH: 60%	time: 18 hours; temperature: 37.8°C; RH: 60%	time: 18 hours; temperature: 37.8°C; RH: 60%	time: 18 hours; temperature: 37.8°C; RH: 60%

POST EXPOSURE CONDITIONING

Note	type: aeration; pressure: 127 mm Hg	type: aeration; pressure: 127 mm Hg	type: aeration; pressure: 127 mm Hg	type: aeration; pressure: 127 mm Hg	type: aeration; pressure: 127 mm Hg	type: aeration; pressure: 127 mm Hg	type: aeration; pressure: 127 mm Hg	type: aeration; pressure: 127 mm Hg
Temperature (°C)	32.2	32.2	32.2	32.2	32.2	32.2	32.2	32.2

POST EXPOSURE CONDITIONING II

Note	type: ambient conditions	type: ambient conditions	type: ambient conditions	type: ambient conditions	type: ambient conditions	type: ambient conditions	type: ambient conditions	type: ambient conditions
Time (hours)	168	1344	168	1344	168	1344	168	1344

PROPERTIES RETAINED (%)

Tensile Strength @ Yield	104.2 (io)	102.6 (io)	95.3 (io)	84 (io)	92 (io)	86.1 (io)	87.5 (io)	76.8 (io)
Elongation	40 (bn)	40 (bn)	40 (bn)	40 (bn)	80 (bn)	60 (bn)	60 (bn)	40 (bn)
Modulus	100.7 (gw)	89.4 (gw)	93.5 (gw)	87.4 (gw)	85.6 (gw)	98.7 (gw)	93.7 (gw)	95.5 (gw)
Dart Impact (total energy)	100 (dz)	161.5 (dz)	138.5 (dz)	138.5 (dz)	107.7 (dz)	153.8 (dz)	130.8 (dz)	138.5 (dz)
Dart Impact (peak energy)	110 (da)	160 (da)	160 (da)	140 (da)	120 (da)	160 (da)	130 (da)	140 (da)

SURFACE and APPEARANCE

Yellowness Index	1.12	0.88	1.16	1.12	1	0.94	1.27	1.31
Haze (%)	2.07	2.11	3.36	2.34	1.44	1.54	4.57	4.09
Transmittance (%)	92	91	92	92	92	93	92	92

TABLE 21: Effect of Ethylene Oxide Sterilization on Acrylic

Material Family	ACRYLIC RESIN			
Material Supplier/Name	CONTINENTAL ACRYCAL MP CP61			
Material Note	transparent	transparent	transparent	transparent
Reference No.	6	6	6	6

EXPOSURE CONDITIONS

Type	Ethylene Oxide	Ethylene Oxide	Ethylene Oxide	Ethylene Oxide
Details	12% EtO and 88% Freon	12% EtO and 88% Freon	8.6% EtO and 91.4% HCFC-124	8.6% EtO and 91.4% HCFC-124
Concentration	600 mg/l	600 mg/l		
Number of Cycles	1	1	1	1
Note	RH: 60%; test lab: Ethox Corp.	RH: 60%; test lab: Ethox Corp.	RH: 60%; test lab: Ethox Corp.	RH: 60%; test lab: Ethox Corp.
Temperature (°C)	48.9	48.9	48.9	48.9
Time (hours)	6	6	6	6

PRE EXPOSURE CONDITIONING

Preconditioning Note	time: 18 hours; temperature: 37.8°C; RH: 60%	time: 18 hours; temperature: 37.8°C; RH: 60%	time: 18 hours; temperature: 37.8°C; RH: 60%	time: 18 hours; temperature: 37.8°C; RH: 60%

POST EXPOSURE CONDITIONING

Note	type: aeration; note: 10 air changes per hour	type: aeration; note: 30 air changes per hour	type: aeration; note: 10 air changes per hour	type: aeration; note: 30 air changes per hour
Temperature (°C)	32.2	54.4	32.2	54.4

RESIDUALS (ppm)

Residuals Determined	ethylene oxide	ethylene oxide	ethylene oxide	ethylene oxide
Little or No Aeration	187	187	198	198
17 hour Aeration		42		77
24 hour Aeration	82	28	105	36
48 hour Aeration	47	25	81	35
72 hour Aeration	34		65	

GRAPH 10: Gamma Radiation Dose vs. Tensile Strength of Acrylic Resin

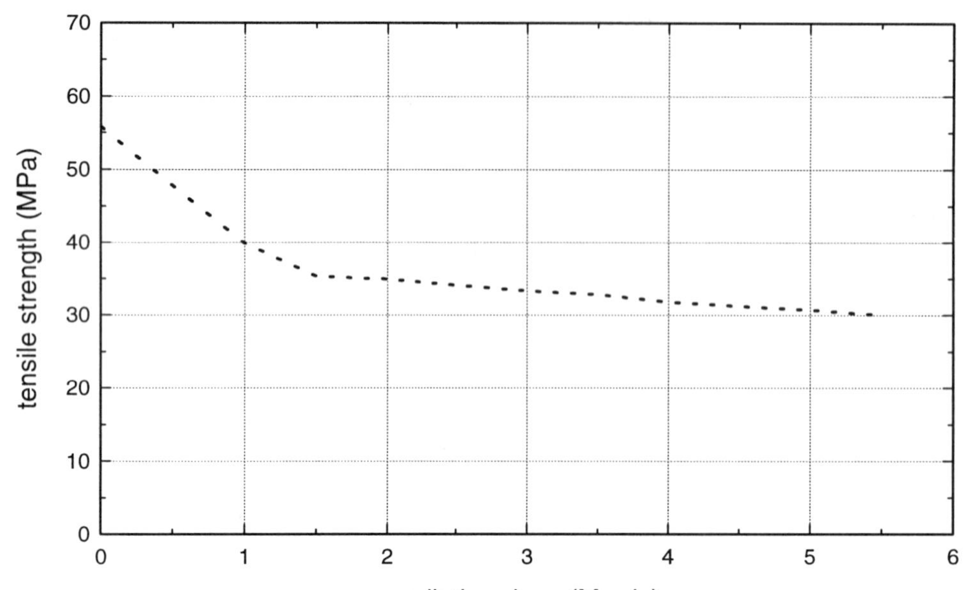

··············	Novacor Acrylic (100% PMMA; transpar.)
Reference No.	106

GRAPH 11: Gamma Radiation Dose vs. Elongation of Acrylic Resin

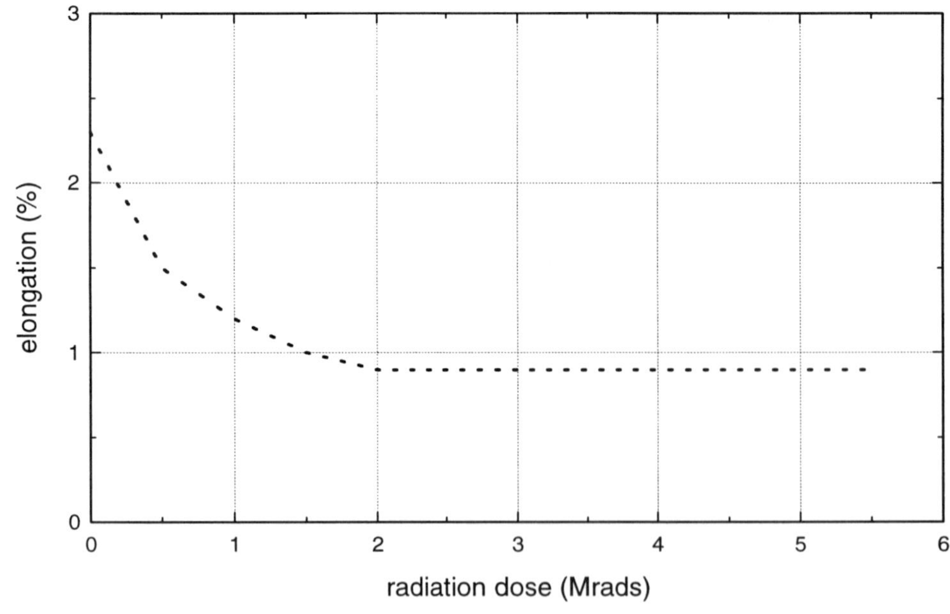

··············	Novacor Acrylic (100% PMMA; transpar.)
Reference No.	106

GRAPH 12: Gamma Radiation Dose vs. Notched Izod Impact Strength of Acrylic Resin

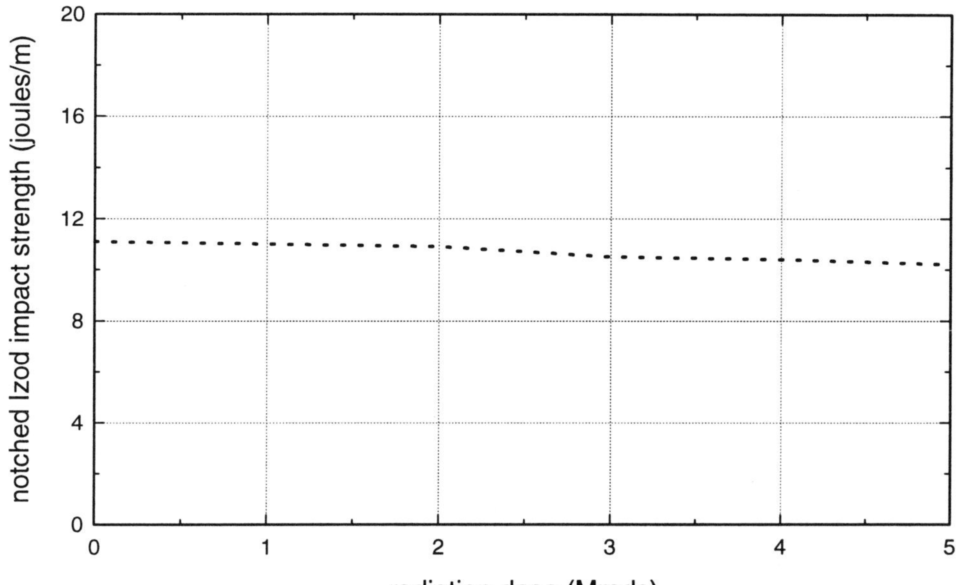

....	Novacor Acrylic (100% PMMA; transpar.)
Reference No.	106

GRAPH 13: Gamma Radiation Dose vs. Yellowness Index of Acrylic Resin

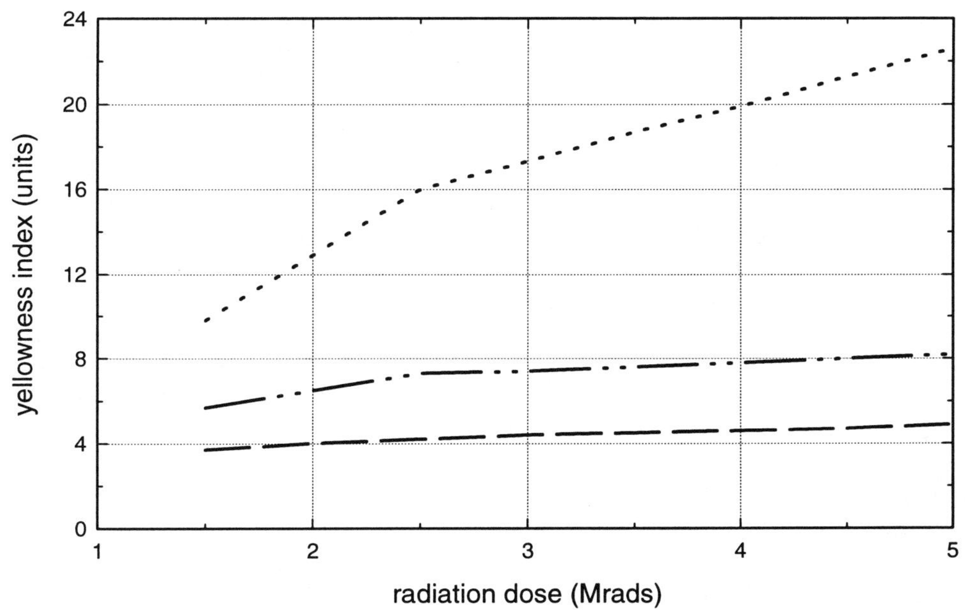

....	Rohm & Haas Plexiglas V-Grade Acrylic; post exposure time: 0 days; thickness: 3.2 mm
—··—··	Rohm & Haas Plexiglas V-Grade Acrylic; post exposure time: 56 days; post exposure temperature: 23°C; thickness: 3.2 mm
— — — —	Rohm & Haas Plexiglas V-Grade Acrylic; post exposure time: 84 days; post exposure temperature: 23°C; thickness: 3.2 mm
Reference No.	90

GRAPH 14: Gamma Radiation Dose vs. Yellowness Index of Acrylic Resin

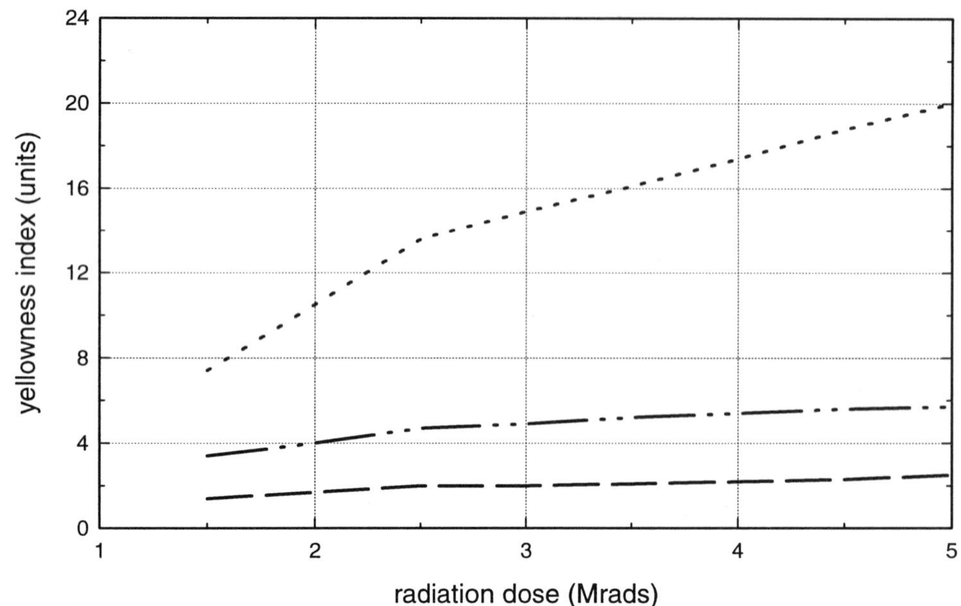

....	Rohm & Haas Plexiglas DR-G Acrylic; post exposure time: 0 days; thickness: 3.2 mm
—·—··	Rohm & Haas Plexiglas DR-G Acrylic; post exposure time: 56 days; post exposure temperature: 23°C; thickness: 3.2 mm
— — — —	Rohm & Haas Plexiglas DR-G Acrylic; post exposure time: 84 days; post exposure temperature: 23°C; thickness: 3.2 mm
Reference No.	90

GRAPH 15: Gamma Radiation Dose vs. Yellowness Index of Acrylic Resin

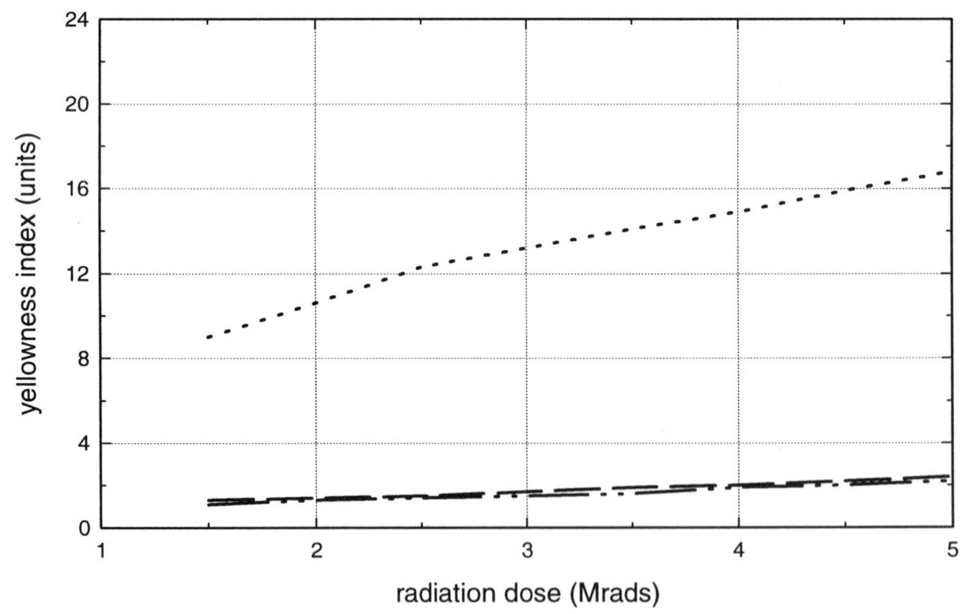

....	Rohm & Haas Plexiglas HFI-10G Acrylic; post exposure time: 0 days; thickness: 3.2 mm
—·—··	Rohm & Haas Plexiglas HFI-10G Acrylic; post exposure time: 56 days; post exposure temperature: 23°C; thickness: 3.2 mm
— — — —	Rohm & Haas Plexiglas HFI-10G Acrylic; post exposure time: 84 days; post exposure temperature: 23°C; thickness: 3.2 mm
Reference No.	90

GRAPH 16: Gamma Radiation Dose vs. Yellowness Index of Acrylic Resin

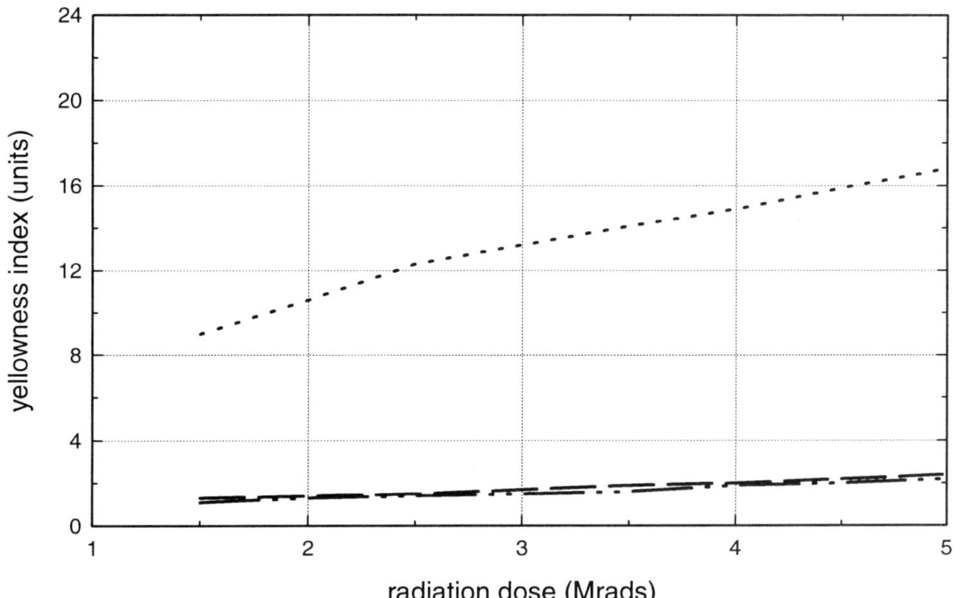

....	Rohm & Haas Plexiglas HFI-7G Acrylic; post exposure time: 0 days; thickness: 3.2 mm
—··—··	Rohm & Haas Plexiglas HFI-7G Acrylic; post exposure time: 56 days; post exposure temperature: 23°C; thickness: 3.2 mm
– – – –	Rohm & Haas Plexiglas HFI-7G Acrylic; post exposure time: 84 days; post exposure temperature: 23°C; thickness: 3.2 mm
Reference No.	90

GRAPH 17: Gamma Radiation Dose vs. Yellowness Index of Acrylic Resin

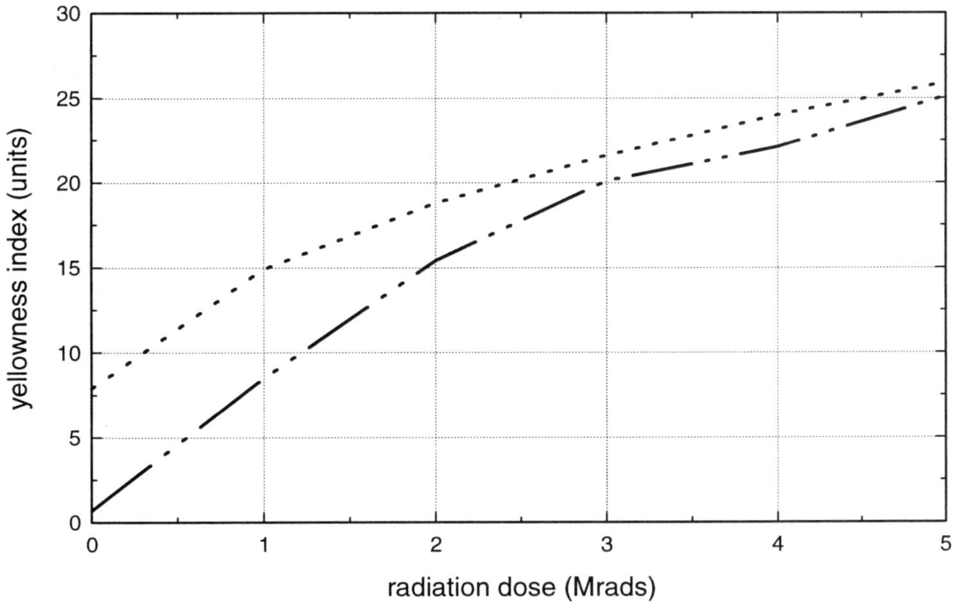

....	Acrylic
—··—··	Acrylic
Reference No.	106

GRAPH 18: **Post Gamma Radiation Exposure Time vs. Yellowness Index of Acrylic Resin**

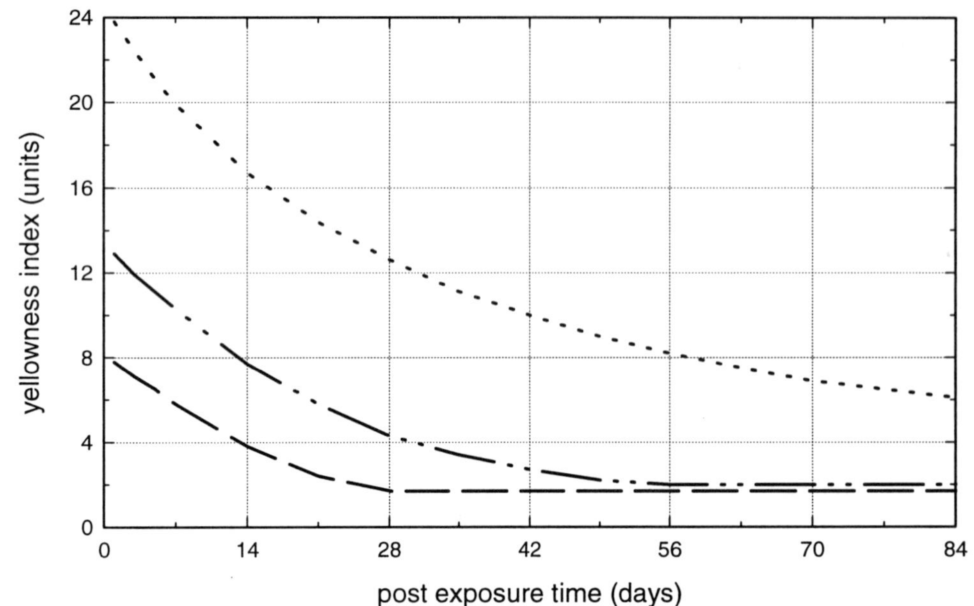

....	Rohm & Haas Plexiglas HFI-10G Acrylic; 2.5 Mrads; post exposure temperature: 23 degrees C; thickness: 6.4 mm
—·—··	Rohm & Haas Plexiglas HFI-10G Acrylic; 2.5 Mrads; post exposure temperature: 23 degrees C; thickness: 3.2 mm
— — — —	Rohm & Haas Plexiglas HFI-10G Acrylic; 2.5 Mrads; post exposure temperature: 23 degrees C; thickness: 1.5 mm
Reference No.	91

GRAPH 19: **Post Gamma Radiation Exposure Time vs. Yellowness Index of Acrylic Resin**

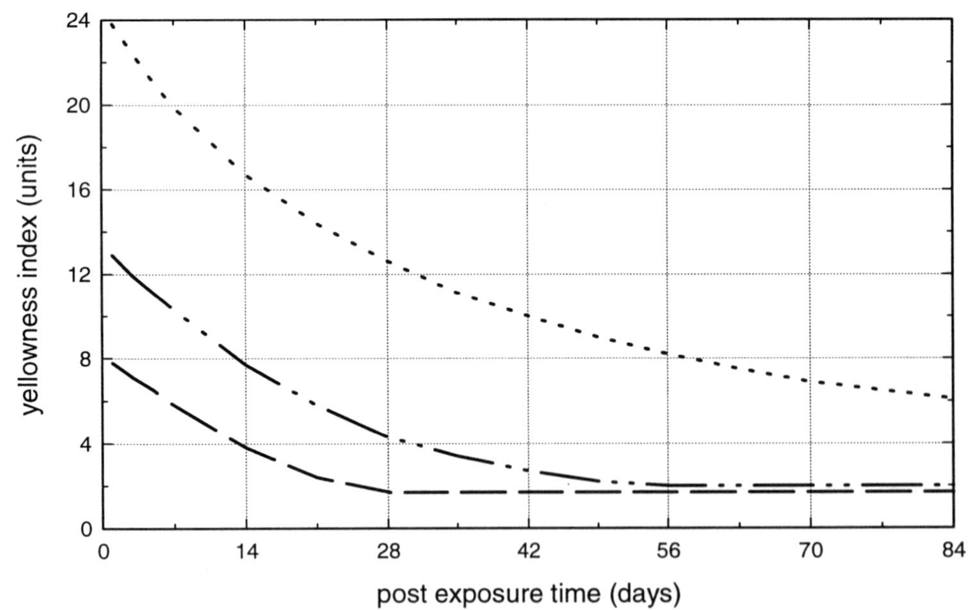

....	Rohm & Haas Plexiglas HFI-7G Acrylic; 2.5 Mrads; post exposure temperature: 23 degrees C; thickness: 6.4 mm
—·—··	Rohm & Haas Plexiglas HFI-7G Acrylic; 2.5 Mrads; post exposure temperature: 23 degrees C; thickness: 3.2 mm
— — — —	Rohm & Haas Plexiglas HFI-7G Acrylic; 2.5 Mrads; post exposure temperature: 23 degrees C; thickness: 1.5 mm
Reference No.	91

GRAPH 20: **Post Gamma Radiation Exposure Time vs. Yellowness Index of Acrylic Resin**

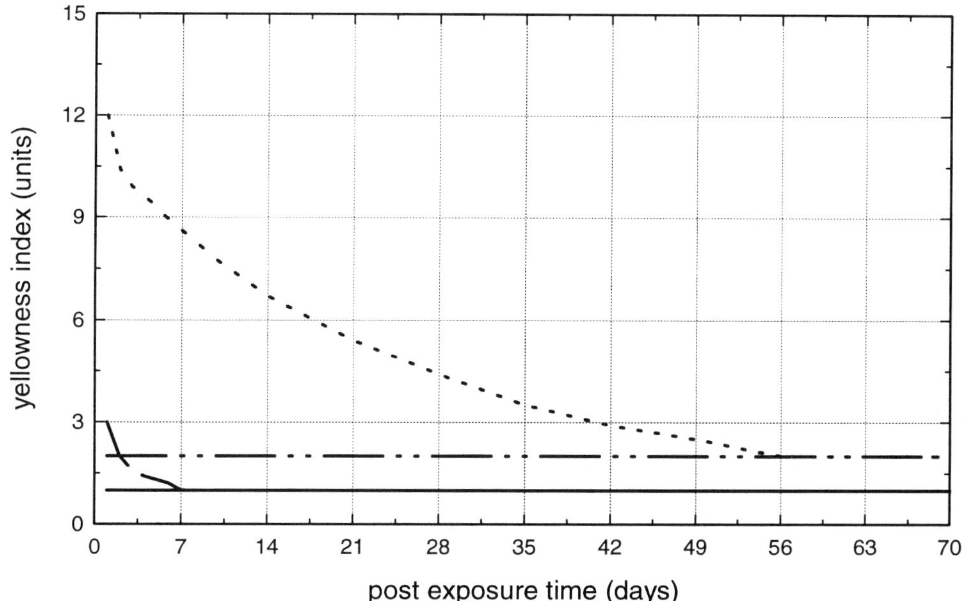

....	Rohm & Haas Plexiglas HFI-7G Acrylic; 5 Mrad
—··—··	Rohm & Haas Plexiglas HFI-7G Acrylic; before exposure
– – – –	Rohm & Haas Plexiglas SG-7 Acrylic; 5 Mrad
———	Rohm & Haas Plexiglas SG-7 Acrylic; before exposure
Reference No.	22

GRAPH 21: **Post Gamma Radiation Exposure Time vs. Yellowness Index of Acrylic Resin**

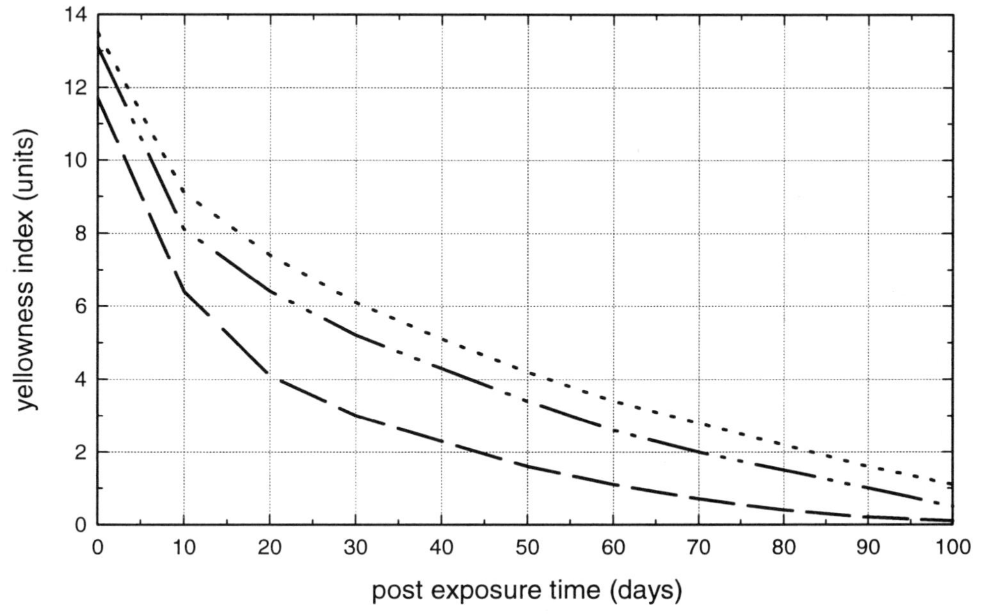

....	ICI Perspex CP924G Acrylic; 5 Mrad; thickness: 3.2 mm; test method: ASTM D1925; post exposure temperature: 25°C
—··—··	ICI Perspex CP927G Acrylic; 5 Mrad; thickness: 3.2 mm; test method: ASTM D1925; post exposure temperature: 25°C
– – – –	ICI Perspex CP1000IG Acrylic; 5 Mrad; thickness: 3.2 mm; test method: ASTM D1925; post exposure temperature: 25°C
Reference No.	74

GRAPH 22: Post Gamma Radiation Exposure Time vs. Yellowness Index of Acrylic Resin

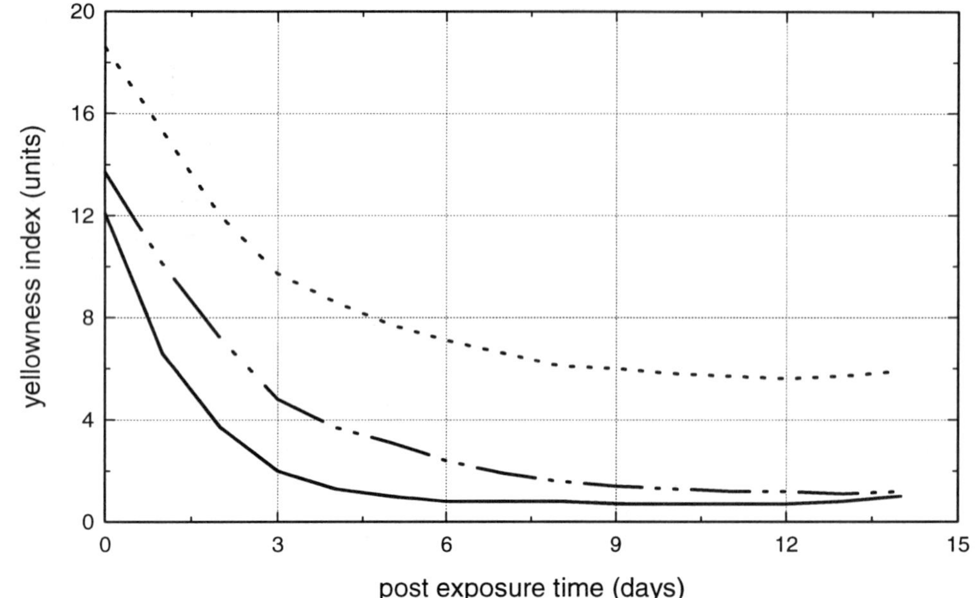

···············	Rohm & Haas Plexiglas V-Grade Acrylic; 2.5 Mrad; thickness: 3.2 mm; post exposure temperature: 60°C
—··—··	Rohm & Haas Plexiglas DR-G Acrylic; 2.5 Mrad; thickness: 3.2 mm; post exposure temperature: 60°C
— — —	Rohm & Haas Plexiglas HFI-10G Acrylic; 2.5 Mrad; thickness: 3.2 mm; post exposure temperature: 60°C
———	Rohm & Haas Plexiglas MI-7G Acrylic; 2.5 Mrad; thickness: 3.2 mm; post exposure temperature: 60°C
Reference No.	90

GRAPH 23: Post Gamma Radiation Exposure Time vs. Yellowness Index of Acrylic Resin

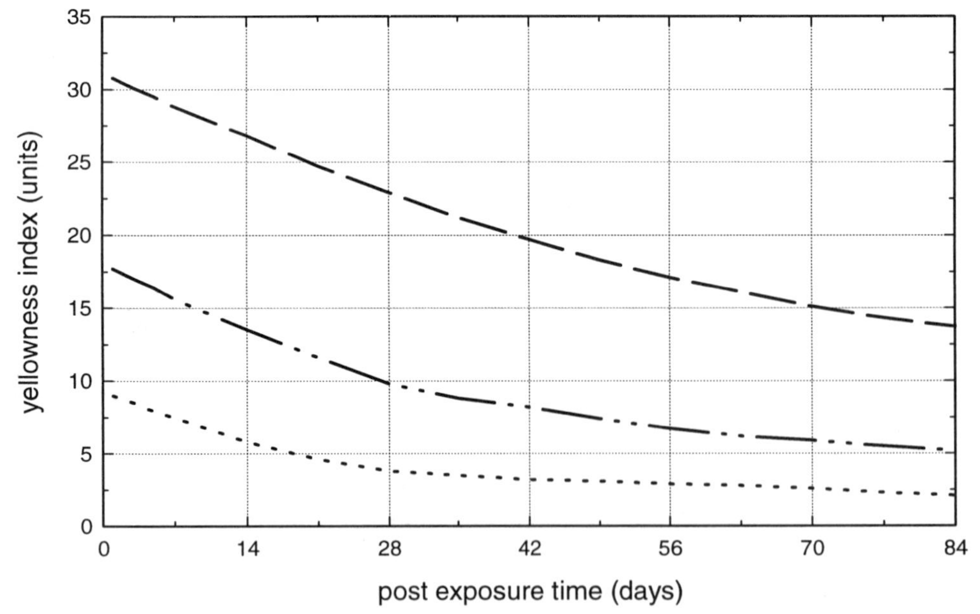

···············	Rohm & Haas Plexiglas V-Grade Acrylic; 2.5 Mrads; post exposure temperature: 23°C; thickness: 1.5 mm
—··—··	Rohm & Haas Plexiglas V-Grade Acrylic; 2.5 Mrads; post exposure temperature: 23°C; thickness: 3.2 mm
— — —	Rohm & Haas Plexiglas V-Grade Acrylic; 2.5 Mrads; post exposure temperature: 23°C; thickness: 6.4 mm
Reference No.	90

GRAPH 24: **Post Gamma Radiation Exposure Time vs. Yellowness Index of Acrylic Resin**

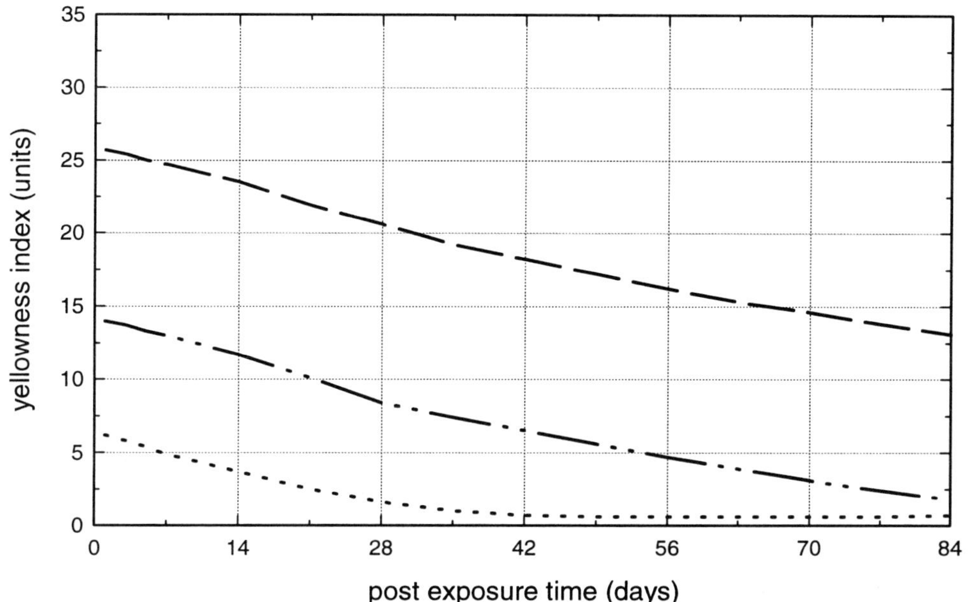

···············	Rohm & Haas Plexiglas DR-G Acrylic; 2.5 Mrads; post exposure temperature: 23°C; thickness: 1.5 mm
—··—··	Rohm & Haas Plexiglas DR-G Acrylic; 2.5 Mrads; post exposure temperature: 23°C; thickness: 3.2 mm
— — — —	Rohm & Haas Plexiglas DR-G Acrylic; 2.5 Mrads; post exposure temperature: 23°C; thickness: 6.4 mm
Reference No.	90

GRAPH 25: **Post Gamma Radiation Exposure Time vs. Yellowness Index of Acrylic Resin**

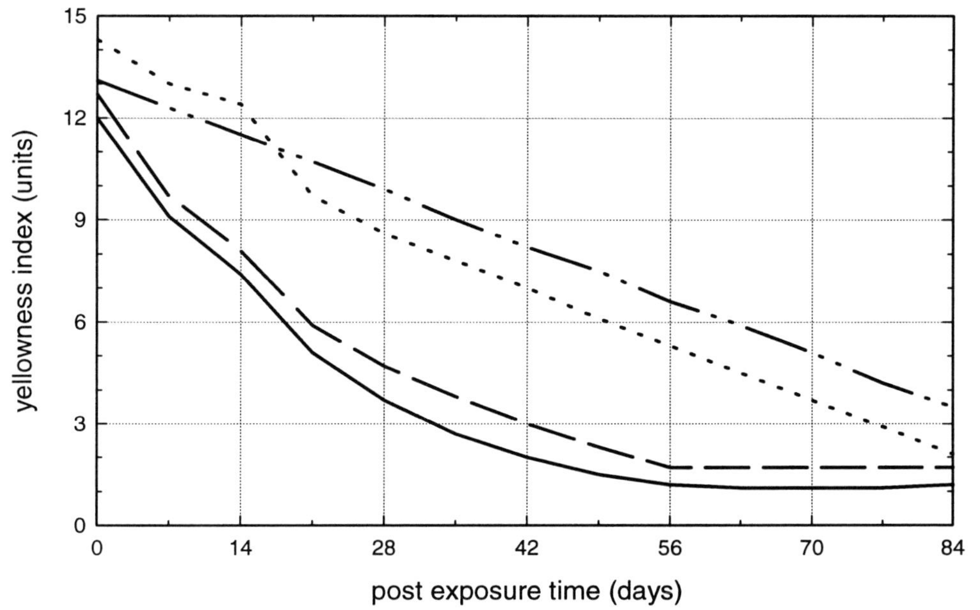

···············	Rohm & Haas Plexiglas DR-G Acrylic; 2.5 Mrads; post exposure temperature: 23°C; thickness: 3.2 mm
—··—··	Rohm & Haas Plexiglas V-Grade Acrylic; 2.5 Mrads; post exposure temperature: 23°C; thickness: 3.2 mm
— — — —	Rohm & Haas Plexiglas HFI-7G Acrylic; 2.5 Mrads; post exposure temperature: 23°C; thickness: 3.2 mm
———	Rohm & Haas Plexiglas HFI-10G Acrylic; 2.5 Mrads; post exposure temperature: 23°C; thickness: 3.2 mm
Reference No.	91

GRAPH 26: **Post Gamma Radiation Exposure Time vs. Yellowness Index of Acrylic Resin**

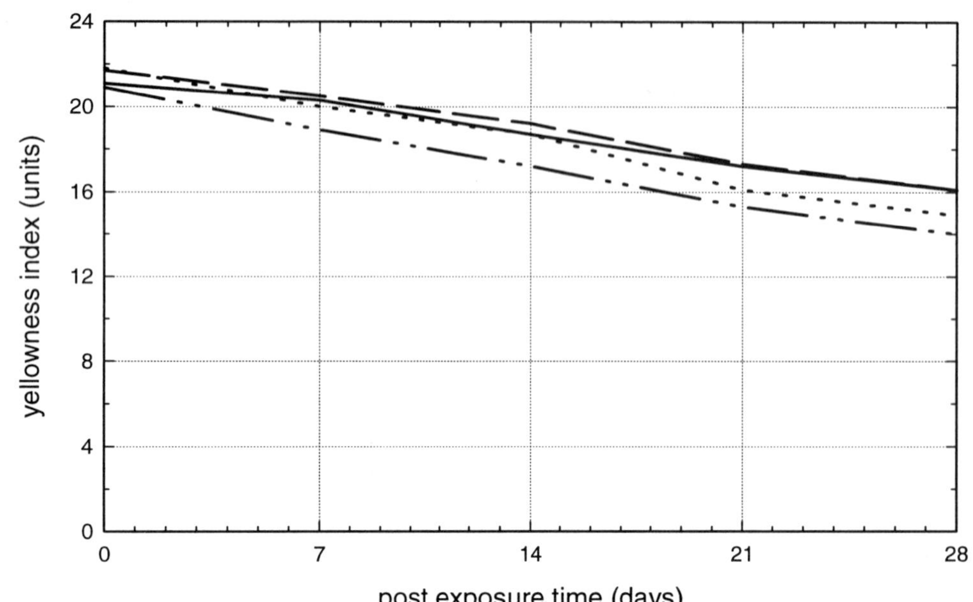

···············	Acrylic (impact modified); 3 Mrads, light storage
—·—··	Acrylic (gen. purp. grade); 3 Mrads, light storage
– – – –	Acrylic (impact modified); 3 Mrads, dark storage
———	Acrylic (gen. purp. grade); 3 Mrads, dark storage
Reference No.	106

GRAPH 27: **Post Gamma Radiation Exposure Time vs. Yellowness Index of Acrylic Resin**

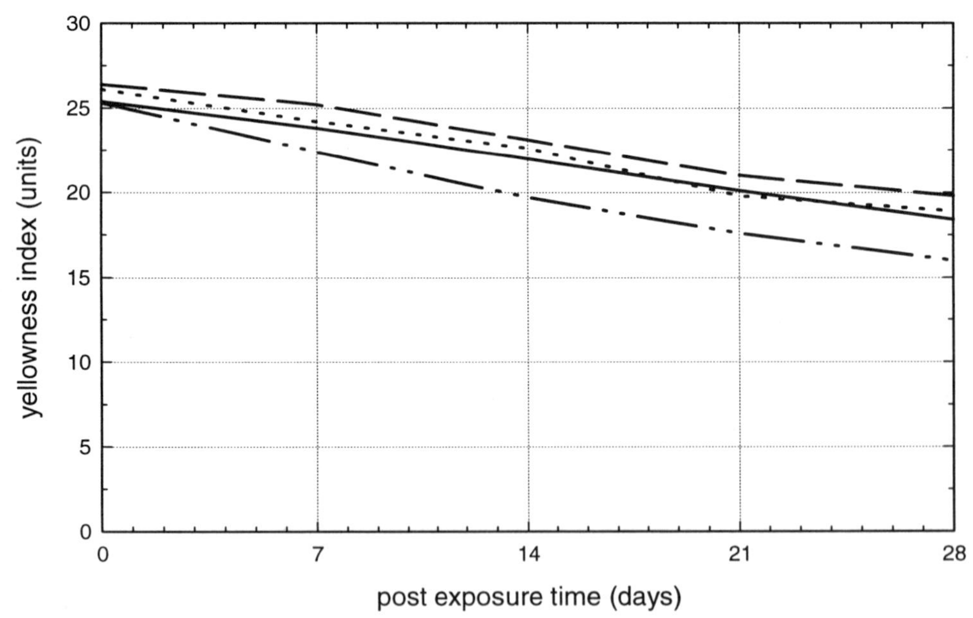

···············	Acrylic (impact modified); 5 Mrads, light storage
—·—··	Acrylic (gen. purp. grade); 5 Mrads, light storage
– – – –	Acrylic (impact modified); 5 Mrads, dark storage
———	Acrylic (gen. purp. grade); 5 Mrads, dark storage
Reference No.	106

GRAPH 28: **Post Gamma Radiation Exposure Temperature vs. Yellowness Index of Acrylic Resin**

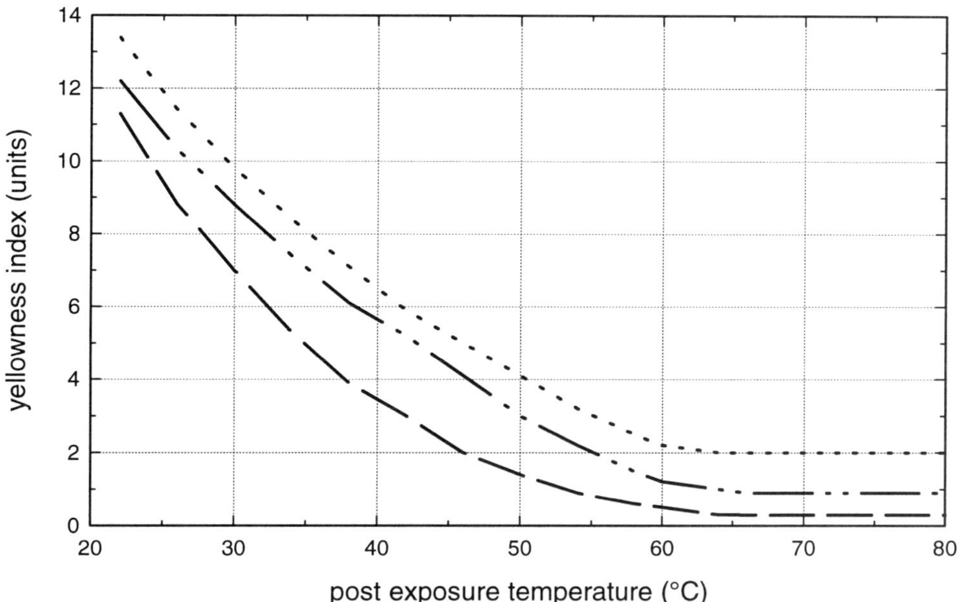

···············	ICI Perspex CP924G Acrylic; 5 Mrad; thickness: 3.2 mm; post exposure time: 1 day; test method: ASTM D1925
—··—··	ICI Perspex CP927G Acrylic; 5 Mrad; thickness: 3.2 mm; post exposure time: 1 day; test method: ASTM D1925
— — — —	ICI Perspex CP1000IG Acrylic; 5 Mrad; thickness: 3.2 mm; post exposure time: 1 day; test method: ASTM D1925
Reference No.	74

GRAPH 29: **Post Gamma Radiation Exposure Temperature vs. Yellowness Index of Acrylic Resin**

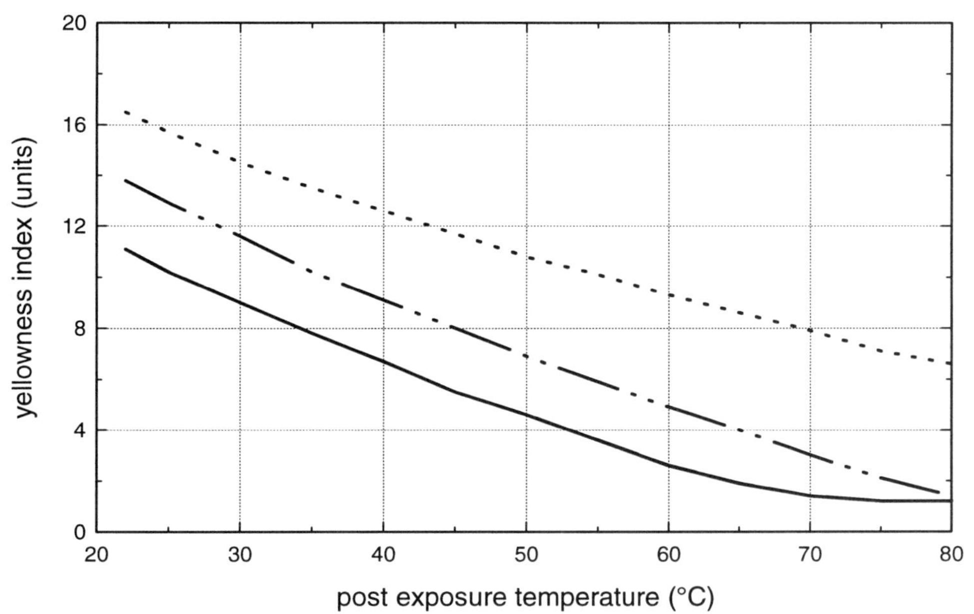

···············	Rohm & Haas Plexiglas V-Grade Acrylic; 2.5 Mrads; post exposure time: 3 days; thickness: 3.2 mm
—··—··	Rohm & Haas Plexiglas DR-G Acrylic; 2.5 Mrads; post exposure time: 3 days; thickness: 3.2 mm
— — — —	Rohm & Haas Plexiglas HFI-7G Acrylic; 2.5 Mrads; post exposure time: 3 days; thickness: 3.2 mm
————	Rohm & Haas Plexiglas HFI-10G Acrylic; 2.5 Mrads; post exposure time: 3 days; thickness: 3.2 mm
Reference No.	91

GRAPH 30: Electron Beam Radiation Dose vs. Tensile Strength of Acrylic Resin

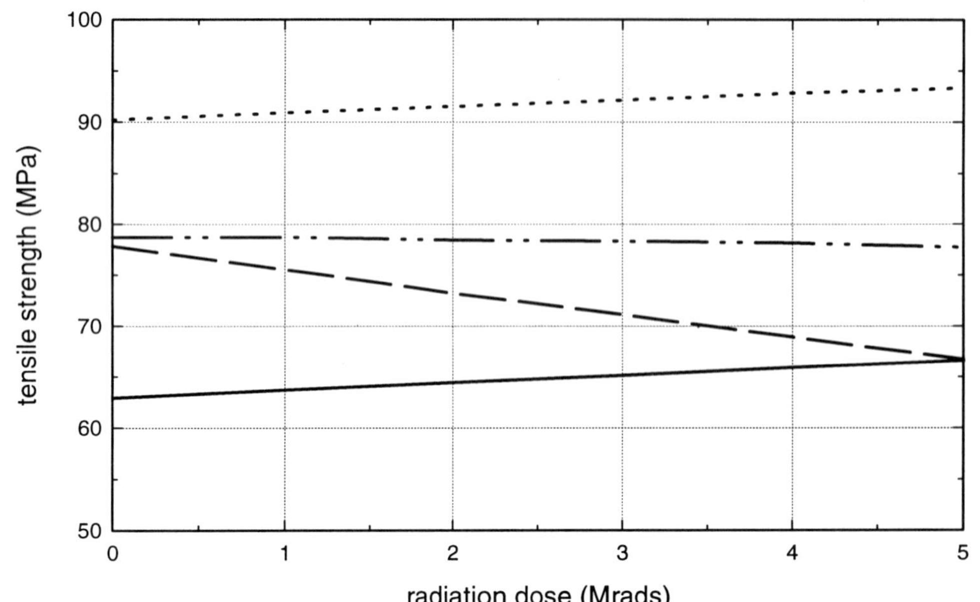

....	ICI Perspex CP924G Acrylic; test method: ASTM D638
—··—··	ICI Perspex CP927G Acrylic; test method: ASTM D638
– – – –	ICI Perspex CP927GHF Acrylic; test method: ASTM D638
———	ICI Perspex CP1000IG Acrylic; test method: ASTM D638
Reference No.	72

GRAPH 31: Electron Beam Radiation Dose vs. Flexural Strength of Acrylic Resin

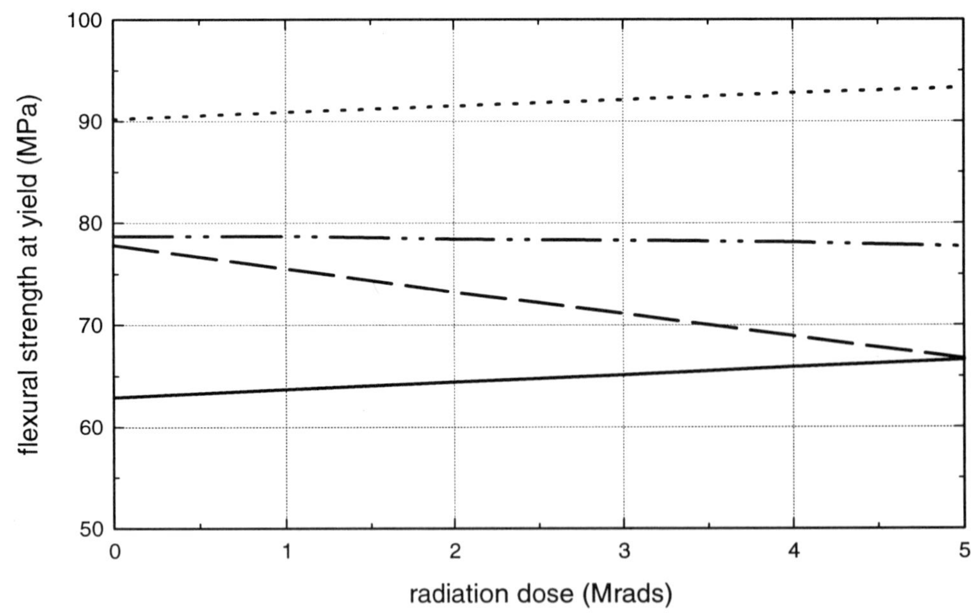

....	ICI Perspex CP924G Acrylic; test method: ASTM D790
—··—··	ICI Perspex CP927G Acrylic; test method: ASTM D790
– – – –	ICI Perspex CP927GHF Acrylic; test method: ASTM D790
———	ICI Perspex CP1000IG Acrylic; test method: ASTM D790
Reference No.	72

GRAPH 32: Electron Beam Radiation Dose vs. Impact Strength of Acrylic Resin

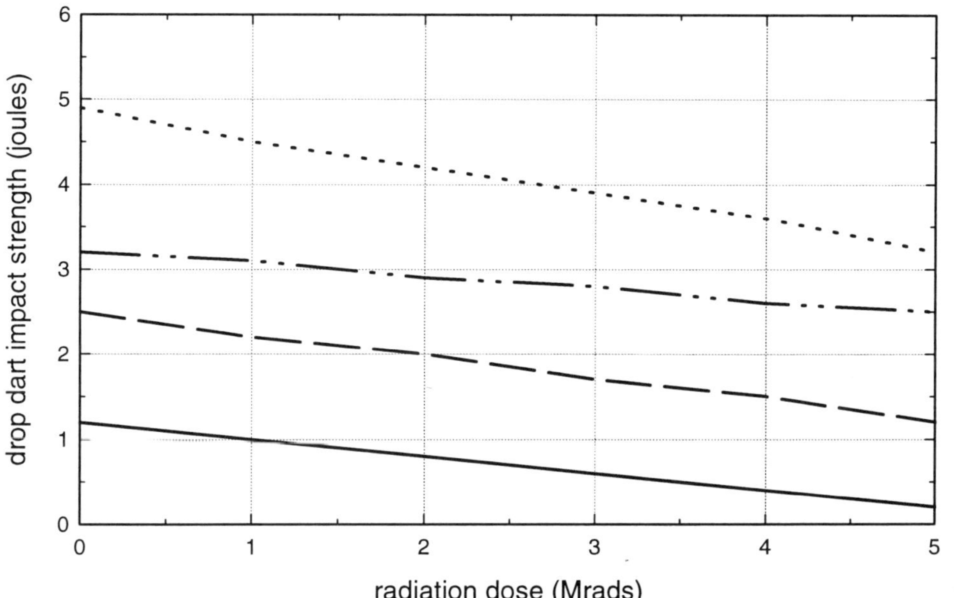

···············	ICI Perspex CP1000IG Acrylic; thickness: 3.2 mm; test method: ASTM D3029 GB; dart weight: 1814 grams
—··—··	ICI Perspex CP927GHF Acrylic; thickness: 3.2 mm; test method: ASTM D3029 GB; dart weight: 1814 grams
— — — —	ICI Perspex CP927G Acrylic; thickness: 3.2 mm; test method: ASTM D3029 GB; dart weight: 1814 grams
———	ICI Perspex CP924G Acrylic; thickness: 3.2 mm; test method: ASTM D3029 GB; dart weight: 1814 grams
Reference No.	72

GRAPH 33: Electron Beam Radiation Dose vs. Yellowness Index of Acrylic Resin

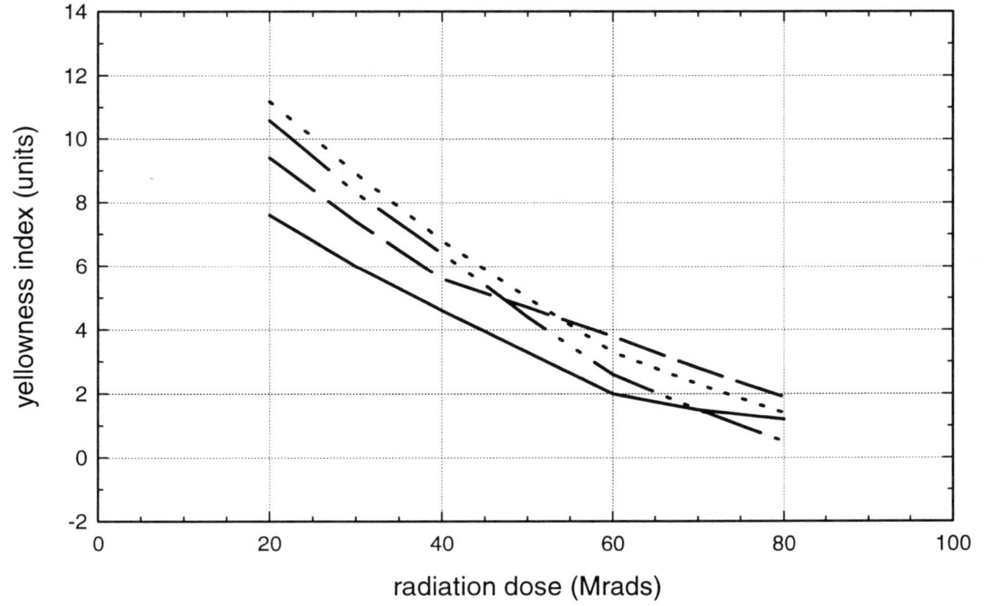

···············	ICI Perspex CP927G Acrylic; thickness: 3.2 mm; test method: ASTM D1925
—··—··	ICI Perspex CP924G Acrylic; thickness: 3.2 mm; test method: ASTM D1925
— — — —	ICI Perspex CP927GHF Acrylic; thickness: 3.2 mm; test method: ASTM D1925
———	ICI Perspex CP1000IG Acrylic; thickness: 3.2 mm; test method: ASTM D1925
Reference No.	72

GRAPH 34: Post Electron Beam Radiation Exposure Time vs. Yellowness Index of Acrylic Resin

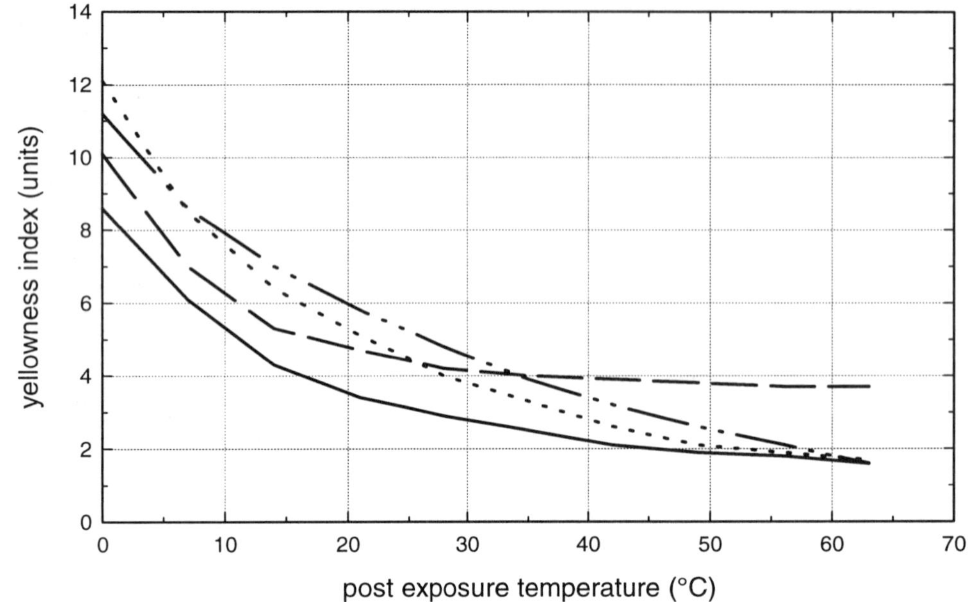

···············	ICI Perspex CP927G Acrylic; 5 Mrad; thickness: 3.2 mm; test method: ASTM D1925; post exposure temperature: 25°C
—··—··	ICI Perspex CP924G Acrylic; 5 Mrad; thickness: 3.2 mm; test method: ASTM D1925; post exposure temperature: 25°C
– – – –	ICI Perspex CP927GHF Acrylic; 5 Mrad; thickness: 3.2 mm; test method: ASTM D1925; post exposure temperature: 25°C
———	ICI Perspex CP1000IG Acrylic; 5 Mrad; thickness: 3.2 mm; test method: ASTM D1925; post exposure temperature: 25°C
Reference No.	72

GRAPH 35: Post Electron Beam Radiation Exposure Temperature vs. Yellowness Index of Acrylic Resin

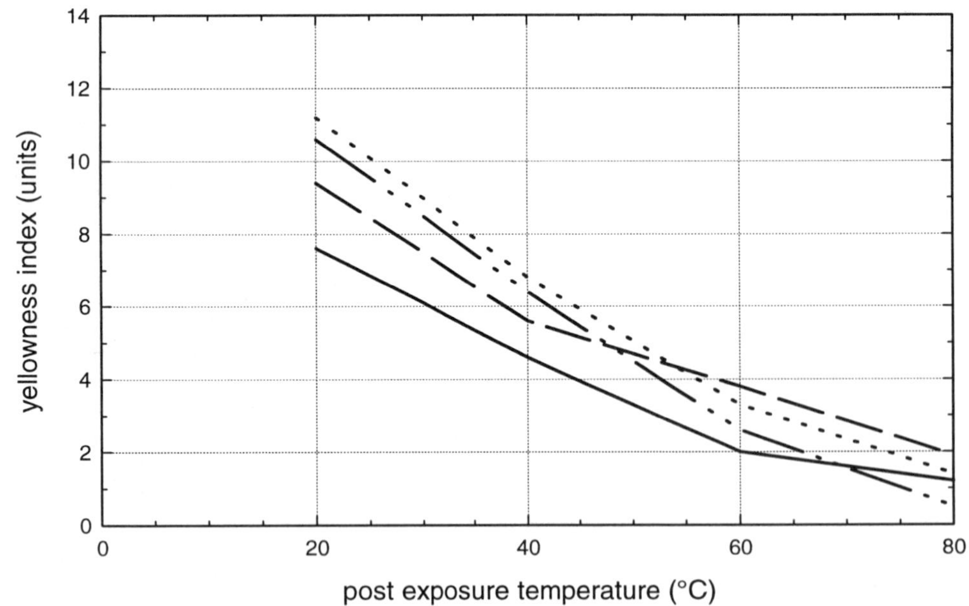

···············	ICI Perspex CP927G Acrylic; 5 Mrad; thickness: 3.2 mm; post exposure time: 1 day; test method: ASTM D1925
—··—··	ICI Perspex CP924G Acrylic; 5 Mrad; thickness: 3.2 mm; post exposure time: 1 day; test method: ASTM D1925
– – – –	ICI Perspex CP927GHF Acrylic; 5 Mrad; thickness: 3.2 mm; post exposure time: 1 day; test method: ASTM D1925
———	ICI Perspex CP1000IG Acrylic; 5 Mrad; thickness: 3.2 mm; post exposure time: 1 day; test method: ASTM D1925
Reference No.	72

GRAPH 36: **Beta Radiation Dose vs. Tensile Strength of Acrylic Resin**

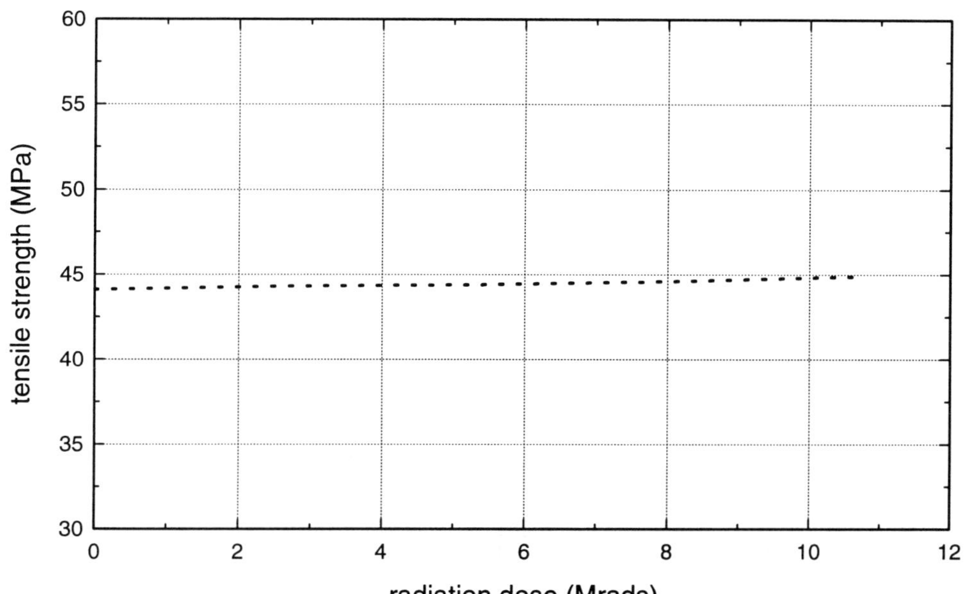

....	Acrylic
Reference No.	105

GRAPH 37: **Beta Radiation Dose vs. Tensile Modulus of Acrylic Resin**

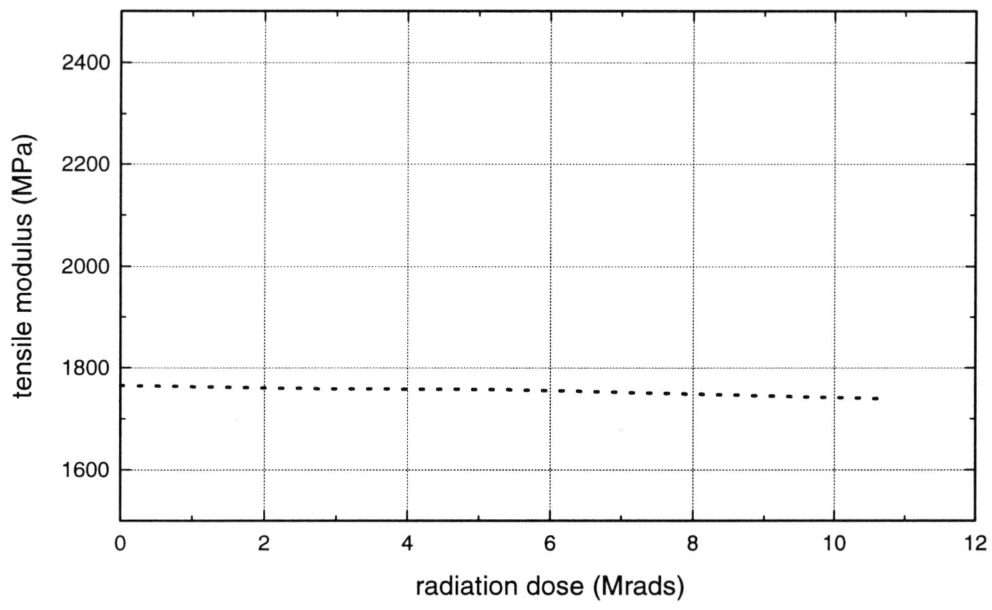

....	Acrylic
Reference No.	105

GRAPH 38: Beta Radiation Dose vs. Notched Izod Impact Strength of Acrylic Resin

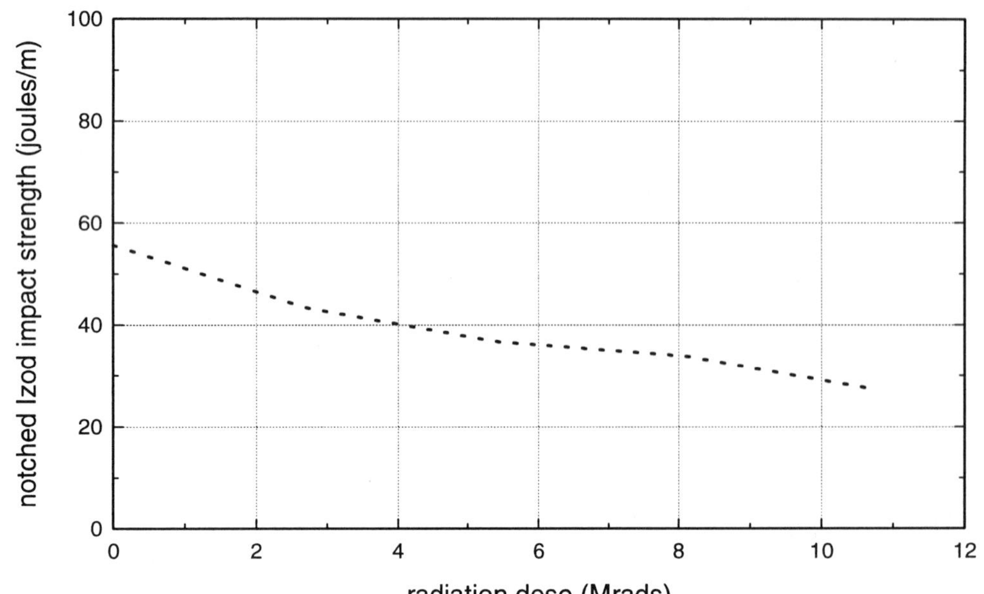

....	Acrylic
Reference No.	105

GRAPH 39: Beta Radiation Dose vs. Yellowness Index of Acrylic Resin

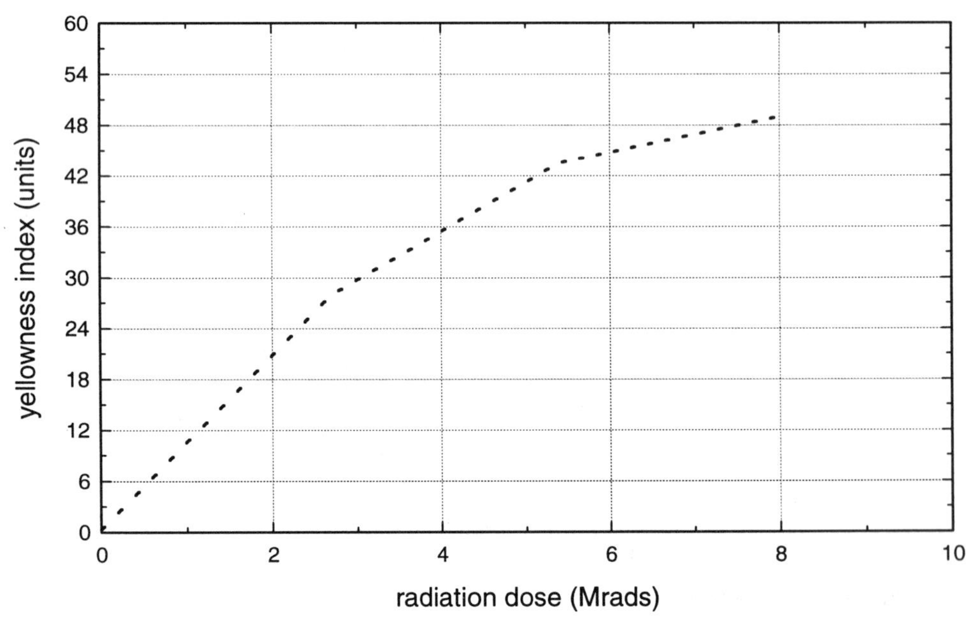

....	Acrylic
Reference No.	105

GRAPH 40: Post Beta Radiation Exposure Time vs. Yellowness Index of Acrylic Resin

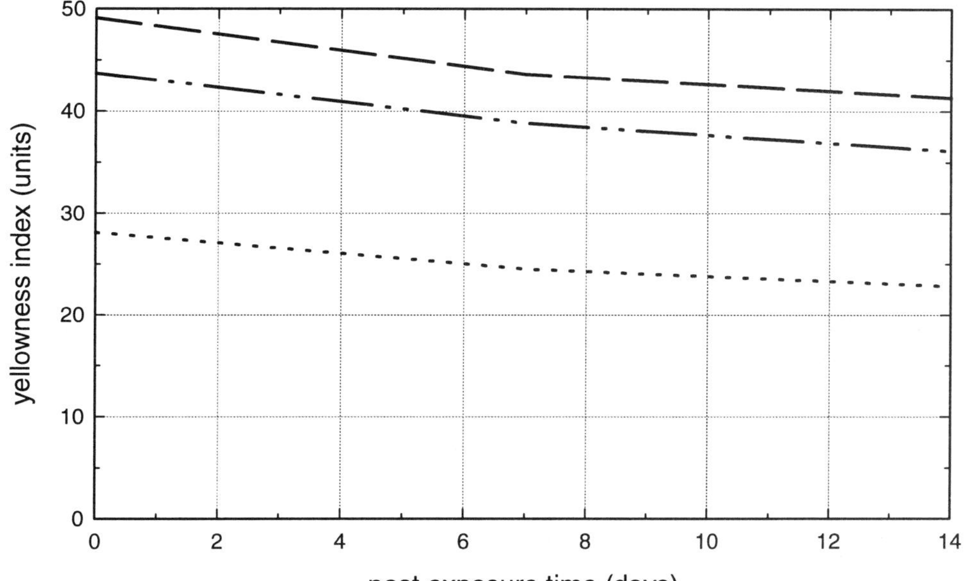

....	Acrylic; 2.7 Mrads
—··—··	Acrylic; 5.4 Mrads
— — — —	Acrylic; 8.1 Mrads
Reference No.	105

Gamma Radiation Resistance

Novacor Chemicals: Acrylic Copolymer (composition: 20% methyl methacrylate, 80% styrene; features: transparent); **Acrylic Copolymer** (composition: 30% methyl methacrylate, 70% styrene; features: transparent); **Acrylic Copolymer** (composition: 60% methyl methacrylate, 40% styrene; features: transparent)

Acrylic (methyl methacrylate) with various levels of styrene (0 to 80%) were irradiated. Tensile strength, tensile elongation, and Izod were all tested, as well as color, shortly after and five weeks after exposure. The acrylic rapidly lost one third of its tensile strength, and ultimately lost 44%. The three copolymers maintained their original tensile strength, even through 6.0 Mrads of exposure. Tensile elongation followed a similar pattern. All materials start out with 2-3% elongation, but the acrylic resin lost 50% of its elongation, even at a very low radiation dosage, while the copolymers showed good retention over the entire exposure range. The Izod impact strengths of the copolymers were initially stronger than the acrylic resin; there was no appreciable loss of any impact strengths. Generally the yellowing increased with the amount of radiation and the methyl methacrylate content of the polymer. A decrease in color intensity occurred over time in all cases, resulting in the three copolymers exposed at 1.27 and 2.80 Mrads being nearly equal to the unexposed specimens. However the acrylic resin remained light yellow.

Reference: Hauser, Deborah I., Netolicky, William, *The Effects Of Gamma And Ethylene Oxide Sterilization On Styrene Methyl Methacrylate Copolymers And A Toughened Terpolymer,* Medical '91, Ghent, Belgium, conference proceedings - Novacor Chemicals Inc., 1991.

Novacor Chemicals: NAS 30 (features: transparent)

NAS 30 was exposed to 3.0 and 5.0 Mrads and lost no stiffness (flexural modulus). It also showed no change in tensile strength or notched Izod impact. The copolymer yellowed to an index of 7.5 at 5.0 Mrads exposure. After exposure to 3 Mrads it yellowed to an index of 6 and dropped to a level of 2.5 after 4 weeks. After exposure to 5 Mrads it yellowed to an index of 6 and dropped to a level of 2.6 after 4 weeks. There was no material difference in recovery results in light vs. dark.

Reference: Hauser, Deborah I., Netolicky, William, *The Effects Of Gamma And Ethylene Oxide Sterilization On Styrene Methyl Methacrylate Copolymers And A Toughened Terpolymer,* Medical '91, Ghent, Belgium, conference proceedings - Novacor Chemicals Inc., 1991.

Ethylene Oxide (EtO) Resistance

Novacor Chemicals: NAS 30 (features: transparent)

NAS 30 was exposed to both 100% EtO and a mixture of 12% EtO and 88% Freon). There was no effect on the flexural modulus or toughness. However, NAS 30 loses approximately 21% of its tensile strength. Experience shows that annealing will greatly reduce molded-in stress, thereby significantly reducing the effect that EtO has on NAS 30 resin. When exposed to the EtO/Freon mixture there was no loss in flexural modulus, however the tensile strength of the NAS 30 copolymer drops off by 7%.

Reference: Hauser, Deborah I., Netolicky, William, *The Effects Of Gamma And Ethylene Oxide Sterilization On Styrene Methyl Methacrylate Copolymers And A Toughened Terpolymer,* Medical '91, Ghent, Belgium, conference proceedings - Novacor Chemicals Inc., 1991.

TABLE 22: Effect of Gamma Radiation Sterilization on Acrylic Copolymer

Material Family	ACRYLIC COPOLYMER							
Material Supplier/Name	NOVACOR							
Material Note	20% methyl methacrylate, 80% styrene; transparent	20% methyl methacrylate, 80% styrene; transparent	20% methyl methacrylate, 80% styrene; transparent	20% methyl methacrylate, 80% styrene; transparent	30% methyl methacrylate, 70% styrene; transparent	30% methyl methacrylate, 70% styrene; transparent	30% methyl methacrylate, 70% styrene; transparent	30% methyl methacrylate, 70% styrene; transparent
Reference No.	106	106	106	106	106	106	106	106
EXPOSURE CONDITIONS								
Type	Gamma Radiation	Gamma Radiation	Gamma Radiation	Gamma Radiation	Gamma Radiation	Gamma Radiation	Gamma Radiation	Gamma Radiation
Radiation Dose (Mrads)	1.27	2.8	3.55	5.43	1.27	2.8	3.55	5.43
POST EXPOSURE CONDITIONING								
Time (hours)	0	0	0	0	0	0	0	0
SURFACE and APPEARANCE								
Yellowness Index Note	light yellow	light yellow	light yellow	light yellow	light yellow	light yellow	light yellow	yellow

TABLE 23: Effect of Gamma Radiation Sterilization on Acrylic Copolymer

Material Family	ACRYLIC COPOLYMER							
Material Supplier/Name	NOVACOR							
Material Note	60% methyl methacrylate, 40% styrene; transparent	60% methyl methacrylate, 40% styrene; transparent	60% methyl methacrylate, 40% styrene; transparent	60% methyl methacrylate, 40% styrene; transparent	20% methyl methacrylate, 80% styrene; transparent	20% methyl methacrylate, 80% styrene; transparent	20% methyl methacrylate, 80% styrene; transparent	20% methyl methacrylate, 80% styrene; transparent
Reference No.	106	106	106	106	106	106	106	106
EXPOSURE CONDITIONS								
Type	Gamma Radiation	Gamma Radiation	Gamma Radiation	Gamma Radiation	Gamma Radiation	Gamma Radiation	Gamma Radiation	Gamma Radiation
Radiation Dose (Mrads)	1.27	2.8	3.55	5.43	1.27	2.8	3.55	5.43
POST EXPOSURE CONDITIONING								
Time (hours)	0	0	0	0	840	840	840	840
SURFACE and APPEARANCE								
Yellowness Index Note	light yellow	yellow	dark yellow	dark yellow	almost water white	almost water white	almost water white	almost water white

TABLE 24: Effect of Gamma Radiation Sterilization on Acrylic Copolymer

Material Family	ACRYLIC COPOLYMER							
Material Supplier/Name	NOVACOR							
Material Note	30% methyl methacrylate, 70% styrene; transparent	30% methyl methacrylate, 70% styrene; transparent	30% methyl methacrylate, 70% styrene; transparent	30% methyl methacrylate, 70% styrene; transparent	60% methyl methacrylate, 40% styrene; transparent	60% methyl methacrylate, 40% styrene; transparent	60% methyl methacrylate, 40% styrene; transparent	60% methyl methacrylate, 40% styrene; transparent
Reference No.	106	106	106	106	106	106	106	106
EXPOSURE CONDITIONS								
Type	Gamma Radiation	Gamma Radiation	Gamma Radiation	Gamma Radiation	Gamma Radiation	Gamma Radiation	Gamma Radiation	Gamma Radiation
Radiation Dose (Mrads)	1.27	2.8	3.55	5.43	1.27	2.8	3.55	5.43
POST EXPOSURE CONDITIONING								
Time (hours)	840	840	840	840	840	840	840	840
SURFACE and APPEARANCE								
Yellowness Index Note	almost water white	almost water white	almost water white	very light yellow	almost water white	almost water white	light yellow	light yellow

TABLE 25: Effect of Gamma Radiation Sterilization on Novacor NAS 30 Acrylic Copolymer

Material Family	ACRYLIC COPOLYMER	
Material Supplier/Name	NOVACOR NAS 30	
Material Note	transparent	transparent
Reference No.	106	106
EXPOSURE CONDITIONS		
Type	Gamma Radiation	Gamma Radiation
Radiation Dose (Mrads)	3	5
PROPERTIES RETAINED (%)		
Tensile Strength	96 (he)	96 (he)
Modulus	100 (bz)	100 (bz)
SURFACE and APPEARANCE		
Δ Yellowness Index	5 (kt)	6 (kt)

TABLE 26: Effect of Ethylene Oxide Sterilization on Novacor NAS 30 Acrylic Copolymer

Material Family	ACRYLIC COPOLYMER	
Material Supplier/Name	NOVACOR NAS 30	
Material Note	transparent	transparent
Reference No.	106	106
EXPOSURE CONDITIONS		
Type	Ethylene Oxide	Ethylene Oxide
Details	100% EtO	12% EtO and 88% Freon
Concentration	1000 mg/l	1000 mg/l
Number of Cycles	1	1
Temperature (°C)	54.4	54.4
Time (hours)	4-6	4-6
PROPERTIES RETAINED (%)		
Tensile Strength	80 (he)	92 (he)
Modulus	100 (bz)	100 (bz)
Notched Izod Impact	93 (fm)	93 (fm)

GRAPH 41: Gamma Radiation Dose vs. Tensile Strength of Acrylic Copolymer

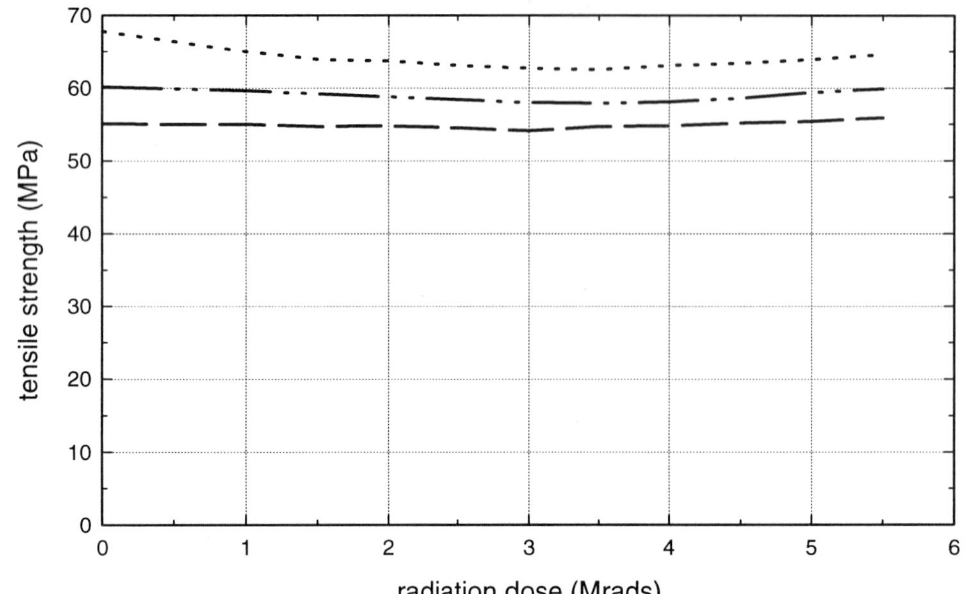

···············	Novacor Acrylic Copol. (60% PMMA, 40% styrene; transpar.)
—·–·—	Novacor Acrylic Copol. (30% PMMA, 70% styrene; transpar.)
— — —	Novacor Acrylic Copol. (20% PMMA, 80% styrene; transpar.)
Reference No.	106

GRAPH 42: Gamma Radiation Dose vs. Elongation of Acrylic Copolymer

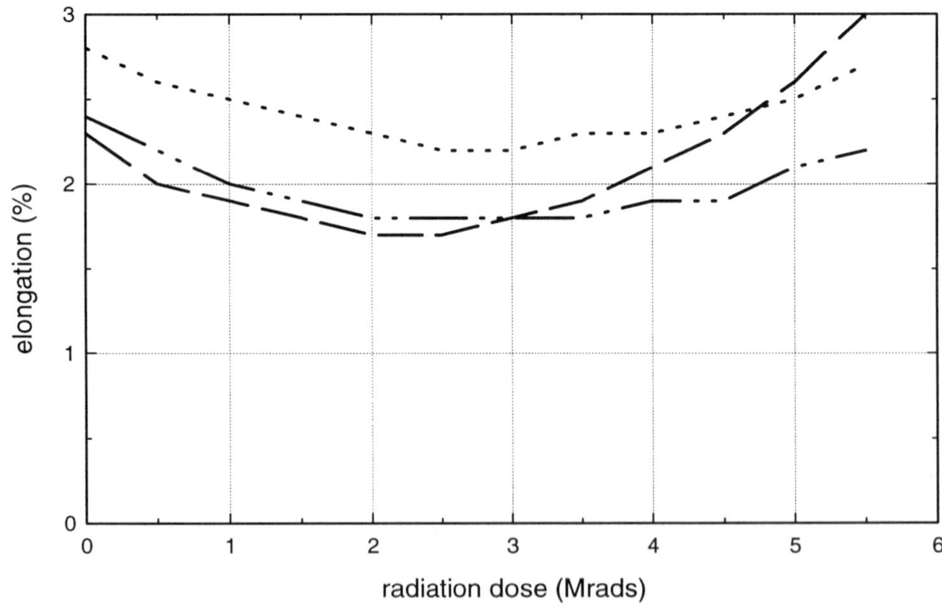

···············	Novacor Acrylic Copol. (60% PMMA, 40% styrene; transpar.)
—·–·—	Novacor Acrylic Copol. (30% PMMA, 70% styrene; transpar.)
— — —	Novacor Acrylic Copol. (20% PMMA, 80% styrene; transpar.)
Reference No.	106

GRAPH 43: Gamma Radiation Dose vs. Notched Izod Impact Strength of Acrylic Copolymer

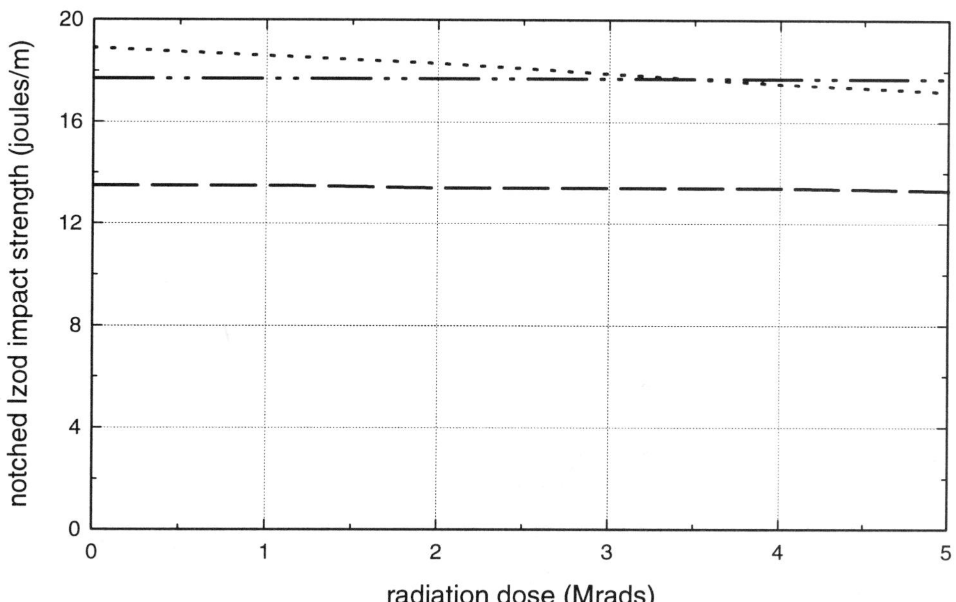

....	Novacor Acrylic Copol. (20% PMMA, 80% styrene; transpar.)
—··—··	Novacor Acrylic Copol. (30% PMMA, 70% styrene; transpar.)
— — — —	Novacor Acrylic Copol. (60% PMMA, 40% styrene; transpar.)
Reference No.	106

GRAPH 44: Gamma Radiation Dose vs. Yellowness Index of Acrylic Copolymer

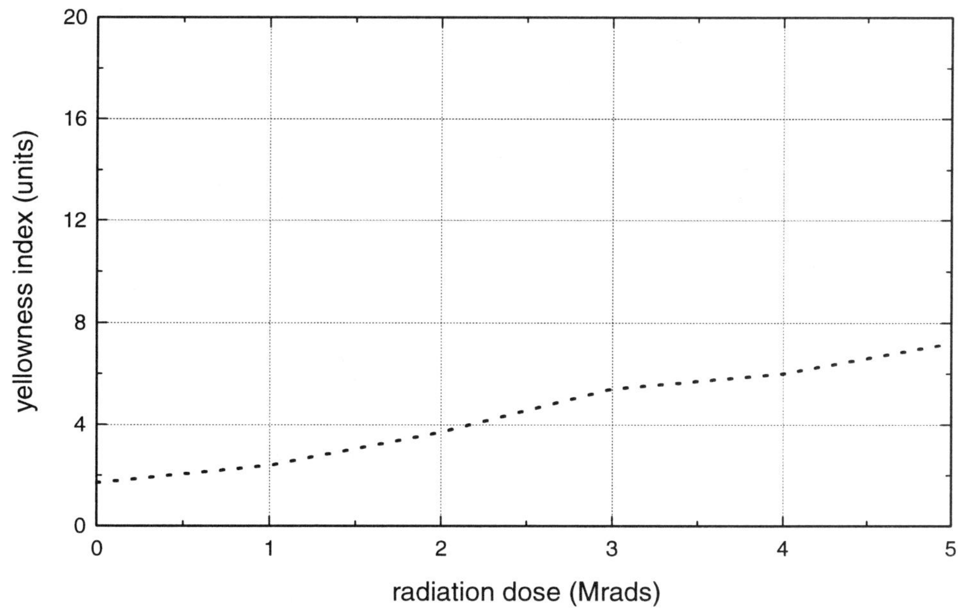

....	Novacor NAS 30 Acrylic Copol. (transpar.)
Reference No.	106

GRAPH 45: Post Gamma Radiation Exposure Time vs. Yellowness Index of Acrylic Copolymer

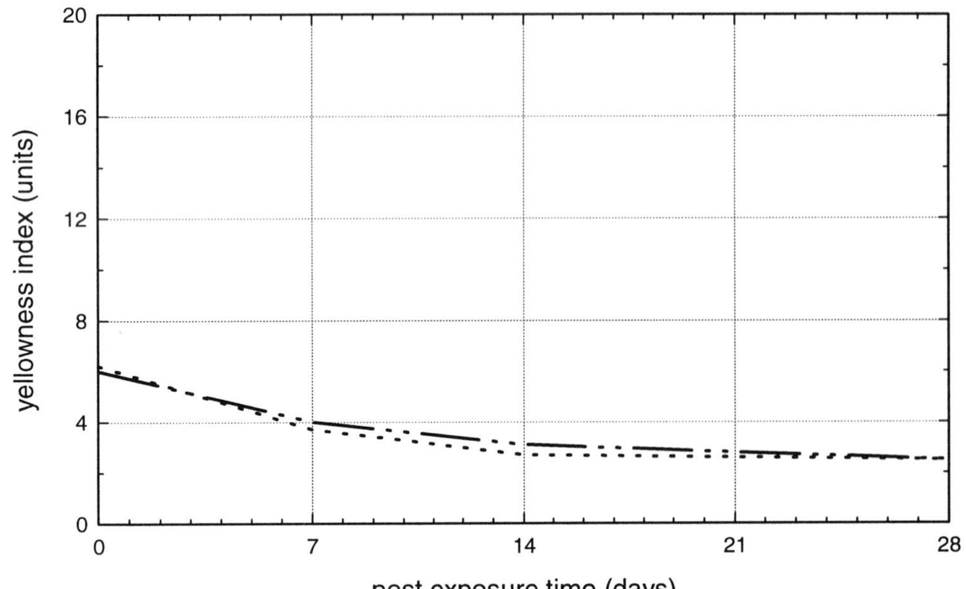

....	Novacor NAS 30 Acrylic Copol. (transpar.); 3 Mrads, light storage
—··—··	Novacor NAS 30 Acrylic Copol. (transpar.); 3 Mrads, dark storage
Reference No.	106

GRAPH 46: Post Gamma Radiation Exposure Time vs. Yellowness Index of Acrylic Copolymer

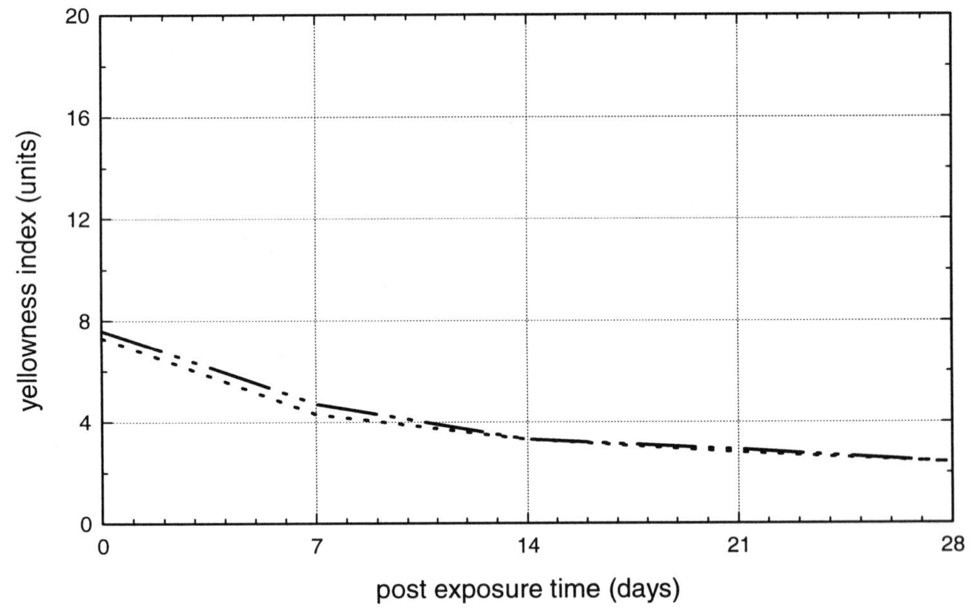

....	Novacor NAS 30 Acrylic Copol. (transpar.); 5 Mrads, light storage
—··—··	Novacor NAS 30 Acrylic Copol. (transpar.); 5 Mrads, dark storage
Reference No.	106

Gamma Radiation Resistance

Novacor Chemicals: Zylar 90 (features: transparent)

Zylar 90, when exposed to 3.0 and 5.0 Mrads, lost no stiffness (flexural modulus). There was no appreciable change in tensile strength or notched Izod impact. The yellowness index changed to 20.6 with 5.0 Mrads exposure. With 3.0 Mrads exposure yellowness reached 16 and dropped to an index of 6 after 4 weeks. Samples exhibited no significant difference in recovery in the light vs. in the dark.

Reference: Hauser, Deborah I., Netolicky, William, *The Effects Of Gamma And Ethylene Oxide Sterilization On Styrene Methyl Methacrylate Copolymers And A Toughened Terpolymer,* Medical '91, Ghent, Belgium, conference proceedings - Novacor Chemicals Inc., 1991.

Novacor Chemicals: Zylar 93-546 (features: transparent, impact modified); **Zylar 94-568** (features: transparent, impact modified); **Zylar ST 561** (features: transparent, high impact); **Zylar ST 94-560** (features: transparent, high impact); **Zylar ST 94-562** (features: transparent, high impact)

Zylar 94 series resins were tested at both 3.5 and 7.0 Mrads exposure to gamma radiation after 7 months aging. Seven Mrads exposure was chosen with the expectation that parts would be exposed to two sterilizing cycles of 3.5 Mrads each. There was no change in the physical properties. Samples retained tensile strength, tensile elongation, and notched Izod impact strength at both exposure levels. The yellowness index was measured and there was little change in appearance.

Reference: Haines, Jim, Hauser, Debbie, *Zylar Clear Alloys For Medical Applications,* supplier technical report - Novacor Chemicals Inc.

Beta Radiation Resistance

Novacor Chemicals: Zylar 93-546 (features: transparent, impact modified); **Zylar 94-568** (features: transparent, impact modified); **Zylar ST 561** (features: transparent, high impact); **Zylar ST 94-560** (features: transparent, high impact); **Zylar ST 94-562** (features: transparent, high impact)

Beta sterilization is increasing in importance in Europe. Samples were tested at beta radiation levels of 2.7 to 10.8 Mrads. Physical properties were retained after 10.8 Mrads. As with gamma exposure, the Zylar resins did not change color to a large degree. Yellowness decreased after two weeks, without regard to storage in light or dark.

Reference: Haines, Jim, Hauser, Debbie, *Zylar Clear Alloys For Medical Applications,* supplier technical report - Novacor Chemicals Inc.

Ethylene Oxide (EtO) Resistance

Novacor Chemicals: Zylar 90 (features: transparent)

Zylar 90 (styrene methyl methacrylate butadiene terpolymer) was exposed to both 100% EtO and a mixture of 12% EtO and 88% Freon). On exposure to EtO there was no effect on the flexural modulus. Zylar 90 loses approximately 7% of its Izod impact value. Exposure to the EtO/Freon mixture shows no loss in flexural modulus, however the Izod impact of Zylar 90 drops off by 9%.

Reference: Hauser, Deborah I., Netolicky, William, *The Effects Of Gamma And Ethylene Oxide Sterilization On Styrene Methyl Methacrylate Copolymers And A Toughened Terpolymer,* Medical '91, Ghent, Belgium, conference proceedings - Novacor Chemicals Inc., 1991.

Novacor Chemicals: Zylar 93-546 (features: transparent, impact modified); **Zylar 94-568** (features: transparent, impact modified); **Zylar ST 561** (features: transparent, high impact); **Zylar ST 94-560** (features: transparent, high impact); **Zylar ST 94-562** (features: transparent, high impact)

Zylar 93-546 resin lost approximately 10% of its notched Izod impact strength after EtO exposure, and 25% after exposure to a mixture of EtO (12%) and Freon (88%). Neither tensile modulus nor flexural modulus were affected. Other Zylar grades (94-568, ST 94-560, ST 94-561, ST 94-562) appeared essentially unaffected

Reference: Haines, Jim, Hauser, Debbie, *Zylar Clear Alloys For Medical Applications,* supplier technical report - Novacor Chemicals Inc.

TABLE 27: Effect of Gamma Radiation Sterilization on Acrylic Terpolymer

Material Family	ACRYLIC TERPOLYMER							
Material Supplier/Name	NOVACOR ZYLAR 90		NOVACOR ZYLAR 93-546		NOVACOR ZYLAR 94-568		NOVACOR ZYLAR ST 94-560	
Material Note	transparent	transparent	transparent, impact modified	transparent, impact modified	transparent, impact modified	transparent, impact modified	transparent, high impact	transparent, high impact
Reference No.	106	106	105	105	105	105	105	105
EXPOSURE CONDITIONS								
Type	Gamma Radiation	Gamma Radiation	Gamma Radiation	Gamma Radiation	Gamma Radiation	Gamma Radiation	Gamma Radiation	Gamma Radiation
Radiation Dose (Mrads)	3	5	3.5	7	3.5	7	3.5	7
PROPERTIES RETAINED (%)								
Tensile Strength	100 (he)	100 (he)	100 (he)	96 (he)	96 (he)	96 (he)	96 (he)	96 (he)
Elongation			92 (ai)	90 (ai)	86 (ai)	82 (ai)	92 (ai)	86 (ai)
Modulus	100 (bz)	100 (bz)						
Notched Izod Impact	100 (fm)	100 (fm)	95 (fm)	100 (fm)	83 (fm)	86 (fm)	86 (fm)	83 (fm)
SURFACE and APPEARANCE								
Δ Yellowness Index	25.5 (kt)	19 (kt)	1.5 (kt)	2.6 (kt)	1.2 (kt)	2.4 (kt)	1.6 (kt)	3.1 (kt)

TABLE 28: Effect of Ethylene Oxide Sterilization on Acrylic Terpolymer

Material Family	ACRYLIC TERPOLYMER	
Material Supplier/Name	NOVACOR ZYLAR 90	
Material Note	transparent	transparent
Reference No.	106	106
EXPOSURE CONDITIONS		
Type	Ethylene Oxide	Ethylene Oxide
Details	100% EtO	12% EtO and 88% Freon
Concentration	1000 mg/l	1000 mg/l
Number of Cycles	1	1
Temperature (°C)	54.4	54.4
Time (hours)	4-6	4-6
PROPERTIES RETAINED (%)		
Tensile Strength	100 (he)	100 (he)
Modulus	100 (bz)	100 (bz)

TABLE 29: Effect of Ethylene Oxide Sterilization on Acrylic Terpolymer

Material Family	ACRYLIC TERPOLYMER				
Material Supplier/Name	Novacor Zylar 93-546	Novacor Zylar 94-568	Novacor Zylar ST 94-560	Novacor Zylar ST 561	Novacor Zylar ST 94-562
Material Note	transparent, impact modified	transparent, impact modified	transparent, high impact	transparent, high impact	transparent, high impact
Reference No.	105	105	105	105	105
EXPOSURE CONDITIONS					
Type	Ethylene Oxide	Ethylene Oxide	Ethylene Oxide	Ethylene Oxide	Ethylene Oxide
Details	100% EtO	100% EtO	100% EtO	100% EtO	100% EtO
Number of Cycles	1	1	1	1	1
PROPERTIES RETAINED (%)					
Notched Izod Impact	90 (fm)	100 (fm)	100 (fm)	100 (fm)	98 (fm)

Material Family	ACRYLIC TERPOLYMER				
Material Supplier/Name	Novacor Zylar 93-546	Novacor Zylar 94-568	Novacor Zylar ST 94-560	Novacor Zylar ST 561	Novacor Zylar ST 94-562
Material Note	transparent, impact modified	transparent, impact modified	transparent, high impact	transparent, high impact	transparent, high impact
Reference No.	105	105	105	105	105
EXPOSURE CONDITIONS					
Type	Ethylene Oxide	Ethylene Oxide	Ethylene Oxide	Ethylene Oxide	Ethylene Oxide
Details	12% EtO and 88% Freon	12% EtO and 88% Freon	12% EtO and 88% Freon	12% EtO and 88% Freon	12% EtO and 88% Freon
Number of Cycles	1	1	1	1	1
PROPERTIES RETAINED (%)					
Notched Izod Impact	75 (fm)	100 (fm)	100 (fm)	100 (fm)	97 (fm)

GRAPH 47: Gamma Radiation Dose vs. Tensile Strength of Acrylic Terpolymer

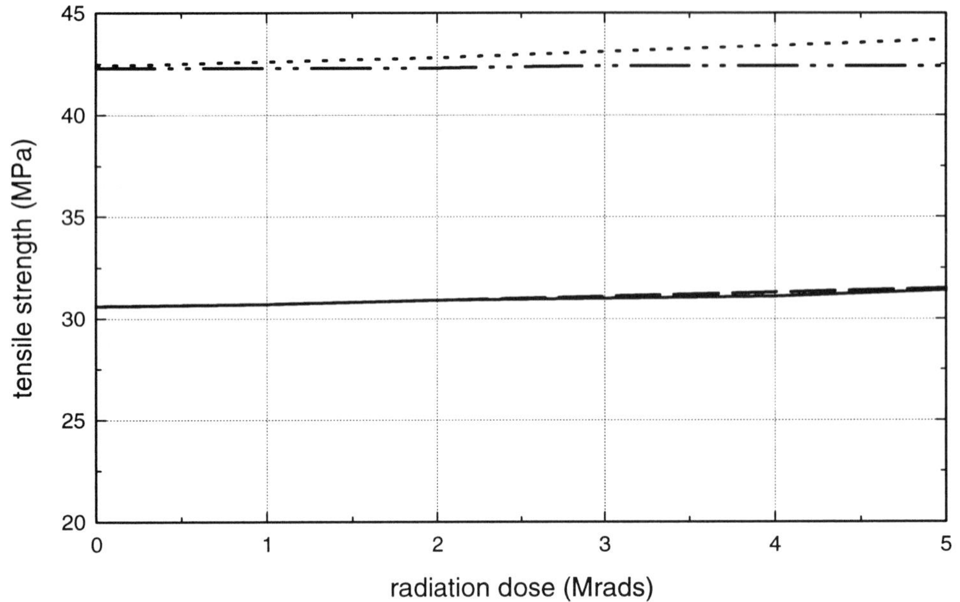

....	Novacor Zylar 94-568 Acrylic Terpol.; post exposure time: 0 days; thickness: 3.2 mm
—··—··	Novacor Zylar 94-568 Acrylic Terpol.; post exposure time: 212 days; thickness: 3.2 mm
— — —	Novacor Zylar ST 94-560 Acrylic Terpol.; post exposure time: 0 days; thickness: 3.2 mm
———	Novacor Zylar ST 94-560 Acrylic Terpol.; post exposure time: 212 days; thickness: 3.2 mm
Reference No.	105

GRAPH 48: Gamma Radiation Dose vs. Tensile Strength of Acrylic Terpolymer

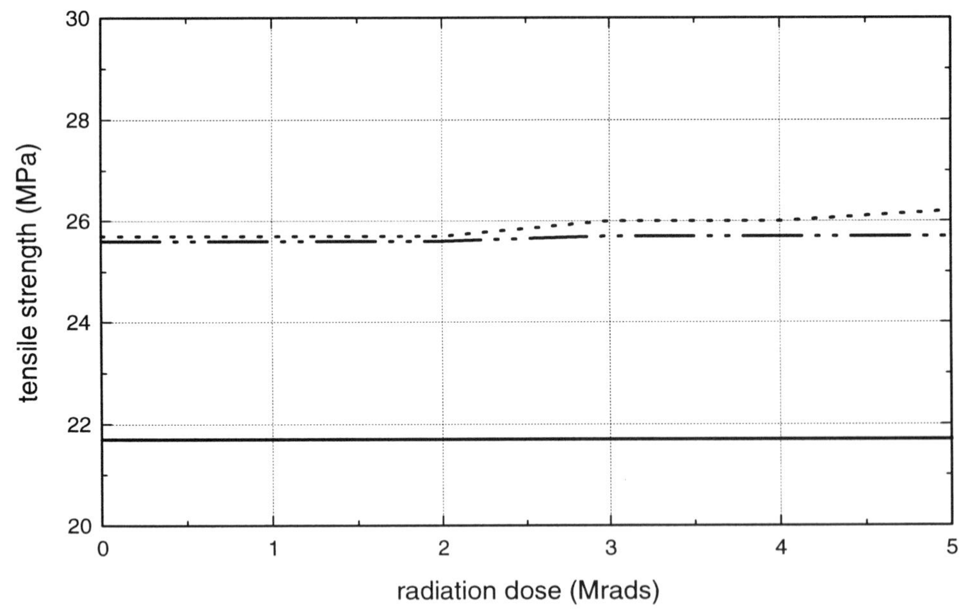

....	Novacor Zylar ST 561 Acrylic Terpol.; post exposure time: 0 days; thickness: 3.2 mm
—··—··	Novacor Zylar ST 561 Acrylic Terpol.; post exposure time: 212 days; thickness: 3.2 mm
———	Novacor Zylar ST 94-562 Acrylic Terpol.; post exposure time: 0 days; thickness: 3.2 mm
———	Novacor Zylar ST 94-562 Acrylic Terpol.; post exposure time: 212 days; thickness: 3.2 mm
Reference No.	105

GRAPH 49: Gamma Radiation Dose vs. Elongation of Acrylic Terpolymer

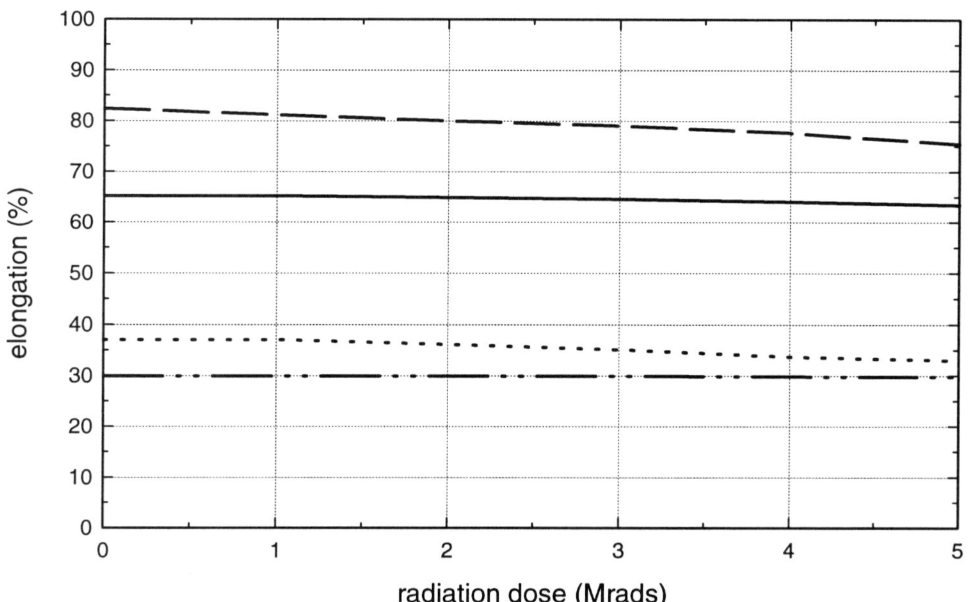

···········	Novacor Zylar 94-568 Acrylic Terpol.; post exposure time: 0 days; thickness: 3.2 mm
—·—··	Novacor Zylar 94-568 Acrylic Terpol.; post exposure time: 212 days; thickness: 3.2 mm
– – – –	Novacor Zylar ST 94-560 Acrylic Terpol.; post exposure time: 0 days; thickness: 3.2 mm
———	Novacor Zylar ST 94-560 Acrylic Terpol.; post exposure time: 212 days; thickness: 3.2 mm
Reference No.	105

GRAPH 50: Gamma Radiation Dose vs. Elongation of Acrylic Terpolymer

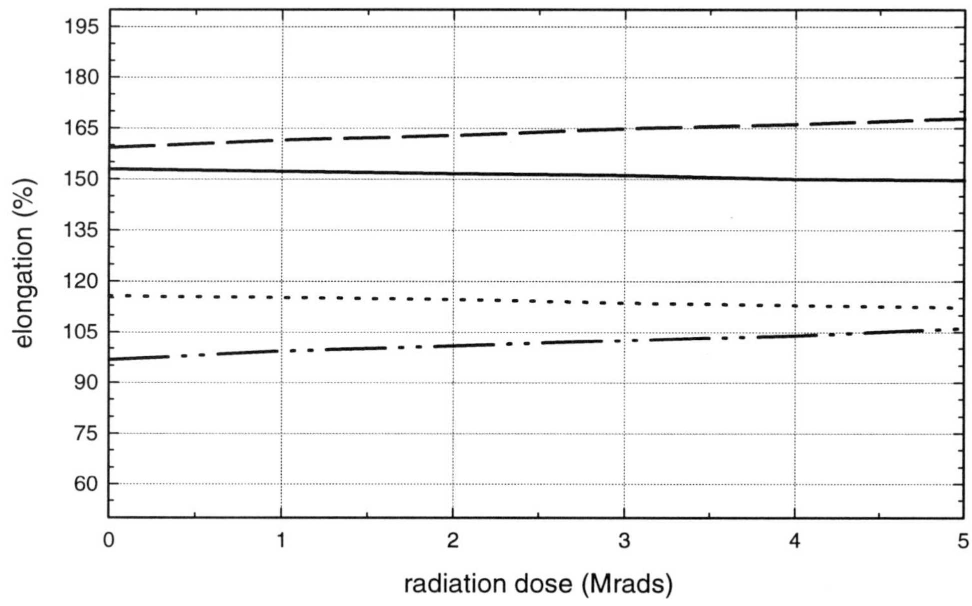

···········	Novacor Zylar ST 561 Acrylic Terpol.; post exposure time: 0 days; thickness: 3.2 mm
—·—··	Novacor Zylar ST 561 Acrylic Terpol.; post exposure time: 212 days; thickness: 3.2 mm
– – – –	Novacor Zylar ST 94-562 Acrylic Terpol.; post exposure time: 0 days; thickness: 3.2 mm
———	Novacor Zylar ST 94-562 Acrylic Terpol.; post exposure time: 212 days; thickness: 3.2 mm
Reference No.	105

GRAPH 51: **Gamma Radiation Dose vs. Notched Izod Impact Strength of Acrylic Terpolymer**

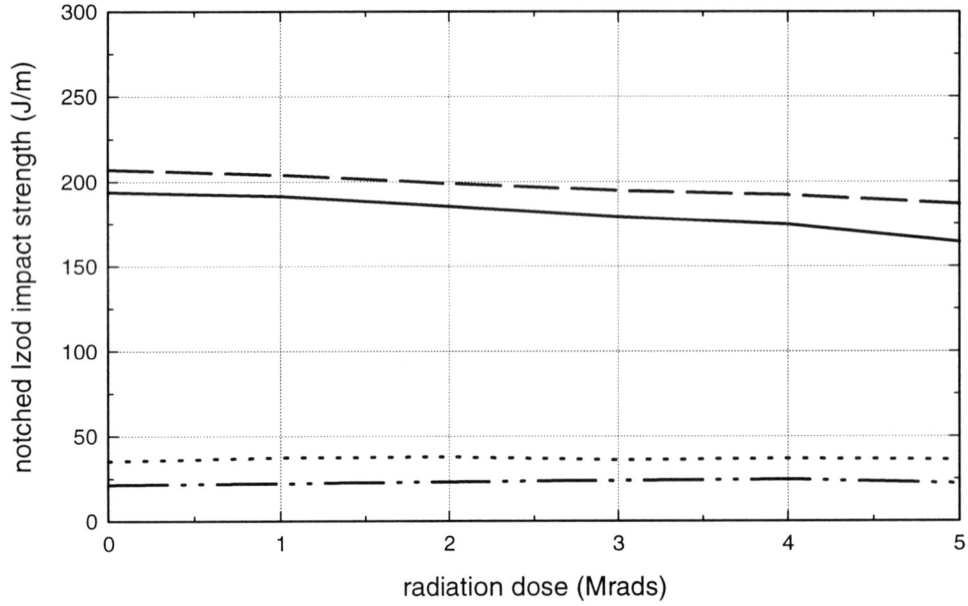

···············	Novacor Zylar 94-568 Acrylic Terpol.; post exposure time: 0 days; thickness: 3.2 mm
—··—··	Novacor Zylar 94-568 Acrylic Terpol.; post exposure time: 212 days; thickness: 3.2 mm
— — —	Novacor Zylar ST 94-560 Acrylic Terpol.; post exposure time: 0 days; thickness: 3.2 mm
———	Novacor Zylar ST 94-560 Acrylic Terpol.; post exposure time: 212 days; thickness: 3.2 mm
Reference No.	105

GRAPH 52: **Gamma Radiation Dose vs. Notched Izod Impact Strength of Acrylic Terpolymer**

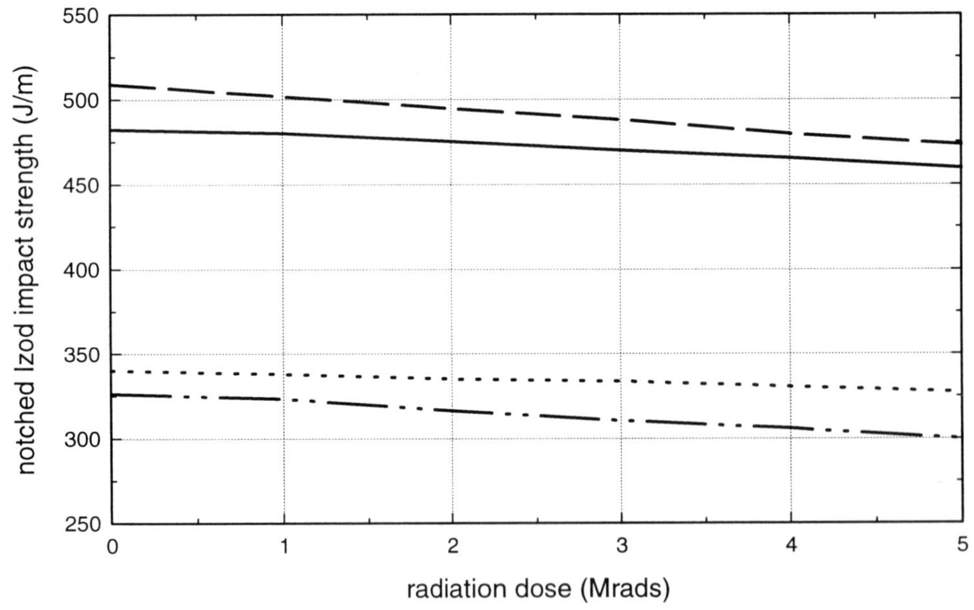

···············	Novacor Zylar ST 561 Acrylic Terpol.; post exposure time: 0 days; thickness: 3.2 mm
—··—··	Novacor Zylar ST 561 Acrylic Terpol.; post exposure time: 212 days; thickness: 3.2 mm
— — —	Novacor Zylar ST 94-562 Acrylic Terpol.; post exposure time: 0 days; thickness: 3.2 mm
———	Novacor Zylar ST 94-562 Acrylic Terpol.; post exposure time: 212 days; thickness: 3.2 mm
Reference No.	105

GRAPH 53: Gamma Radiation Dose vs. Yellowness Index of Acrylic Terpolymer

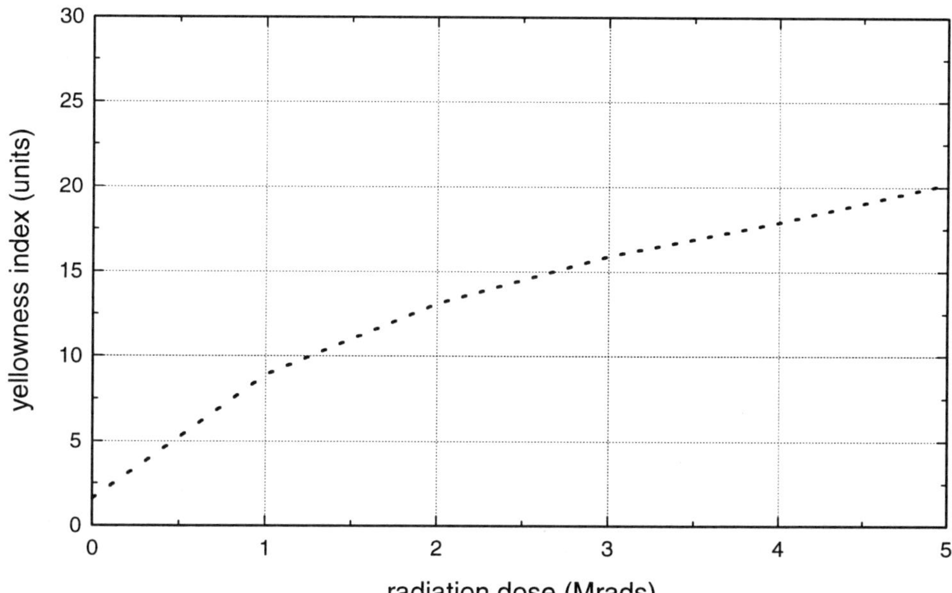

....	Novacor Zylar 90 Acrylic Terpol. (transpar.)
Reference No.	106

GRAPH 54: Post Gamma Radiation Exposure Time vs. Yellowness Index of Acrylic Terpolymer

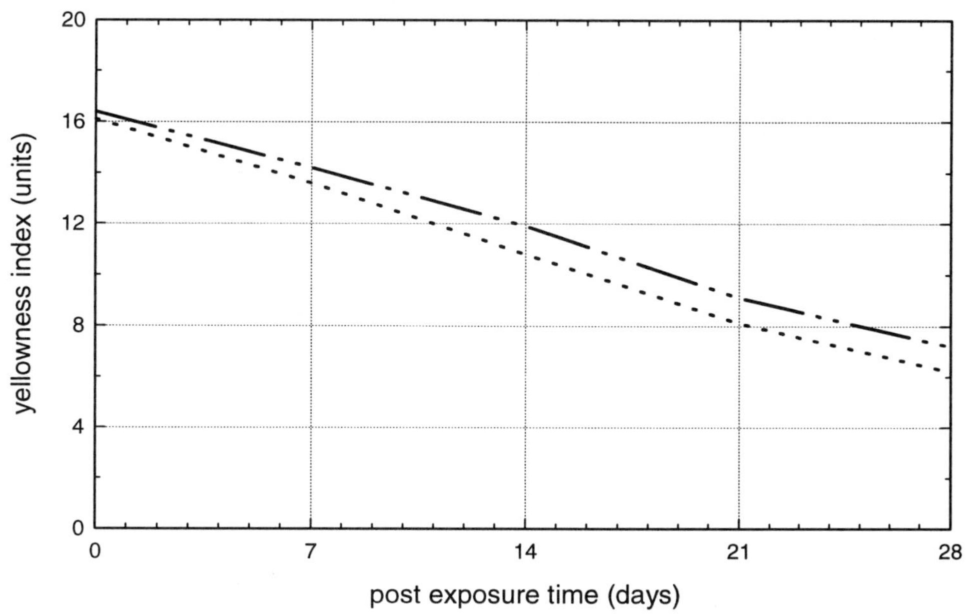

....	Novacor Zylar 90 Acrylic Terpol. (transpar.); 3 Mrads, light storage
—··—··	Novacor Zylar 90 Acrylic Terpol. (transpar.); 3 Mrads, dark storage
Reference No.	106

GRAPH 55: Post Gamma Radiation Exposure Time vs. Yellowness Index of Acrylic Terpolymer

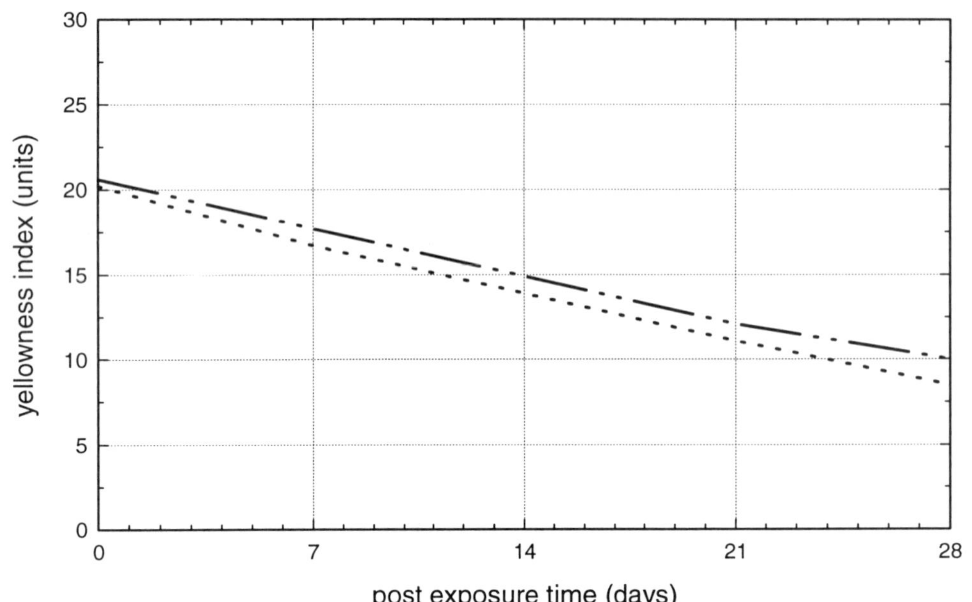

....	Novacor Zylar 90 Acrylic Terpol. (transpar.); 5 Mrads, light storage
—··—··	Novacor Zylar 90 Acrylic Terpol. (transpar.); 5 Mrads, dark storage
Reference No.	106

GRAPH 56: Beta Radiation Dose vs. Tensile Strength of Acrylic Terpolymer

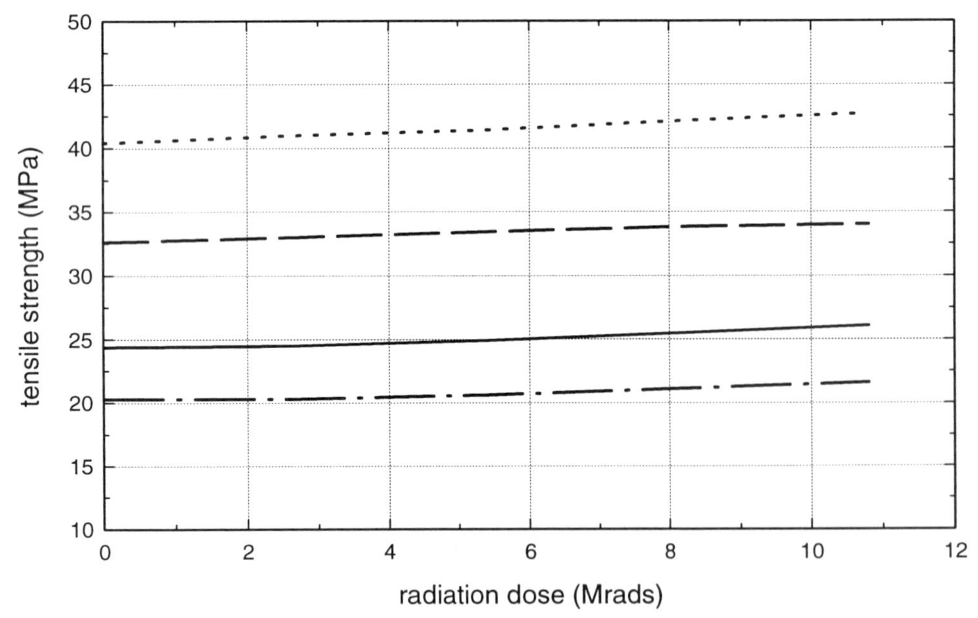

....	Novacor Zylar 94-568 Acrylic Terpol.
—··—··	Novacor Zylar 93-546 Acrylic Terpol.
— — —	Novacor Zylar ST 94-560 Acrylic Terpol.
———	Novacor Zylar ST 561 Acrylic Terpol.
—·—·—	Novacor Zylar ST 94-562 Acrylic Terpol.
Reference No.	105

GRAPH 57: **Beta Radiation Dose vs. Tensile Modulus of Acrylic Terpolymer**

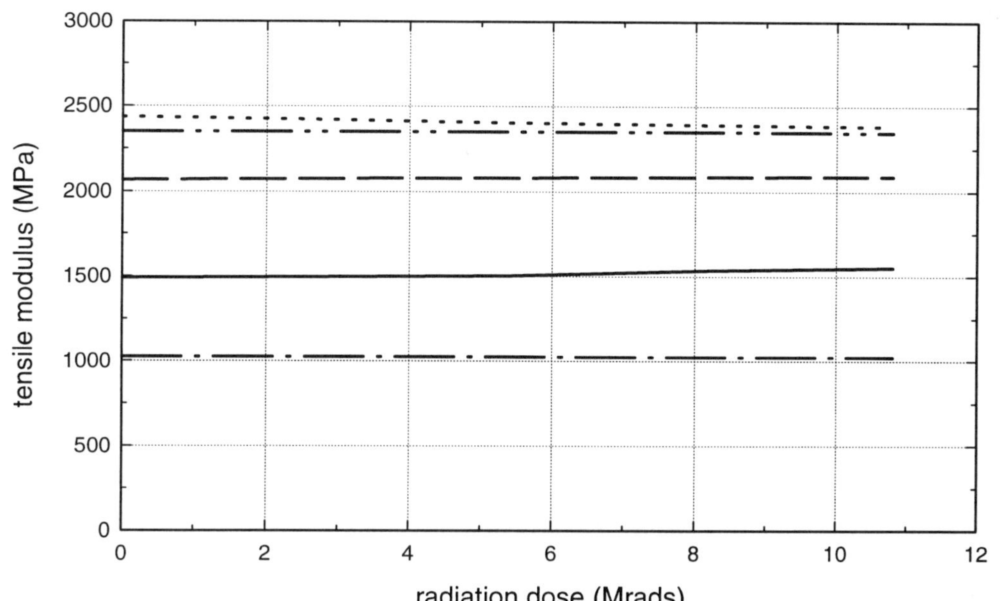

	Novacor Zylar 94-568 Acrylic Terpol.
	Novacor Zylar 93-546 Acrylic Terpol.
	Novacor Zylar ST 94-560 Acrylic Terpol.
	Novacor Zylar ST 561 Acrylic Terpol.
	Novacor Zylar ST 94-562 Acrylic Terpol.
Reference No.	105

GRAPH 58: **Beta Radiation Dose vs. Notched Izod Impact Strength of Acrylic Terpolymer**

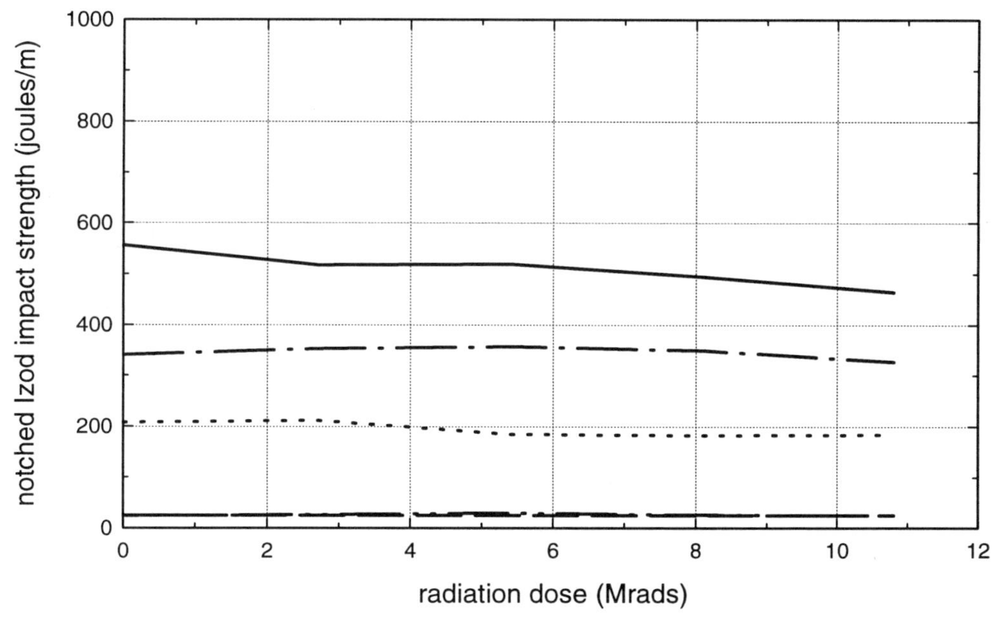

	Novacor Zylar ST 94-560 Acrylic Terpol.
	Novacor Zylar 94-568 Acrylic Terpol.
	Novacor Zylar 93-546 Acrylic Terpol.
	Novacor Zylar ST 94-562 Acrylic Terpol.
	Novacor Zylar ST 561 Acrylic Terpol.
Reference No.	105

GRAPH 59: Beta Radiation Dose vs. Yellowness Index of Acrylic Terpolymer

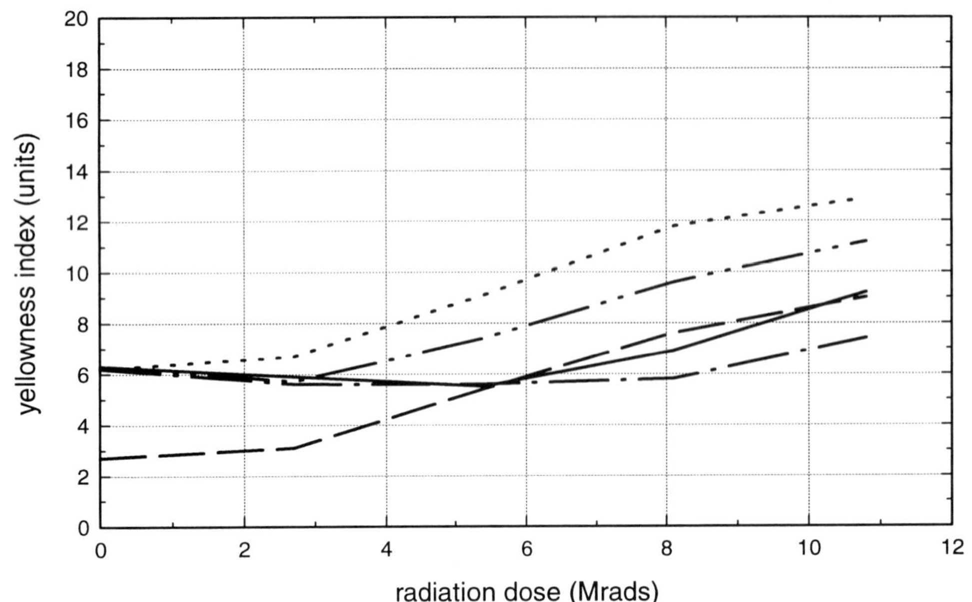

GRAPH 60: Post Beta Radiation Exposure Time vs. Yellowness Index of Acrylic Terpolymer

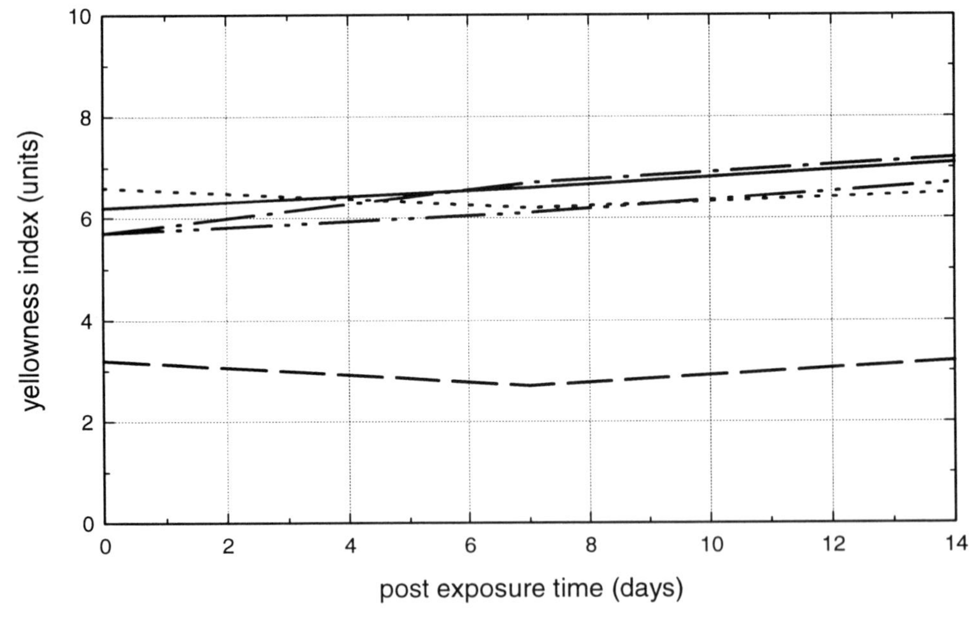

GRAPH 61: Post Beta Radiation Exposure Time vs. Yellowness Index of Acrylic Terpolymer

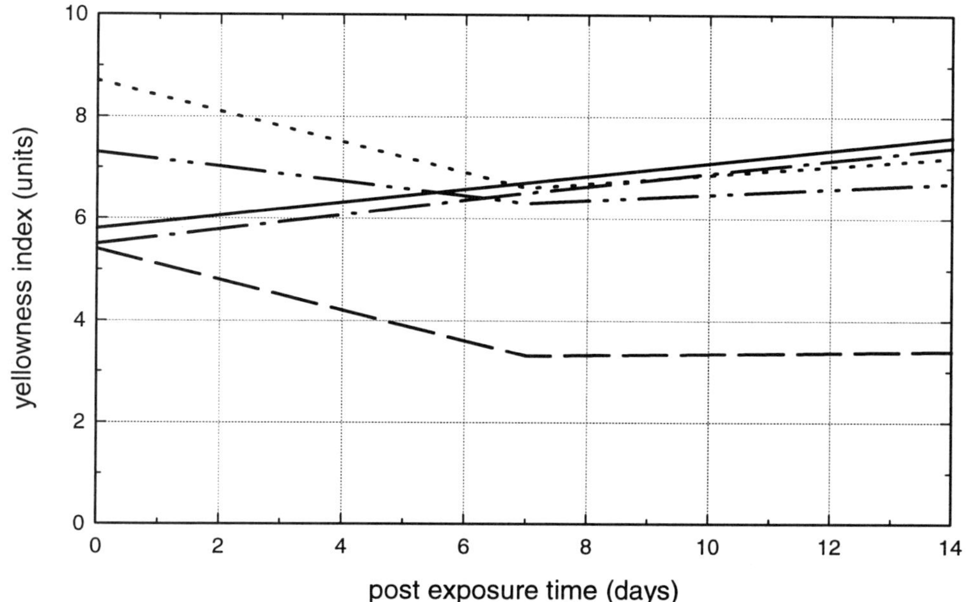

....	Novacor Zylar 94-568 Acrylic Terpol.; 5.4 Mrads
—·—··	Novacor Zylar ST 94-560 Acrylic Terpol.; 5.4 Mrads
— — —	Novacor Zylar 93-546 Acrylic Terpol.; 5.4 Mrads
———	Novacor Zylar ST 561 Acrylic Terpol.; 5.4 Mrads
—·—·—	Novacor Zylar ST 94-562 Acrylic Terpol.; 5.4 Mrads
Reference No.	105

GRAPH 62: Post Beta Radiation Exposure Time vs. Yellowness Index of Acrylic Terpolymer

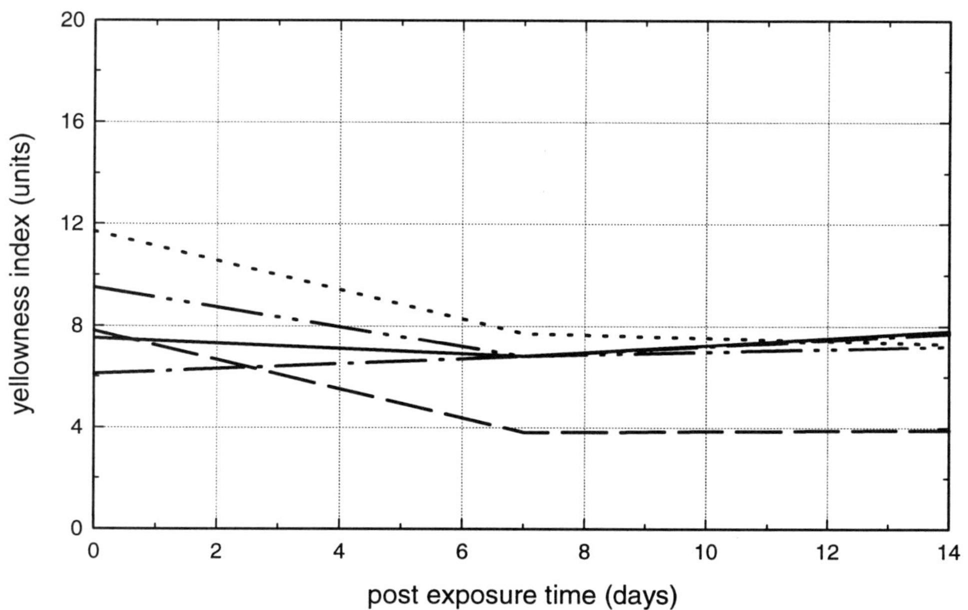

....	Novacor Zylar 94-568 Acrylic Terpol.; 8.1 Mrads
—·—··	Novacor Zylar ST 94-560 Acrylic Terpol.; 8.1 Mrads
— — —	Novacor Zylar 93-546 Acrylic Terpol.; 8.1 Mrads
———	Novacor Zylar ST 561 Acrylic Terpol.; 8.1 Mrads
—·—·—	Novacor Zylar ST 94-562 Acrylic Terpol.; 8.1 Mrads
Reference No.	105

GRAPH 63: Post Beta Radiation Exposure Time vs. Yellowness Index of Acrylic Terpolymer

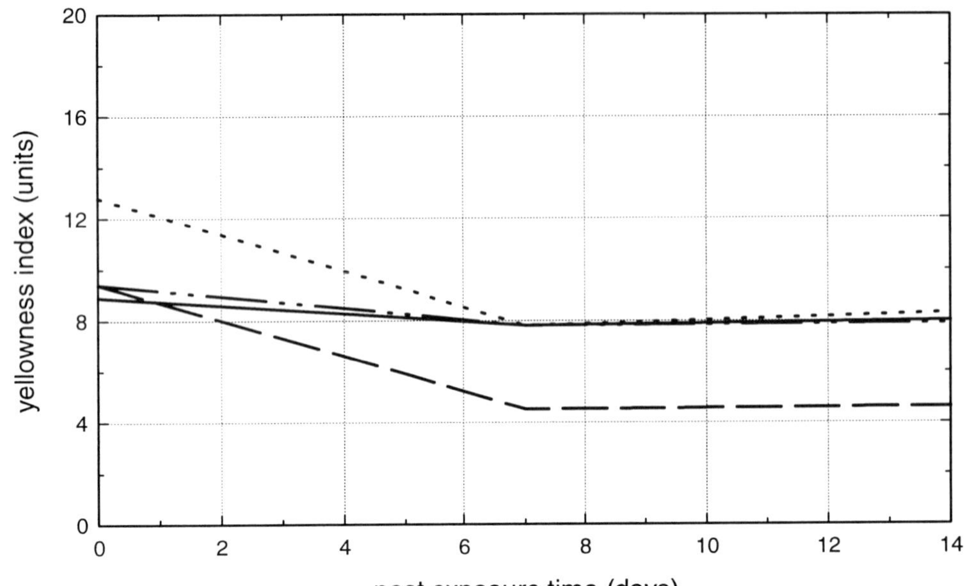

Gamma Radiation Resistance

BP Chemicals: Barex 210 (features: transparent)

Barex 210 turned amber most likely by crosslinking and formation of double bonds. The Barex 210 demonstrated low transmission and high yellowness index values.

Reference: Atakent, A. I., *Gamma Radiation Study,* supplier technical report - Cyro Industries, 1990.

Gamma Radiation Resistance

Eval Company: Eval EP-E (form: film; features: transparent, barrier properties); **Eval EP-F** (form: film; features: transparent, barrier properties)

Eval resins maintain their physical properties after being irradiated by 5.0 Mrads of either gamma or beta rays. The only appreciable effect being a reduction in elongation at break. No discoloration after irradiation was noticed.

Reference: *Effects Of Radiation On Eval Resins,* supplier design guide (Technical Bulletin No. 140) - Eval Company Of America.

Steam Resistance

Eval Company: Eval EP-E (form: film; features: transparent, barrier properties); **Eval EP-F** (form: film; features: transparent, barrier properties)

When Eval composite films are subjected to boiling sterilization, the gas barrier properties of the film are temporarily reduced. The films recover their original properties with time. The time to restore the original barrier properties depends on the polymers and thickness utilized as the outer layer. When polyamide is used for the outside layer the original gas barrier properties are completely restored in one day. It should be emphasized that in composite structures with Eval film employed as the intermediate layer, the Eval polymer never undergoes structural changes caused by the boiling process. The deterioration of the polymer's gas barrier properties is temporary and due to moisture absorption during boiling. As time passes, the moisture evaporates and the gas barrier properties are restored.

Reference: *Kuraray Eval Resin,* supplier design guide (5-2,000-507) - KurarayCo., Ltd.

TABLE 30: Effect of Gamma Radiation Sterilization on Ethylene Vinyl Alcohol Copolymer

Material Family	ETHYLENE VINYL ALCOHOL COPOLYMER			
Material Supplier/Name	EVAL EP-F		EVAL EP-E	
Material Note	film; transparent, barrier properties; thickness: 0.015 mm	film; transparent, barrier properties; thickness: 0.015 mm	film; transparent, barrier properties; thickness: 0.02 mm	film; transparent, barrier properties; thickness: 0.02 mm
Reference No.	64	64	64	64
EXPOSURE CONDITIONS				
Type	Gamma Radiation	Gamma Radiation	Gamma Radiation	Gamma Radiation
Radiation Dose (Mrads)	2	5	2	5
PROPERTIES RETAINED (%)				
Tensile Strength	91.7 (hn)	82.5 (hn)	73.7 (ho)	57.5 (ho)
Tensile Strength (transverse)	105 (hq)	160 (hq)	91.7 (hl)	144.5 (hl)
Tensile Strength @ Yield	110.4 (ic)	105.2 (ic)	96.2 (id)	62.3 (id)
Tensile Strength @ Yield (transverse)	111.5 (if)		90.9 (ig)	
Elongation	91.3 (am)	69.6 (am)	117.4 (an)	56.5 (an)
Elongation (transverse)	113.6 (ao)	7.7 (ao)	35.7 (ap)	3.6 (ap)

TABLE 31: Effect of Beta Radiation Sterilization on Ethylene Vinyl Alcohol Copolymer

Material Family	ETHYLENE VINYL ALCOHOL COPOLYMER	
Material Supplier/Name	EVAL EP-F	EVAL EP-E
Material Note	film; transparent, barrier properties; thickness: 0.015 mm	film; transparent, barrier properties; thickness: 0.02 mm
Reference No.	64	64
EXPOSURE CONDITIONS		
Type	Beta Radiation	Beta Radiation
Radiation Dose (Mrads)	5	5
PROPERTIES RETAINED (%)		
Tensile Strength	91.7 (hn)	78.7 (ho)
Tensile Strength (transverse)	95 (hq)	88.8 (hl)
Tensile Strength @ Yield	101.7 (ic)	94.3 (id)
Tensile Strength @ Yield (transverse)	101.7 (if)	90.9 (ig)
Elongation	95.7 (am)	113 (an)
Elongation (transverse)	40.9 (ao)	42.9 (ap)

Gamma Radiation Resistance

3M: Kel-F 81

Kel-F PCTFE has the ability to perform satisfactorily while absorbing large doses of gamma radiation. Kel-F 81 PCTFE will suffer only minimal physical property loss at a dosage level of 18 Mrads (dose rate of 0.07 Mrads/hour in air at 25°C).

Reference: *Kel-F 81 PCTFE Engineering Manual,* supplier design guide (98-0211-5944-1 (120.5) DPI) - 3M Industrial Chemical Products Division, 1990.

TABLE 32: Effect of Gamma Radiation Sterilization on Polychlorotrifluoroethylene

Material Family	POLYCHLOROTRIFLUOROETHYLENE	
Material Supplier/Name	3M KEL-F 81	
Reference No.	96	96

EXPOSURE CONDITIONS

Type	Gamma Radiation	Gamma Radiation
Radiation Dose (Mrads)	16	24
Note	dose rate: 0.07 Mrads/hr in air	dose rate: 0.07 Mrads/hr in air
Temperature (degrees C)	25	25

PROPERTIES RETAINED (%)

Tensile Strength	70 (he)	60 (he)
Tensile Strength @ Yield	70 (ib)	110 (ib)
Elongation	80 (ai)	80 (ai)
Modulus	115 (gs)	110 (gs)

Gamma Radiation Resistance

ICI: Fluon

Harrington and Giberson, in a study of the decline in the tensile strength and elongation of PTFE when exposed to gamma radiation, showed that irradiation in a vacuum was less damaging than irradiation in air. This point was confirmed by Wall and Florin. The more recent work by Monnet and Bensa gives further data on the effect of radiation dose on mechanical properties. They found that as little as 0.01 to 0.1 Mrad dose can affect mechanical properties. One Mrad has a measurable effect and 2 to 3 Mrads in air reduces strength by 40 to 75%. Four Mrads reduces tensile strength to 2% of the original.

Reference: *Physical Properties Of Unfilled And Filled Polytetrafluoroethylene,* supplier design guide (Technical Service Note F12/13) - ICI PLC, 1981.

TABLE 33: Effect of Gamma Radiation Sterilization on Polytetrafluoroethylene

Material Family	POLYTETRAFLUOROETHYLENE				
Material Supplier/Name	ICI FLUON				
Reference No.	71	71	71	71	71

EXPOSURE CONDITIONS

Type	Gamma Radiation	Gamma Radiation	Gamma Radiation	Gamma Radiation	Gamma Radiation
Note	dose: 2.4 ev/g x 10^{-20} in air	dose: 4.1 ev/g x 10^{-20} in air	dose: 0.7 ev/g x 10^{-20} in vacuum	dose: 4.1 ev/g x 10^{-20} in vacuum	dose: 32.0 ev/g x 10^{-20} in vacuum

PROPERTIES RETAINED (%)

Tensile Strength	2 (he)	0 (he)	73 (he)	51 (he)	43 (he)

Radiation Resistance

Atochem: Kynar

The different grades of KYNAR homopolymer are readily crosslinked and do not degrade when irradiated with moderate doses of high energy electron or gamma radiation. The efficiency of crosslinking is influenced by the grade, i.e., molecular weight variations are important. KYNAR fabricated products utilizing radiation technology are heat-shrinkable tubing and insulated wire capable of withstanding high temperatures.

Reference: *Kynar Polyvinylidene Fluoride,* supplier technical report (PL705-Rev4-1-91) - Atochem North America, Inc., 1991.

Gamma Radiation Resistance

Atochem: Foraflon 1000HD (features: general purpose grade)

Tests with intensive dosages have shown that Foraflon 1000HD can withstand exposure to 20 Mrads without undergoing any change other than an increase in yield strength and a reduction in elongation at break. The phenomenon of increase in hardness continues up to 40 Mrads. Then up to about 100 Mrads, two phenomena are in competition: degradation and crosslinking, which bring about a relative stability in mechanical properties, and at the same time some browning, which becomes very considerable above 100 Mrads. Above 100 Mrads, the degradation becomes progressive and rapid.

Reference: *Foraflon PVDF,* supplier design guide (694.E/07.87/20) - Atochem S. A., 1987.

TABLE 34: Effect of Electron Beam Radiation Sterilization on Polyvinylidene Fluoride Thermoplastic Elastomer

Material Family	POLYVINYLIDENE FLUORIDE THERMOPLASTIC ELASTOMER			
Material Supplier/Name	ATOCHEM KYNAR FLEX 2800			
Reference No.	7	7	7	7
EXPOSURE CONDITIONS				
Type	Electron Beam Radiation	Electron Beam Radiation	Electron Beam Radiation	Electron Beam Radiation
Radiation Dose (Mrads)	2	4	8	20
PROPERTIES RETAINED (%)				
Tensile Strength	100 (kp)	104.3 (kp)	100 (kp)	113.3 (kp)
Modulus	106.4 (hd)	106.4 (hd)	101.6 (hd)	104.8 (hd)

GRAPH 64: Gamma Radiation Dose vs. Tensile Strength at Yield of Polyvinylidene Fluoride

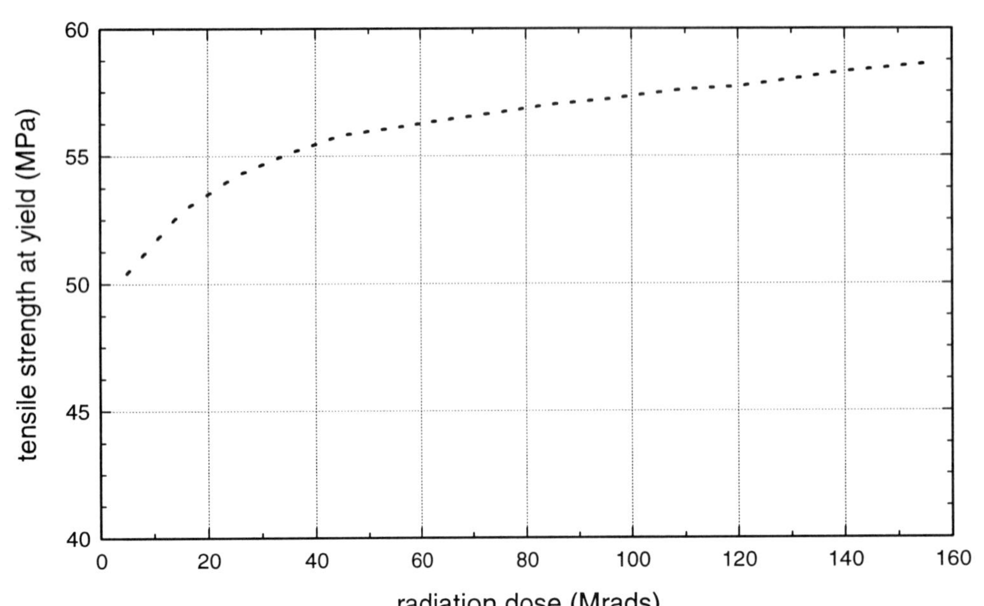

....	Atochem Foraflon 1000HD PVDF (gen. purp. grade)
Reference No.	89

GRAPH 65: Gamma Radiation Dose vs. Elongation at Break of Polyvinylidene Fluoride

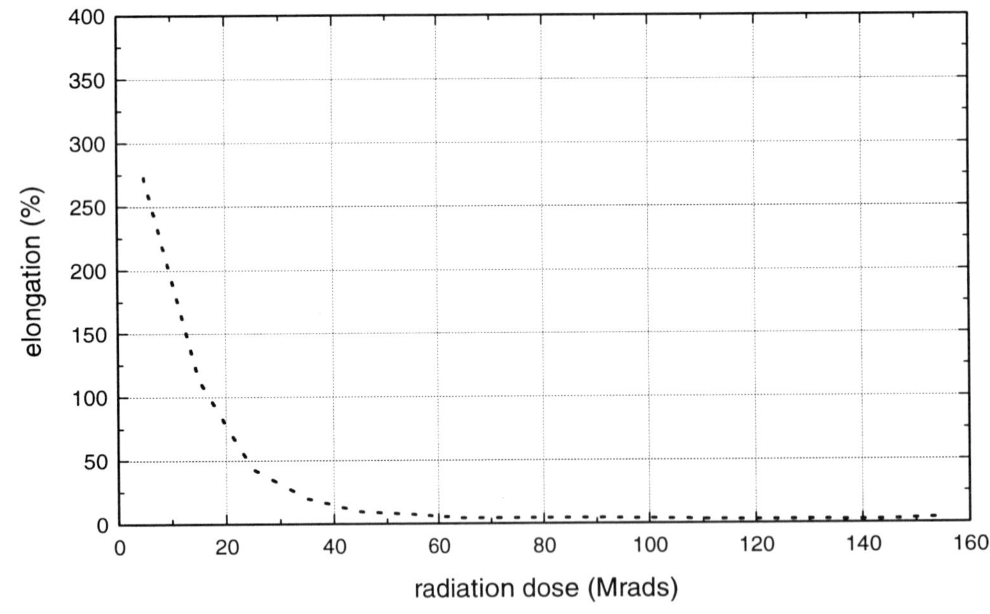

....	Atochem Foraflon 1000HD PVDF (gen. purp. grade)
Reference No.	89

Gamma Radiation Resistance

GE Plastics: Noryl HP (features: medical grade); **Noryl PX1404**

Properties of Noryl HP and Noryl PX1404 are retained through 5 megarads of gamma radiation.

Reference: *Sterilization Comparison,* supplier marketing literature (JRW/3/91) - General Electric Plastics, 1991.

GE Plastics: Noryl HP (features: medical grade)

Healthcare grades of Noryl HP resins withstand gamma radiation sterilization with virtually no loss of properties. Even after exposure to 2.5 Mrads of gamma radiation, theresins' mechanical performance is virtually unaffected. The color shift was small (0.6 yellowness index) with gamma exposure. An important benefit in applications such as labware, surgical instruments and equipment subjected to frequent abuse.

Reference: *GE Plastics Medical Applications,* supplier marketing literature (NBM-100 (1/89) RTB) - General Electric Company, 1989.

Ethylene Oxide (EtO) Resistance

GE Plastics: Noryl HP (features: medical grade)

Properties of Noryl HP resins are unchanged to 10 ethylene oxide sterilization cycles.

Reference: *Sterilization Comparison,* supplier marketing literature (JRW/3/91) - General Electric Plastics, 1991.

GE Plastics: Noryl HP (features: medical grade)

Healthcare grades of Noryl HP resins withstand ethylene oxide sterilization with virtually no loss of properties.

Reference: *GE Plastics Medical Applications,* supplier marketing literature (NBM-100 (1/89) RTB) - General Electric Company, 1989.

Steam Resistance

GE Plastics: Noryl PX1404

After 2000 autoclave cycles (Hi-Vac @ 132°C), Noryl PX1404 retains 50% of its impact strength.

Sterilization Comparison, supplier marketing literature (JRW/3/91) - General Electric Plastics, 1991.

GE Plastics: Noryl GFN3-701 (composition: 30% glass fiber reinforced)

After 4000 autoclave cycles (Hi-Vac @ 132°C), Noryl GFN3-701 retains 54% of its impact strength and 80% of its tensile strength.

Sterilization Comparison, supplier marketing literature (ACB/3/93) - General Electric Plastics, 1993.

GE Plastics: Noryl HP (features: medical grade)

Noryl HP resins are not recommended for steam autoclave sterilization.

Sterilization Comparison, supplier marketing literature (JRW/3/91) - General Electric Plastics, 1991.

GE Plastics: Noryl

One of the properties common to all Noryl resins is hydrolytic stability. Repeated steam autoclaving results in only small decreases in impact and flexural properties during the first few cycles. Thereafter, no further change in toughness or stiffness is observed even after 50 autoclaving cycles.

Reference: *Noryl Extrusion Resins,* supplier design guide (CDX-265) - General Electric Company.

TABLE 35: Effect of Gamma Radiation Sterilization on Modified Polyphenylene Oxide

Material Family	**MODIFIED POLYPHENYLENE ETHER**
Material Supplier/Name	**GE NORYL HP**
Material Note	medical grade
Reference No.	49

EXPOSURE CONDITIONS

Type	Gamma Radiation
Radiation Dose (Mrads)	2.5

PROPERTIES RETAINED (%)

Tensile Strength	101 (hm)
Elongation	102 (ak)
Notched Izod Impact	94 (fm)

TABLE 36: Effect of Ethylene Oxide Sterilization on Modified Polyphenylene Oxide

Material Family	**MODIFIED POLYPHENYLENE OXIDE**					
Material Supplier/Name	**GE NORYL HP41**					
Material Note	medical grade	medical grade	medical grade	medical grade	medical grade	medical grade
Reference No.	108	108	108	108	108	108

EXPOSURE CONDITIONS

Type	Ethylene Oxide	Ethylene Oxide	Ethylene Oxide	Ethylene Oxide	Ethylene Oxide	Ethylene Oxide
Details	100% EtO	100% EtO	100% EtO	100% EtO	100% EtO	100% EtO
Concentration	550 mg/l	550 mg/l	550 mg/l	550 mg/l	550 mg/l	550 mg/l
Time (hours)	5	5	5	5	5	5

PRE EXPOSURE CONDITIONING

Preconditioning Note	type: pre vacuum; pressure: 660-711 mm Hg; RH: 45-60%; dwell time: 15 minutes	type: pre vacuum; pressure: 660-711 mm Hg; RH: 45-60%; dwell time: 15 minutes	type: pre vacuum; pressure: 660-711 mm Hg; RH: 45-60%; dwell time: 15 minutes	type: pre vacuum; pressure: 660-711 mm Hg; RH: 45-60%; dwell time: 15 minutes	type: pre vacuum; pressure: 660-711 mm Hg; RH: 45-60%; dwell time: 15 minutes	type: pre vacuum; pressure: 660-711 mm Hg; RH: 45-60%; dwell time: 15 minutes

POST EXPOSURE CONDITIONING

Note	type: aeration; note: mechanical with fan	type: aeration; note: mechanical with fan	type: aeration; note: mechanical with fan	type: aeration; note: ambient conditions	type: aeration; note: ambient conditions	type: aeration; note: ambient conditions
Temperature (°C)	49	49	49	49	49	49

RESIDUALS (ppm)

Residuals Determined	ethylene oxide	ethylene chlorohydrin	ethylene glycol	ethylene oxide	ethylene chlorohydrin	ethylene glycol
24 hour Aeration	198	none detected	37	313	none detected	66
168 hour Aeration	51			138		
336 hour Aeration	38		20	116		32

Gamma Radiation Resistance

EMS-American Grilon: Grilamid TR55 (features: transparent, amorphous)

The values for tensile strength seemed not to change much over the range of dose levels (0.0 to 15.0 Mrads); at break the values decreased by more than 10% at 15 Mrads. The ultimate elongation follows the same trend as tensile strength at break. The flexural strength and modulus seem to increase over the whole range. Notched Izod stayed at a fairly level pace, and tensile impact declined slightly. After irradiation of 15 Mrads, all of the properties are within specification. The color changes with irradiation; yellowness mostly. This change is noticeable, but at normal dosages, should not cause a problem.

Reference: *Laboratory Report - Gamma Sterilization Of Grilamid TR55,* supplier technical report - EMS-American Grilon Inc., 1984.

TABLE 37: Effect of Gamma Radiation Sterilization on Nylon 12

Material Family	NYLON 12					
Material Supplier/Name	EMSER GRILAMID TR55					
Material Note	transparent, amorphous	transparent, amorphous	transparent, amorphous	transparent, amorphous	transparent, amorphous	transparent, amorphous
Reference No.	61	61	61	61	61	61
EXPOSURE CONDITIONS						
Type	Gamma Radiation	Gamma Radiation	Gamma Radiation	Gamma Radiation	Gamma Radiation	Gamma Radiation
Radiation Dose (Mrads)	2.83	5.71	7.51	10	12	14.87
Note	test lab: Isomedix, Inc.	test lab: Isomedix, Inc.	test lab: Isomedix, Inc.	test lab: Isomedix, Inc.	test lab: Isomedix, Inc.	test lab: Isomedix, Inc.
PROPERTIES RETAINED (%)						
Tensile Strength	99.1 (hs)	94.3 (hs)	102.6 (hs)	91 (hs)	89 (hs)	91 (hs)
Tensile Strength @ Yield	98.9 (ih)	98.8 (ih)	99.7 (ih)	100 (ih)	99.6 (ih)	99.5 (ih)
Elongation	107.9 (as)	107.3 (as)	98.8 (as)	102.4 (as)	93.9 (as)	96.3 (as)
Elongation @ Yield	100 (be)	100 (be)	100 (be)	100 (be)	100 (be)	100 (be)
Flexural Strength	99.6 (cl)	100.4 (cl)	100.4 (cl)	101.3 (cl)	100 (cl)	102.6 (cl)
Modulus	101.8 (cb)	95.8 (cb)	105.4 (cb)	101.8 (cb)	102.7 (cb)	103 (cb)
Notched Izod Impact	100 (fo)	101.2 (fo)	92.2 (fo)	100 (fo)	109.6 (fo)	96.4 (fo)
Gardner Impact	116.2 (gr)	102.4 (gr)	96.9 (gr)	114 (gr)	96.2 (gr)	88.4 (gr)
SURFACE and APPEARANCE						
ΔL Color	-0.06 (ab)	-0.15 (ab)	-0.03 (ab)	0.26 (ab)	0.17 (ab)	2.62 (ab)
Δa Color	-0.2 (n)	-0.44 (n)	-0.61 (n)	-0.98 (n)	-1.38 (n)	-0.77 (n)
Δb Color	0.72 (s)	1.33 (s)	1.72 (s)	2.83 (s)	3.86 (s)	6.61 (s)

GRAPH 66: Gamma Radiation Dose vs. Tensile Strength of Nylon 12

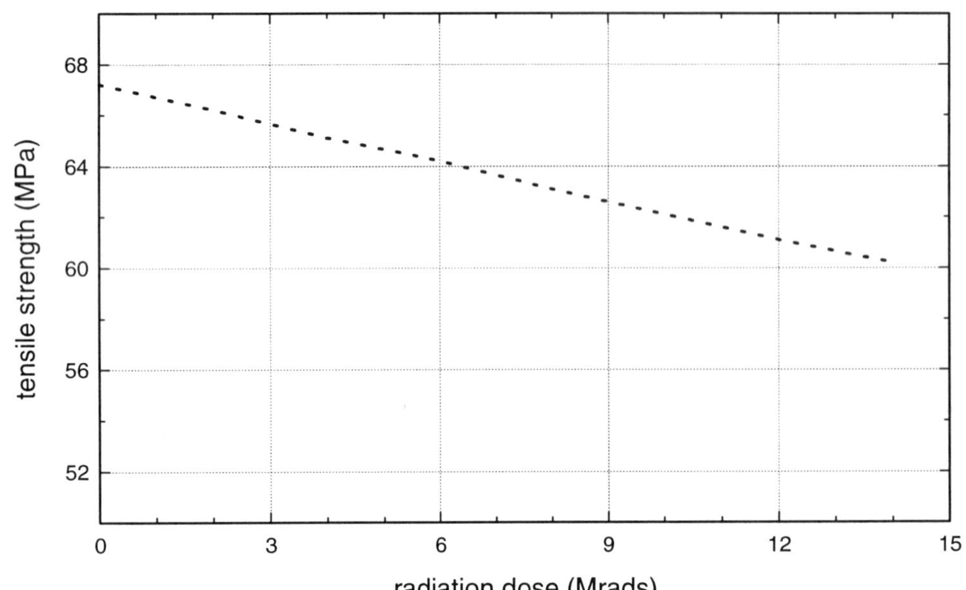

....	Emser Grilamid TR55 Nylon 12 (transp., amorphous)
Reference No.	61

GRAPH 67: Gamma Radiation Dose vs. Elongation at Break of Nylon 12

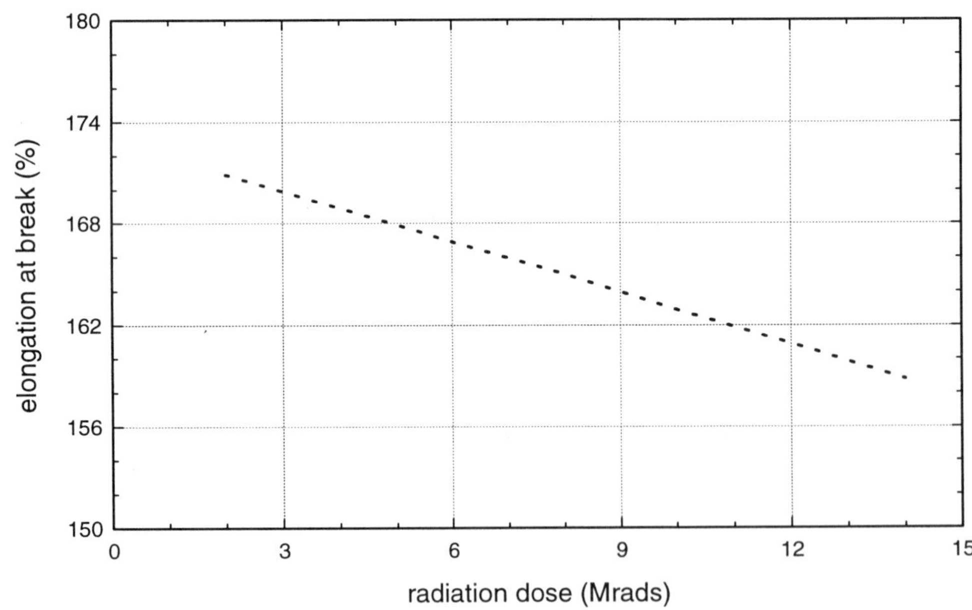

....	Emser Grilamid TR55 Nylon 12 (transp., amorphous)
Reference No.	61

GRAPH 68: Gamma Radiation Dose vs. Flexural Yield Strength of Nylon 12

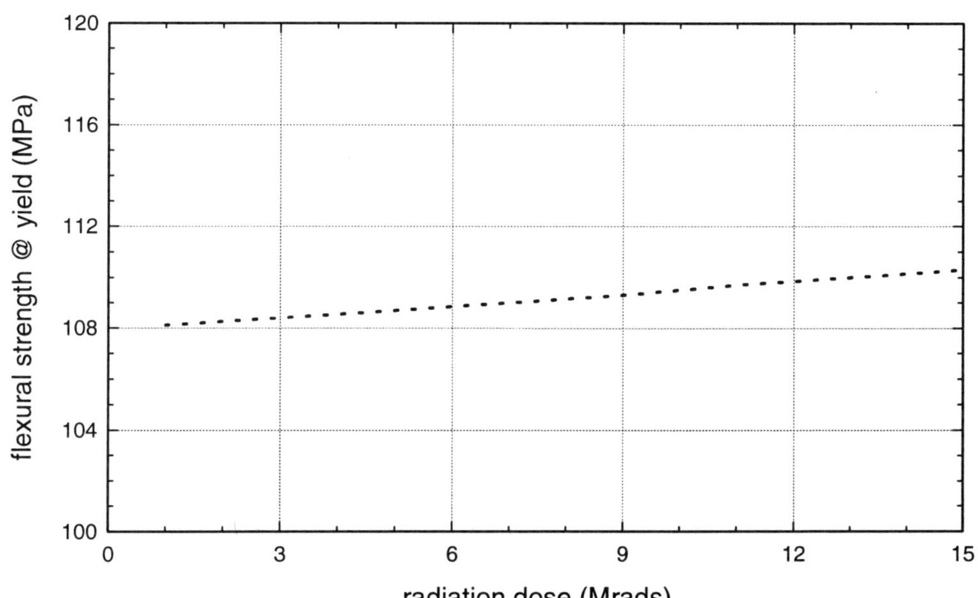

GRAPH 69: Gamma Radiation Dose vs. Flexural Modulus of Nylon 12

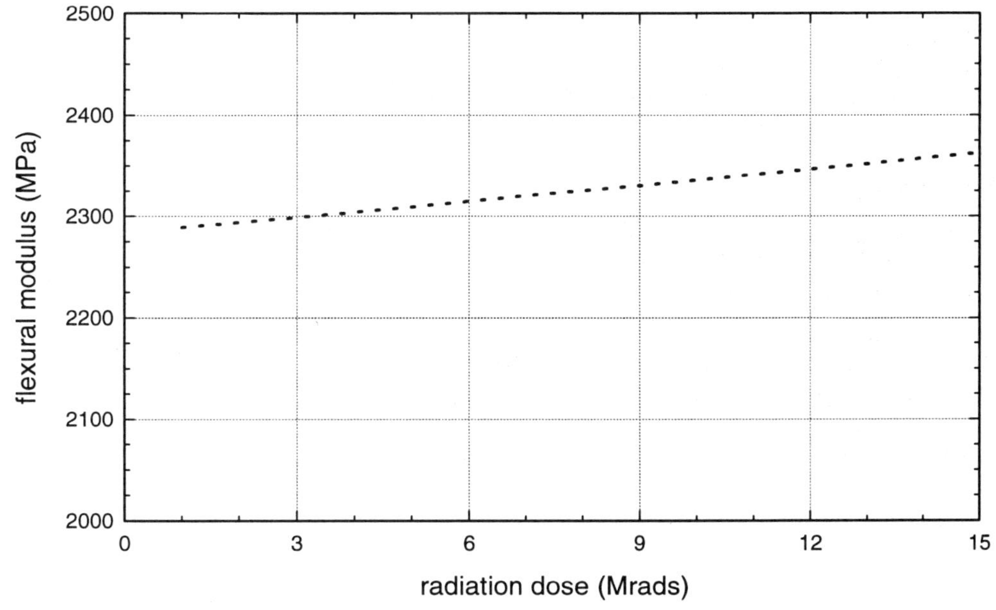

GRAPH 70: **Gamma Radiation Dose vs. Notched Izod Impact Strength of Nylon 12**

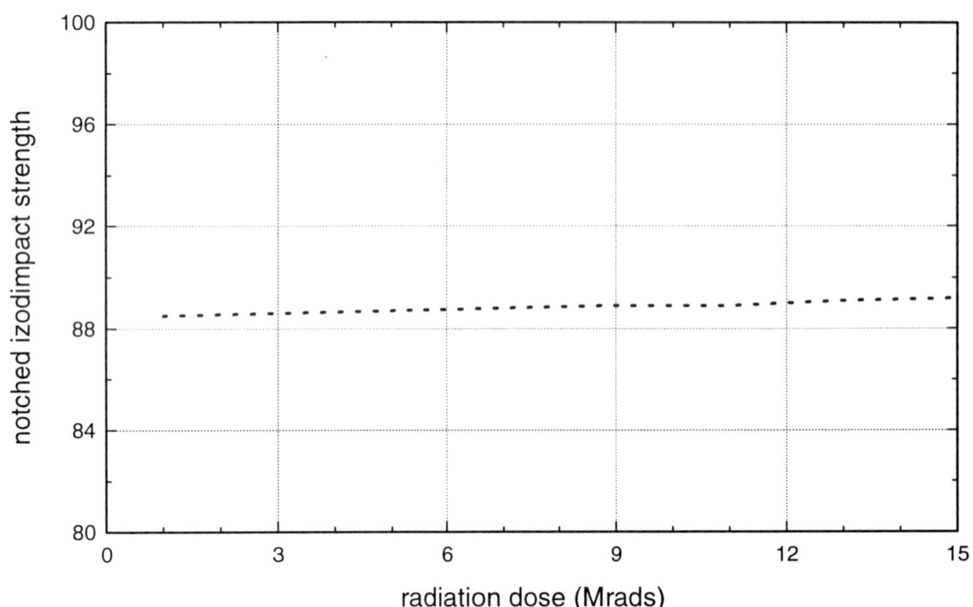

···············	Emser Grilamid TR55 Nylon 12 (transp., amorphous)
Reference No.	61

GRAPH 72: **Gamma Radiation Dose vs. Color Coordinate a of Nylon 12**

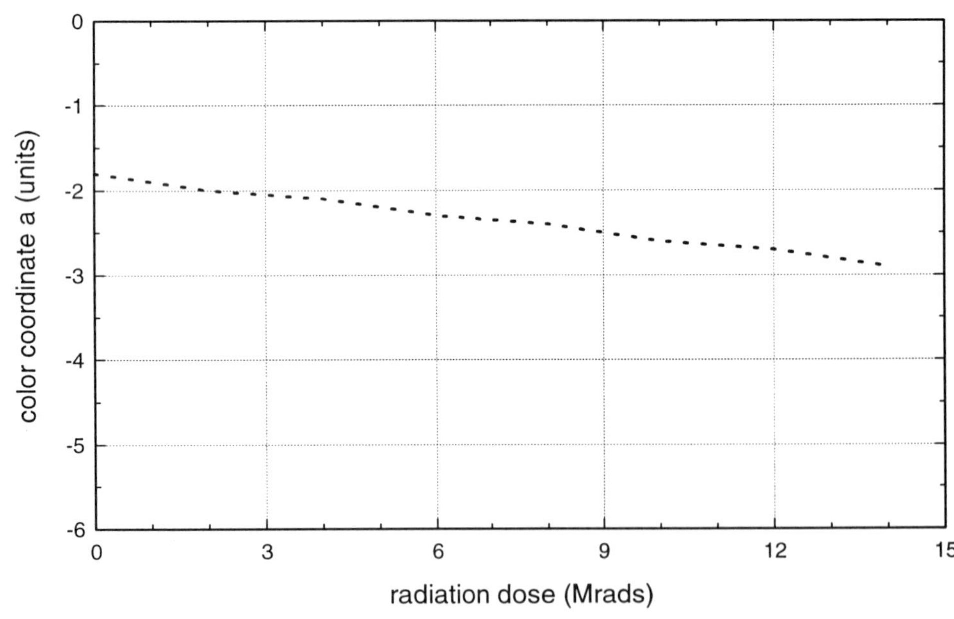

···············	Emser Grilamid TR55 Nylon 12 (transp., amorphous)
Reference No.	61

GRAPH 73: Gamma Radiation Dose vs. Color Coordinate b of Nylon 12

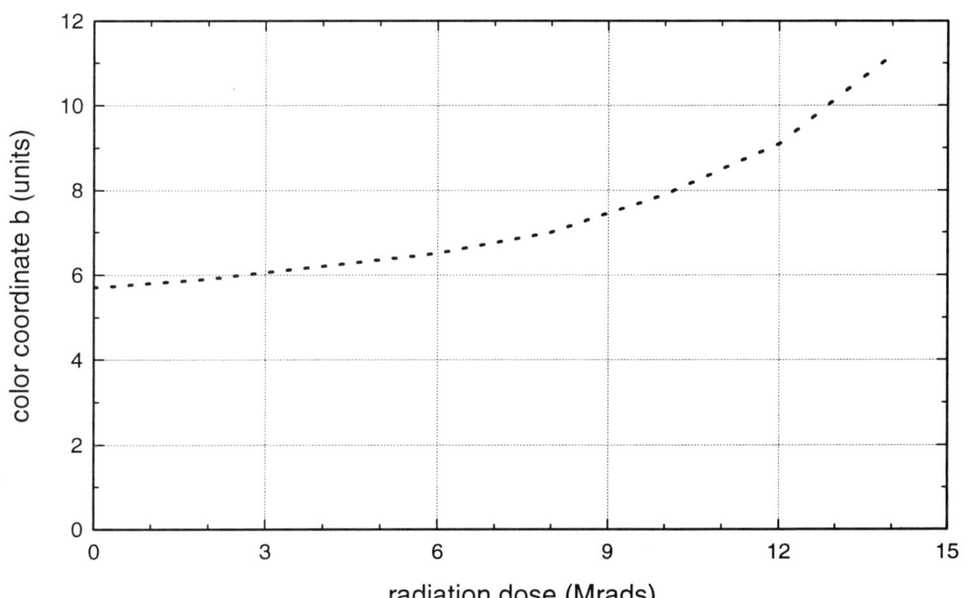

....	Emser Grilamid TR55 Nylon 12 (transp., amorphous)
Reference No.	61

GRAPH 74: Gamma Radiation Dose vs. Color Coordinate L of Nylon 12

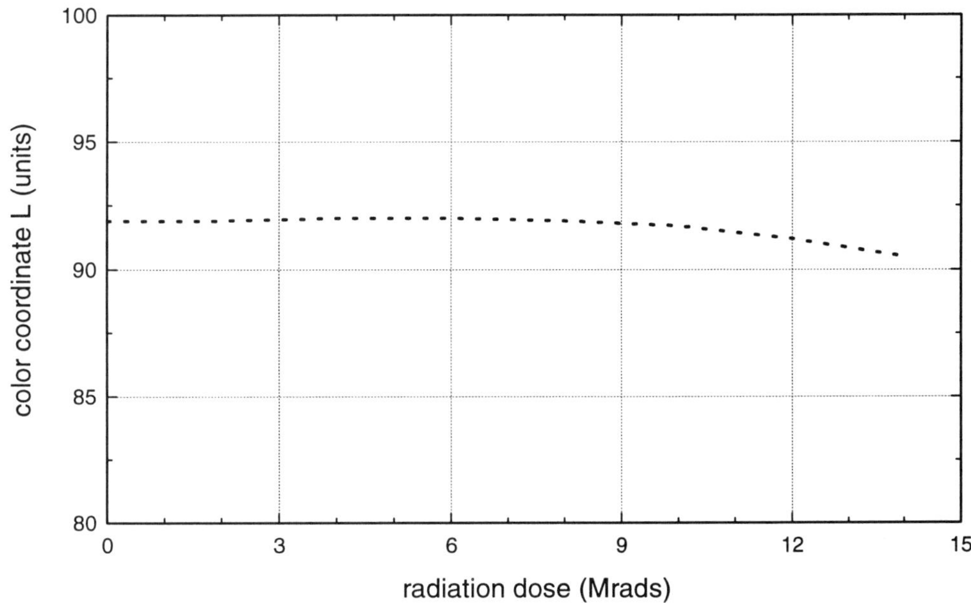

....	Emser Grilamid TR55 Nylon 12 (transp., amorphous)
Reference No.	61

Gamma Radiation Resistance

BASF: Ultramid B

Unreinforced nylon is a thermoplastic with average resistance to radiation. High energy radiation alters the properties of unreinforced Ultramid resins to different extents. Some properties are changed by a medium dose, and others hardly alter even at high doses. In the range extending up to 10,000 kJ/kg (1000 Mrads), the dielectric properties, i.e. the dielectric strength, the dissipation factor, and the resistance to tracking are hardly influenced.

The glass reinforced resins, including those containing a flame retardant, are extremely resistant to radiation. For instance, exposure to a dose of 2000 kJ/kg (200 Mrads) impairs the impact strength by only 15 to 30%. Ultramid articles sterilized with 25 kJ/kg (2.5 Mrads) gamma rays do not suffer any impairment of the mechanical properties. Uncolored or white parts assume a slight yellow tinge.

Reference: *Ultramid Nylon Resins Product Line, Properties, Processing,* supplier design guide (B 568/1e/4.91) - BASF Corporation, 1991.

Steam Resistance

Miles: Durethan

Certain grades of Durethan nylon can be used under standard autoclaving conditions (i.e., 121°C for 15 to 30 minutes).

Reference: *Medical Milestones,* supplier marketing literature (KU-F-2019(10F)/ 112/6/92) - Miles Inc., 1992.

Gamma Radiation Resistance

BASF: Ultramid A

Unreinforced nylon is a thermoplastic with average resistance to radiation. High energy radiation alters the properties of unreinforced Ultramid resins to different extents. Some properties are changed by a medium dose, and others hardly alter even at high doses. In the range extending up to 10,000 kJ/kg (1000 Mrads), the dielectric properties, i.e. the dielectric strength, the dissipation factor, and the resistance to tracking are hardly influenced.

The glass reinforced resins, including those containing a flame retardant, are extremely resistant to radiation. For instance, exposure to a dose of 2000 kJ/kg (200 Mrads) impairs the impact strength by only 15 to 30%. Ultramid articles sterilized with 25 kJ/kg (2.5 Mrads) gamma rays do not suffer any impairment of the mechanical properties. Uncolored or white parts assume a slight yellow tinge.

Reference: *Ultramid Nylon Resins Product Line, Properties, Processing,* supplier design guide (B 568/1e/4.91) - BASF Corporation, 1991.

Steam Resistance

Miles: Durethan

Certain grades of Durethan nylon can be used under standard autoclaving conditions (i.e., 121°C for 15 to 30 minutes).

Reference: *Medical Milestones,* supplier marketing literature (KU-F-2019(10F)/ 112/6/92) - Miles Inc., 1992.

TABLE 38: Effect of Ethylene Oxide Sterilization on Nylon 66

Material Family	NYLON 66							
Material Supplier/Name	DUPONT ZYTEL 101							
Reference No.	6	6	6	6	6	6	6	6
EXPOSURE CONDITIONS								
Type	Ethylene Oxide	Ethylene Oxide	Ethylene Oxide	Ethylene Oxide	Ethylene Oxide	Ethylene Oxide	Ethylene Oxide	Ethylene Oxide
Details	12% EtO and 88% Freon	12% EtO and 88% Freon	12% EtO and 88% Freon	12% EtO and 88% Freon	8.6% EtO and 91.4% HCFC-124	8.6% EtO and 91.4% HCFC-124	8.6% EtO and 91.4% HCFC-124	8.6% EtO and 91.4% HCFC-124
Concentration	600 mg/l	600 mg/l	600 mg/l	600 mg/l				
Number of Cycles	1	1	2	2	1	1	2	2
Note	RH: 60%; test lab: Ethox Corp.	RH: 60%; test lab: Ethox Corp.	RH: 60%; test lab: Ethox Corp.	RH: 60%; test lab: Ethox Corp.	RH: 60%; test lab: Ethox Corp.	RH: 60%; test lab: Ethox Corp.	RH: 60%; test lab: Ethox Corp.	RH: 60%; test lab: Ethox Corp.
Temperature (°C)	48.9	48.9	48.9	48.9	48.9	48.9	48.9	48.9
Time (hours)	6	6	6	6	6	6	6	6
PRE EXPOSURE CONDITIONING								
Preconditioning Note	time: 18 hours; temperature: 37.8°C; RH: 60%	time: 18 hours; temperature: 37.8°C; RH: 60%	time: 18 hours; temperature: 37.8°C; RH: 60%	time: 18 hours; temperature: 37.8°C; RH: 60%	time: 18 hours; temperature: 37.8°C; RH: 60%	time: 18 hours; temperature: 37.8°C; RH: 60%	time: 18 hours; temperature: 37.8°C; RH: 60%	time: 18 hours; temperature: 37.8°C; RH: 60%
POST EXPOSURE CONDITIONING								
Note	type: aeration; pressure: 127 mm Hg	type: aeration; pressure: 127 mm Hg	type: aeration; pressure: 127 mm Hg	type: aeration; pressure: 127 mm Hg	type: aeration; pressure: 127 mm Hg	type: aeration; pressure: 127 mm Hg	type: aeration; pressure: 127 mm Hg	type: aeration; pressure: 127 mm Hg
Temperature (°C)	32.2	32.2	32.2	32.2	32.2	32.2	32.2	32.2
POST EXPOSURE CONDITIONING II								
Note	type: ambient conditions	type: ambient conditions	type: ambient conditions	type: ambient conditions	type: ambient conditions	type: ambient conditions	type: ambient conditions	type: ambient conditions
Time (hours)	168	1344	168	1344	168	1344	168	1344
PROPERTIES RETAINED (%)								
Tensile Strength @ Yield	100.3 (il)	101.6 (il)	96.6 (il)	97.4 (il)	94 (il)	102.3 (il)	100.5 (il)	100.7 (il)
Elongation	72 (bn)	62 (bn)	50 (bn)	64.5 (bn)	59 (bn)	85.5 (bn)	78.5 (bn)	65.5 (bn)
Modulus	87.1 (gu)	102.9 (gu)	79.4 (gu)	85.3 (gu)	74.3 (gu)	97.9 (gu)	89.8 (gu)	93.8 (gu)
Dart Impact (total energy)	98.5 (ea)	123.1 (ea)	101.5 (ea)	104.6 (ea)	100 (ea)	73.8 (ed)	101.5 (ea)	120 (ea)
Dart Impact (peak energy)	73.8 (cu)	92.9 (cu)	78.6 (cu)	81 (cu)	73.8 (cu)	64.3 (cy)	76.2 (cu)	92.9 (cu)
SURFACE and APPEARANCE								
ΔE Color	0.61	0.5	0.7	0.84	0.54	0.65	0.59	0.51

TABLE 39: Effect of Ethylene Oxide Sterilization on Nylon 66

Material Family	NYLON 66			
Material Supplier/Name	DUPONT ZYTEL 101			
Reference No.	6	6	6	6
EXPOSURE CONDITIONS				
Type	Ethylene Oxide	Ethylene Oxide	Ethylene Oxide	Ethylene Oxide
Details	12% EtO and 88% Freon	12% EtO and 88% Freon	8.6% EtO and 91.4% HCFC-124	8.6% EtO and 91.4% HCFC-124
Concentration	600 mg/l	600 mg/l		
Number of Cycles	1	1	1	1
Note	RH: 60%; test lab: Ethox Corp.	RH: 60%; test lab: Ethox Corp.	RH: 60%; test lab: Ethox Corp.	RH: 60%; test lab: Ethox Corp.
Temperature (°C)	48.9	48.9	48.9	48.9
Time (hours)	6	6	6	6
PRE EXPOSURE CONDITIONING				
Preconditioning Note	time: 18 hours; temperature: 37.8°C; RH: 60%	time: 18 hours; temperature: 37.8°C; RH: 60%	time: 18 hours; temperature: 37.8°C; RH: 60%	time: 18 hours; temperature: 37.8°C; RH: 60%
POST EXPOSURE CONDITIONING				
Note	type: aeration; note: 10 air changes per hour	type: aeration; note: 30 air changes per hour	type: aeration; note: 10 air changes per hour	type: aeration; note: 30 air changes per hour
Temperature (°C)	32.2	54.4	32.2	54.4
RESIDUALS (ppm)				
Residuals Determined	ethylene oxide	ethylene oxide	ethylene oxide	ethylene oxide
Little or No Aeration	65	65	77	77
17 hour Aeration	13	3	54	16
24 hour Aeration	5	1	44	9
48 hour Aeration	2		34	7
72 hour Aeration	<2		17	

TABLE 40: Effect of Steam Sterilization on Nylon 66

Material Family	NYLON 66	
Material Supplier/Name	DUPONT ZYTEL 122L	
Material Note	hydrolysis resistant	hydrolysis resistant
Reference No.	68	68

EXPOSURE CONDITIONS

Type	Steam	Steam
Number of Cycles	1	1
Temperature (°C)	120	120
Time (hours)	200	400

PROPERTIES RETAINED (%)

Tensile Strength	101.9 (he)	97.1 (he)
Elongation	36.7 (ai)	29.3 (ai)

GRAPH 75: Electron Beam Radiation Dose vs. Tensile Strength Retained of Nylon 66

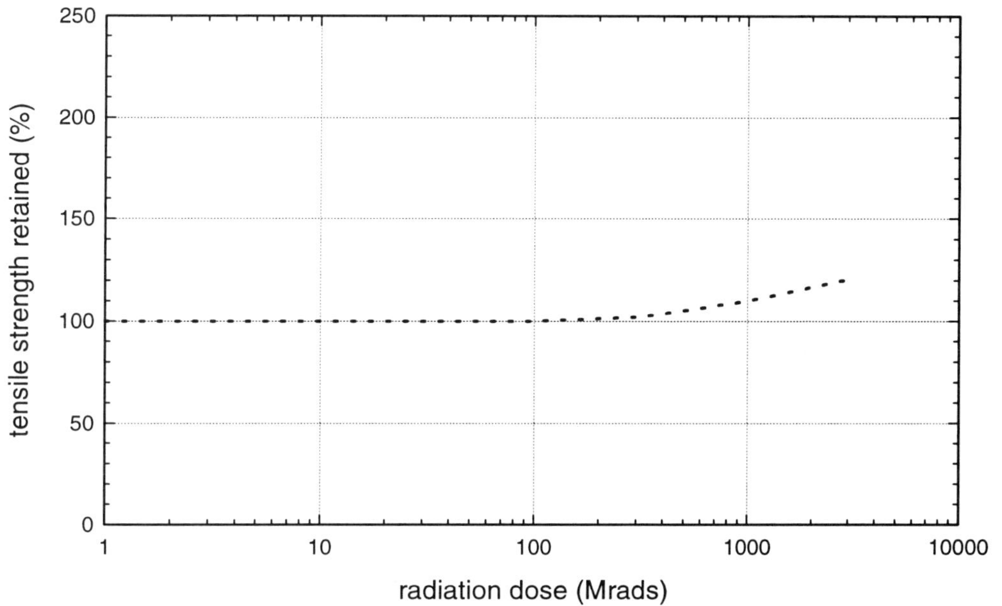

....	BASF Ultramid A3 Nylon 66 (high flow); 2 MeV electron beams; rate: 0.5 Mrad/min.
Reference No.	93

GRAPH 76: Electron Beam Radiation Dose vs. Elongation at Break of Nylon 66

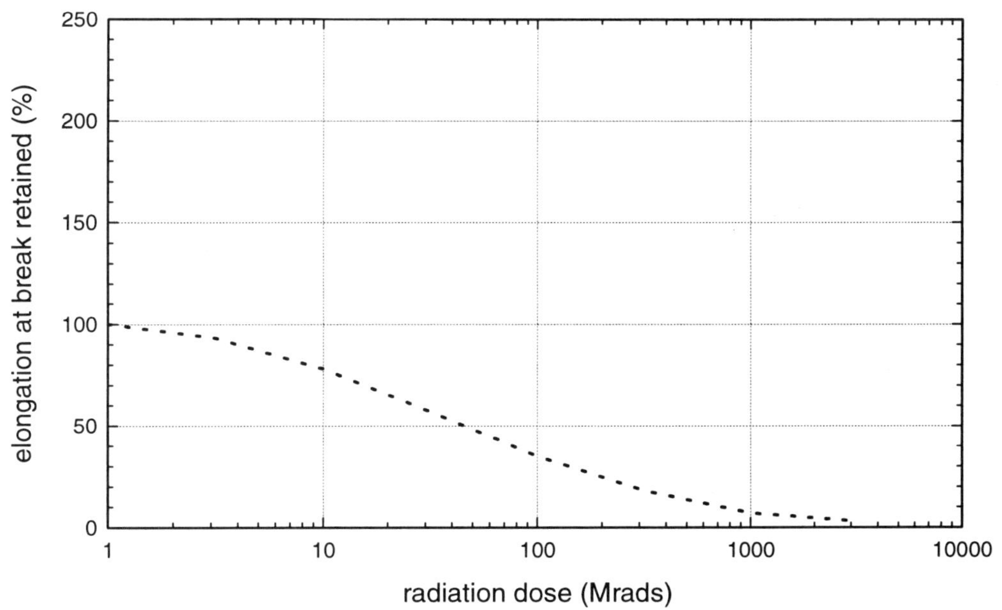

....	BASF Ultramid A3 Nylon 66 (high flow); 2 MeV electron beams; rate: 0.5 Mrad/min.
Reference No.	93

GRAPH 77: Electron Beam Radiation Dose vs. Elastic Modulus Retained of Nylon 66

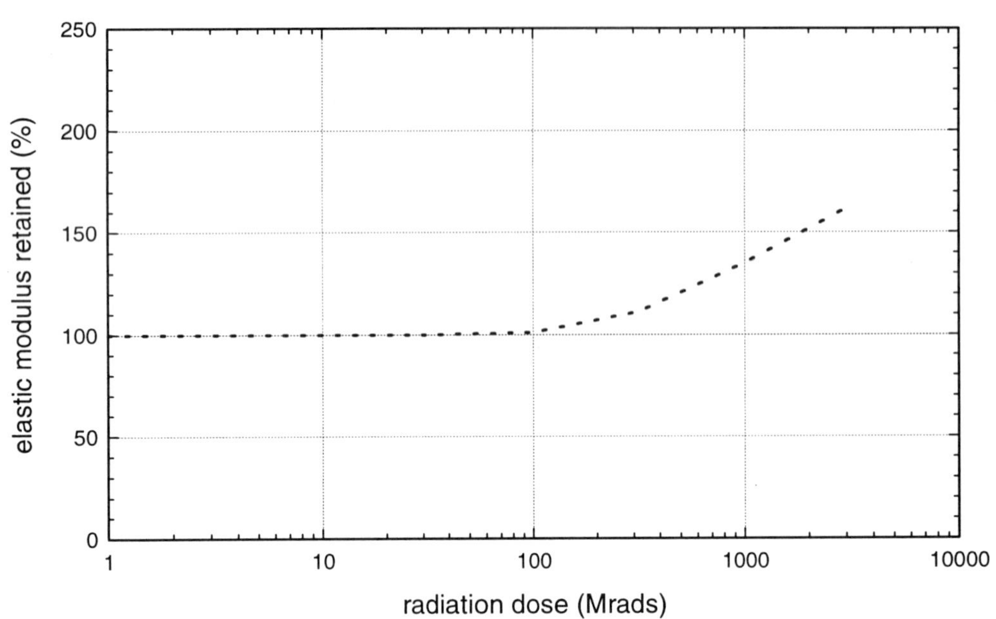

....	BASF Ultramid A3 Nylon 66 (high flow); 2 MeV electron beams; rate: 0.5 Mrad/min.
Reference No.	93

GRAPH 78: Electron Beam Radiation Dose vs. Impact Strength of Nylon 66

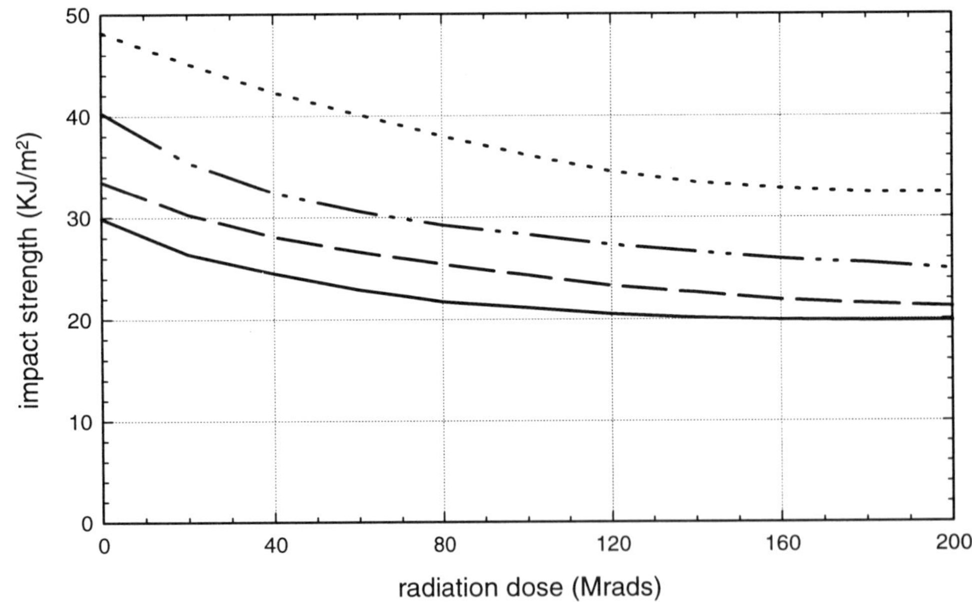

....	BASF Ultramid A3WG7 Nylon 66 (35% glass fiber; high flow, heat stabil.); 2 MeV electron beams; rate: 0.5 Mrad/min.
—··—··	BASF Ultramid A3XG7 Nylon 66 (35% glass fiber; high flow, flame retard.); 2 MeV electron beams; rate: 0.5 Mrad/min.
— — —	BASF Ultramid A3WG5 Nylon 66 (25% glass fiber; high flow, heat stabil.); 2 MeV electron beams; rate: 0.5 Mrad/min.
———	BASF Ultramid A3XG5 Nylon 66 (25% glass fiber; high flow, flame retard.); 2 MeV electron beams; rate: 0.5 Mrad/min.
Reference No.	93

GRAPH 79: **Electron Beam Radiation Dose vs. Impact Strength Retained of Nylon 66**

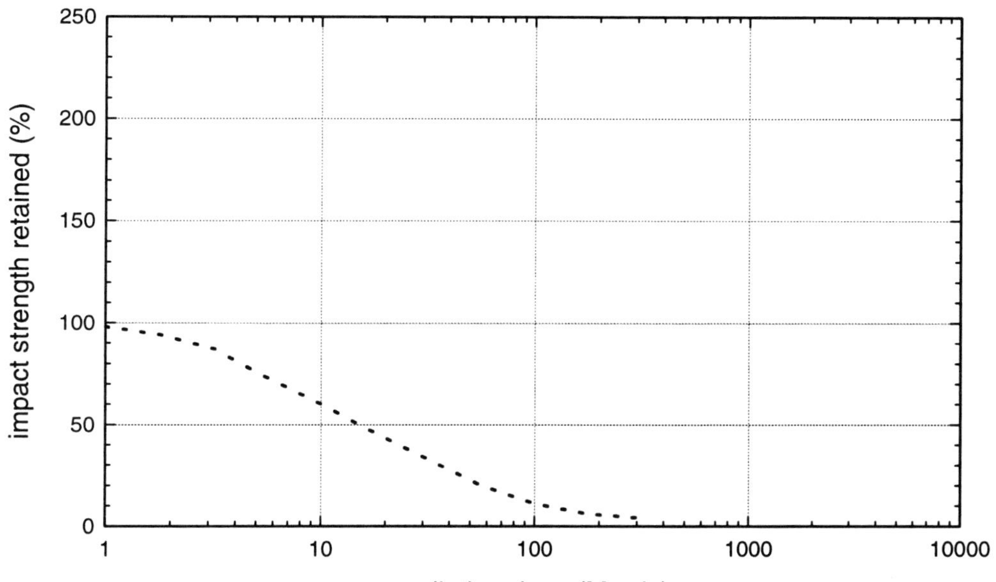

....	BASF Ultramid A3 Nylon 66 (high flow); 2 MeV electron beams; rate: 0.5 Mrad/min.
Reference No.	93

Radiation Resistance

Dow Chemical: Calibre 2061-15-FC850122 (features: gamma radiation stabilized); **Calibre MegaRad 2081-15-FC030006** (features: gamma radiation stabilized, transparent); **Calibre MegaRad 2081-15-FC030105** (features: gamma radiation stabilized, transparent); **Calibre MegaRad 2081-15-FC030116** (features: gamma radiation stabilized, transparent)

Physical properties were unaffected by radiation sterilization and easily withstood up to 10 Mrads of either gamma or E-beam radiation. The transparent resins have an initial purple tint which changes to smoke grey after exposure to high energy radiation. The color tint and radiation dose level must be coordinated or the resulting resin will be purple (too much tint for the radiation level) or yellow (too much irradiation for tint package).

No visible color differences were found in the gamma exposed samples versus the E-beam samples. Polycarbonate resins are very sensitive to photo-bleaching; fluorescent light greatly increased the color reversal of the samples.

Reference: Hermanson, Nancy J., Steffens, John F., *The Physical and Visual Property Changes in Thermoplastic Resins After Exposure to High Energy Sterilization - Gamma Versus Electron Beam,* ANTEC 1993, conference proceedings - Society of Plastics Engineers, 1993.

Gamma Radiation Resistance

GE Plastics: Lexan GR (features: gamma radiation stabilized, transparent, medical grade)

Lexan GR resins retain their properties to 10 megarads of gamma radiation. There is a slight color shift.

Reference: *Sterilization Comparison,* supplier marketing literature (ACB/3/93) - General Electric Plastics, 1993.

GE Plastics: Lexan HP (features: transparent, medical grade)

Lexan HP resins retain their properties to 10 megarads of gamma radiation. There is yellowing at 2 megarads.

Reference: *Sterilization Comparison,* supplier marketing literature (ACB/3/93) - General Electric Plastics, 1993.

GE Plastics: Lexan HPS (features: gamma radiation stabilized, transparent, medical grade)

Lexan HPS resins retain their properties to 10 megarads of gamma radiation. Samples are clear at 2.5 megarads and light grey at 5 megarads.

Reference: *Sterilization Comparison,* supplier marketing literature (ACB/3/93) - General Electric Plastics, 1993.

Dow Chemical: Calibre 2060 (features: transparent)

The physical properties of PC are well maintained up to 10.0 Mrads of gamma radiation. PC will discolor upon irradiation. The optical properties bleach back in time and are accelerated by fluorescent light exposure. Yellowness index values increased five-fold between 2.5 and 10.0 Mrads.

Reference: Sturdevant, Marianne F., *Sterilization Compatibility of Rigid Thermoplastic Materials,* supplier technical report (301-1548) - Dow Chemical Company, 1988.

Dow Chemical: (features: transparent, developmental material)

The physical properties of PC are well maintained up to 10.0 Mrads of gamma radiation. AROMATIC ESTERCARBONATE will generally discolor more and remain permanent as compared to other polycarbonates. No difference is seen when comparing the physical properties of the irradiated samples versus those stored in complete darkness.

Reference: Sturdevant, Marianne F., *Sterilization Compatibility of Rigid Thermoplastic Materials,* supplier technical report (301-1548) - Dow Chemical Company, 1988.

Dow Chemical: (features: transparent, natural resin)

The physical properties of polycarbonate are very stable to gamma radiation and can easily withstand levels up to 10 Mrads of radiation. It tends to discolor more than styrenic resins. However, the reversal or fading of this gamma induced discoloration can be accelerated by exposure to fluorescent or mild UV light. Color reversal will occur within one month of irradiation.

Reference: Sturdevant, Marianne F., *The Long-term Effects of Ethylene Oxide and Gamma Radiation Sterilization on the Properties of Rigid Thermoplastic Materials,* ANTEC 1990, conference proceedings - Society of Plastics Engineers, 1990.

GE Plastics: Lexan 141 (features: transparent)

Polycarbonate specimens underwent a change in flow, increasing some 25-fold at all three radiation levels. This suggests that chain scission occurred as the principal or sole mechanism of molecular change at 2.5 Mrads, but was accompanied by crosslinking at higher dosages. Polycarbonate discolors during irradiation. The level of discoloration increases with the increase in radiation dosage. Because of higher initial color, polysulfone parts are always of deeper color than those of polycarbonate. Polycarbonate parts show a greater rate of change after sterilization than do parts made from polysulfone. There is no evidence of change in mechanical properties at any of the three irradiation levels. Tensile strength, elongation, modulus, and notched Izod impact values remain essentially unchanged. This is surprising, in view of the chain scission noted above, and indicates that such molecular changes are not sufficient to affect these properties. There is no significant change in chemical resistance following irradiation. No dimensional changes occurred.

Reference: *Gamma Irradiaiton Studies On Polysulfone And Polycarbonate,* supplier technical report (project no.: 214M01; file no.: 7845-R) - Amoco Performance Products, Inc., 1982.

GE Plastics: (features: transparent)

Gamma radiation up to 10.0 Mrads has little effect on physical properties. The molecular weight of samples tested did not change significantly. However, samples were found to discolor. Gamma stable PC was also tested and demonstrated over ten percent improvement in yellowness index.

Reference: Paulino, J. C., Baccaro, L. E., *Effects Of Sterilization Procedures On Physical Properties of Engineering Thermoplastics,* supplier technical report - General Electric Company.

GE Plastics: Lexan (features: transparent)

Polycarbonate is much more resistant to gamma irradiation than most polymers. The primary effect of gamma irradiation on polycarbonate is chain scission (chain breakage). Fortunately, the high chain stiffness of polycarbonate makes it very difficult for the two ends of the chain to move apart. Hence, many of the broken chains will recombine. In addition, the aromatic nature of polycarbonate gives it other opportunities to absorb the energy from the gamma photon, rather than just by breaking the polymer chain. From a retention of mechanical properties point of view, polycarbonate has always been considered one of the most resistant materials for gamma irradiation.

Reference: *Lexan GR Resin Gamma Resistant Product Information Book,* supplier marketing literature (NBM-110) - General Electric Company.

GE Plastics: Lexan GR (features: gamma radiation stabilized, transparent)

After gamma irradiation Lexan GR will reach its dark aging equilibrium color shift after several days.

Reference: *Lexan GR Resin Gamma Resistant Product Information Book,* supplier marketing literature (NBM-110) - General Electric Company.

GE Plastics: Lexan HP (features: transparent, medical grade); **Lexan HPS** (features: gamma radiation stabilized, transparent, medical grade)

Lexan HP resins are a family of medical-grade polycarbonates developed to improve consistency in processing and part performance. They offer exceptional strength in a wide range of controlled viscosities. Low-viscosity grades provide superior flow for thin-wall and intricate designs; high viscosity grades are ideal for molding thick sections without sinks. Lexan HPS resins are special radiation-stabilized grades engineered for improved color stability. They offer the same cleanliness, consistency, physical and thermal properties as Lexan HP resin, but with enhanced stability to color shift under gamma radiation. The color shift of Lexan HPS resin after 2.5 Mrads of ionizing radiation is only 40% that of unstabilized Lexan HP resin. Lexan resin complies with FDA/USP VI. Typical Lexan resin applications include syringes, kidney dialysis devices, blood oxygenators and other medical laboratory devices and equipment.

Reference: *GE Plastics Medical Applications,* supplier marketing literature (NBM-100 (1/89) RTB) - General Electric Company, 1989.

Miles: Makrolon (features: transparent)

Miles offers Makrolon polycarbonate with a proprietary formulation that enables molded parts to be sterilized by gamma radiation with minimal apparent yellowing. Three different tints are offered. The tints are designed to impart a neutral density, transparent gray shade to the molded part post sterilization. The rate and extent to which stabilized polycarbonates reach a neutral color post sterilization depends chiefly on whether the material is stored in light or darkness. In contrast to samples kept in the dark, light accelerates the rate and more completely reverses the color changes. Tint 1117 is available in several different resin grades and simply masks the yellowing which occurs as a result of sterilization. In contrast to tint 1117, tints 1118 and 1119 contain a radiation stabilizer. Tint 1118 is useful for devices which will be packed in dark shipping containers, irradiated, and remain unexposed to light until the end product is actually used. Tint 1119 is good for items which will be packed for shipment in light (or for those which are packaged in dark containers, but later exposed to light). For example, blister packed products are often sterilized in their shipping cartons and remain in darkness until they are unpacked at their point of purchase for retail shelf display.

Irradiation may cause some materials to lose physical properties. Although general purpose polycarbonate grades yellow significantly when exposed to gamma radiation, physical properties of the material are retained (within the radiation exposure limits required for medical device sterilization). After 5.0 Mrads of exposure - the dosage of Cobalt 60 to which many medical apparatus are subjected for sterilization - test specimens essentially retained their physical properties.

Reference: *Medical Milestones,* supplier marketing literature (KU-F-2019(10F)/ 112/6/92) - Miles Inc., 1992.

Miles: Makrolon Rx2530 (features: gamma radiation stabilized, transparent, high flow)

Makrolon Rx-2530 (radiation stabilized, high flow polycarbonate) can be considered for applications requiring minimum color change after sterilizing doses of gamma radiation. Rx-2530 has proprietary technology for gamma stabilization which reduces yellowing by 60%. The yellowing from gamma sterilization fades with time after exposure. In non stabilized polycarbonate several weeks are needed to reach equilibrium. This equilibrium is accelerated by exposure to light. Rx-2530 will reach equilibrium when stored in the dark. When formulated with a tint, it will have a neutral color that is stable with dark storage after sterilizing doses of gamma radiation.

Reference: *Product Information Makrolon Rx-2350,* supplier marketing literature (201 (1/89)) - Miles Inc., 1989.

Polycarbonate (features: gamma radiation stabilized, transparent); **Polycarbonate** (features: transparent, general purpose grade)

Both the general purpose polycarbonate and the gamma grade polycarbonate (PC-"G") retained their stiffness (flexural modulus), tensile strength, and notched Izod impact properties at 3.0 and 5.0 Mrads exposure. After 3.0 Mrads exposure both polycarbonates had a yellowness index of approximately 6. Within 3 weeks the color index for both polycarbonates dropped into the negative range, meaning that they became blue again. Five Mrads of gamma irradiation caused a color change of the PC-G resin to a 20 index, and 36.5 for the general purpose polycarbonate. Recovery, in both instances, took place in the light. In the dark the result was much different. At 3.0 Mrads the gamma grade polycarbonate remained slightly yellow and the general purpose grade was at an index of 8. At 5.0 Mrads the general purpose PC remained quite yellow and the gamma grade recovered to an index of 12 vs. 2.1 in the light.

Reference: Hauser, Deborah I., Netolicky, William, *The Effects Of Gamma And Ethylene Oxide Sterilization On Styrene Methyl Methacrylate Copolymers And A Toughened Terpolymer,* Medical '91, Ghent, Belgium, conference proceedings - Novacor Chemicals Inc., 1991.

Electron Beam Radiation

GE Plastics: Lexan GR (features: gamma radiation stabilized, transparent)

Lexan GR resin has not been tested with electron beam irradiation, and is not recommended for electron beam irradiation. Despite the claims of electron beam proponents, it is the view of GE Plastics that electron beam irradiation is not the same as gamma irradiation. These two methods differ in two significant areas. Gamma irradiation is much more penetrating than electron beam irradiation. This means that doses within a commercial sample are very uniform. Electron beams are highly absorbed by all polymers. Hence, the dosage variation within a single device can be significant with electron beam irradiation. Second, high energy electron beams can induce radioactivity when the energy of the beam gets into the 10 MeV range. Contract electron beam irradiation sources have been quoted to be in the range of 5 to 10 MeV, which is of particular concern because the depth of penetration of electrons is controlled by the beam energy. High beam energies are needed to obtain commercially useful sterilization.

Reference: *Lexan GR Resin Gamma Resistant Product Information Book,* supplier marketing literature (NBM-110) - General Electric Company.

Miles: Makrolon (features: transparent)

The physical properties of Makrolon polycarbonate are not appreciably affected by electron beam sterilization. E-beam radiation is useful for delivering and controlling precise radiation doses and sterilizing materials which may be degraded by alternate sterilization techniques.

Reference: *Medical Milestones,* supplier marketing literature (KU-F-2019(10F)/ 112/6/92) - Miles Inc., 1992.

Beta Radiation Resistance

Polycarbonate (features: transparent)

Beta sterilization is increasing in importance in Europe. Samples of polycarbonate were tested at beta radiation levels of 2.7 to 10.8 Mrads. Physical properties were retained after 10.8 Mrads. The material became very yellow. After two weeks of storage there was a reduction in yellowness. Storage in light caused a greater reduction in yellowness as compared to storage in dark.

Reference: Haines, Jim, Hauser, Debbie, *Zylar Clear Alloys For Medical Applications,* supplier technical report - Novacor Chemicals Inc.

Ethylene Oxide (EtO) Resistance

GE Plastics: Lexan HP (features: transparent, medical grade); **Lexan HPS** (features: gamma radiation stabilized, transparent, medical grade)

Properties of Lexan HP and Lexan HPS resins are unchanged to 50 ethylene oxide sterilization cycles.

Reference: *Sterilization Comparison,* supplier marketing literature (ACB/3/93) - General Electric Plastics, 1993.

Dow Chemical: Calibre 2060 (features: transparent)

The physical property performance is unaffected by EtO exposure. Exposure of up to five cycles of ethylene oxide gas did not affect the notched Izod impact strength. The tensile yield strength was also unaffected by multiple exposures. The first day following exposure, residual EtO levels measured over 500 ppm. After 20 days the residual level dropped to below 200 ppm.

Reference: Sturdevant, Marianne F., *Sterilization Compatibility of Rigid Thermoplastic Materials,* supplier technical report (301-1548) - Dow Chemical Company, 1988.

Dow Chemical: (features: transparent, developmental material)

The physical property performance of aromatic ester carbonate (AEC) is unaffected by EtO exposure. Exposure of up to five cycles of ethylene oxide gas did not affect the notched Izod impact strength. The tensile yield strength was also unaffected by multiple exposures. The first day following exposure, residual EtO levels measured over 500 ppm. AEC does not get below 200 ppm for 30 days and retains 126 ppm after 63 days.

Reference: Sturdevant, Marianne F., *Sterilization Compatibility of Rigid Thermoplastic Materials,* supplier technical report (301-1548) - Dow Chemical Company, 1988.

Dow Chemical: (features: transparent, natural resin)

Polycarbonate withstands normal ethylene oxide sterilization conditions. However, care should be taken to minimize excessive or multiple exposure to ethylene oxide as it may cause embrittlement and chemical attack leading to stress cracking. Five repeated sterilization cycles caused some embrittlement. The embrittlement is seen as a loss of tensile elongation at break and a decrease in instrumented dart impact energy. After multiple EtO cycles, the embrittlement appears to compound with time. The elongation and instrumented impact strengths at six months and one year were significantly less than what was observed at two weeks after sterilization.

Reference: Sturdevant, Marianne F., *The Long-term Effects of Ethylene Oxide and Gamma Radiation Sterilization on the Properties of Rigid Thermoplastic Materials,* ANTEC 1990, conference proceedings - Society of Plastics Engineers, 1990.

Dow Chemical: Calibre 2060-10 (features: transparent)

A very slight decrease is seen in the tensile yield properties after exposure to EtO sterilization. Using both CFC-12 and HCFC-124 as carrier gases showed similar results. The same is also true for the instrumented dart impact results; EtO sterilization does not affect the practical toughness. Yellowness index, light transmission and percent haze are not affected by EtO sterilization regardless of the carrier used.

Reference: Hermanson, Nancy J., *Effects Of Alternate Carriers Of Ethylene Oxide Sterilant On Thermoplastics,* supplier technical report (301-02018) - Dow Chemical Company.

GE Plastics: (features: transparent)

After one day of mechanical aeration (aeration with fan), levels of ethylene oxide and ethylene glycol were less than 180 ppm. Levels of ethylene clorohydren were non-detectable (i.e., less than 4 ppm) at one, seven and fourteen days. Ethylene oxide and ethylene glycol levels were less than 40 ppm after fourteen days of mechanical aeration.

The ultimate elongation decreased slightly. However the yield elongation didn't change significantly. This is attributed to surface effects which are magnified after the material passes the yield point. Overall mechanical properties did not change significantly.

Reference: Paulino, J. C., Baccaro, L. E., *Effects Of Sterilization Procedures On Physical Properties of Engineering Thermoplastics,* supplier technical report - General Electric Company.

GE Plastics: Lexan GR (features: gamma radiation stabilized, transparent)

Lexan GR has been tested and is recommended for EtO sterilization.

Reference: *Lexan GR Resin Gamma Resistant Product Information Book,* supplier marketing literature (NBM-110) - General Electric Company.

Miles: Makrolon (features: transparent)

Today, ethylene oxide is the most widely used sterilization method. Miles engineering thermoplastics can be EtO sterilized with either the common mixture of EtO and chlorinated fluorocarbon (CFC) or the preferred mixture of EtO and carbon dioxide (CO_2). Makrolon polycarbonate retains its physical properties and clarity after repeated cycles of EtO sterilization. After 5 cycles of EtO/CO_2 sterilization there is no measurable change in the physical properties. The impact strength of test specimens remained unchanged after 50 sterilization cycles with pure EtO at 55°C, where each cycle lasted up to 6 hours. To avoid degradation of material properties, temperature during sterilization should not exceed 65°C. Also, processing parameters and mold design should be optimized to minimize the part stress levels which may be influenced by the diluent (i.e., CFC). Degassing is required following sterilization with EtO. EtO sterilized devices must be quarantined and degassed until very low levels of EtO are obtained. Degassing times will vary depending upon part complexity, but it is reasonable to assume that complex parts with large surface areas will typically require longer degassing times.

Reference: *Medical Milestones,* supplier marketing literature (KU-F-2019(10F)/ 112/6/92) - Miles Inc., 1992.

Steam Resistance

GE Plastics: Lexan HP (features: transparent, medical grade); **Lexan HPS** (features: gamma radiation stabilized, transparent, medical grade)

After 1 to 3 autoclave cycles (Hi-Vac @ 132°C), Lexan HP and Lexan HPS resins have limited utility.

Sterilization Comparison, supplier marketing literature (ACB/3/93) - General Electric Plastics, 1993.

GE Plastics: Lexan GR (features: gamma radiation stabilized, transparent, medical grade)

Lexan GR resins are not recommended for steam autoclave sterilization.

Sterilization Comparison, supplier marketing literature (ACB/3/93) - General Electric Plastics, 1993.

GE Plastics: (features: transparent)

PC bars and disks were tested for physical property retention after a series of cycles. A gravity displacement cycle was used. PC showed retention of tensile and yield strength.

Paulino, J. C., Baccaro, L. E., *Effects Of Sterilization Procedures On Physical Properties of Engineering Thermoplastics,* supplier technical report - General Electric Company.

GE Plastics: Lexan 8040 (features: transparent)

Lexan 8040 film allows autoclaving with no loss of clarity or structural integrity and no stress-whitening or pinholes.

Lexan 8040 Film For Medical Packaging - Comparison Data, supplier marketing literature (SP-2013 (7/88) RTB) - General Electric Company, 1988.

GE Plastics: Lexan GR (features: gamma radiation stabilized, transparent)

Standard polycarbonate is marginal for steam sterilization in an autoclave, with good performance only achieved using high viscosity resins in low residual stress parts for one autoclave cycle. Thus, standard grades of polycarbonate are autoclaveable for disposable items, but not recommended for reusable devices. Lexan GR has a lesser window for autoclaving. In its highest melt viscosity, lowest melt flow, Lexan GR may be autoclaveable at 121°C, if no residual stresses are present. Fixturing may be required to maintain dimensional tolerances. Lexan GR is not recommended for steam autoclave sterilization.

Lexan GR Resin Gamma Resistant Product Information Book, supplier marketing literature (NBM-110) - General Electric Company.

Dow Chemical: Calibre (features: transparent)

Steam autoclave will sometimes cause hydrolysis to occur to such a degree that the physical properties, particularly the toughness and clarity of a polycarbonate part, are seriously altered. Thus, any polycarbonate part designed for service involving exposure in an autoclave should be tested in the use environment to determine if degradation occurs.

Calibre Engineering Thermoplastics Basic Design Manual, supplier design guide (301-1040-1288) - Dow Chemical Company, 1988.

Miles: Makrolon (features: transparent)

For non critical applications, autoclaving is an acceptable method of sterilizing parts made from Makrolon polycarbonate. For critical applications (i.e., those in which the maintenance of physical properties and/or clarity are vital). Sterilization temperatures for parts made of Makrolon polycarbonate must not exceed 121°C or the parts could deform. Although Makrolon polycarbonate resins can be treated many times with boiling water or steam at 121°C, permanent immersion of parts in water above 60°C or in steam causes loss of material properties and must be avoided. Parts molded from polycarbonate resins should be protected from damage by substances such as alkaline corrosion inhibitors which are frequently added to boiler feed water. Condensed steam should not be allowed to accumulate inside plastic objects (e.g., baby bottles) by incorrectly positioning them in autoclaves or steam sterilizers as this may also cause damage to the parts.

Medical Milestones, supplier marketing literature (KU-F-2019(10F)/ 112/6/92) - Miles Inc., 1992.

Miles: Apec HT (features: transparent, high heat grade)

Parts made from Apec HT resins can be subjected to repeated steam sterilizations at higher than the standard temperature of 121°C (e.g., up to 143°C) thus reducing the amount of time required to sterilize. Although Apec polycarbonate resins can be treated many times with boiling water or steam at 121°C, permanent immersion of parts in water above 60°C or in steam causes loss of material properties and must be avoided.

Reference: *Medical Milestones,* supplier marketing literature (KU-F-2019(10F)/ 112/6/92) - Miles Inc., 1992.

TABLE 41: Effect of Gamma Radiation Sterilization on Polycarbonate

Material Family	POLYCARBONATE					
Material Supplier/Name	DOW					
Material Note	transparent, natural resin	transparent, natural resin	transparent, natural resin	transparent, natural resin	transparent, natural resin	transparent, natural resin
Reference No.	5	5	5	5	5	5
EXPOSURE CONDITIONS						
Type	Gamma Radiation	Gamma Radiation	Gamma Radiation	Gamma Radiation	Gamma Radiation	Gamma Radiation
Details	source: Cobalt 60	source: Cobalt 60	source: Cobalt 60	source: Cobalt 60	source: Cobalt 60	source: Cobalt 60
Radiation Dose (Mrads)	2.5	2.5	2.5	10	10	10
Note	test lab: Radiations Sterilizers Inc.	test lab: Radiations Sterilizers Inc.	test lab: Radiations Sterilizers Inc.	test lab: Radiations Sterilizers Inc.	test lab: Radiations Sterilizers Inc.	test lab: Radiations Sterilizers Inc.
POST EXPOSURE CONDITIONING						
Note	type: storage in dark	type: storage in dark	type: storage in dark	type: storage in dark	type: storage in dark	type: storage in dark
Temperature (°C)	21	21	21	21	21	21
Time (hours)	336	4368	8760	336	4368	8760
PROPERTIES RETAINED (%)						
Tensile Strength	100.8 (hw)	97.3 (hw)	93.7 (hw)	101.1 (hw)	97.7 (hw)	97.6 (hw)
Tensile Strength @ Yield	98.3 (il)	98.2 (il)	99.1 (il)	97.1 (il)	97.3 (il)	97.3 (il)
Elongation	103.3 (ax)	95.9 (ax)	86.2 (ax)	104.9 (ax)	100 (ax)	100 (ax)
Dart Impact (total energy)	99.7 (ff)	93.7 (ff)	83.9 (ff)	94.6 (ff)	90.4 (ff)	87 (ff)
Notched Izod Impact	99.4 (ft)	105 (ft)	98.8 (ft)	72 (ft)	95 (ft)	91.9 (ft)

TABLE 42: Effect of Gamma Radiation Sterilization on Polycarbonate

Material Family	POLYCARBONATE			
Material Supplier/Name	MILES MAKROLON FCR2458		GE LEXAN HP	GE LEXAN HPS
Material Note	transparent	transparent	transparent, medical grade	gamma radiation stabilized, transparent, medical grade
Reference No.	82	82	49	49
EXPOSURE CONDITIONS				
Type	Gamma Radiation	Gamma Radiation	Gamma Radiation	Gamma Radiation
Radiation Dose (Mrads)	2.5	5	2.5	2.5
PROPERTIES RETAINED (%)				
Tensile Strength	97 (hm)	98 (hm)	91 (hm)	91 (hm)
Tensile Strength @ Yield	100 (ib)	100 (ib)		
Elongation	100 (ak)	95.8 (ak)	98 (ak)	98 (ak)
Notched Izod Impact	98.7 (fn)	101.3 (fn)	94 (fm)	94 (fm)
Heat Deflection Temperature	98.8 (k)	98.1 (k)		

TABLE 43: Effect of Gamma Radiation Sterilization on Polycarbonate

Material Family	POLYCARBONATE							
Material Supplier/Name	GE LEXAN HPS1125		GE LEXAN HPS1136		GE LEXAN HPS1124		GE LEXAN HP1125	
Material Note	gamma radiation stabilized, transparent, medical grade, tinted	gamma radiation stabilized, transparent, medical grade, tinted	gamma radiation stabilized, transparent, medical grade, tinted	gamma radiation stabilized, transparent, medical grade, tinted	gamma radiation stabilized, transparent, medical grade, tinted	gamma radiation stabilized, transparent, medical grade, tinted	transparent, medical grade, tinted	transparent, medical grade, tinted
Reference No.	49	49	49	49	49	49	49	49
EXPOSURE CONDITIONS								
Type	Gamma Radiation	Gamma Radiation	Gamma Radiation	Gamma Radiation	Gamma Radiation	Gamma Radiation	Gamma Radiation	Gamma Radiation
Radiation Dose (Mrads)	2.5	2.5	2.5	2.5	2.5	2.5	2.5	2.5
POST EXPOSURE CONDITIONING								
Note		type: storage in dark		type: storage in dark		type: storage in dark		type: storage in dark
Time (hours)	0	336	0	336	0	336	0	336
SURFACE and APPEARANCE								
Δ Yellowness Index	10 (ku)	6 (ku)	10 (ku)	6 (ku)	10 (ku)	6 (ku)	22 (ku)	15 (ku)

TABLE 44: Effect of Gamma Radiation Sterilization on Polycarbonate

Material Family	POLYCARBONATE							
Material Supplier/Name	DOW CALIBRE 2060		DOW		DOW CALIBRE 2060		DOW	
Material Note	transparent	transparent	transparent, developmental material	transparent, developmental material	transparent	transparent	transparent, developmental material	transparent, developmental material
Reference No.	3	3	3	3	3	3	3	3
EXPOSURE CONDITIONS								
Type	Gamma Radiation	Gamma Radiation	Gamma Radiation	Gamma Radiation	Gamma Radiation	Gamma Radiation	Gamma Radiation	Gamma Radiation
Details	source: Cobalt 60	source: Cobalt 60	source: Cobalt 60	source: Cobalt 60	source: Cobalt 60	source: Cobalt 60	source: Cobalt 60	source: Cobalt 60
Radiation Dose (Mrads)	2.5	10	2.5	10	2.5	10	2.5	10
POST EXPOSURE CONDITIONING								
Note	type: storage under fluorescent light	type: storage under fluorescent light	type: storage under fluorescent light	type: storage under fluorescent light	type: storage in dark	type: storage in dark	type: storage in dark	type: storage in dark
Temperature (°C)	21	21	21	21	21	21	21	21
Time (hours)	336	336	336	336	336	336	336	336
PROPERTIES RETAINED (%)								
Tensile Strength @ Yield	98.9 (ii)	97.7 (ii)	100 (ii)	98.9 (ii)	98.9 (ii)	97.7 (ii)	98.9 (ii)	97.9 (ii)
Notched Izod Impact	96.3 (fp)	88.8 (fp)	98.6 (fp)	104.2 (fp)	99.4 (fp)	72 (fp)	101.4 (fp)	97.2 (fp)

TABLE 45: Effect of Gamma Radiation Sterilization on Polycarbonate

Material Family	POLYCARBONATE			
Material Note	gamma radiation stabilized, transparent	gamma radiation stabilized, transparent	transparent, general purpose grade	transparent, general purpose grade
Reference No.	106	106	106	106
EXPOSURE CONDITIONS				
Type	Gamma Radiation	Gamma Radiation	Gamma Radiation	Gamma Radiation
Radiation Dose (Mrads)	3	5	3	5
PROPERTIES RETAINED (%)				
Tensile Strength	100 (he)	100 (he)	99 (he)	99 (he)
Modulus			100 (bz)	100 (bz)
Notched Izod Impact			96 (fm)	96 (fm)
SURFACE and APPEARANCE				
Δ Yellowness Index	16 (kt)	30 (kt)	30 (kt)	50 (kt)

TABLE 46: Effect of Gamma Radiation Sterilization on Polycarbonate

Material Family	POLYCARBONATE			
Material Supplier/Name	DOW CALIBRE MEGARAD 2081-15-FC030105			
Material Note	gamma radiation stabilized, transparent	gamma radiation stabilized, transparent	gamma radiation stabilized, transparent	gamma radiation stabilized, transparent
Reference No.	1	1	1	1
EXPOSURE CONDITIONS				
Type	Gamma Radiation	Gamma Radiation	Gamma Radiation	Gamma Radiation
Radiation Dose (Mrads)	2.5	2.5	10	10
Note	test lab: SteriGenics	test lab: SteriGenics	test lab: SteriGenics	test lab: SteriGenics
POST EXPOSURE CONDITIONING				
Note	type: aging	type: aging	type: aging	type: aging
Time (hours)	168	1344	168	1344
PROPERTIES RETAINED (%)				
Tensile Strength @ Yield	100 (ir)	109.7 (ir)	98.4 (ir)	104.8 (ir)
Elongation	102.3 (bq)	106.2 (bq)	96.9 (bq)	100.8 (bq)
Flexural Strength	102 (cm)	94 (cm)	102 (cm)	101 (cm)
Modulus	108 (ha)	96.6 (ha)	105.9 (ha)	101.3 (ha)
Flexural Modulus	103.3 (ch)	108.2 (ch)	107.4 (ch)	109.5 (ch)
Dart Impact (total energy)	92.6 (ei)	102.9 (ei)	76.5 (ei)	102.9 (ei)
Dart Impact (peak energy)	90.6 (dg)	98.4 (dg)	75 (dg)	90.6 (dg)
Notched Izod Impact	105.1 (gh)	101.5 (gh)	97.8 (gh)	98.7 (gh)
Heat Deflection Temperature	100 (m)	99.2 (m)	95.9 (m)	97.5 (m)
Vicat Softening Point	100 (ku)	98.7 (ku)	97.3 (ku)	97.3 (ku)
SURFACE and APPEARANCE				
Yellowness Index	16.9 (kw)	7 (kw)	45.5 (kw)	27.8 (kw)
Haze (%)	0.72	-0.5	1.96	-0.85
Transparency Retained (%)	96.97	121.21	96.97	116.67

TABLE 47: Effect of Gamma Radiation Sterilization on Polycarbonate

Material Family	POLYCARBONATE							
Material Supplier/Name	DOW CALIBRE MEGARAD 2081-15-FC030006				DOW CALIBRE MEGARAD 2081-15-FC030116			
Material Note	gamma radiation stabilized, transparent	gamma radiation stabilized, transparent	gamma radiation stabilized, transparent	gamma radiation stabilized, transparent	gamma radiation stabilized, transparent	gamma radiation stabilized, transparent	gamma radiation stabilized, transparent	gamma radiation stabilized, transparent
Reference No.	1	1	1	1	1	1	1	1

EXPOSURE CONDITIONS

Type	Gamma Radiation	Gamma Radiation	Gamma Radiation	Gamma Radiation	Gamma Radiation	Gamma Radiation	Gamma Radiation	Gamma Radiation
Radiation Dose (Mrads)	2.5	2.5	10	10	2.5	2.5	10	10
Note	test lab: SteriGenics	test lab: SteriGenics	test lab: SteriGenics	test lab: SteriGenics	test lab: SteriGenics	test lab: SteriGenics	test lab: SteriGenics	test lab: SteriGenics

POST EXPOSURE CONDITIONING

Note	type: aging	type: aging	type: aging	type: aging	type: aging	type: aging	type: aging	type: aging
Time (hours)	168	1344	168	1344	168	1344	168	1344

SURFACE and APPEARANCE

Yellowness Index	11.56 (kw)	8.13 (kw)	41.86 (kw)	8.6 (kw)	6.22 (kw)	0.52 (kw)	43.72 (kw)	12.27 (kw)
Haze (%)	2.6	1.5	2	3.7	1.7	-1.3	0	-1.4
Transparency Retained (%)	102.7	119.2	94.5	119.2	87.8	108	84	107

TABLE 48: Effect of Gamma Radiation Sterilization on Polycarbonate

Material Family	POLYCARBONATE			
Material Supplier/Name	DOW CALIBRE 2061-15-FC850122			
Material Note	gamma radiation stabilized	gamma radiation stabilized	gamma radiation stabilized	gamma radiation stabilized
Reference No.	1	1	1	1

EXPOSURE CONDITIONS

Type	Gamma Radiation	Gamma Radiation	Gamma Radiation	Gamma Radiation
Radiation Dose (Mrads)	2.5	2.5	10	10
Note	test lab: SteriGenics	test lab: SteriGenics	test lab: SteriGenics	test lab: SteriGenics

POST EXPOSURE CONDITIONING

Note	type: aging	type: aging	type: aging	type: aging
Time (hours)	168	1344	168	1344

SURFACE and APPEARANCE

ΔL Color	-0.7 (ac)	-0.9 (ac)	-1.8 (ac)	-2.6 (ac)
Δa Color	-0.57 (o)	-0.22 (o)	-0.17 (o)	-1.09 (o)
Δb Color	4.3 (t)	3.59 (t)	5.58 (t)	8.25 (t)

TABLE 49: Effect of Gamma Radiation Sterilization on Polycarbonate

Material Family	POLYCARBONATE		
Material Supplier/Name	GE LEXAN 141		
Material Note	transparent	transparent	transparent
Reference No.	43	43	43

EXPOSURE CONDITIONS

Type	Gamma Radiation	Gamma Radiation	Gamma Radiation
Details	source: Cobalt 60	source: Cobalt 60	source: Cobalt 60
Radiation Dose (Mrads)	2.5	4	6
Note	test lab: Isomedix, Inc.	test lab: Isomedix, Inc.	test lab: Isomedix, Inc.

PROPERTIES RETAINED (%)

Tensile Strength	97.9 (he)	91.8 (he)	96.9 (he)
Elongation	61.3 (ai)	83.8 (ai)	81.1 (ai)
Modulus	97.2 (gs)	97.2 (gs)	105.6 (gs)
Notched Izod Impact	92.8 (fm)	103 (fm)	95.8 (fm)
Melt Flow Rate	2600 (a)	2500 (a)	2520 (a)

TABLE 50: Effect of Electron Beam Radiation Sterilization on Polycarbonate

Material Family	POLYCARBONATE			
Material Supplier/Name	DOW CALIBRE MEGARAD 2081-15-FC030105			
Material Note	gamma radiation stabilized, transparent	gamma radiation stabilized, transparent	gamma radiation stabilized, transparent	gamma radiation stabilized, transparent
Reference No.	1	1	1	1

EXPOSURE CONDITIONS

Type	Electron Beam Radiation	Electron Beam Radiation	Electron Beam Radiation	Electron Beam Radiation
Radiation Dose (Mrads)	2.5	2.5	10	10
Note	test lab: E-Beam Services, Inc.	test lab: E-Beam Services, Inc.	test lab: E-Beam Services, Inc.	test lab: E-Beam Services, Inc.

POST EXPOSURE CONDITIONING

Note	type: aging	type: aging	type: aging	type: aging
Time (hours)	168	1344	168	1344

PROPERTIES RETAINED (%)

Tensile Strength @ Yield	98.4 (ir)	101.6 (ir)	116.1 (ir)	101.6 (ir)
Elongation	93.1 (bq)	83.8 (bq)	91.5 (bq)	95.4 (bq)
Flexural Strength	102 (cm)	96 (cm)	101 (cm)	99 (cm)
Modulus	101.3 (ha)	98.7 (ha)	117.7 (ha)	103.8 (ha)
Flexural Modulus	104.5 (ch)	106.2 (ch)	105.3 (ch)	109.1 (ch)
Dart Impact (total energy)	100 (ei)	95.6 (ei)	92.6 (ei)	95.6 (ei)
Dart Impact (peak energy)	100 (dg)	96.9 (dg)	87.5 (dg)	87.5 (dg)
Notched Izod Impact	103 (gh)	92.2 (gh)	101.5 (gh)	99.3 (gh)
Heat Deflection Temperature	100 (m)	99.2 (m)	94.2 (m)	95 (m)
Vicat Softening Point	98.7 (ku)	98 (ku)	97.3 (ku)	98 (ku)

SURFACE and APPEARANCE

Δ Yellowness Index	11.1 (kw)	7.1 (kw)	43.6 (kw)	28.1 (kw)
Haze (%)	0.98	-0.57	0.86	0.21
Transparency Retained (%)	107.58	119.7	92.42	113.64

TABLE 51: Effect of Electron Beam Radiation Sterilization on Polycarbonate

Material Family	POLYCARBONATE							
Material Supplier/Name	DOW CALIBRE MEGARAD 2081-15-FC030006				DOW CALIBRE MEGARAD 2081-15-FC030116			
Material Note	gamma radiation stabilized, transparent	gamma radiation stabilized, transparent	gamma radiation stabilized, transparent	gamma radiation stabilized, transparent	gamma radiation stabilized, transparent	gamma radiation stabilized, transparent	gamma radiation stabilized, transparent	gamma radiation stabilized, transparent
Reference No.	1	1	1	1	1	1	1	1
EXPOSURE CONDITIONS								
Type	Electron Beam Radiation	Electron Beam Radiation	Electron Beam Radiation	Electron Beam Radiation	Electron Beam Radiation	Electron Beam Radiation	Electron Beam Radiation	Electron Beam Radiation
Radiation Dose (Mrads)	2.5	2.5	10	10	2.5	2.5	10	10
Note	test lab: E-Beam Services, Inc.	test lab: E-Beam Services, Inc.	test lab: E-Beam Services, Inc.	test lab: E-Beam Services, Inc.	test lab: E-Beam Services, Inc.	test lab: E-Beam Services, Inc.	test lab: E-Beam Services, Inc.	test lab: E-Beam Services, Inc.
POST EXPOSURE CONDITIONING								
Note	type: aging	type: aging	type: aging	type: aging	type: aging	type: aging	type: aging	type: aging
Time (hours)	168	1344	168	1344	168	1344	168	1344
SURFACE and APPEARANCE								
Δ Yellowness Index	11.54 (kw)	8.71 (kw)	11.12 (kw)	8.84 (kw)	3.75 (kw)	0.73 (kw)	36.67 (kw)	20.91 (kw)
Haze (%)	2.5	3.1	1.9	1.2	0.8	-0.6	0.1	-1.4
Transparency Retained (%)	106.8	117.8	106.8	119.2	97	108	82	104

TABLE 52: Effect of Electron Beam Radiation Sterilization on Polycarbonate

Material Family	Polycarbonate	Polycarbonate
Material Supplier/Name	Miles Makrolon	Miles Makrolon
Material Note	transparent	transparent
Reference No.	82	82
EXPOSURE CONDITIONS		
Type	Electron Beam Radiation	Electron Beam Radiation
Radiation Dose (Mrads)	2.5	5
PROPERTIES RETAINED (%)		
Tensile Strength	107.6 (hm)	104.3 (hm)
Tensile Strength @ Yield	100 (ib)	98.9 (ib)
Elongation	96 (ak)	96 (ak)
Notched Izod Impact	103.6 (fn)	97.6 (fn)
Heat Deflection Temperature	99.6 (k)	101.2 (k)

TABLE 53: Effect of Electron Beam Radiation Sterilization on Polycarbonate

Material Family	POLYCARBONATE			
Material Supplier/Name	DOW CALIBRE 2061-15-FC850122			
Material Note	gamma radiation stabilized	gamma radiation stabilized	gamma radiation stabilized	gamma radiation stabilized
Reference No.	1	1	1	1

EXPOSURE CONDITIONS

Type	Electron Beam Radiation	Electron Beam Radiation	Electron Beam Radiation	Electron Beam Radiation
Radiation Dose (Mrads)	2.5	2.5	10	10
Note	test lab: E-Beam Services, Inc.	test lab: E-Beam Services, Inc.	test lab: E-Beam Services, Inc.	test lab: E-Beam Services, Inc.

POST EXPOSURE CONDITIONING

Note	type: aging	type: aging	type: aging	type: aging
Time (hours)	168	1344	168	1344

SURFACE and APPEARANCE

Δ E Color	-1.3 (ac)	-0.7 (ac)	-3.6 (ac)	-2.5 (ac)
Δ L Color	-1.5 (o)	-0.88 (o)	-1.78 (o)	-1.36 (o)
Δ a Color	5.1 (t)	3.83 (t)	8.77 (t)	7.74 (t)

TABLE 54: Effect of Ethylene Oxide Sterilization on Polycarbonate

Material Family	POLYCARBONATE					
Material Supplier/Name	DOW					
Material Note	transparent, natural resin	transparent, natural resin	transparent, natural resin	transparent, natural resin	transparent, natural resin	transparent, natural resin
Reference No.	5	5	5	5	5	5

EXPOSURE CONDITIONS

Type	Ethylene Oxide	Ethylene Oxide	Ethylene Oxide	Ethylene Oxide	Ethylene Oxide	Ethylene Oxide
Details	12% EtO and 88% Freon	12% EtO and 88% Freon	12% EtO and 88% Freon	12% EtO and 88% Freon	12% EtO and 88% Freon	12% EtO and 88% Freon
Concentration	660 mg/l	660 mg/l	660 mg/l	660 mg/l	660 mg/l	660 mg/l
Number of Cycles	1	1	1	5	5	5
Note	RH: 60%; test lab: Ethox Corp.	RH: 60%; test lab: Ethox Corp.	RH: 60%; test lab: Ethox Corp.	RH: 60%; test lab: Ethox Corp.	RH: 60%; test lab: Ethox Corp.	RH: 60%; test lab: Ethox Corp.
Temperature (°C)	49	49	49	49	49	49
Time (hours)	≥ 6	≥ 6	≥ 6	≥ 6	≥ 6	≥ 6

PRE EXPOSURE CONDITIONING

Preconditioning Note	time: 8 hours; temperature: 37.8°C; RH: 60%	time: 8 hours; temperature: 37.8°C; RH: 60%	time: 8 hours; temperature: 37.8°C; RH: 60%	time: 8 hours; temperature: 37.8°C; RH: 60%	time: 8 hours; temperature: 37.8°C; RH: 60%	time: 8 hours; temperature: 37.8°C; RH: 60%

POST EXPOSURE CONDITIONING

Note	type: evacuation; pressure: 127 mm Hg	type: evacuation; pressure: 127 mm Hg	type: evacuation; pressure: 127 mm Hg	type: evacuation; pressure: 127 mm Hg	type: evacuation; pressure: 127 mm Hg	type: evacuation; pressure: 127 mm Hg

POST EXPOSURE CONDITIONING II

Note	type: aeration	type: aeration	type: aeration	type: aeration	type: aeration	type: aeration
Temperature (°C)	32.2	32.2	32.2	32.2	32.2	32.2
Time (hours)	≥ 16	≥ 16	≥ 16	≥ 16	≥ 16	≥ 16

POST EXPOSURE CONDITIONING III

Note	type: storage in dark	type: storage in dark	type: storage in dark	type: storage in dark	type: storage in dark	type: storage in dark
Temperature (°C)	21	21	21	21	21	21
Time (hours)	336	4368	8760	336	4368	8760

PROPERTIES RETAINED (%)

Tensile Strength	102.9 (hx)	98.2 (hx)	93.5 (hx)	78.8 (hx)	76.4 (hx)	75.8 (hx)
Tensile Strength @ Yield	100 (ij)	99.2 (ij)	99 (ij)	99.6 (ij)	100 (ij)	99.3 (ij)
Elongation	102.4 (ay)	98.4 (ay)	90.2 (ay)	44.7 (ay)	18.7 (ay)	20.3 (ay)
Dart Impact (total energy)	93.9 (fh)	93.3 (fh)	94.3 (fh)	89.8 (fe)	64.5 (fe)	62.1 (fe)
Notched Izod Impact	98.8 (fv)	98.1 (fv)	102.5 (fv)	101.2 (fv)	102.5 (fv)	101.2 (fv)

TABLE 55: Effect of Ethylene Oxide Sterilization on Polycarbonate

Material Family	POLYCARBONATE							
Material Supplier/Name	DOW CALIBRE 2060-10							
Material Note	transparent	transparent	transparent	transparent	transparent	transparent	transparent	transparent
Reference No.	6	6	6	6	6	6	6	6

EXPOSURE CONDITIONS

Type	Ethylene Oxide	Ethylene Oxide	Ethylene Oxide	Ethylene Oxide	Ethylene Oxide	Ethylene Oxide	Ethylene Oxide	Ethylene Oxide
Details	12% EtO and 88% Freon	12% EtO and 88% Freon	12% EtO and 88% Freon	12% EtO and 88% Freon	8.6% EtO and 91.4% HCFC-124	8.6% EtO and 91.4% HCFC-124	8.6% EtO and 91.4% HCFC-124	8.6% EtO and 91.4% HCFC-124
Concentration	600 mg/l	600 mg/l	600 mg/l	600 mg/l				
Number of Cycles	1	1	2	2	1	1	2	2
Note	RH: 60%; test lab: Ethox Corp.	RH: 60%; test lab: Ethox Corp.	RH: 60%; test lab: Ethox Corp.	RH: 60%; test lab: Ethox Corp.	RH: 60%; test lab: Ethox Corp.	RH: 60%; test lab: Ethox Corp.	RH: 60%; test lab: Ethox Corp.	RH: 60%; test lab: Ethox Corp.
Temperature (°C)	48.9	48.9	48.9	48.9	48.9	48.9	48.9	48.9
Time (hours)	6	6	6	6	6	6	6	6

PRE EXPOSURE CONDITIONING

Preconditioning Note	time: 18 hours; temperature: 37.8°C; RH: 60%	time: 18 hours; temperature: 37.8°C; RH: 60%	time: 18 hours; temperature: 37.8°C; RH: 60%	time: 18 hours; temperature: 37.8°C; RH: 60%	time: 18 hours; temperature: 37.8°C; RH: 60%	time: 18 hours; temperature: 37.8°C; RH: 60%	time: 18 hours; temperature: 37.8°C; RH: 60%	time: 18 hours; temperature: 37.8°C; RH: 60%

POST EXPOSURE CONDITIONING

Note	type: aeration; pressure: 127 mm Hg	type: aeration; pressure: 127 mm Hg	type: aeration; pressure: 127 mm Hg	type: aeration; pressure: 127 mm Hg	type: aeration; pressure: 127 mm Hg	type: aeration; pressure: 127 mm Hg	type: aeration; pressure: 127 mm Hg	type: aeration; pressure: 127 mm Hg
Temperature (°C)	32.2	32.2	32.2	32.2	32.2	32.2	32.2	32.2

POST EXPOSURE CONDITIONING II

Note	type: ambient conditions	type: ambient conditions	type: ambient conditions	type: ambient conditions	type: ambient conditions	type: ambient conditions	type: ambient conditions	type: ambient conditions
Time (hours)	168	1344	168	1344	168	1344	168	1344

PROPERTIES RETAINED (%)

Tensile Strength @ Yield	99.4 (il)	95.5 (il)	99.8 (il)	95.7 (il)	97.1 (il)	97.7 (il)	99.9 (il)	95.4 (il)
Elongation	87.8 (bm)	84.9 (bm)	65.5 (bm)	48.2 (bm)	85.6 (bm)	89.9 (bm)	91.4 (bm)	91.4 (bm)
Modulus	89 (gu)	104.1 (gu)	103.1 (gu)	100.6 (gu)	90.3 (gu)	101.9 (gu)	96.9 (gu)	83.1 (gu)
Dart Impact (total energy)	91.5 (dr)	105.1 (dr)	100 (dr)	98.3 (dr)	98.3 (dr)	103.4 (dr)	91.5 (dr)	106.8 (dr)
Dart Impact (peak energy)	91.4 (eq)	100 (eq)	96.6 (eq)	93.1 (eq)	93.1 (eq)	100 (eq)	87.9 (eq)	101.7 (eq)

SURFACE and APPEARANCE

Yellowness Index	1.61	1.33	1.52	1.43	1.77	1.48	1.58	1.66
Haze (%)	3.07	3.12	6.25	5.23	1.99	2.93	4.27	6.44
Transmittance (%)	88	88	87	88	88	88	88	87

TABLE 56: Effect of Ethylene Oxide Sterilization on Polycarbonate

Material Family	POLYCARBONATE			
Material Supplier/Name	DOW CALIBRE 2060-10			
Material Note	transparent	transparent	transparent	transparent
Reference No.	6	6	6	6

EXPOSURE CONDITIONS

Type	Ethylene Oxide	Ethylene Oxide	Ethylene Oxide	Ethylene Oxide
Details	12% EtO and 88% Freon	12% EtO and 88% Freon	8.6% EtO and 91.4% HCFC-124	8.6% EtO and 91.4% HCFC-124
Concentration	600 mg/l	600 mg/l		
Number of Cycles	1	1	1	1
Note	RH: 60%; test lab: Ethox Corp.	RH: 60%; test lab: Ethox Corp.	RH: 60%; test lab: Ethox Corp.	RH: 60%; test lab: Ethox Corp.
Temperature (°C)	48.9	48.9	48.9	48.9
Time (hours)	6	6	6	6

PRE EXPOSURE CONDITIONING

Preconditioning Note	time: 18 hours; temperature: 37.8°C; RH: 60%	time: 18 hours; temperature: 37.8°C; RH: 60%	time: 18 hours; temperature: 37.8°C; RH: 60%	time: 18 hours; temperature: 37.8°C; RH: 60%

POST EXPOSURE CONDITIONING

Note	type: aeration; note: 10 air changes per hour	type: aeration; note: 30 air changes per hour	type: aeration; note: 10 air changes per hour	type: aeration; note: 30 air changes per hour
Temperature (°C)	32.2	54.4	32.2	54.4

RESIDUALS (ppm)

Residuals Determined	ethylene oxide	ethylene oxide	ethylene oxide	ethylene oxide
Little or No Aeration	1011	1011	1249	1249
24 hour Aeration	625	319	378	458
48 hour Aeration		227		328
72 hour Aeration	275	216	363	235
168 hour Aeration	167		253	

TABLE 57: Effect of Ethylene Oxide Sterilization on Polycarbonate

Material Family	POLYCARBONATE			
Material Supplier/Name	DOW CALIBRE 2060		DOW	
Material Note	transparent	transparent	transparent, developmental material	transparent, developmental material
Reference No.	3	3	3	3

EXPOSURE CONDITIONS

Type	Ethylene Oxide	Ethylene Oxide	Ethylene Oxide	Ethylene Oxide
Details	12% EtO and 88% Freon	12% EtO and 88% Freon	12% EtO and 88% Freon	12% EtO and 88% Freon
Concentration	660 mg/l	660 mg/l	660 mg/l	660 mg/l
Number of Cycles	1	5	1	5
Note	RH: 60%	RH: 60%	RH: 60%	RH: 60%
Temperature (°C)	49	49	49	49
Time (hours)	≥ 6	≥ 6	≥ 6	≥ 6

PRE EXPOSURE CONDITIONING

Preconditioning Note	time: 8 hours; temperature: 37.8°C; RH: 60%	time: 8 hours; temperature: 37.8°C; RH: 60%	time: 8 hours; temperature: 37.8°C; RH: 60%	time: 8 hours; temperature: 37.8°C; RH: 60%

POST EXPOSURE CONDITIONING

Note	type: evacuation; pressure: 127 mm Hg	type: evacuation; pressure: 127 mm Hg	type: evacuation; pressure: 127 mm Hg	type: evacuation; pressure: 127 mm Hg

POST EXPOSURE CONDITIONING II

Note	type: aeration	type: aeration	type: aeration	type: aeration
Temperature (°C)	32	32	32	32
Time (hours)	≥ 16	≥ 16	≥ 16	≥ 16

POST EXPOSURE CONDITIONING III

Note	type: storage in dark; RH: 50%	type: storage in dark; RH: 50%	type: storage in dark; RH: 50%	type: storage in dark; RH: 50%
Temperature (°C)	21	21	21	21
Time (hours)	336	336	336	336

PROPERTIES RETAINED (%)

Tensile Strength @ Yield	100 (ii)	100 (ii)	98.9 (ii)	98.9 (ii)
Notched Izod Impact	98.8 (fp)	101.2 (fp)	102.8 (fp)	111.3 (fp)

TABLE 58: Effect of Ethylene Oxide Sterilization on Polycarbonate

Material Family	POLYCARBONATE	
Material Supplier/Name	DOW CALIBRE 2060	DOW
Material Note	transparent	transparent, developmental material
Reference No.	3	3

EXPOSURE CONDITIONS

Type	Ethylene Oxide	Ethylene Oxide
Details	12% EtO and 88% Freon	12% EtO and 88% Freon
Concentration	660 mg/l	660 mg/l
Number of Cycles	1	1
Note	RH: 60%	RH: 60%
Temperature (°C)	49	49
Time (hours)	≥ 6	≥ 6

PRE EXPOSURE CONDITIONING

Preconditioning Note	time: 8 hours; temperature: 37.8°C; RH: 60%	time: 8 hours; temperature: 37.8°C; RH: 60%

POST EXPOSURE CONDITIONING

Note	type: evacuation; pressure: 127 mm Hg	type: evacuation; pressure: 127 mm Hg

POST EXPOSURE CONDITIONING II

Note	type: aeration	type: aeration
Temperature (°C)	32	32

RESIDUALS (ppm)

Residuals Determined	ethylene oxide	ethylene oxide
24 hour Aeration	514	877
72 hour Aeration	335	449
168 hour Aeration	290	362
744 hour Aeration	202	270
792 hour Aeration		207
816 hour Aeration	99	
840 hour Aeration		175
864 hour Aeration	97	
888 hour Aeration		126

TABLE 59: Effect of Ethylene Oxide Sterilization on Polycarbonate

Material Family	POLYCARBONATE					
Material Supplier/Name	MILES MAKROLON FCR-2458		MILES MAKROLON 2608			
Material Note	transparent	transparent	transparent	transparent	gamma radiation stabilized, transparent	gamma radiation stabilized, transparent
Reference No.	81	81	81	81	106	106
EXPOSURE CONDITIONS						
Type	Ethylene Oxide	Ethylene Oxide	Ethylene Oxide	Ethylene Oxide	Ethylene Oxide	Ethylene Oxide
Details	12% EtO and 88% Freon	12% EtO and 88% Freon	10% EtO and 90% CO_2	10% EtO and 90% CO_2	100% EtO	12% EtO and 88% Freon
Concentration					1000 mg/l	1000 mg/l
Number of Cycles	1	5	1	5	1	1
Note	RH: 60%; pressure: 0.051 MPa	RH: 60%; pressure: 0.051 MPa	RH: 45%; pressure: 0.158 MPa	RH: 45%; pressure: 0.158 MPa		
Temperature (°C)	48.9	48.9	48.9	48.9	54.4	54.4
Time (hours)	6	6	10	10	4-6	4-6
PROPERTIES RETAINED (%)						
Tensile Strength	95 (hm)	85 (hm)	102.9 (hm)	103.9 (hm)	100 (he)	100 (he)
Tensile Strength @ Yield	100 (ib)	100 (ib)	102.2 (ib)	101.1 (ib)		
Elongation	95.8 (ak)	79.2 (ak)	100 (ak)	100 (ak)		
Modulus					100 (bz)	100 (bz)
Notched Izod Impact	97.4 (fn)	95.4 (fn)	97.4 (fn)	95.5 (fn)	100 (fm)	99 (fm)
Heat Deflection Temperature	99.2 (j)	97.7 (j)				

TABLE 60: Effect of Ethylene Oxide Sterilization on Polycarbonate

Material Family	POLYCARBONATE		
Material Supplier/Name	MILES MAKROLON 2608		
Material Note	transparent	transparent	transparent
Reference No.	81	81	81

EXPOSURE CONDITIONS

Type	Ethylene Oxide	Ethylene Oxide	Ethylene Oxide
Number of Cycles	1	1	1

RESIDUALS (ppm)

Residuals Determined	ethylene oxide	ethylene chlorohydrin	ethylene glycol
24 hour Aeration	119		
168 hour Aeration	19	0	0
336 hour Aeration	8	0	0

TABLE 61: Effect of Ethylene Oxide Sterilization on Polycarbonate

Material Family	Polycarbonate
Material Supplier/Name	GE
Material Note	transparent
Reference No.	46

EXPOSURE CONDITIONS

Type	Ethylene Oxide
Details	100% EtO
Concentration	550 mg/l
Number of Cycles	5
Note	test lab: Sterilization Technical Services, Inc.
Temperature (°C)	54.4
Time (hours)	5

PRE EXPOSURE CONDITIONING

Preconditioning Note	type: pre vacuum; pressure: 660-711 mm Hg; RH: 45-60%; dwell time: 15 minutes

PROPERTIES RETAINED (%)

Tensile Strength @ Yield	112 (ib)
Elongation @ Yield	107 (bd)
Modulus	103 (gs)

TABLE 62: Effect of Ethylene Oxide Sterilization on Polycarbonate

Material Family	POLYCARBONATE					
Material Supplier/Name	**GE LEXAN 141**	**GE LEXAN HP3**	**GE LEXAN SP1310**	**GE LEXAN 141**	**GE LEXAN HP3**	**GE LEXAN SP1310**
Material Note	transparent	transparent, medical grade	transparent, random copolymer	transparent	transparent, medical grade	transparent, random copolymer
Reference No.	107	107	107	107	107	107
EXPOSURE CONDITIONS						
Type	Ethylene Oxide	Ethylene Oxide	Ethylene Oxide	Ethylene Oxide	Ethylene Oxide	Ethylene Oxide
Details	100% EtO	100% EtO	100% EtO	100% EtO	100% EtO	100% EtO
Concentration	800 mg/l	800 mg/l	800 mg/l	800 mg/l	800 mg/l	800 mg/l
Temperature (°C)	51.7	51.7	51.7	51.7	51.7	51.7
Time (hours)	8	8	8	8	8	8
PRE EXPOSURE CONDITIONING						
Preconditioning Note	type: pre vacuum; pressure: 660-711 mm Hg; RH: 45-60%; dwell time: 15 minutes	type: pre vacuum; pressure: 660-711 mm Hg; RH: 45-60%; dwell time: 15 minutes	type: pre vacuum; pressure: 660-711 mm Hg; RH: 45-60%; dwell time: 15 minutes	type: pre vacuum; pressure: 660-711 mm Hg; RH: 45-60%; dwell time: 15 minutes	type: pre vacuum; pressure: 660-711 mm Hg; RH: 45-60%; dwell time: 15 minutes	type: pre vacuum; pressure: 660-711 mm Hg; RH: 45-60%; dwell time: 15 minutes
POST EXPOSURE CONDITIONING						
Note	type: aeration; note: ambient conditions	type: aeration; note: ambient conditions	type: aeration; note: ambient conditions	type: aeration; note: ambient conditions	type: aeration; note: ambient conditions	type: aeration; note: ambient conditions
Temperature (°C)	23	23	23	23	23	23
RESIDUALS (ppm)						
Residuals Determined	ethylene chlorohydrin	ethylene chlorohydrin	ethylene chlorohydrin	ethylene glycol	ethylene glycol	ethylene glycol
336 hour Aeration	<3	<3	<3	<20	<24	<21

TABLE 63: Effect of Ethylene Oxide Sterilization on Polycarbonate

Material Family	POLYCARBONATE					
Material Supplier/Name	**GE LEXAN 141**	**GE LEXAN HP3**	**GE LEXAN SP1310**	**GE LEXAN 141**	**GE LEXAN HP3**	**GE LEXAN SP1310**
Material Note	transparent	transparent, medical grade	transparent, random copolymer	transparent	transparent, medical grade	transparent, random copolymer
Reference No.	107	107	107	107	107	107
EXPOSURE CONDITIONS						
Type	Ethylene Oxide	Ethylene Oxide	Ethylene Oxide	Ethylene Oxide	Ethylene Oxide	Ethylene Oxide
Details	100% EtO	100% EtO	100% EtO	100% EtO	100% EtO	100% EtO
Concentration	800 mg/l	800 mg/l	800 mg/l	800 mg/l	800 mg/l	800 mg/l
Temperature (°C)	51.7	51.7	51.7	51.7	51.7	51.7
Time (hours)	8	8	8	8	8	8
PRE EXPOSURE CONDITIONING						
Preconditioning Note	type: pre vacuum; pressure: 660-711 mm Hg; RH: 45-60%; dwell time: 15 minutes	type: pre vacuum; pressure: 660-711 mm Hg; RH: 45-60%; dwell time: 15 minutes	type: pre vacuum; pressure: 660-711 mm Hg; RH: 45-60%; dwell time: 15 minutes	type: pre vacuum; pressure: 660-711 mm Hg; RH: 45-60%; dwell time: 15 minutes	type: pre vacuum; pressure: 660-711 mm Hg; RH: 45-60%; dwell time: 15 minutes	type: pre vacuum; pressure: 660-711 mm Hg; RH: 45-60%; dwell time: 15 minutes
POST EXPOSURE CONDITIONING						
Note	type: aeration; note: mechanical with fan	type: aeration; note: mechanical with fan	type: aeration; note: mechanical with fan	type: aeration; note: mechanical with fan	type: aeration; note: mechanical with fan	type: aeration; note: mechanical with fan
Temperature (°C)	49	49	49	49	49	49
RESIDUALS (ppm)						
Residuals Determined	ethylene chlorohydrin	ethylene chlorohydrin	ethylene chlorohydrin	ethylene glycol	ethylene glycol	ethylene glycol
336 hour Aeration	<3	<3	<3	<25	<21	<23

TABLE 64: Effect of Ethylene Oxide Sterilization on Polycarbonate

Material Family	POLYCARBONATE					
Material Supplier/Name	GE LEXAN HPS3			GE LEXAN HP3		
Material Note	gamma radiation stabilized, transparent, medical grade	gamma radiation stabilized, transparent, medical grade	gamma radiation stabilized, transparent, medical grade	transparent, medical grade	transparent, medical grade	transparent, medical grade
Reference No.	108	108	108	108	108	108

EXPOSURE CONDITIONS

Type	Ethylene Oxide	Ethylene Oxide	Ethylene Oxide	Ethylene Oxide	Ethylene Oxide	Ethylene Oxide
Details	100% EtO	100% EtO	100% EtO	100% EtO	100% EtO	100% EtO
Concentration	550 mg/l	550 mg/l	550 mg/l	550 mg/l	550 mg/l	550 mg/l
Time (hours)	5	5	5	5	5	5

PRE EXPOSURE CONDITIONING

Preconditioning Note	type: pre vacuum; pressure: 660-711 mm Hg; RH: 45-60%; dwell time: 15 minutes	type: pre vacuum; pressure: 660-711 mm Hg; RH: 45-60%; dwell time: 15 minutes	type: pre vacuum; pressure: 660-711 mm Hg; RH: 45-60%; dwell time: 15 minutes	type: pre vacuum; pressure: 660-711 mm Hg; RH: 45-60%; dwell time: 15 minutes	type: pre vacuum; pressure: 660-711 mm Hg; RH: 45-60%; dwell time: 15 minutes	type: pre vacuum; pressure: 660-711 mm Hg; RH: 45-60%; dwell time: 15 minutes

POST EXPOSURE CONDITIONING

Note	type: aeration; note: mechanical with fan	type: aeration; note: mechanical with fan	type: aeration; note: mechanical with fan	type: aeration; note: mechanical with fan	type: aeration; note: mechanical with fan	type: aeration; note: mechanical with fan
Temperature (°C)	49	49	49	49	49	49

RESIDUALS (ppm)

Residuals Determined	ethylene oxide	ethylene chlorohydrin	ethylene glycol	ethylene oxide	ethylene chlorohydrin	ethylene glycol
24 hour Aeration	245	none detected	21	177	none detected	30
168 hour Aeration	78			54		
336 hour Aeration	56		25	39		<4

TABLE 65: Effect of Ethylene Oxide Sterilization on Polycarbonate

Material Family	POLYCARBONATE					
Material Supplier/Name	GE LEXAN HPS3			GE LEXAN HP3		
Material Note	gamma radiation stabilized, transparent, medical grade	gamma radiation stabilized, transparent, medical grade	gamma radiation stabilized, transparent, medical grade	transparent, medical grade	transparent, medical grade	transparent, medical grade
Reference No.	108	108	108	108	108	108
EXPOSURE CONDITIONS						
Type	Ethylene Oxide	Ethylene Oxide	Ethylene Oxide	Ethylene Oxide	Ethylene Oxide	Ethylene Oxide
Details	100% EtO	100% EtO	100% EtO	100% EtO	100% EtO	100% EtO
Concentration	550 mg/l	550 mg/l	550 mg/l	550 mg/l	550 mg/l	550 mg/l
Time (hours)	5	5	5	5	5	5
PRE EXPOSURE CONDITIONING						
Preconditioning Note	type: pre vacuum; pressure: 660-711 mm Hg; RH: 45-60%; dwell time: 15 minutes	type: pre vacuum; pressure: 660-711 mm Hg; RH: 45-60%; dwell time: 15 minutes	type: pre vacuum; pressure: 660-711 mm Hg; RH: 45-60%; dwell time: 15 minutes	type: pre vacuum; pressure: 660-711 mm Hg; RH: 45-60%; dwell time: 15 minutes	type: pre vacuum; pressure: 660-711 mm Hg; RH: 45-60%; dwell time: 15 minutes	type: pre vacuum; pressure: 660-711 mm Hg; RH: 45-60%; dwell time: 15 minutes
POST EXPOSURE CONDITIONING						
Note	type: aeration; note: ambient conditions	type: aeration; note: ambient conditions	type: aeration; note: ambient conditions	type: aeration; note: ambient conditions	type: aeration; note: ambient conditions	type: aeration; note: ambient conditions
Temperature (°C)	49	49	49	49	49	49
RESIDUALS (ppm)						
Residuals Determined	ethylene oxide	ethylene chlorohydrin	ethylene glycol	ethylene oxide	ethylene chlorohydrin	ethylene glycol
24 hour Aeration	332	none detected	71	360	none detected	76
168 hour Aeration	177			128		
336 hour Aeration	49		33	111		31

GRAPH 80: **Gamma Radiation Dose vs. Yellowness Index of Polycarbonate**

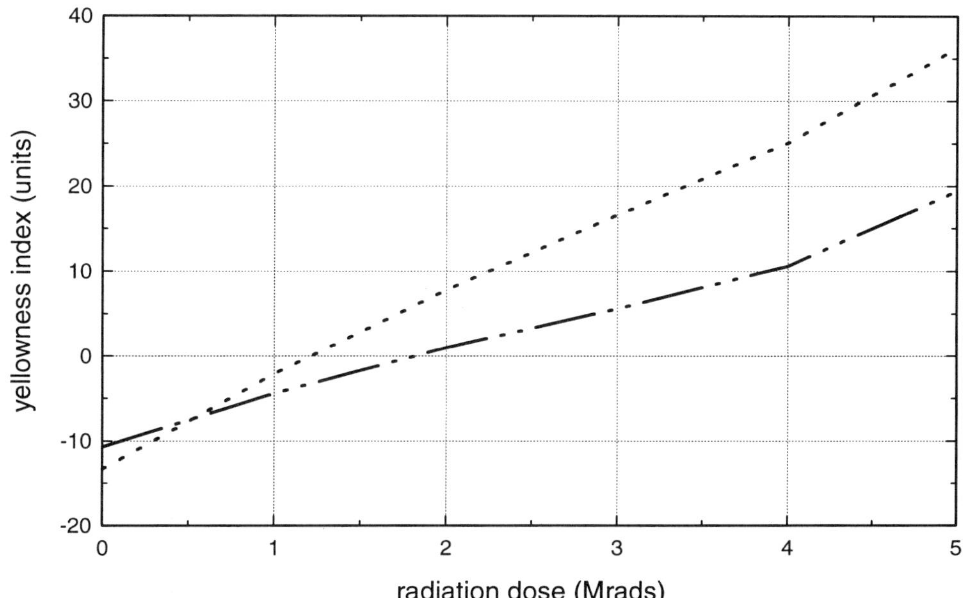

··············	Polycarbonate (gen. purp.)
—··—··	Polycarbonate (gamma rad. stab., transpar.)
Reference No.	106

GRAPH 81: **Post Gamma Radiation Exposure Time vs. Yellowness Index of Polycarbonate**

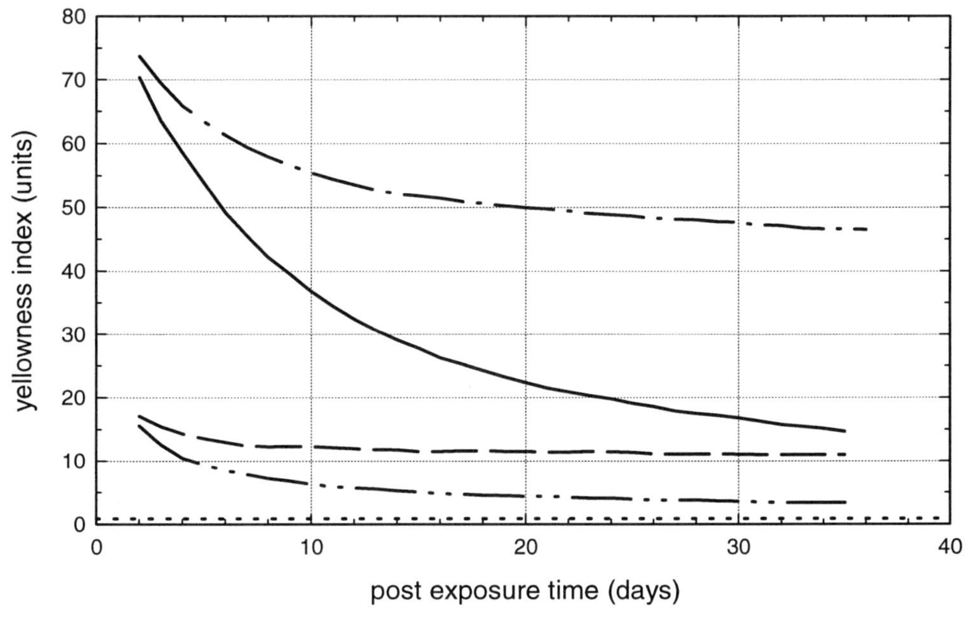

··············	Dow Calibre 2060 Polycarbonate (transpar.); before exposure
—··—··	Dow Calibre 2060 Polycarbonate (transpar.); 2.5 Mrads; light storage
— — —	Dow Calibre 2060 Polycarbonate (transpar.); 2.5 Mrads; dark storage
———	Dow Calibre 2060 Polycarbonate (transpar.); 10 Mrads; light storage
—·—·—	Dow Calibre 2060 Polycarbonate (transpar.); 10 Mrads; dark storage
Reference No.	3

GRAPH 82: **Post Gamma Radiation Exposure Time vs. Yellowness Index of Polycarbonate**

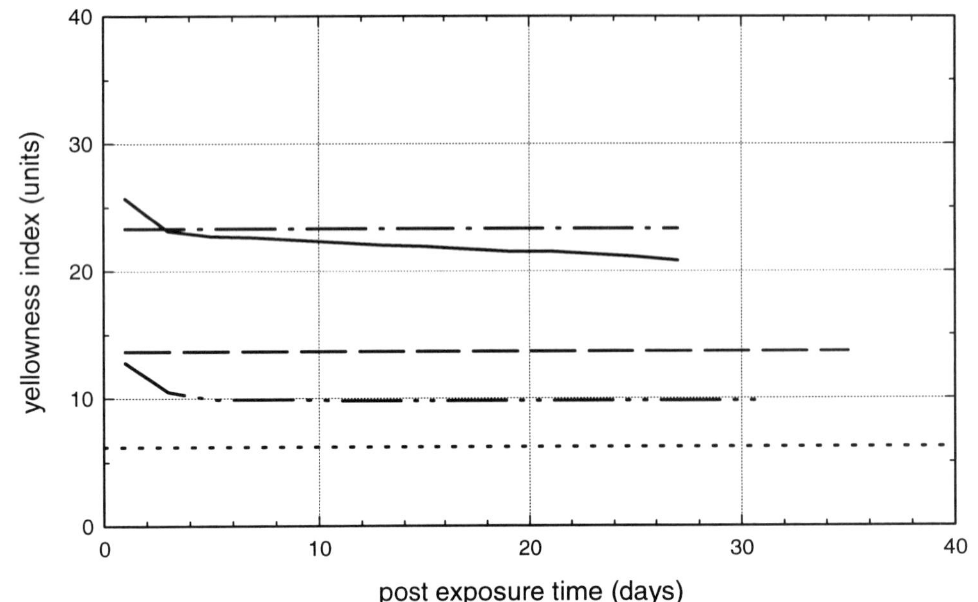

....	Dow Polycarbonate; before exposure
—··—··	Dow Polycarbonate; 2.5 Mrads; light storage
— — — —	Dow Polycarbonate; 2.5 Mrads; dark storage
———	Dow Polycarbonate; 10 Mrads; light storage
—·—·—	Dow Polycarbonate; 10 Mrads; dark storage
Reference No.	3

GRAPH 83: **Post Gamma Radiation Exposure Time vs. Yellowness Index of Polycarbonate**

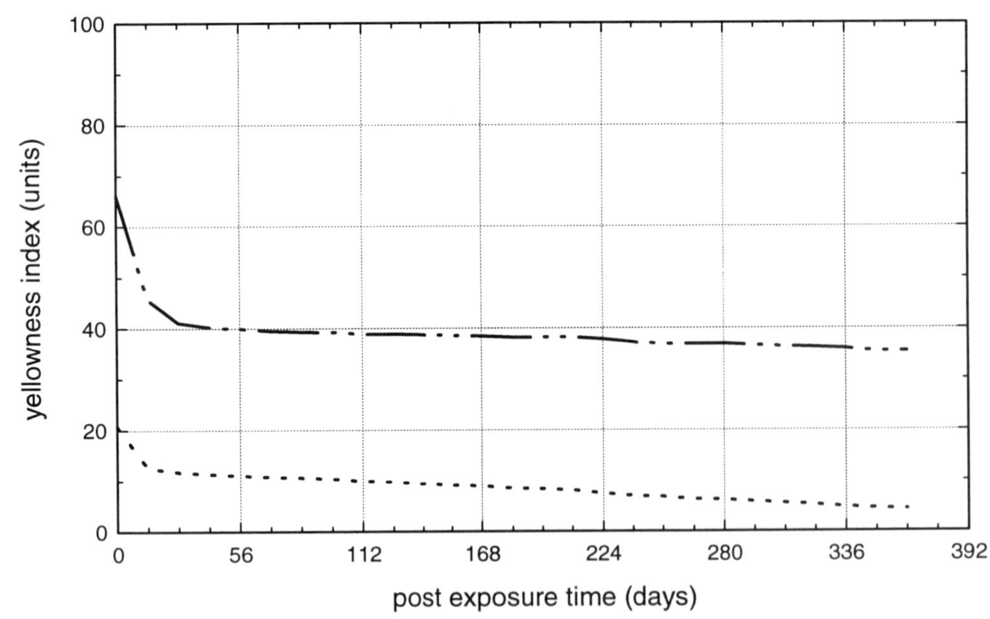

....	Dow Polycarbonate; 2.5 Mrad
—··—··	Dow Polycarbonate; 10 Mrad
Reference No.	5

GRAPH 84: **Post Gamma Radiation Exposure Time vs. Yellowness Index of Polycarbonate**

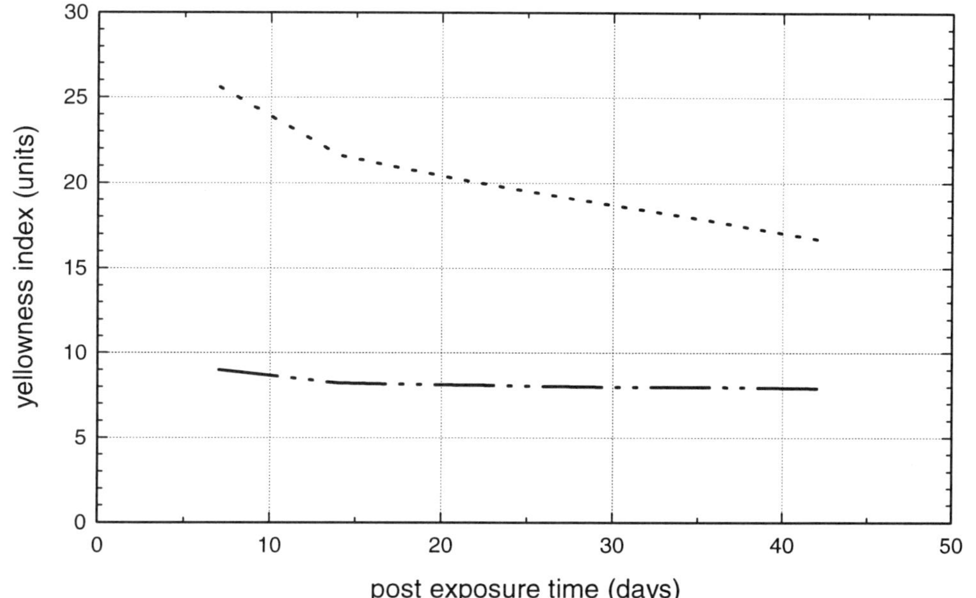

···············	Miles Makrolon Polycarbonate (transpar.); 3.5 Mrads; dark storage; thickness: 2.54 mm
—··—··	Miles Makrolon Rx2530 Polycarbonate; 3.5 Mrads; dark storage; thickness: 2.54 mm
Reference No.	83

GRAPH 85: **Post Gamma Radiation Exposure Time vs. Yellowness Index of Polycarbonate**

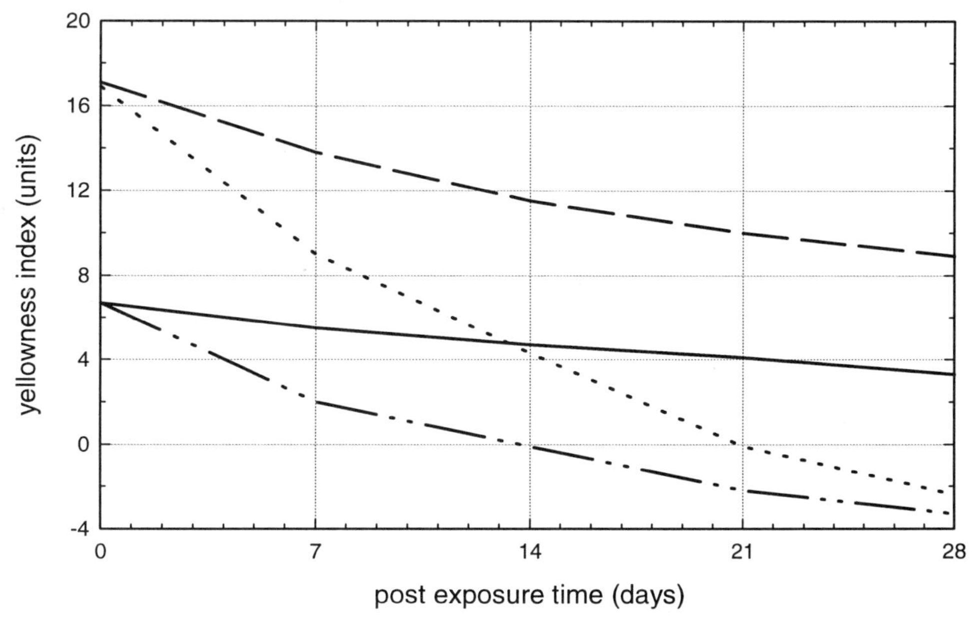

···············	Polycarbonate; 3 Mrads, light storage
—··—··	Polycarbonate (gamma rad. stab., transpar.); 3 Mrads, light storage
— — —	Polycarbonate; 3 Mrads, dark storage
———	Polycarbonate (gamma rad. stab., transpar.); 3 Mrads, dark storage
Reference No.	106

GRAPH 86: **Post Gamma Radiation Exposure Time vs. Yellowness Index of Polycarbonate**

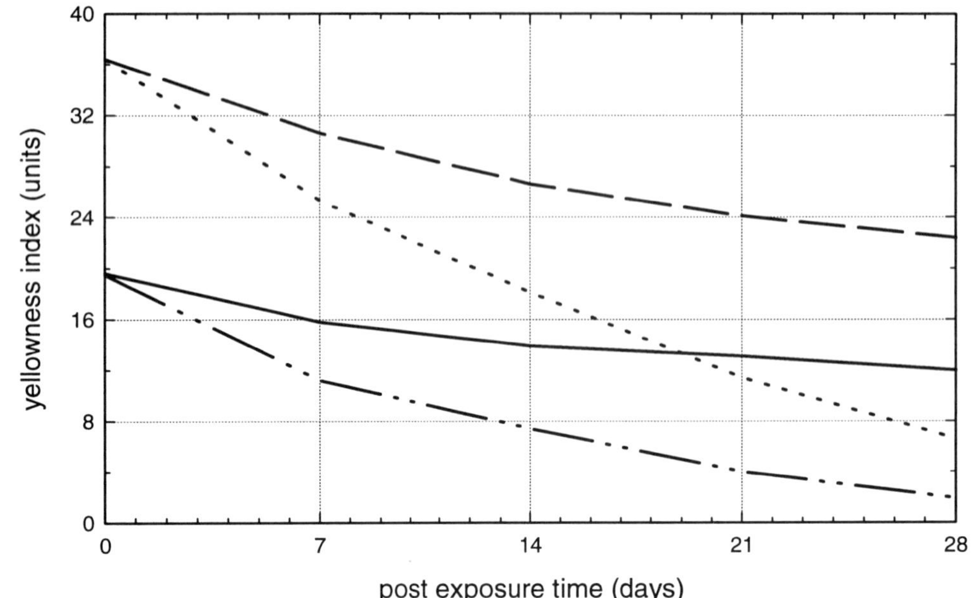

GRAPH 87: **Post Gamma Radiation Exposure Time vs. Percent Light Transmission of Polycarbonate**

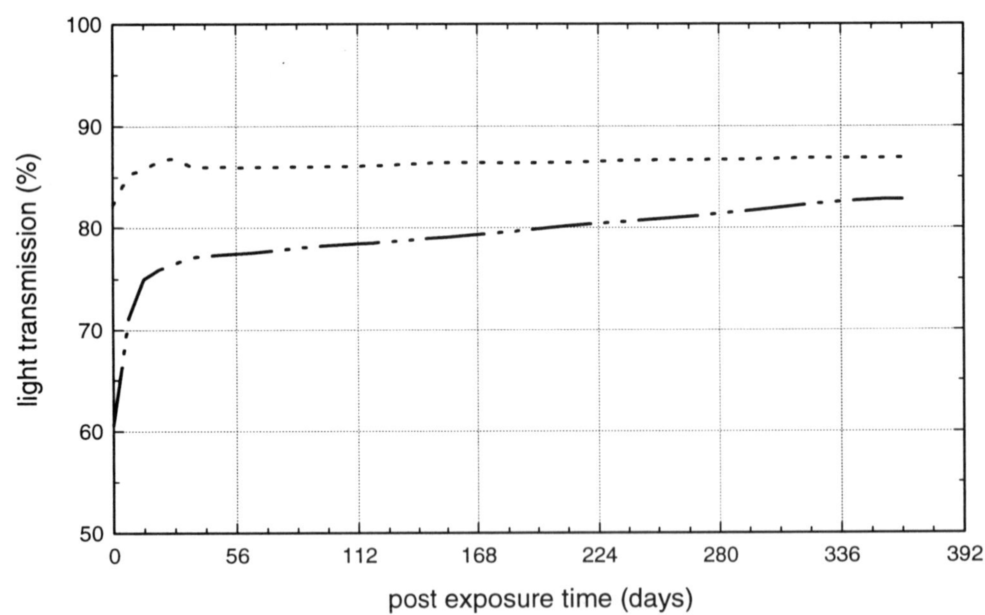

GRAPH 88: Beta Radiation Dose vs. Tensile Strength of Polycarbonate

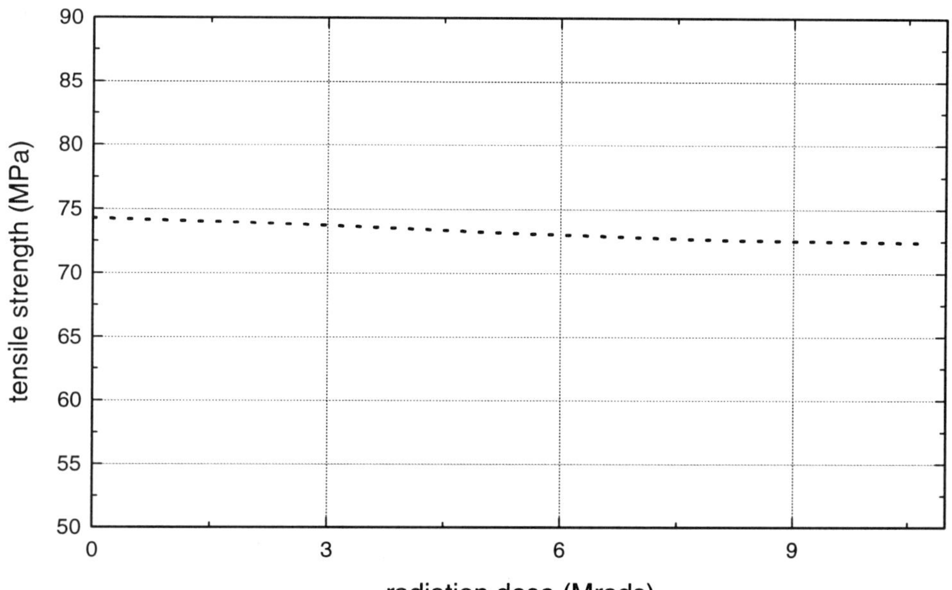

....	Polycarbonate (transpar.)
Reference No.	105

GRAPH 89: Beta Radiation Dose vs. Tensile Modulus of Polycarbonate

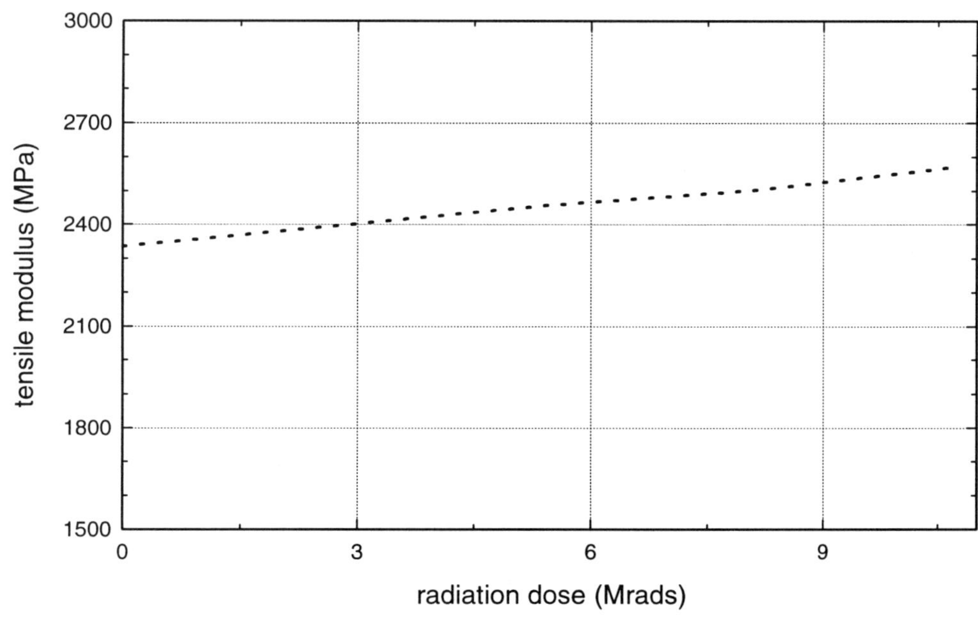

....	Polycarbonate (transpar.)
Reference No.	105

GRAPH 90: Beta Radiation Dose vs. Notched Izod Impact Strength of Polycarbonate

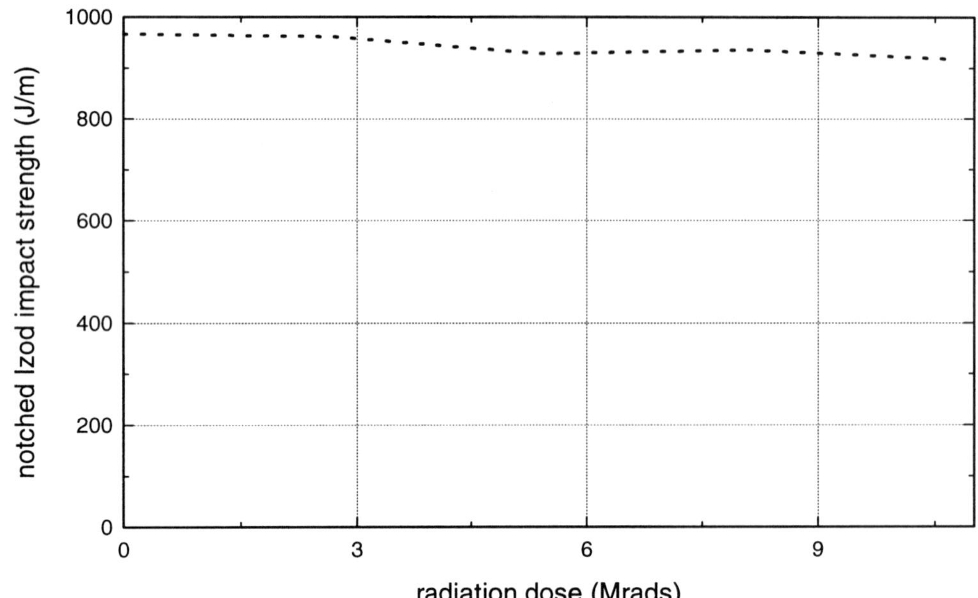

....	Polycarbonate (transpar.)
Reference No.	105

GRAPH 91: Beta Radiation Dose vs. Yellowness Index of Polycarbonate

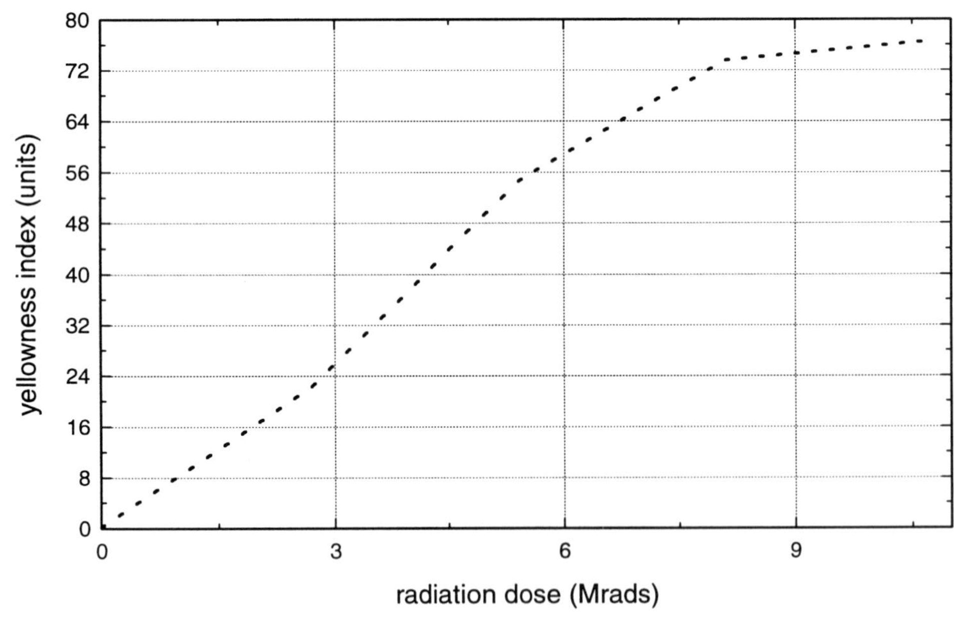

....	Polycarbonate (transpar.)
Reference No.	105

GRAPH 92: Post Beta Radiation Exposure Time vs. Yellowness Index of Polycarbonate

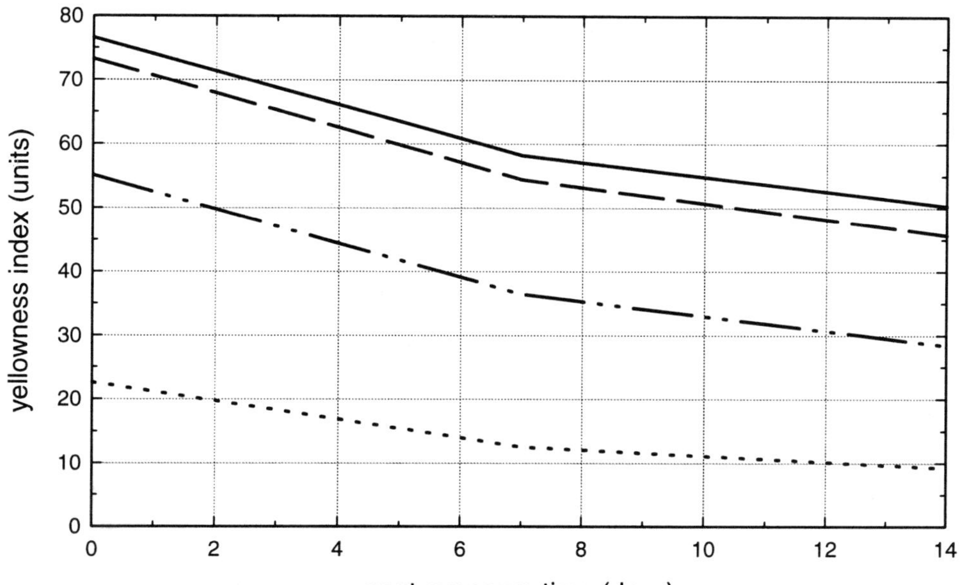

....	Polycarbonate (transpar.); 2.7 Mrads
—··—··	Polycarbonate (transpar.); 5.4 Mrads
– – – –	Polycarbonate (transpar.); 8.1 Mrads
———	Polycarbonate (transpar.); 10.8 Mrads
Reference No.	105

Gamma Radiation Resistance

GE Plastics: Lexan PPC (features: transparent, high heat grade)

Lexan PPC resins retain their properties through 10 megarads of gamma radiation. At 5 megarads, limited yellowing occurs.

Reference: *Sterilization Comparison,* supplier marketing literature (JRW/3/91) - General Electric Plastics, 1991.

Ethylene Oxide (EtO) Resistance

GE Plastics: Lexan PPC (features: transparent, high heat grade)

Properties of Lexan PPC resins are unchanged to 50 ethylene oxide sterilization cycles.

Reference: *Sterilization Comparison,* supplier marketing literature (JRW/3/91) - General Electric Plastics, 1991.

Steam Resistance

GE Plastics: Lexan PPC (features: transparent, high heat grade)

After 200 autoclave cycles (Hi-Vac @ 132°C), Lexan PPC resins retain 92% of their impact strength.

Reference: *Sterilization Comparison,* supplier marketing literature (JRW/3/91) - General Electric Plastics, 1991.

GRAPH 93: **Number of Steam Sterilization Cycles vs. Drop Dart Impact Strength of Polyester Carbonate**

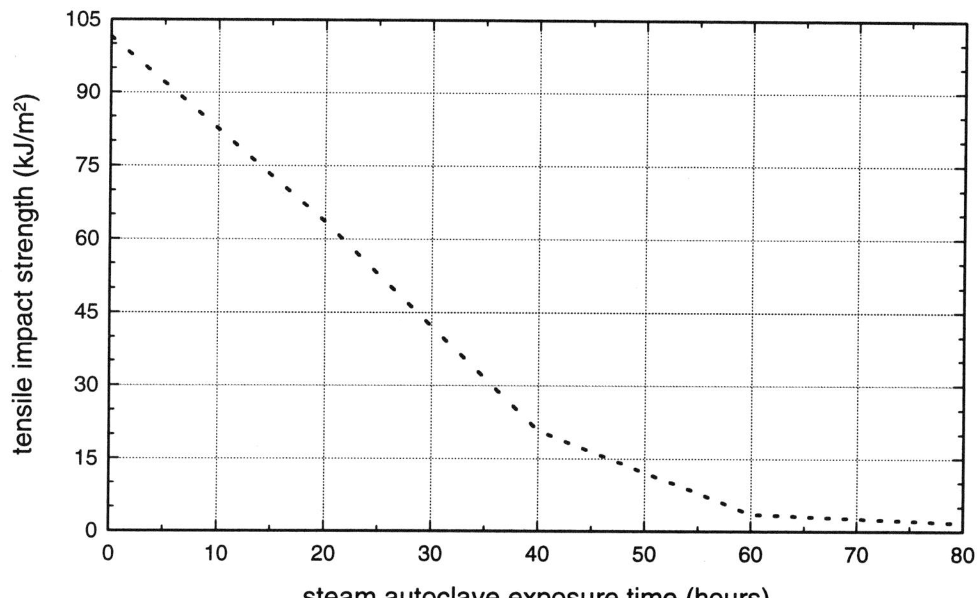

....	Polyester Carbonate; 30 min. @ 132°C per cycle
Reference No.	17

Gamma Radiation Resistance

DuPont: (form: film)

Gamma sterilization can cause chain scission and radical recombination in polyester film. These reactions can randomize the amorphous domains and destroy the crystalline structure, resulting in lower glass transition temperature (Tg) and melting temperature (Tm). The Tg was around 73°C for control and gamma sterilized samples. A minimal change of Tg among control and gamma sterilized samples indicated that no serious chain scission occurred during irradiation and the amorphous domain structure was basically unchanged. Therefore, Tg among these samples were comparable. Tm of polyester film was observed around 256°C for all samples. There was no statistical difference in Tm for both 4 mil and 5 mil samples after irradiation. The properties of polyester remained similar after gamma sterilization.

The thermogravimetric analysis (TGA) curves of non sterilized and gamma sterilized polyester samples were similar to each other. For all the samples, the onset temperature (To) was around 452°C and the first derivative curve peaked around 480°C.

There was no significant difference in the tensile strength along the machine direction for samples of both thicknesses. Although the tensile strength along the machine direction was insensitive to the sterilization method, the tensile strength along the transverse direction was somewhat sensitive. The tensile strength of 4 mil samples was reduced after gamma sterilization, while the tensile strength of 5 mil sample increased. These apparently contradictory results might be partly explained by the thickness of the samples. Chain scission and radical recombination might occur during gamma sterilization. In the case that more chain scission occurs than recombination, the tensile strength would reduce. On the other hand, if effective radical recombination occurs, the tensile strength would increase. In this case, the radical recombination appeared to be more effective for the thick samples, therefore resulting in higher tensile strength. These changes were within a 10% range of the control sample. Therefore it is reasonable to assume that these are acceptable sterilization methods.

Reference: Hu, C. B., Ma, M. T., Mcintyre, J., Nguyen, D., Myers, K.E., *The Effect Of Steam And Gamma Sterilization On The Thermal And Mechanical Properties Of Polyester Film,* ANTEC 1994, conference proceedings - Society of Plastics Engineers, 1994.

Steam Resistance

DuPont: (form: film)

For steam sterilization, hydrolysis and/or annealing can occur on polyester film. If hydrolysis occurs to a significant extent, the amorphous domain and crystalline structures would be affected greatly, resulting in changes of the glass transition temperature (Tg) and the crystalline melting temperature (Tm). If annealing occurs , the local motion of the molecules in the polymer chain could result in a better domain and crystalline structures. Since the previous amorphous domain and crystalline structure would likely become better organized, the Tg and Tm might shift to a higher temperature.

The Tg was around 73°C for control and around 150°C for all steam sterilized samples. Among steam sterilized samples, Tg was higher as the number of steam cycles increased. A major shifting of Tg was observed for all samples. The degree of shifting appeared to somewhat depend on the steam cycle. For example, Tg of the sample with 3 normal cycles was higher than with 1 normal cycle, Tg of the sample with 3 normal cycles and 1 flash cycle was higher than that with 3 normal cycles only. However, there was no statistical difference of Tg between samples treated with 5 normal and 3 flash cycles and 3 normal and 1 flash cycle. This observation indicated that the amorphous domain reorganized after the first steam sterilization cycle and got better after repeated steam cycles. The equilibrium domain structure was probably reached between 3 normal and 1 flash cycles and 5 normal and 3 flash cycles. The thickness of polyester did not have any effect on this phenomenon since the same pattern was observed for 4 mils and 5 mils of thickness.

Tm of polyester film was observed around 256°C for all samples. A slightly higher Tm was found for the 4 mil sample after 5 normal and 3 flash steam cycles and no statistical difference was observed for all other samples, regardless of the sterilization conditions. As for the 5 mil samples, the Tm of all samples were statistically equivalent. These results indicated that no serious hydrolysis was identified. On the other hand, the slightly higher Tm of 4 mil polyester sample after extensive steam sterilization cycles suggested the possible occurrence of annealing in this sample after repeated steam sterilization. A better crystalline structure without hydrolysis usually would not deteriorate the polymer properties. The properties of polyester remained similar after gamma and steam sterilization.

The thermogravimetric analysis (TGA) curves of non sterilized and steam sterilized polyester samples were similar to each other. For all the samples, the onset temperature (To) was around 452°C and the first derivative curve peaked around 480°C. Slightly higher temperatures were observed for steam sterilized samples. Again, the slightly higher To and characteristic temperature (Tp) might be due to the better domain structure of polyester film after repeated annealing, resulting in a slightly more stable material.

There was no significant difference in the tensile strength along the machine direction for samples of both thicknesses.

All types of steam sterilization in this study caused reversible annealing (morphological change), however, the thermal stability of the polyester material in general did not change.

Reference: Hu, C. B., Ma, M. T., Mcintyre, J., Nguyen, D., Myers, K.E., *The Effect Of Steam And Gamma Sterilization On The Thermal And Mechanical Properties Of Polyester Film,* ANTEC 1994, conference proceedings - Society of Plastics Engineers, 1994.

TABLE 66: Effect of Gamma Radiation Sterilization on Polyester

Material Family	POLYESTER			
Material Supplier/Name	DUPONT			
Material Note	film; thickness: 0.102 mm	film; thickness: 0.102 mm	film; thickness: 0.127 mm	film; thickness: 0.127 mm
Reference No.	24	24	24	24

EXPOSURE CONDITIONS

Type	Gamma Radiation	Gamma Radiation	Gamma Radiation	Gamma Radiation
Radiation Dose (Mrads)	3.2-3.6	6.4-7.2	3.2-3.6	6.4-7.2
Note		specimens exposed to two 3.2-3.6 megarads cycles		specimens exposed to two 3.2-3.6 megarads cycles

PROPERTIES RETAINED (%)

Tensile Strength	105 (hg)	106.3 (hg)	99 (hh)	100.8 (hh)
Tensile Strength (transverse)	96.6 (hj)	93.2 (hj)	104.2 (hk)	106.2 (hk)
Glass Transition Temperature	98.3 (cs)	97.9 (cs)	100 (cs)	101.4 (cs)
Melting Point	100.2 (f)	100.3 (f)	100 (f)	100 (f)

TABLE 67: Effect of Steam Sterilization on Polyester

Material Family	POLYESTER							
Material Supplier/Name	DUPONT							
Material Note	film; thickness: 0.102 mm	film; thickness: 0.102 mm	film; thickness: 0.102 mm	film; thickness: 0.102 mm	film; thickness: 0.127 mm	film; thickness: 0.127 mm	film; thickness: 0.127 mm	film; thickness: 0.127 mm
Reference No.	24	24	24	24	24	24	24	24

EXPOSURE CONDITIONS

Type	Steam	Steam	Steam	Steam	Steam	Steam	Steam	Steam
Number of Cycles	1	3	3	5	1	3	3	5
Note	normal steam cycle	normal steam cycle	normal steam cycle	normal steam cycle	normal steam cycle	normal steam cycle	normal steam cycle	normal steam cycle
Temperature (°C)	121	121	121	121	121	121	121	121
Time (hours)	0.77	0.77	0.77	0.77	0.77	0.77	0.77	0.77

EXPOSURE CONDITIONS II

Number of Cycles			1	3			1	3
Note			flash steam cycle	flash steam cycle			flash steam cycle	flash steam cycle
Temperature (°C)			132	132			132	132
Time (hours)			0.25	0.25			0.25	0.25

PROPERTIES RETAINED (%)

Tensile Strength	106.7 (hg)	102.1 (hg)	103.7 (hg)	101.6 (hg)	102.3 (hh)	97.1 (hh)	97.7 (hh)	99.4 (hh)
Tensile Strength (transverse)	100.8 (hj)	101.7 (hj)	100.7 (hj)	97.6 (hj)	98 (hk)	101.2 (hk)	97.1 (hk)	96.2 (hk)
Glass Transition Temperature	194.8 (cs)	200.1 (cs)	206.4 (cs)	210.5 (cs)	199 (cs)	203.7 (cs)	213.3 (cs)	216 (cs)
Melting Point	100.2 (f)	100.3 (f)	100.3 (f)	100.7 (f)	99.7 (f)	100.1 (f)	100 (f)	100.4 (f)

Gamma Radiation Resistance

Eastman Performance Plastics: Ektar DN001 (features: transparent, high CHDM (cyclohexane dimethanol) content)

Copolyesters are more inherently stable to gamma irradiation than most polymer systems. There is no significant change in physical properties and a very slight color change following gamma radiation exposure. This contrasts with the loss in mechanical strength and/or color shift found in many plastics.

Reference: Goulder, G. K., Seymour, R. W., *Polyesters And Copolyesters For Use In Medical Applications,* supplier technical report - Eastman Performace Plastics.

Beta Radiation Resistance

Copolyester (features: transparent)

Beta sterilization is increasing in importance in Europe. Samples of a copolyester blend were tested at beta radiation levels of 2.7 to 10.8 Mrads. Physical properties were retained after 10.8 Mrads. In contrast to gamma sterilization, beta sterilization generates more heat. At 10.8 Mrads the surface of the copolyester blend began to adhere to the packaging. The material became quite yellow. After two weeks of storage there was some reduction in yellowness. Storage in light or dark did not have a meaningful impact.

Reference: Haines, Jim, Hauser, Debbie, *Zylar Clear Alloys For Medical Applications,* supplier technical report - Novacor Chemicals Inc.

Ethylene Oxide (EtO) Resistance

Eastman Performance Plastics: Ektar DN001 (features: transparent, high CHDM (cyclohexane dimethanol) content)

No significant property changes were observed.

Reference: Goulder, G. K., Seymour, R. W., *Polyesters And Copolyesters For Use In Medical Applications,* supplier technical report - Eastman Performace Plastics.

TABLE 68: Effect of Gamma Radiation Sterilization on Polyester Copolymer

Material Family	POLYESTER COPOLYMER	
Material Supplier/Name	EASTMAN EKTAR DN001	
Material Note	transparent, high CHDM (cyclohexane dimethanol) content	transparent, high CHDM (cyclohexane dimethanol) content
Reference No.	4	4

EXPOSURE CONDITIONS

Type	Gamma Radiation	Gamma Radiation
Radiation Dose (Mrads)	2.5	5

SURFACE and APPEARANCE

Δb Color	0.4 (w)	0.45 (w)

TABLE 69: Effect of Ethylene Oxide Sterilization on Polyester Copolymer

Material Family	**Polyester Copolymer**
Material Supplier/Name	**Eastman Ektar DN001**
Material Note	transparent, high CHDM (cyclohexane dimethanol) content
Reference No.	4

EXPOSURE CONDITIONS

Type	Ethylene Oxide
Temperature (°C)	54.4
Time (hours)	3

EXPOSURE CONDITIONS II

Number of Cycles	1

POST EXPOSURE CONDITIONING

Note	type: aeration
Temperature (°C)	54.4
Time (hours)	10

PROPERTIES RETAINED (%)

Tensile Strength	100 (hs)
Tensile Strength @ Yield	104.5 (ih)
Elongation	100 (as)
Elongation @ Yield	106.5 (be)
Flexural Strength	100 (cl)
Modulus	100 (cb)
Notched Izod Impact	100 (gi)
Unnotched Izod Impact	100 (kr)
Heat Deflection Temperature	100 (l)

SURFACE and APPEARANCE

Hardness Units Change	R0 (gm)
Haze Retained (%)	100
Transmittance Retained (%)	100
Δa Color	-0.27 (q)
Δb Color	-0.25 (x)

GRAPH 94: **Beta Radiation Dose vs. Tensile Strength of Polyester Copolymer**

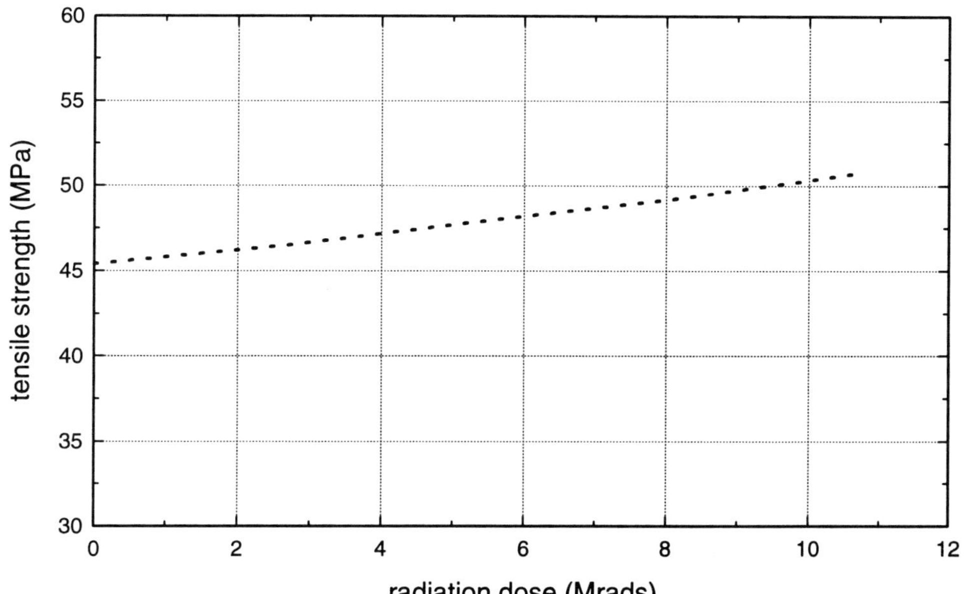

....	Polyester Copol. (transpar.)
Reference No.	105

GRAPH 95: **Beta Radiation Dose vs. Tensile Modulus of Polyester Copolymer**

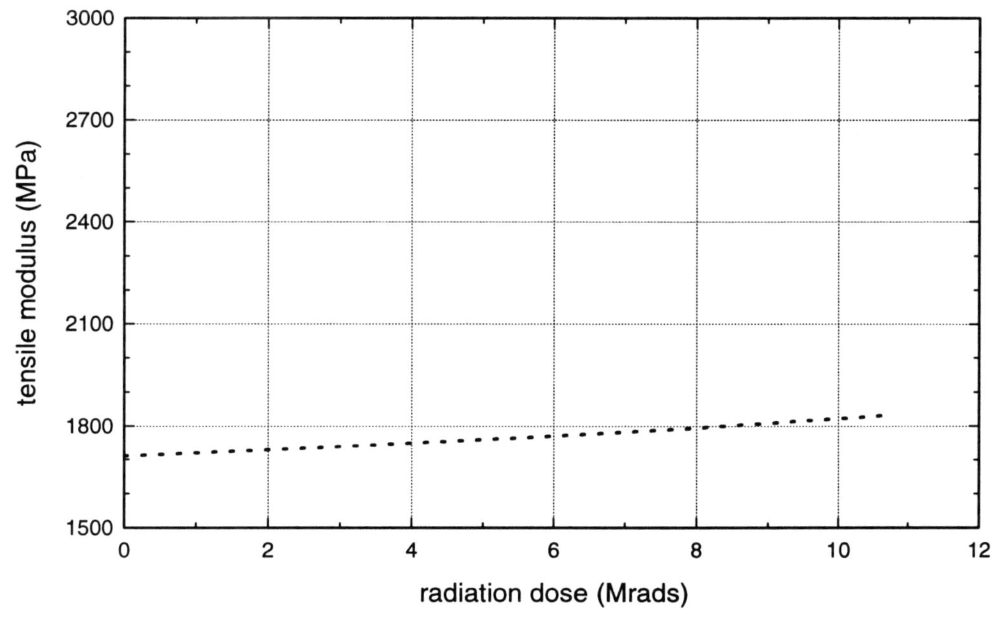

....	Polyester Copol. (transpar.)
Reference No.	105

GRAPH 96: Beta Radiation Dose vs. Yellowness Index of Polyester Copolymer

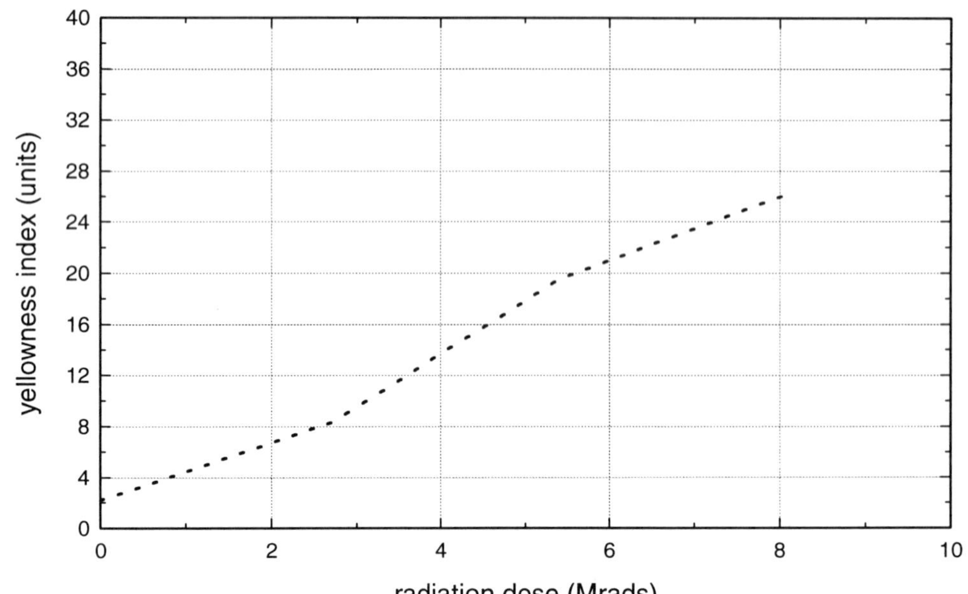

....	Polyester Copol. (transpar.)
Reference No.	105

GRAPH 97: Post Beta Radiation Exposure Time vs. Yellowness Index of Polyester Copolymer

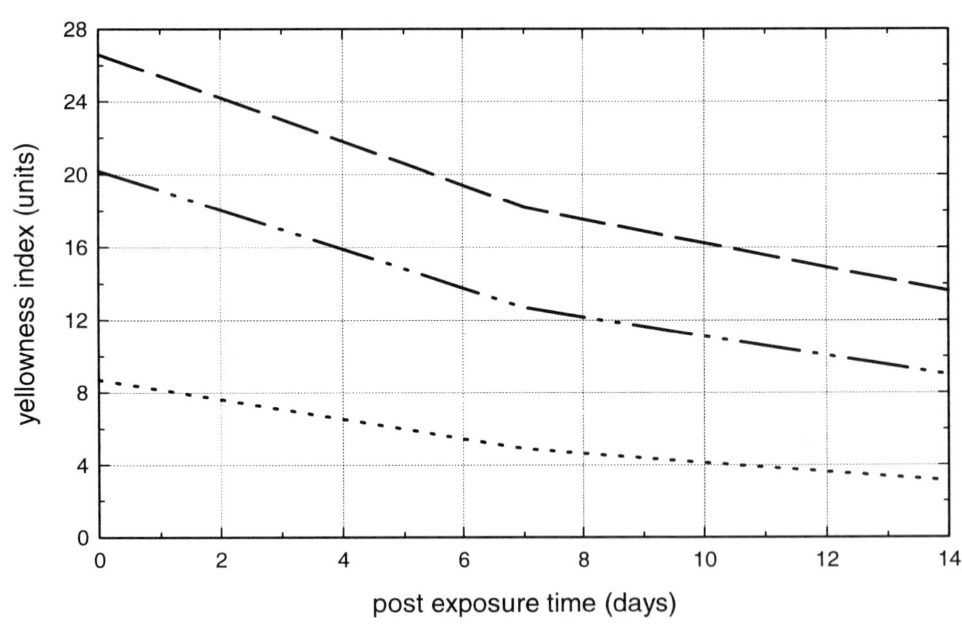

....	Polyester Copol. (transpar.); 2.7 Mrads
—··—··	Polyester Copol. (transpar.); 5.4 Mrads
— — — —	Polyester Copol. (transpar.); 8.1 Mrads
Reference No.	105

Gamma Radiation Resistance

BASF: Ultradur (features: transparent)

Ultradur PBT is resistant to gamma radiation (2.5 Mrads over 6 hours), however after irradiation there may be a slight yellowing but no reduction in the impact resistance. In general, thermoplastic polyesters show no significant loss in properties below dosages of 90 Mrads.

Reference: Johnson, James A., supplier written correspondence - BASF Corporation, 1994.

GE Plastics:

Gamma radiation up to 10.0 Mrads has little effect on physical properties. The molecular weight of samples tested did not change significantly.

Reference: Paulino, J. C., Baccaro, L. E., *Effects Of Sterilization Procedures On Physical Properties of Engineering Thermoplastics,* supplier technical report - General Electric Company.

GE Plastics: Valox HP (features: medical grade)

Valox HP resins retain their properties through 2.5 megarads of gamma radiation.

Reference: *Sterilization Comparison,* supplier marketing literature (ACB/3/93) - General Electric Plastics, 1993.

GE Plastics: Valox HP (features: medical grade)

Valox HP is a crystalline thermoplastic polyester that can withstand gamma radiation sterilization. Exposure to 2.5 Mrads of gamma radiation has no negative effects on the materials' mechanical properties: elongation remains the same, and tensile strength actually increases by 17%.

Reference: *GE Plastics Medical Applications,* supplier marketing literature (NBM-100 (1/89) RTB) - General Electric Company, 1989.

Ethylene Oxide (EtO) Resistance

GE Plastics: Valox HP (features: medical grade)

Properties of Valox HP resins are unchanged to 100 ethylene oxide sterilization cycles.

Reference: *Sterilization Comparison,* supplier marketing literature (ACB/3/93) - General Electric Plastics, 1993.

GE Plastics: Valox HP (features: medical grade)

Crystalline thermoplastic polyester materials with excellent lubricity and chemical resistance, Valox HP resins withstand ethylene oxide sterilization. The naturally opaque materials have chemical resistance superior to most amorphous materials, making them well suited for applications requiring resistance to substances such as lipids and hydrocarbons. In addition, their lubricity makes them well suited for moving parts such as intravenous spikes and stopcock hubs which must contact and move with other parts without being damaged.

Reference: *GE Plastics Medical Applications,* supplier marketing literature (NBM-100 (1/89) RTB) - General Electric Company, 1989.

BASF: Ultradur (features: transparent)

Ultradur PBT is resistant to EtO sterilization, based on 40-100 volume percent for a maximum of 8 hours at 30 to 70°C.

Reference: Johnson, James A., supplier written correspondence - BASF Corporation, 1994.

GE Plastics:

After one day of mechanical aeration (aeration with fan) levels of ethylene oxide and ethylene glycol were less than 60 ppm. Levels of ethylene clorohydren were non-detectable (i.e., less than 4 ppm) at one, seven and fourteen days. Ethylene oxide and ethylene glycol levels were less than 40 ppm after fourteen days of mechanical aeration (aeration with fan). The ultimate elongation decreased slightly. However the yield elongation didn't change significantly. This is attributed to surface effects which are magnified after the material passes the yield point. Overall mechanical properties did not change significantly.

Reference: Paulino, J. C., Baccaro, L. E., *Effects Of Sterilization Procedures On Physical Properties of Engineering Thermoplastics,* supplier technical report - General Electric Company.

Steam Resistance

BASF: Ultradur (features: transparent)

PBT is not generally recommended for steam sterilization because of the possibility of hydrolytic degradation. However, if the temperature is below 60°C and the number of sterilization cycles is low, PBT parts may perform satisfactorily but practical tests under the appropriate conditions are required to verify this.

Johnson, James A., supplier written correspondence - BASF Corporation, 1994.

GE Plastics:

The heat deflection temperature of PBT is less than the autoclave cycle causing the material to deform.

Paulino, J. C., Baccaro, L. E., *Effects Of Sterilization Procedures On Physical Properties of Engineering Thermoplastics,* supplier technical report - General Electric Company.

GE Plastics: Valox HP (features: medical grade)

Valox HP resins are not recommended for steam autoclave sterilization.

Reference: *Sterilization Comparison,* supplier marketing literature (ACB/3/93) - General Electric Plastics, 1993.

TABLE 70: Effect of Gamma Radiation Sterilization on Polybutylene Terephthalate

Material Family	POLYBUTYLENE TEREPHTHALATE	
Material Supplier/Name	**GE Valox HP**	**GE Valox HP260**
Material Note	medical grade	medical grade
Reference No.	49	49

EXPOSURE CONDITIONS

Type	Gamma Radiation	Gamma Radiation
Radiation Dose (Mrads)	2.5	2.5

PROPERTIES RETAINED (%)

Tensile Strength	117 (hm)	117 (hm)
Elongation	103 (ak)	103 (ak)
Notched Izod Impact	110 (fm)	110 (fm)

TABLE 71: Effect of Ethylene Oxide Sterilization on Polybutylene Terephthalate

Material Family	POLYBUTYLENE TEREPHTHALATE						
Material Supplier/Name	GE	GE VALOX HP210			GE VALOX HP260		
Material Note		medical grade	medical grade	medical grade	medical grade	medical grade	medical grade
Reference No.	46	108	108	108	108	108	108

EXPOSURE CONDITIONS

Type	Ethylene Oxide	Ethylene Oxide	Ethylene Oxide	Ethylene Oxide	Ethylene Oxide	Ethylene Oxide	Ethylene Oxide
Details	100% EtO	100% EtO	100% EtO	100% EtO	100% EtO	100% EtO	100% EtO
Concentration	550 mg/l	550 mg/l	550 mg/l	550 mg/l	550 mg/l	550 mg/l	550 mg/l
Number of Cycles	5						
Note	test lab: Sterilization Technical Services, Inc.						
Temperature (°C)	54.4						
Time (hours)	5	5	5	5	5	5	5

PRE EXPOSURE CONDITIONING

Preconditioning Note	type: pre vacuum; pressure: 660-711 mm Hg; RH: 45-60%; dwell time: 15 minutes	type: pre vacuum; pressure: 660-711 mm Hg; RH: 45-60%; dwell time: 15 minutes	type: pre vacuum; pressure: 660-711 mm Hg; RH: 45-60%; dwell time: 15 minutes	type: pre vacuum; pressure: 660-711 mm Hg; RH: 45-60%; dwell time: 15 minutes	type: pre vacuum; pressure: 660-711 mm Hg; RH: 45-60%; dwell time: 15 minutes	type: pre vacuum; pressure: 660-711 mm Hg; RH: 45-60%; dwell time: 15 minutes	type: pre vacuum; pressure: 660-711 mm Hg; RH: 45-60%; dwell time: 15 minutes

POST EXPOSURE CONDITIONING

Note		type: aeration; note: mechanical with fan	type: aeration; note: mechanical with fan	type: aeration; note: mechanical with fan	type: aeration; note: mechanical with fan	type: aeration; note: mechanical with fan	type: aeration; note: mechanical with fan
Temperature (°C)		49	49	49	49	49	49

RESIDUALS (ppm)

Residuals Determined		ethylene oxide	ethylene chlorohydrin	ethylene glycol	ethylene oxide	ethylene chlorohydrin	ethylene glycol
24 hour Aeration		37	<1	<4	33	<1	<4
168 hour Aeration		8			12		
336 hour Aeration		5		<4	5		<4

PROPERTIES RETAINED (%)

Tensile Strength @ Yield	120 (ib)						
Elongation @ Yield	113 (bd)						
Modulus	101 (gs)						

TABLE 72: Effect of Ethylene Oxide Sterilization on Polybutylene Terephthalate

Material Family	POLYBUTYLENE TEREPHTHALATE					
Material Supplier/Name	GE VALOX HP210			GE VALOX HP260		
Material Note	medical grade	medical grade	medical grade	medical grade	medical grade	medical grade
Reference No.	108	108	108	108	108	108
EXPOSURE CONDITIONS						
Type	Ethylene Oxide	Ethylene Oxide	Ethylene Oxide	Ethylene Oxide	Ethylene Oxide	Ethylene Oxide
Details	100% EtO	100% EtO	100% EtO	100% EtO	100% EtO	100% EtO
Concentration	550 mg/l	550 mg/l	550 mg/l	550 mg/l	550 mg/l	550 mg/l
Time (hours)	5	5	5	5	5	5
PRE EXPOSURE CONDITIONING						
Preconditioning Note	type: pre vacuum; pressure: 660-711 mm Hg; RH: 45-60%; dwell time: 15 minutes	type: pre vacuum; pressure: 660-711 mm Hg; RH: 45-60%; dwell time: 15 minutes	type: pre vacuum; pressure: 660-711 mm Hg; RH: 45-60%; dwell time: 15 minutes	type: pre vacuum; pressure: 660-711 mm Hg; RH: 45-60%; dwell time: 15 minutes	type: pre vacuum; pressure: 660-711 mm Hg; RH: 45-60%; dwell time: 15 minutes	type: pre vacuum; pressure: 660-711 mm Hg; RH: 45-60%; dwell time: 15 minutes
POST EXPOSURE CONDITIONING						
Note	type: aeration; note: ambient conditions	type: aeration; note: ambient conditions	type: aeration; note: ambient conditions	type: aeration; note: ambient conditions	type: aeration; note: ambient conditions	type: aeration; note: ambient conditions
Temperature (°C)	49	49	49	49	49	49
RESIDUALS (ppm)						
Residuals Determined	ethylene oxide	ethylene chlorohydrin	ethylene glycol	ethylene oxide	ethylene chlorohydrin	ethylene glycol
24 hour Aeration	100	<1	35	111	<1	42
168 hour Aeration	53			73		
336 hour Aeration	48		28	62		25

Gamma Radiation Resistance

Eastman Performance Plastics: Ektar DN003

Gamma irradiation with typical doses of 2.5 and 5.0 Mrads does not affect physical properties of the material. Color change in Ektar DN003 is virtually undetectable.

Reference: *Ektar Polymers For Healthcare Products,* supplier marketing literature (P/MD-11) - Eastman Performance Plastics, 1991.

Ethylene Oxide (EtO) Resistance

Eastman Performance Plastics: Ektar DN003

Physical properties and color of Ektar polymers are virtually unaffected by ethylene oxide sterilization.

Reference: *Ektar Polymers For Healthcare Products,* supplier marketing literature (P/MD-11) - Eastman Performance Plastics, 1991.

Chemical Sterilants

Eastman Performance Plastics: Ektar DN003

IV solutions, betadine, bleach and other chemicals common to the healthcare environment have virtually no effect on Ektar DN003. At typical assembly strains, these materials are stress crack resistant to most solutions.

Reference: *Ektar Polymers For Healthcare Products,* supplier marketing literature (P/MD-11) - Eastman Performance Plastics, 1991.

Gamma Radiation Resistance

Lustro Plastics: Ultros PETG (form: film; features: transparent, medical grade)

Ultros PETG accepts radiation sterilization.

Reference: *Ultros PETG Copolyester For Film And Sheet For Medical Device Packaging,* supplier marketing literature - Lustro Plastics Co.

Eastman Performance Plastics: Ektar GN002

Gamma irradiation with typical doses of 2.5 and 5.0 Mrads does not affect physical properties of the material. Color change in Ektar GN002 is virtually undetectable.

Reference: *Ektar Polymers For Healthcare Products,* supplier marketing literature (P/MD-11) - Eastman Performance Plastics, 1991.

Eastman: Kodar PETG (features: transparent)

Kodar PETG exhibits good resistance to gamma radiation. Slight yellowing which tended to clear up in three weeks was observed. The benzene ring in these materials protect the structure from a breakdown.

Reference: Atakent, A. I., *Gamma Radiation Study,* supplier technical report - Cyro Industries, 1990.

Beta Radiation Resistance

Polyester PETG (features: transparent)

Beta sterilization is increasing in importance in Europe. Samples of PETG were tested at beta radiation levels of 2.7 to 10.8 Mrads. Physical properties were retained after 10.8 Mrads. In contrast to gamma sterilization, beta sterilization generates more heat. At 8.1 Mrads the surface of the PETG began to melt to the protective package. The material became yellow. After two weeks of storage there was some reduction in yellowness. Storage in light or dark did not have a meaningful impact.

Reference: Haines, Jim, Hauser, Debbie, *Zylar Clear Alloys For Medical Applications,* supplier technical report - Novacor Chemicals Inc.

Ethylene Oxide (EtO) Resistance

Lustro Plastics: Ultros PETG (form: film; features: transparent, medical grade)

Ultros PETG accepts ethylene oxide sterilization.

Reference: *Ultros PETG Copolyester For Film And Sheet For Medical Device Packaging,* supplier marketing literature - Lustro Plastics Co.

Eastman Performance Plastics: Ektar GN002

Physical properties and color of Ektar polymers are virtually unaffected by ethylene oxide sterilization.

Reference: *Ektar Polymers For Healthcare Products,* supplier marketing literature (P/MD-11) - Eastman Performance Plastics, 1991.

Chemical Sterilants

Eastman Performance Plastics: Ektar GN002

IV solutions, betadine, bleach and other chemicals common to the healthcare environment have virtually no effect on Ektar GN002. At typical assembly strains, these materials are stress crack resistant to most solutions.

Reference: *Ektar Polymers For Healthcare Products,* supplier marketing literature (P/MD-11) - Eastman Performance Plastics, 1991.

GRAPH 98: Beta Radiation Dose vs. Tensile Strength of Polycyclohexylenedimethylene Ethylene Terephthalate

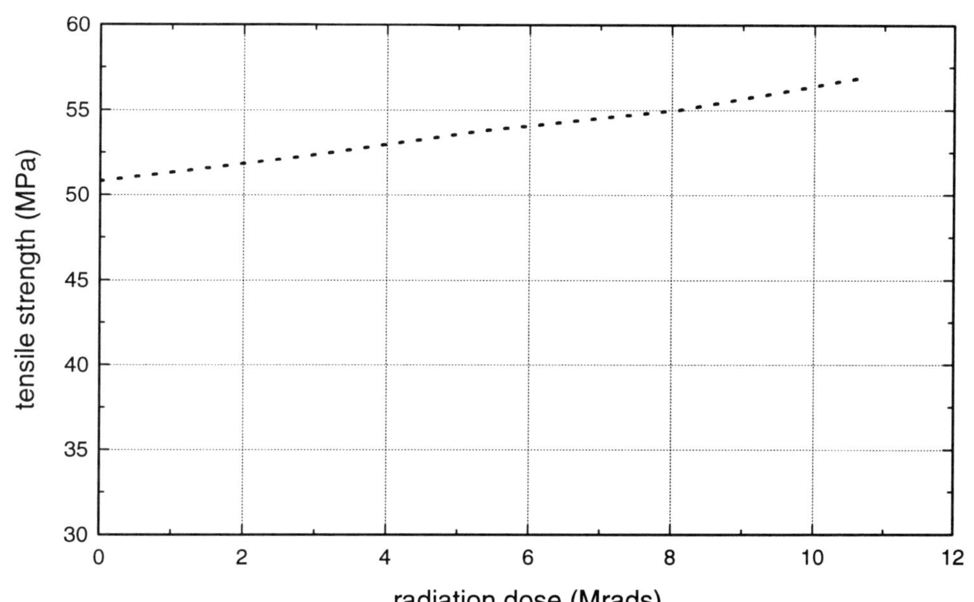

....	Polyester - PETG (transpar.)
Reference No.	105

GRAPH 99: Beta Radiation Dose vs. Tensile Modulus of Polycyclohexylenedimethylene Ethylene Terephthalate

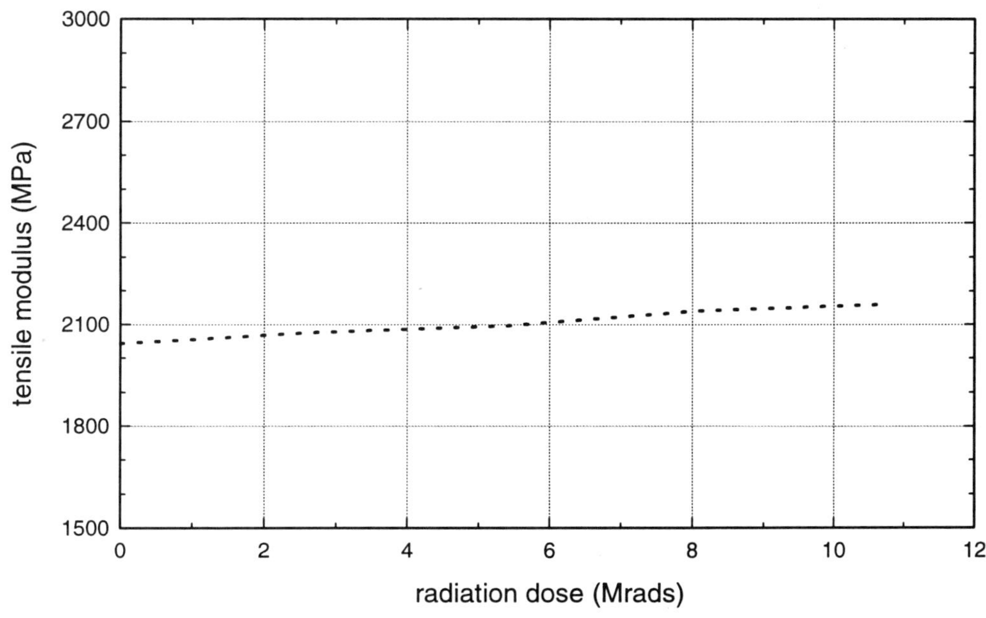

....	Polyester - PETG (transpar.)
Reference No.	105

GRAPH 100: Beta Radiation Dose vs. Notched Izod Impact Strength of Polycyclohexylenedimethylene Ethylene Terephthalate

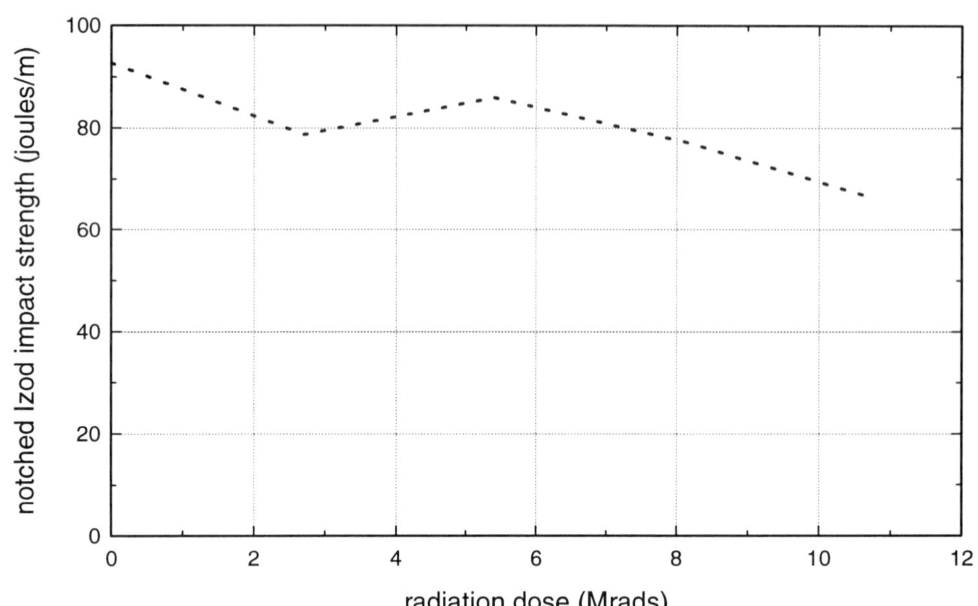

....	Polyester - PETG (transpar.)
Reference No.	105

GRAPH 101: Beta Radiation Dose vs. Yellowness Index of Polycyclohexylenedimethylene Ethylene Terephthalate

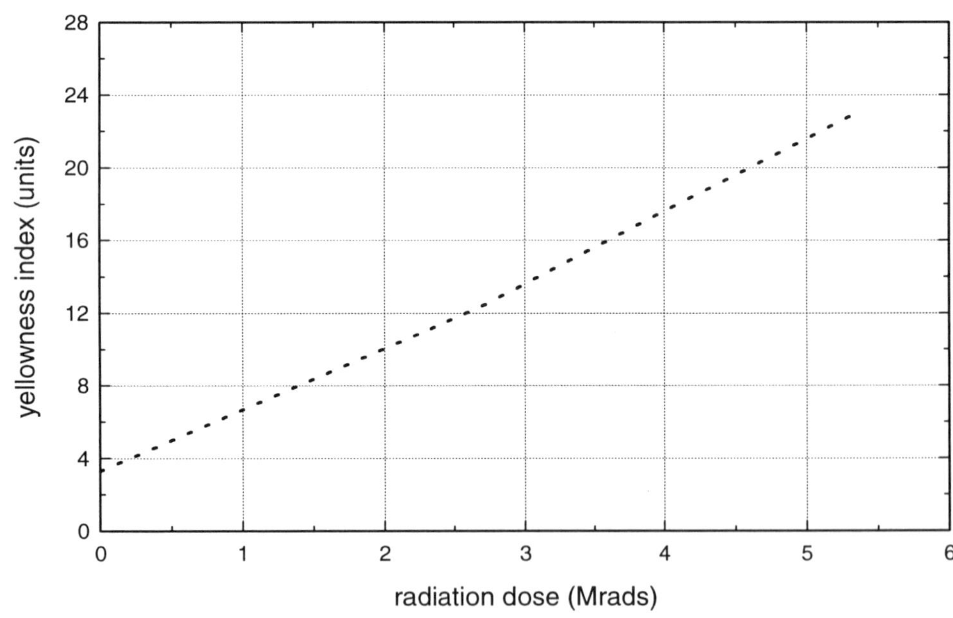

....	Polyester - PETG (transpar.)
Reference No.	105

<u>GRAPH 102:</u> Post Beta Radiation Exposure Time vs. Yellowness Index of Polycyclohexylenedimethylene Ethylene Terephthalate

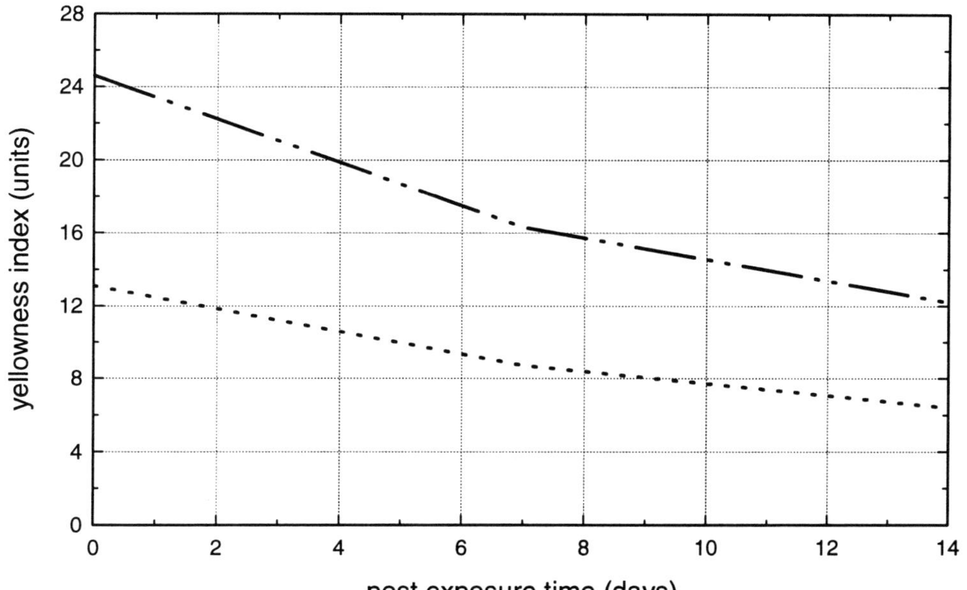

···············	Polyester - PETG (transpar.); 2.7 Mrads
—··—··	Polyester - PETG (transpar.); 5.4 Mrads
Reference No.	105

Gamma Radiation Resistance

Hoechst AG: Vectra A950

Vectra B950 (composition: copolyester amide)

Vectra has excellent resistance to gamma rays. Moldings made from Vectra basic grades A950 and B950 withstand doses of 5 x 103 kJ/kg (500 Mrads) without significant deterioration in mechanical properties.

Reference: *Vectra Polymer Materials,* supplier design guide (B 121 BR E 9102/014) - Hoechst, 1991.

Steam Resistance

Hoechst AG: Vectra A130 (composition: 30% glass fiber reinforced); **Vectra A625** (composition: 25% graphite filled); **Vectra A950**

Vectra has good resistance to hydrolysis. Prolonged exposure in hot water and steam at high temperatures(121°C, 2 bar, up to 1000 hours) leads to gradual hydrolytic degradation. The glass fiber reinforced grades exhibit a more severe decline in mechanical properties, as do many other glass fiber reinforced polymers, owing to capillary action at the glass fibre/polymer interfaces (wick effect). Vectra A625 has particularly good resistance to hydrolysis. Under selected test conditions, virtually no change in tensile strength or elastic modulus occurred after 1000 hours.

Reference: *Vectra Polymer Materials,* supplier design guide (B 121 BR E 9102/014) - Hoechst, 1991.

GRAPH 103: Steam Exposure Time vs. Tensile Strength Retained of Liquid Crystal Polymer

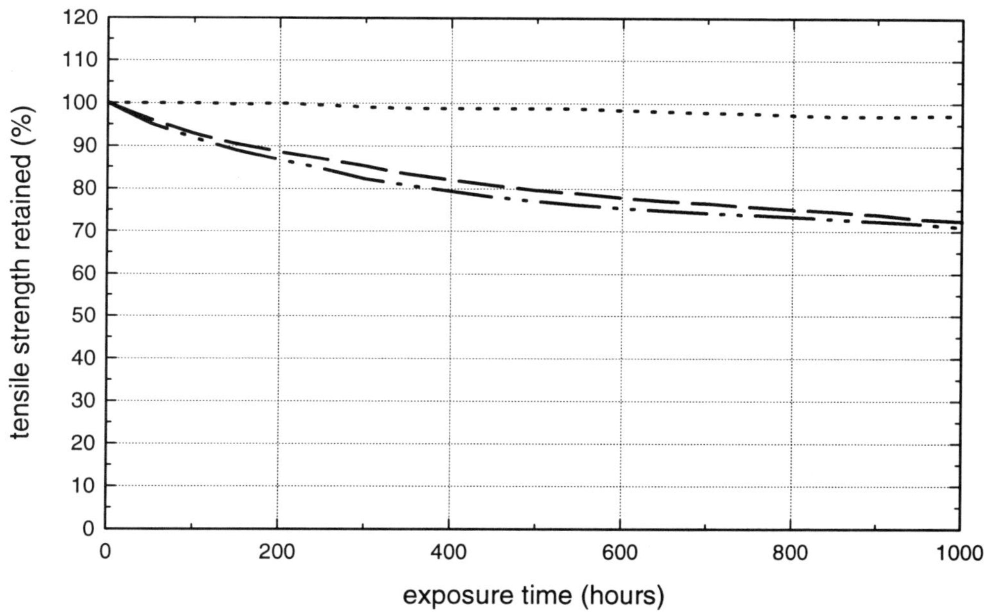

....	Hoechst AG Vectra A625 Liquid Crystal Polymer (25% graphite filled); exposure temperature: 121°C; pressure: 0.2 MPa
—·—··	Hoechst AG Vectra A950 Liquid Crystal Polymer; exposure temperature: 121°C; pressure: 0.2 MPa
– – – –	Hoechst AG Vectra A130 Liquid Crystal Polymer (30% glass fiber); exposure temperature: 121°C; pressure: 0.2 MPa
Reference No.	70

GRAPH 104: Steam Exposure Time vs. Tensile Modulus Retained of Liquid Crystal Polymer

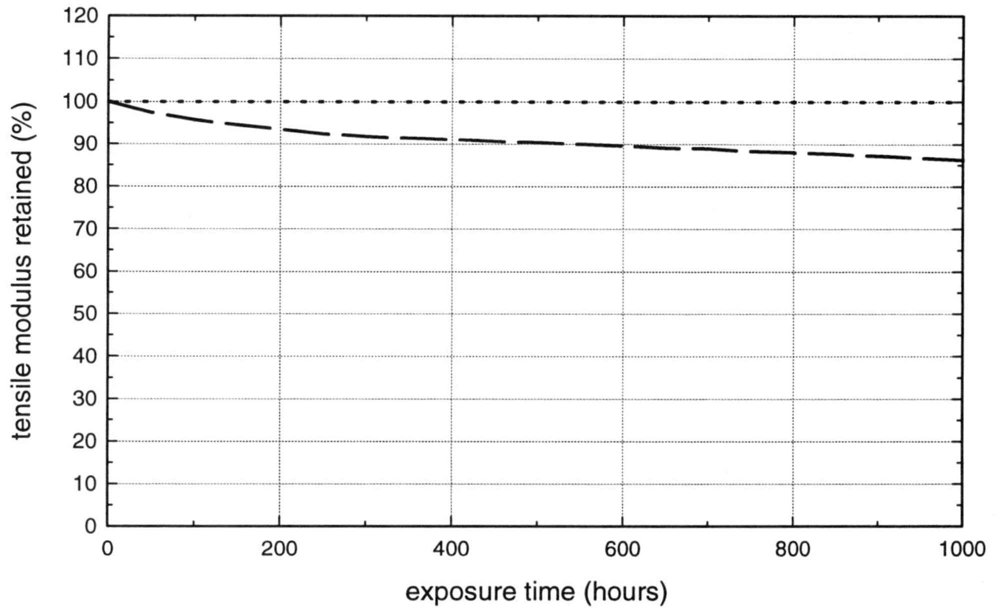

....	Hoechst AG Vectra A625 Liquid Crystal Polymer (25% graphite filled); exposure temperature: 121°C; pressure: 0.2 MPa
....	Hoechst AG Vectra A950 Liquid Crystal Polymer; exposure temperature: 121°C; pressure: 0.2 MPa
– – – –	Hoechst AG Vectra A130 Liquid Crystal Polymer (30% glass fiber); exposure temperature: 121°C; pressure: 0.2 MPa
Reference No.	70

Radiation Resistance

Rhone Poulenc: Kinel

Parts molded in Kinel are not deteriorated following exposure to a dose of 10,000 Mrads radiation.

Reference: *Kinel Polyimide Compounds, Properties, Applications,* supplier design guide - Rhone Poulenc.

DuPont: Vespel SP

Vespel parts have outstanding resistance to high energy, ionizing radiation. A test sample prepared from unfilled SP polyimide, irradiated in a 2MEV Van de Graff beam at an intensity of 10 watts/cm2 (absorbed dose of 4 x 108 Rads), showed no significant change in tensile strength, elongation or appearance. After 1500 hours in the Brookhaven Pile at 175°C, giving an absorbed dose of about 109 Rads, the SP polyimide sample was embrittled but still form stable.

Reference: *Exploring Vespel Territory - The Properties of Du Pont Vespel Parts,* supplier design guide (E-26800) - Du Pont Company.

Ube: Upilex (form: film)

Compared with other films, Upilex is highly resistant to irradiation. Upilex appears to have a life several times longer than conventional polyimide films when exposed to radiation, based on testing conducted by the Japan Atomic Energy Research Institute.

Reference: *Ube Ultra-High Heat-Resistant Polimide Film Upilex,* supplier marketing literature - Ube Industries, Ltd.

Gamma Radiation Resistance

DuPont: Vespel SP-1 (composition: unfilled); Vespel SP-21 (composition: 15 wt% graphite); Vespel SP-22 (composition: 40 wt% graphite)

At exposure levels up to and including 100 Mrads of gamma radiation, Vespel bars displayed less than 1.0% weight loss. This result is the mean of a group of treated samples compared to bars receiving no exposure.

Vespel tensile bars showed less than 6.5% loss in strength up to the maximum exposure level of 100 Mrads.

All three Vespel formulations showed small changes in percent elongation at break. The most significant loss (19.2% from controls) occurred at the highest level of gamma exposure.

Reference: *Vespel Parts And Radiation,* supplier technical report (E-73910) - Du Pont Company, 1989.

Electron Beam Radiation

DuPont: Vespel SP-1 (composition: unfilled); Vespel SP-21 (composition: 15 wt% graphite); Vespel SP-22 (composition: 40 wt% graphite)

At exposure levels up to and including 100 Mrads of electron beam radiation, Vespel bars displayed less than 2.0% weight loss. This result is the mean of a group of treated samples compared to bars receiving no exposure. Vespel tensile bars showed less than 4.5% loss in strength up to the maximum exposure level of 100 Mrads. All three Vespel formulations showed small changes in percent elongation at break. The most significant losses measured less than 15.0% from controls for the highest level of electron beam exposure.

Reference: *Vespel Parts And Radiation,* supplier technical report (E-73910) - Du Pont Company, 1989.

Steam Resistance

DuPont: Vespel SP

SP polyimide parts can be exposed to water up to 100°C, provided the stresses are low enough to take into account the reduced mechanical properties. At 100°C , the tensile strength and elongation of SP polyimide parts are reduced to 45% and 30% of the original values, respectively, in about 500 hours, at which point they level out. Some of the reduced tensile properties caused by moisture absorption may be restored by drying.

Reference: *Exploring Vespel Territory - The Properties of Du Pont Vespel Parts,* supplier design guide (E-26800) - Du Pont Company.

GRAPH 105: Gamma Radiation Dose vs. Tensile Strength Retained of Polyimide

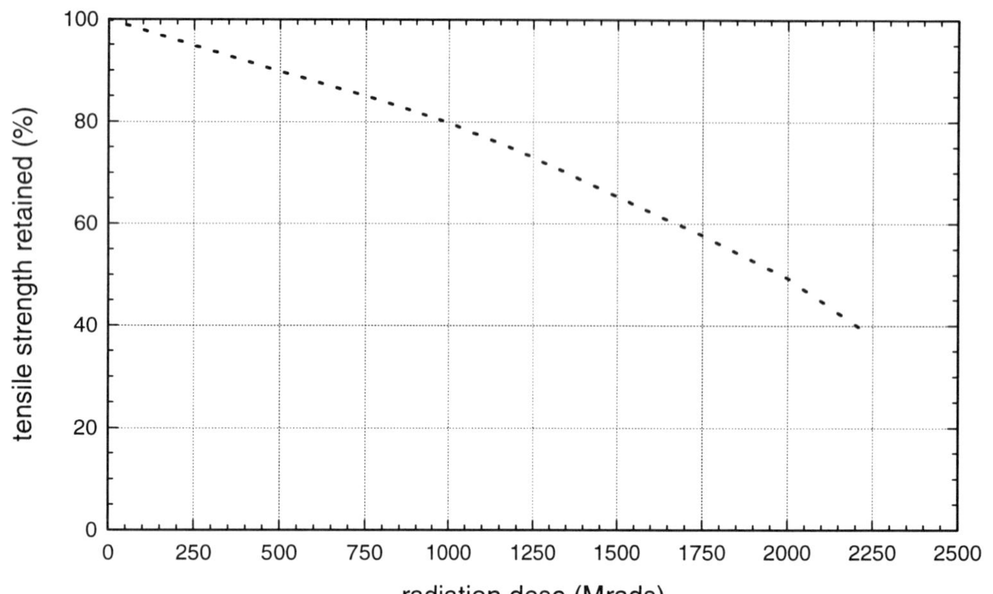

	Ube Upilex R Polyimide (film); Results of tests conducted at the Japan Atomic Energy Research Institute.
Reference No.	97

Gamma Radiation Resistance

Amoco Performance Products: Torlon 4203L

Gamma radiation has a negligible effect on Torlon poly(amide-imide) - only about 5% loss in tensile strength after exposure to 1000 Mrads.

Reference: *Torlon Engineering Polymers / Design Manual,* supplier design guide (F-49893) - Amoco Performance Products.

GRAPH 106: **Gamma Radiation Dose vs. Tensile Strength Retained of Polyamideimide**

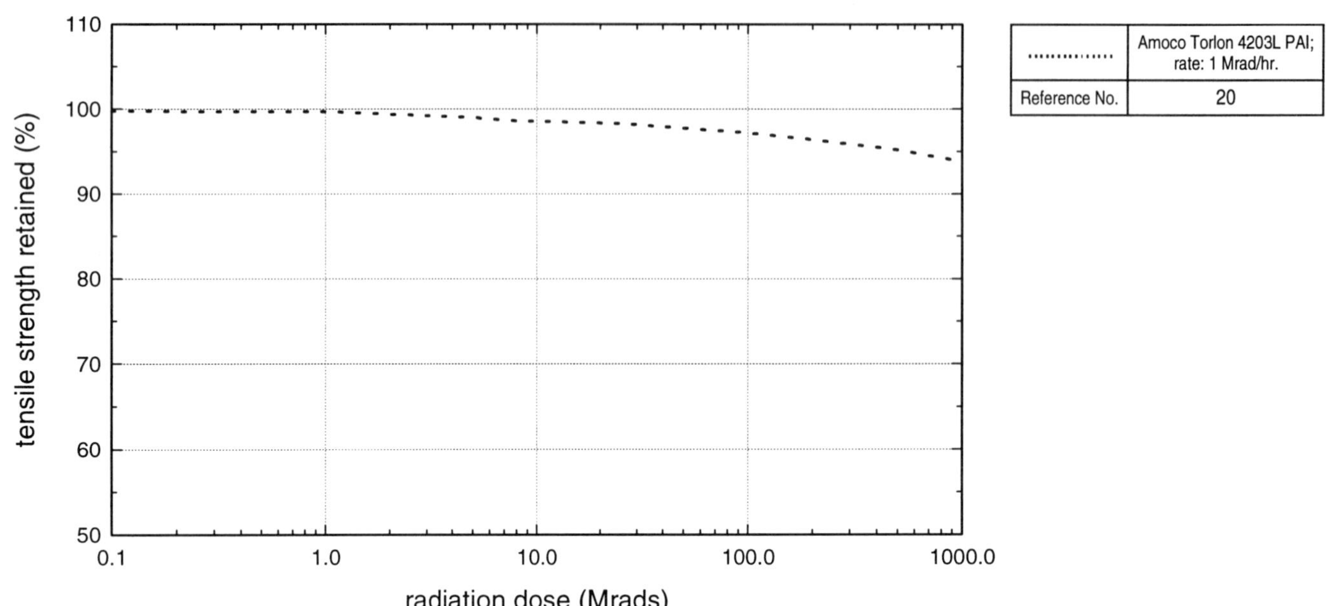

GRAPH 107: **Gamma Radiation Dose vs. Elongation Retained of Polyamideimide**

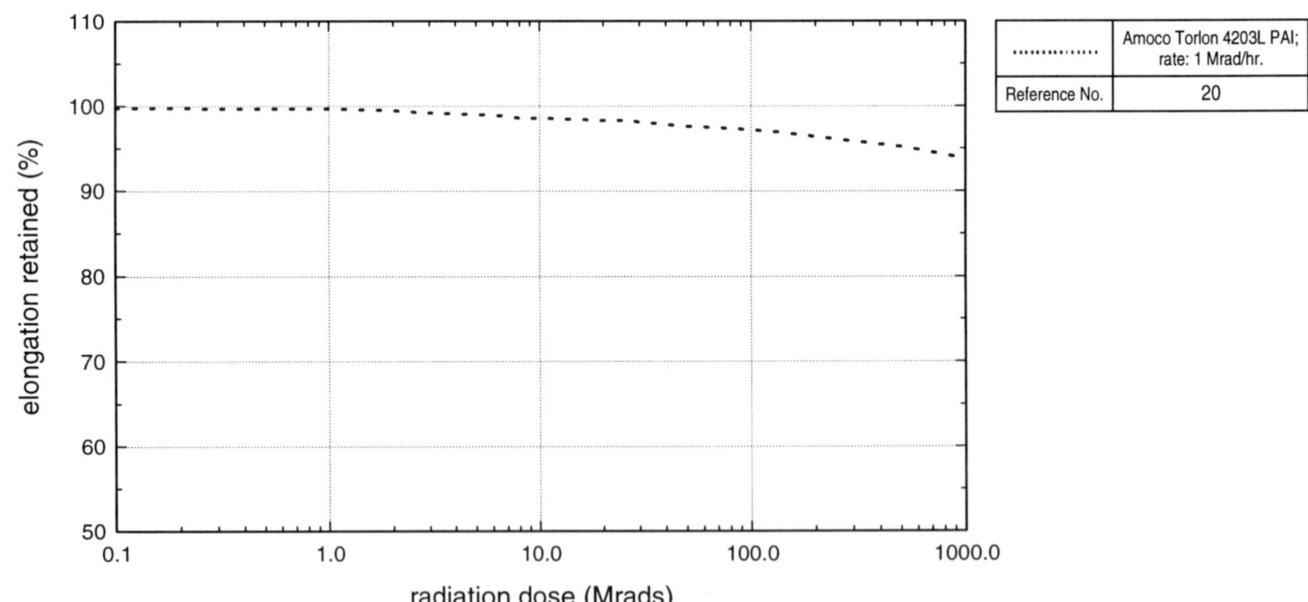

GRAPH 108: Gamma Radiation Dose vs. Flexural Modulus Retained of Polyamideimide

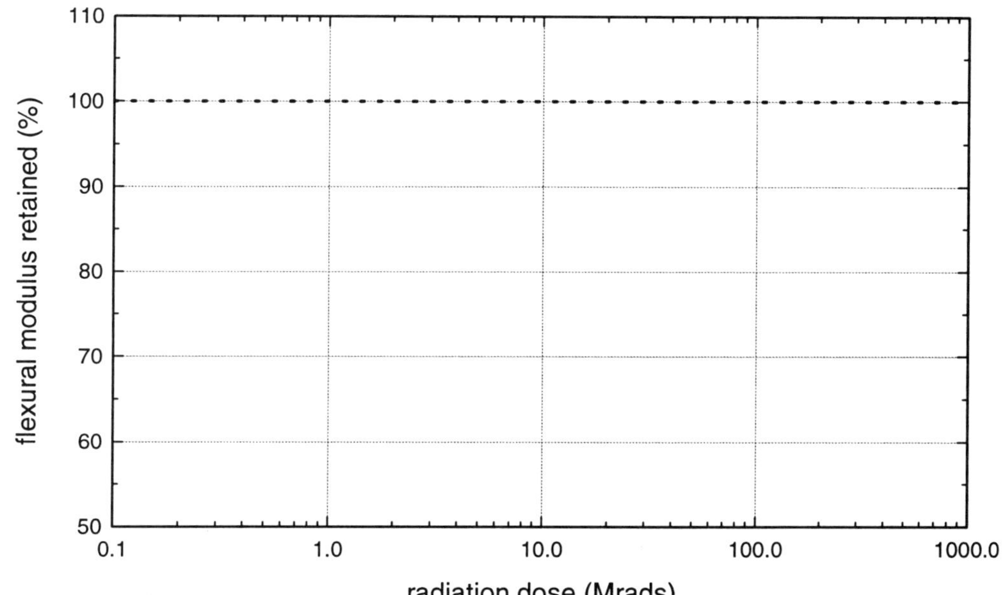

Gamma Radiation Resistance

GE Plastics: (features: transparent, amber tint)

Gamma radiation up to 10.0 Mrads has little effect on physical properties. The molecular weight of samples tested did not change significantly. However, samples were found to discolor.

Reference: Paulino, J. C., Baccaro, L. E., *Effects Of Sterilization Procedures On Physical Properties of Engineering Thermoplastics,* supplier technical report - General Electric Company.

GE Plastics: Ultem 1000 (features: transparent, natural resin, amber tint)

Parts molded of Ultem resin demonstrate resistance to gamma irradiation. A loss of less than 6% tensile strength was observed after cumulative exposure to 500 Mrads at the rate of one Mrad per hour using Cobalt 60.

Reference: Bonifant, Benjamin C., *Designing With Amorphous Thermoplastics,* Medical Device & Diagnostic Industry, trade journal - Cannon Communications, Inc., 1988.

GE Plastics: Ultem 1000 (features: transparent, natural resin, amber tint); **Ultem HP** (features: medical grade)

Ultem 1000 and Ultem HP resins retain their properties through 500 Mrads of gamma radiation.

Reference: *Sterilization Comparison,* supplier marketing literature (ACB/3/93) - General Electric Plastics, 1993.

Ethylene Oxide (EtO) Resistance

GE Plastics: Ultem 1000 (features: transparent, natural resin, amber tint); **Ultem HP** (features: medical grade)

Properties of Ultem 1000 and Ultem HP resins are unchanged to 100 ethylene oxide sterilization cycles.

Reference: *Sterilization Comparison,* supplier marketing literature (ACB/3/93) - General Electric Plastics, 1993.

GE Plastics: (features: transparent, amber tint)

After one day of mechanical aaeration (aeration with fan) levels of ethylene oxide and ethylene glycol were less than 60 ppm. Levels of ethylene clorohydrin were non-detectable (i.e., less than 4 ppm) at one, seven and fourteen days. Ethylene oxide and ethylene glycol levels were less than 40 ppm after fourteen days of mechanical aaeration (aeration with fan). The ultimate elongation decreased slightly. However the yield elongation didn't change significantly. This is attributed to surface effects which are magnified after the material passes the yield point. Overall mechanical properties did not change significantly.

Reference: Paulino, J. C., Baccaro, L. E., *Effects Of Sterilization Procedures On Physical Properties of Engineering Thermoplastics,* supplier technical report - General Electric Company.

GE Plastics: Ultem 1000 (features: transparent, natural resin, amber tint)

Ultem resin has been exposed to repeated cycles of ethylene oxide sterilization using a 3M model 400 Sterivac brand gas sterilizer. After 100 cycles, Ultem 1000 resin retains 100% of its initial tensile strength and more than 75% of its initial tensile elongation. The appearance also remains unaffected. Following a typical hospital sterilization and aeration cycle (2 hours exposure to 100% ethylene oxide gas at 60°C followed by 12 hours of belt aeration at 60°C) samples were analyzed for ethylene oxide retention. The residual amount of ethylene oxide was within the 25 ppm limit suggested by the FDA.

Reference: *Ultem Resin: Advanced Technology For Reusable Medical Devices,* supplier marketing literature (ULT-314A 1/91) RTB) - General Electric Company, 1991.

Steam Resistance

GE Plastics: Ultem 1000 (features: transparent, natural resin, amber tint); **Ultem 1100** (features: transparent, pigmented); **Ultem HP** (features: medical grade)

After 4000 autoclave cycles (Hi-Vac @ 132°C), the tensile strengths of Ultem 1000, Ultem 1100 and Ultem HP resins are maintained.

Sterilization Comparison, supplier marketing literature (ACB/3/93) - General Electric Plastics, 1993.

GE Plastics: (features: transparent, amber tint)

PEI bars and disks were tested for physical property retention after a series of cycles. A prevacuum cycle was used. The test results showed retention of tensile and yield strength, which were retained through 2000 cycles.

Paulino, J. C., Baccaro, L. E., *Effects Of Sterilization Procedures On Physical Properties of Engineering Thermoplastics,* supplier technical report - General Electric Company.

GE Plastics: Ultem 1000 (features: transparent, natural resin, amber tint); **Ultem 1100F** (features: transparent, pigmented)

PEI has excellent retention of tensile strength after 1500 vacuum autoclave cycles. PEI trays showed good resistance to crazing in this environment. In fact, after 500 cycles, no evidence of attack could be seen on the surface; small crazes were noted after 650 cycles. After 1500 cycles in 500 ppm morpholine steam autoclaving, the PEI trays remained serviceable with only a few crazes in very localized areas. Trays molded using highly pigmented PEI, on the other hand, showed no signs of crazing after this number of cycles. PEI absorbs moisture slowly at elevated temperatures. Therefore, PEI develops a less dramatic gradient of moisture content through the thickness of the part during the steam cycle of sterilization than does Polysulfone and polyethersulfone. PEI's greater resistance to chemically accelerated fatigue failure occurs because of lower levels of repeated stress.

Testing was conducted using a commercial vacuum autoclave programmed to provide a 9 minute prevacuum conditioning cycle, a 5 minute sterilization time, and 4 minutes of drying time. Tests were also conducted with and without 50 ppm of morpholine introduced into the feed water system of the vacuum sterilization system. Morpholine is one of the most prevalent boiler feed water additives. Thermoformed trays (representative of medical sterilization trays) were tested and loaded with aluminum bar stock to model the weight of medical instruments.

Reference: Bonifant, Benjamin C., *Designing With Amorphous Thermoplastics,* Medical Device & Diagnostic Industry, trade journal - Cannon Communications, Inc., 1988.

Dry Heat Sterilization

GE Plastics: Ultem HP (features: transparent, medical grade, amber tint)

Ultem HP resins can withstand 160°C of dry heat sterilization.

Reference: *GE Plastics Medical Applications,* supplier marketing literature (NBM-100 (1/89) RTB) - General Electric Company, 1989.

Chemical Sterilants

GE Plastics: Ultem 1000 (features: transparent, natural resin, amber tint)

The effects of several common disinfecting solutions and chemicals on Ultem polyetherimide resin have been tested. Samples were immersed in the solutions and examined for cracks and crazes. Ultem resin exhibits excellent resistance to a wide range of disinfectants and chemicals. Glutaraldehyde, a common ingredient of cold sterilants, is very soluble in water, and regenerates the dialdehyde on vacuum distillation so that it maintains/increases its presence through autoclave cycles. Ultem 1000 resin was exposed at various stress levels at both room temperature and 150°C. The resin performed well at room temperature, even after prolonged exposure (simulating accidental storage in solution). However, exposure at elevated temperature showed that Ultem was attacked. Despite being attacked, it provided much better resistance than polysulfone.

Reference: *Ultem Resin: Advanced Technology For Reusable Medical Devices,* supplier marketing literature (ULT-314A 1/91) RTB) - General Electric Company, 1991.

TABLE 73: Effect of Ethylene Oxide Sterilization on Polyetherimide

Material Family	**POLYETHERIMIDE**
Material Supplier/Name	**GE**
Material Note	transparent, amber tint
Reference No.	46

EXPOSURE CONDITIONS

Type	Ethylene Oxide
Details	100% EtO
Concentration	550 mg/l
Number of Cycles	5
Note	test lab: Sterilization Technical Services, Inc.
Temperature (°C)	54.4
Time (hours)	5

PRE EXPOSURE CONDITIONING

Preconditioning Note	type: pre vacuum; pressure: 660-711 mm Hg; RH: 45-60%; dwell time: 15 minutes

PROPERTIES RETAINED (%)

Tensile Strength @ Yield	107 (ib)
Elongation @ Yield	109 (bd)
Modulus	95 (gs)

TABLE 74: Effect of Ethylene Oxide Sterilization on Polyetherimide

Material Family	POLYETHERIMIDE							
Material Supplier/Name	GE Ultem 1100	GE Ultem 1000	GE Ultem 5011	GE Ultem 5001	GE Ultem 1100	GE Ultem 1000	GE Ultem 5011	GE Ultem 5001
Material Note	transparent, pigmented	transparent, natural resin, amber tint	chemical resistance	chemical resistance	transparent, pigmented	transparent, natural resin, amber tint	chemical resistance	chemical resistance
Reference No.	107	107	107	107	107	107	107	107

EXPOSURE CONDITIONS

Type	Ethylene Oxide	Ethylene Oxide	Ethylene Oxide	Ethylene Oxide	Ethylene Oxide	Ethylene Oxide	Ethylene Oxide	Ethylene Oxide
Details	100% EtO	100% EtO	100% EtO	100% EtO	100% EtO	100% EtO	100% EtO	100% EtO
Concentration	800 mg/l	800 mg/l	800 mg/l	800 mg/l	800 mg/l	800 mg/l	800 mg/l	800 mg/l
Temperature (°C)	51.7	51.7	51.7	51.7	51.7	51.7	51.7	51.7
Time (hours)	8	8	8	8	8	8	8	8

PRE EXPOSURE CONDITIONING

Preconditioning Note	type: pre vacuum; pressure: 660-711 mm Hg; RH: 45-60%; dwell time: 15 minutes	type: pre vacuum; pressure: 660-711 mm Hg; RH: 45-60%; dwell time: 15 minutes	type: pre vacuum; pressure: 660-711 mm Hg; RH: 45-60%; dwell time: 15 minutes	type: pre vacuum; pressure: 660-711 mm Hg; RH: 45-60%; dwell time: 15 minutes	type: pre vacuum; pressure: 660-711 mm Hg; RH: 45-60%; dwell time: 15 minutes	type: pre vacuum; pressure: 660-711 mm Hg; RH: 45-60%; dwell time: 15 minutes	type: pre vacuum; pressure: 660-711 mm Hg; RH: 45-60%; dwell time: 15 minutes	type: pre vacuum; pressure: 660-711 mm Hg; RH: 45-60%; dwell time: 15 minutes

POST EXPOSURE CONDITIONING

Note	type: aeration; note: ambient conditions	type: aeration; note: ambient conditions	type: aeration; note: ambient conditions	type: aeration; note: ambient conditions	type: aeration; note: ambient conditions	type: aeration; note: ambient conditions	type: aeration; note: ambient conditions	type: aeration; note: ambient conditions
Temperature (°C)	23	23	23	23	23	23	23	23

RESIDUALS (ppm)

Residuals Determined	ethylene chlorohydrin	ethylene chlorohydrin	ethylene chlorohydrin	ethylene chlorohydrin	ethylene glycol	ethylene glycol	ethylene glycol	ethylene glycol
336 hour Aeration	<3	<3	<3	<3	<21	<19	<19	<16

TABLE 75: Effect of Ethylene Oxide Sterilization on Polyetherimide

Material Family	POLYETHERIMIDE							
Material Supplier/Name	GE Ultem 1100	GE Ultem 1000	GE Ultem 5011	GE Ultem 5001	GE Ultem 1100	GE Ultem 1000	GE Ultem 5011	GE Ultem 5001
Material Note	transparent, pigmented	transparent, natural resin, amber tint	chemical resistance	chemical resistance	transparent, pigmented	transparent, natural resin, amber tint	chemical resistance	chemical resistance
Reference No.	107	107	107	107	107	107	107	107
EXPOSURE CONDITIONS								
Type	Ethylene Oxide	Ethylene Oxide	Ethylene Oxide	Ethylene Oxide	Ethylene Oxide	Ethylene Oxide	Ethylene Oxide	Ethylene Oxide
Details	100% EtO	100% EtO	100% EtO	100% EtO	100% EtO	100% EtO	100% EtO	100% EtO
Concentration	800 mg/l	800 mg/l	800 mg/l	800 mg/l	800 mg/l	800 mg/l	800 mg/l	800 mg/l
Temperature (°C)	51.7	51.7	51.7	51.7	51.7	51.7	51.7	51.7
Time (hours)	8	8	8	8	8	8	8	8
PRE EXPOSURE CONDITIONING								
Preconditioning Note	type: pre vacuum; pressure: 660-711 mm Hg; RH: 45-60%; dwell time: 15 minutes	type: pre vacuum; pressure: 660-711 mm Hg; RH: 45-60%; dwell time: 15 minutes	type: pre vacuum; pressure: 660-711 mm Hg; RH: 45-60%; dwell time: 15 minutes	type: pre vacuum; pressure: 660-711 mm Hg; RH: 45-60%; dwell time: 15 minutes	type: pre vacuum; pressure: 660-711 mm Hg; RH: 45-60%; dwell time: 15 minutes	type: pre vacuum; pressure: 660-711 mm Hg; RH: 45-60%; dwell time: 15 minutes	type: pre vacuum; pressure: 660-711 mm Hg; RH: 45-60%; dwell time: 15 minutes	type: pre vacuum; pressure: 660-711 mm Hg; RH: 45-60%; dwell time: 15 minutes
POST EXPOSURE CONDITIONING								
Note	type: aeration; note: mechanical with fan	type: aeration; note: mechanical with fan	type: aeration; note: mechanical with fan	type: aeration; note: mechanical with fan	type: aeration; note: mechanical with fan	type: aeration; note: mechanical with fan	type: aeration; note: mechanical with fan	type: aeration; note: mechanical with fan
Temperature (°C)	49	49	49	49	49	49	49	49
RESIDUALS (ppm)								
Residuals Determined	ethylene chlorohydrin	ethylene chlorohydrin	ethylene chlorohydrin	ethylene chlorohydrin	ethylene glycol	ethylene glycol	ethylene glycol	ethylene glycol
336 hour Aeration	<3	<3	<2	<3	<18	<22	<15	<20

TABLE 76: Effect of Ethylene Oxide Sterilization on Polyetherimide

Material Family	POLYETHERIMIDE					
Material Supplier/Name	GE ULTEM 1000					
Material Note	transparent, natural resin, amber tint	transparent, natural resin, amber tint	transparent, natural resin, amber tint	transparent, natural resin, amber tint	transparent, natural resin, amber tint	transparent, natural resin, amber tint
Reference No.	108	108	108	108	108	108
EXPOSURE CONDITIONS						
Type	Ethylene Oxide	Ethylene Oxide	Ethylene Oxide	Ethylene Oxide	Ethylene Oxide	Ethylene Oxide
Details	100% EtO	100% EtO	100% EtO	100% EtO	100% EtO	100% EtO
Concentration	550 mg/l	550 mg/l	550 mg/l	550 mg/l	550 mg/l	550 mg/l
Time (hours)	5	5	5	5	5	5
PRE EXPOSURE CONDITIONING						
Preconditioning Note	type: pre vacuum; pressure: 660-711 mm Hg; RH: 45-60%; dwell time: 15 minutes	type: pre vacuum; pressure: 660-711 mm Hg; RH: 45-60%; dwell time: 15 minutes	type: pre vacuum; pressure: 660-711 mm Hg; RH: 45-60%; dwell time: 15 minutes	type: pre vacuum; pressure: 660-711 mm Hg; RH: 45-60%; dwell time: 15 minutes	type: pre vacuum; pressure: 660-711 mm Hg; RH: 45-60%; dwell time: 15 minutes	type: pre vacuum; pressure: 660-711 mm Hg; RH: 45-60%; dwell time: 15 minutes
POST EXPOSURE CONDITIONING						
Note	type: aeration; note: mechanical with fan	type: aeration; note: mechanical with fan	type: aeration; note: mechanical with fan	type: aeration; note: ambient conditions	type: aeration; note: ambient conditions	type: aeration; note: ambient conditions
Temperature (°C)	49	49	49	49	49	49
RESIDUALS (ppm)						
Residuals Determined	ethylene oxide	ethylene chlorohydrin	ethylene glycol	ethylene oxide	ethylene chlorohydrin	ethylene glycol
24 hour Aeration	58	<1	34	111	<1	41
168 hour Aeration	34			59		
336 hour Aeration	22		19	48		22

TABLE 77: Effect of Steam Sterilization on Polyetherimide

Material Family	POLYETHERIMIDE							
Material Supplier/Name	GE ULTEM 1000							
Material Note	transparent, natural resin, amber tint	transparent, natural resin, amber tint	transparent, natural resin, amber tint	transparent, natural resin, amber tint	transparent, natural resin, amber tint	transparent, natural resin, amber tint	transparent, natural resin, amber tint	transparent, natural resin, amber tint
Reference No.	50	50	50	50	50	50	50	50

EXPOSURE CONDITIONS

Type	Steam Autoclave	Steam Autoclave	Steam Autoclave	Steam Autoclave	Steam Autoclave	Steam Autoclave	Steam Autoclave	Steam Autoclave
Details	no boiler additive	no boiler additive	no boiler additive	no boiler additive	50 ppm morpholine in feed water	50 ppm morpholine in feed water	50 ppm morpholine in feed water	50 ppm morpholine in feed water
Number of Cycles	100	500	1000	1500	100	500	1000	1500
Note	test apparatus: Amsco Eagle Series 2023	test apparatus: Amsco Eagle Series 2023	test apparatus: Amsco Eagle Series 2023	test apparatus: Amsco Eagle Series 2023	test apparatus: Amsco Eagle Series 2023	test apparatus: Amsco Eagle Series 2023	test apparatus: Amsco Eagle Series 2023	test apparatus: Amsco Eagle Series 2023
Temperature (°C)	132	132	132	132	132	132	132	132
Time (hours)	0.083	0.083	0.083	0.083	0.083	0.083	0.083	0.083

PRE EXPOSURE CONDITIONING

Preconditioning Note	dry time: 4 min.	dry time: 4 min.	dry time: 4 min.	dry time: 4 min.	dry time: 4 min.	dry time: 4 min.	dry time: 4 min.	dry time: 4 min.

POST EXPOSURE CONDITIONING

Note	type: prevacuum conditioning; time: 9 min.	type: prevacuum conditioning; time: 9 min.	type: prevacuum conditioning; time: 9 min.	type: prevacuum conditioning; time: 9 min.	type: prevacuum conditioning; time: 9 min.	type: prevacuum conditioning; time: 9 min.	type: prevacuum conditioning; time: 9 min.	type: prevacuum conditioning; time: 9 min.

PROPERTIES RETAINED (%)

Elongation	53 (ak)	50 (ak)	41.7 (ak)	28.3 (ak)	21.7 (ak)	28.3 (ak)	16.7 (ak)	25 (ak)

TABLE 78: Effect of Steam Sterilization on Polyetherimide

Material Family	POLYETHERIMIDE							
Material Supplier/Name	GE ULTEM 1100F							
Material Note	transparent, pigmented	transparent, pigmented	transparent, pigmented	transparent, pigmented	transparent, pigmented	transparent, pigmented	transparent, pigmented	transparent, pigmented
Reference No.	50	50	50	50	50	50	50	50

EXPOSURE CONDITIONS

Type	Steam Autoclave	Steam Autoclave	Steam Autoclave	Steam Autoclave	Steam Autoclave	Steam Autoclave	Steam Autoclave	Steam Autoclave
Details	no boiler additive	no boiler additive	no boiler additive	no boiler additive	50 ppm morpholine in feed water	50 ppm morpholine in feed water	50 ppm morpholine in feed water	50 ppm morpholine in feed water
Number of Cycles	100	500	1000	1500	100	500	1000	1500
Note	test apparatus: Amsco Eagle Series 2023	test apparatus: Amsco Eagle Series 2023	test apparatus: Amsco Eagle Series 2023	test apparatus: Amsco Eagle Series 2023	test apparatus: Amsco Eagle Series 2023	test apparatus: Amsco Eagle Series 2023	test apparatus: Amsco Eagle Series 2023	test apparatus: Amsco Eagle Series 2023
Temperature (°C)	132	132	132	132	132	132	132	132
Time (hours)	0.083	0.083	0.083	0.083	0.083	0.083	0.083	0.083

PRE EXPOSURE CONDITIONING

Preconditioning Note	dry time: 4 min.	dry time: 4 min.	dry time: 4 min.	dry time: 4 min.	dry time: 4 min.	dry time: 4 min.	dry time: 4 min.	dry time: 4 min.

POST EXPOSURE CONDITIONING

Note	type: prevacuum conditioning; time: 9 min.	type: prevacuum conditioning; time: 9 min.	type: prevacuum conditioning; time: 9 min.	type: prevacuum conditioning; time: 9 min.	type: prevacuum conditioning; time: 9 min.	type: prevacuum conditioning; time: 9 min.	type: prevacuum conditioning; time: 9 min.	type: prevacuum conditioning; time: 9 min.

PROPERTIES RETAINED (%)

Elongation	27.5 (ak)	25 (ak)	25 (ak)	17.5 (ak)	20 (ak)	20 (ak)	25 (ak)	22.5 (ak)

GRAPH 109: Gamma Radiation Dose vs. Tensile Strength of Polyetherimide

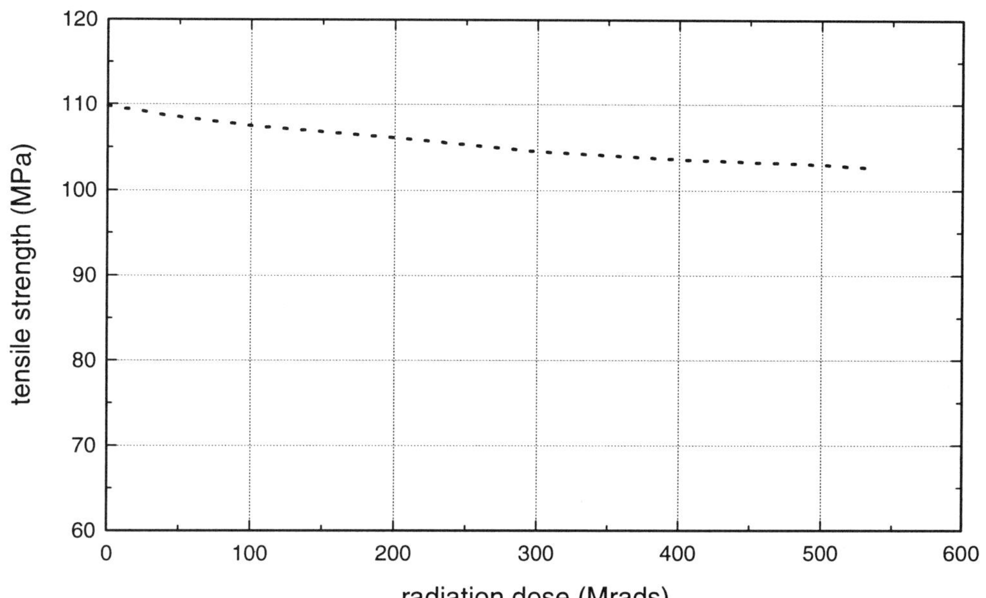

....	GE Ultem 1000 PEI; rate: 1 Mrad/hr.
Reference No.	51

GRAPH 110: Gamma Radiation Dose vs. Elongation at Break of Polyetherimide

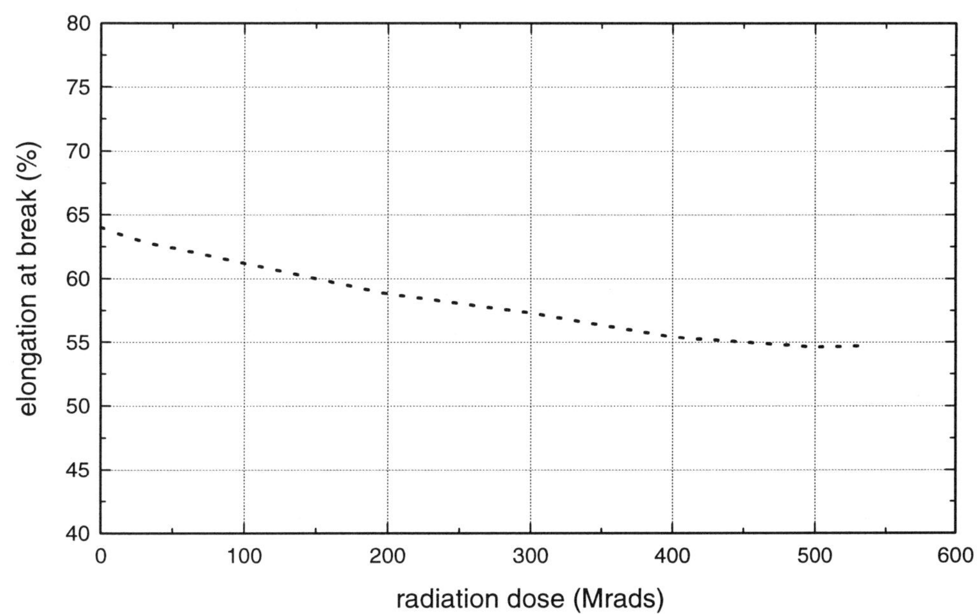

....	GE Ultem 1000 PEI; rate: 1 Mrad/hr.
Reference No.	51

GRAPH 111: Gamma Radiation Dose vs. Elonation at Yield of Polyetherimide

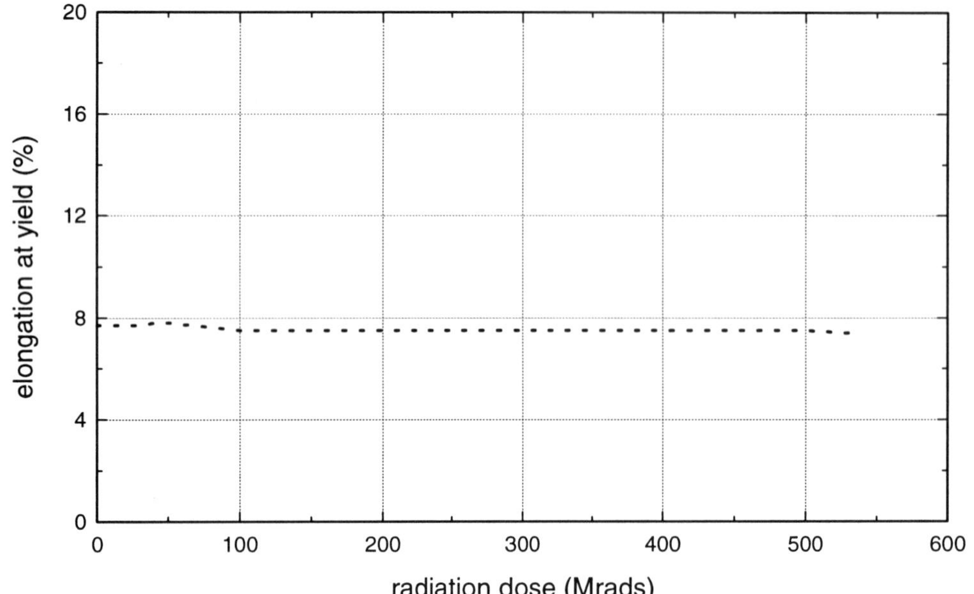

....	GE Ultem 1000 PEI; rate: 1 Mrad/hr.
Reference No.	51

GRAPH 112: Number of Steam Sterilization Cycles vs. Tensile Strength of Polyetherimide

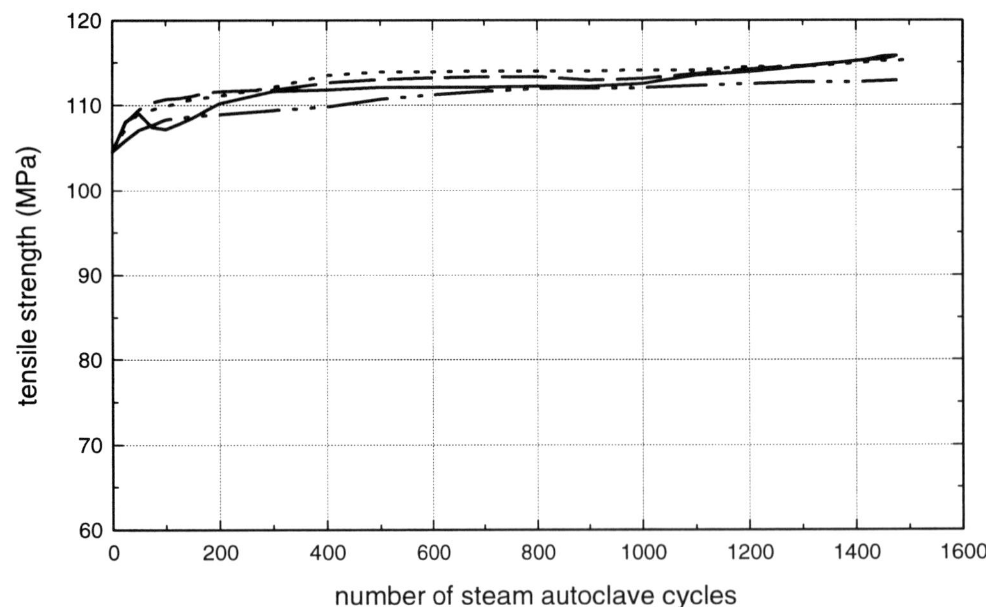

....	GE Ultem 1000 PEI; no boiler additive
—·—··	GE Ultem 1100F PEI; no boiler additive
— — —	GE Ultem 1000 PEI; 50 ppm morpholine
———	GE Ultem 1100F PEI; 50 ppm morpholine
Reference No.	52

GRAPH 113: Number of Steam Sterilization Cycles vs. Tensile Impact Strength of Polyetherimide

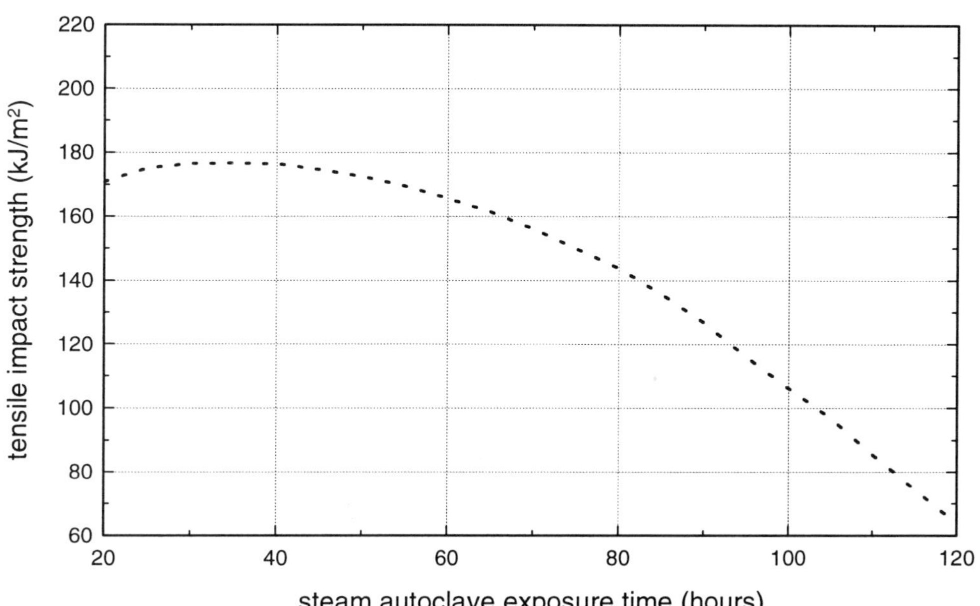

....	PEI; 30 min. @ 132°C per cycle; test method: ASTM D1822
Reference No.	18

GRAPH 114: Number of Steam Sterilization Cycles vs. Drop Dart Impact Strength of Polyetherimide

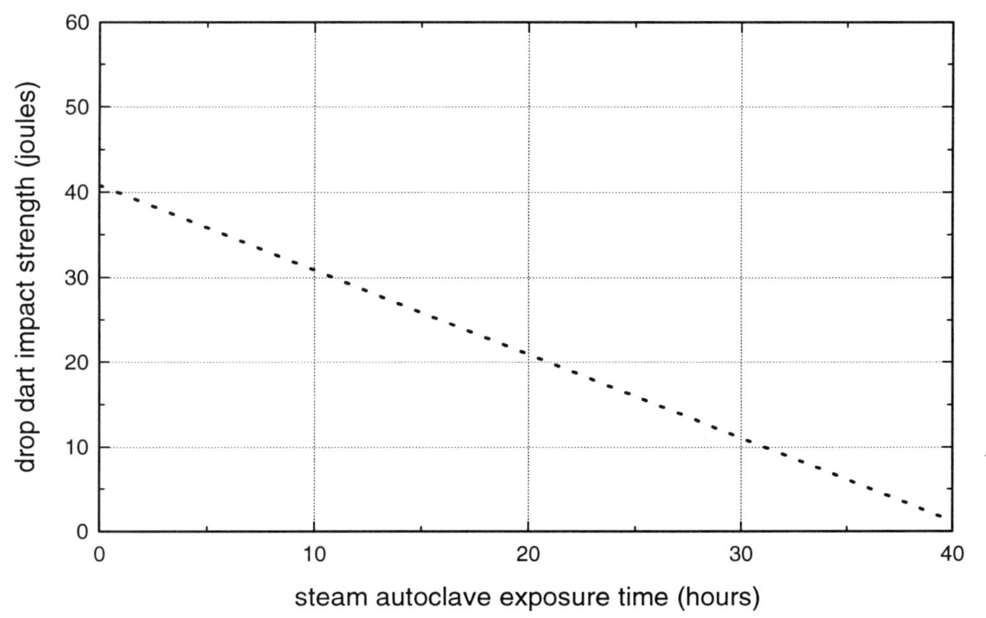

....	PEI; 30 min. @ 132°C per cycle
Reference No.	17

Gamma Radiation Resistance

Victrex USA: Victrex PEK

PEK has excellent resistance to gamma radiation.

Reference: *Victrex PEK Properties And Processing,* supplier design guide (VP2/ October 1987) - ICI Advanced Materials, 1987.

Gamma Radiation Resistance

Victrex USA: Victrex PEEK

Victrex PEEK shows excellent resistance to hard (gamma) irradiation, absorbing over 1000 Mrads of irradiation without suffering significant damage and showing no embrittlement of the polymer. It is believed that PEEK will resist dose levels of well over 10,000 Mrads of particle (alpha or beta) irradiation without significant degradation of properties. Fibre reinforced grades are expected to show even better performance than this. PEEK is significantly more radiation stable than polystyrene, otherwise the most radiation resistant thermoplastic.

Reference: *Victrex PEEK,* supplier design guide (VK2/0586) - ICI Advanced Materials, 1986.

Steam Resistance

Victrex USA: Victrex PEEK

High temperature steam, coupled with chemical sterilization, can be applied with no ill effects.

Reference: *Victrex Polymers For Medical Applications,* supplier technical report - ICI Advanced Materials.

Dry Heat Sterilization

Victrex USA: Victrex PEEK

Hot air sterilization up to 204°C for long periods of time has no dimensional or mechanical effect on the material, thus making it an extremely reliable product when designing for high virulent and especially dangerous environments.

Reference: *Victrex Polymers For Medical Applications,* supplier technical report - ICI Advanced Materials.

TABLE 79: Effect of Steam Sterilization on Polyetheretherketone

Material Family	POLYETHERETHERKETONE				
Material Supplier/Name	VICTREX PEEK 450GL30				
Material Note	30% glass fiber reinforced	30% glass fiber reinforced	30% glass fiber reinforced	30% glass fiber reinforced	30% glass fiber reinforced
Reference No.	79	79	79	79	79

EXPOSURE CONDITIONS

Type	Steam	Steam	Steam	Steam	Steam
Time (hours)	75	350	1000	2000	2500

PROPERTIES RETAINED (%)

Tensile Strength	73.1 (he)	69.4 (he)	67.2 (he)	68.7 (he)	66.4 (he)
Elongation	93.3 (ak)	(ak)	86.7 (ak)	106.7 (ak)	86.7 (ak)

Radiation Resistance

BASF AG: Ultrapek

Ultrapek offers extremely high resistance to gamma radiation over the entire range of temperatures encountered in practice. If an article is irradiated in a heated environment, its resistance towards the combined effect of radiation and oxidation is more significant than its resistance to radiation alone. The tensile strength at break of Ultrapek is unaffected by a radiation dose of 1000 Mrads at 200°C. Ultrapek is more resistant to radiation than polystyrene.

Reference: *Ultrapek Product Line, Properties, Processing,* supplier design guide (B 607 e/10.92) - BASF Aktiengesellschaft, 1992.

Gamma Radiation Resistance

Exxon:

The literature is extensive and is heavily weighted with descriptions of polyethylene as a material which undergoes crosslinking in preference to chain scission upon irradiation. It is generally understood that by whatever mechanism polyethylene reacts following exposure to high energy radiation, it is sufficiently tolerant to moderate doses to allow its use in gamma radiation sterilized medical devices. Specifically designed formulations involving additive packages or other design factors providing tolerance to irradiation are not believed to be required for these applications.

Although there is general agreement that the predominant reaction pathway for polyethylene after irradiation is by crosslinking, there are references to polyethylene degradation by chain scission. Irradiated polyethylene that is characterized as crosslinked is based on the molecular behavior of irradiated polyethylene stored in vacuum. A more realistic scenario is storage in contact with air. While crosslinking reactions may predominate in the absence of air, the possible tendency of irradiated polyethylene exposed to air to oxidatively degrade in a fashion similar to polypropylene has received less attention.

The difference between irradiated polyethylene and polypropylene in respect of air degradation might be very dependent on the resin density, manufacturing process, and irradiation and storage conditions. Some polyethylene might be much closer to polypropylene in this behavior. A further complicating factor is that polypropylene is sometimes described as undergoing crosslinking upon irradiation. This would contradict practical knowledge, if the polypropylene had been stored in air.

The literature pays little attention to the effect of small doses of high energy radiation on the color of polyethylene. The majority of yellowing in lightly irradiated polyolefins is due to phenolic stabilizers in the resin formulation particularly when a hindered amine light stabilizer is also present.

Reference: Portnoy, R. C., *The Response Of Various Polyethylenes To High Energy Radiation,* ANTEC 1994, conference proceedings - Society of Plastics Engineers, 1994.

Radiation Resistance

BASF AG: Lupolen

Lupolen is crosslinked by short term exposure at dose rates higher than 10 Gy/s (=1 krad/s). The dose required depends on the desired degree of crosslinkage and varies between 50 and 300 kGy (5 to 30 Mrads). In the presence of air, long term exposure at dose rates of less than 0.1 Gy/s are responsible for oxidative degradation. Unstabilized polyethylene exposed to a total dose of 20 kGy (= 2 Mrads) is likely to undergo a change in properties, e.g. a decrease in impact strength. Lupolen's resistance to long-term exposure to radiation can be improved by means of the same substances that are used to increase the resistance to high-temperature oxidative degradation, viz. antioxidants.

Reference: *Lupolen, Lucalen Product Line, Properties, Processing,* supplier design guide (B 581 e/(8127) 10.91) - BASF Aktiengesellschaft, 1991.

Gamma Radiation Resistance

Exxon: (density: 0.917; melt index: 12.0 grams/10 min.)

Samples were injection molded into ASTM test parts under standard conditions. Each set of parts was then divided into lots representing the various aging and irradiation conditions. Parts were irradiated at 0, 25, 50, 75 kGy (0, 2.5, 5.0, 7.5 Mrads). One group of parts was then tested immediately after irradiation and another after 21 days aging at 60°C. A third group of parts is in ambient temperature storage for testing at a later date. The ASTM properties tested were the tensile elongation at break and Gardner impact strength at 23°C (for polyolefins a significant fall in Gardner impact strength indicated embrittlement). The Hunter b color of each sample using Gardner discs was also determined. The investigators used their proprietary flex-to-failure test, which they have found to be a sensitive measure of embrittlement due to irradiation. Using this test they determined the deflection at peak flexural load and the mode of flexural failure, either ductile or brittle. Past work pointed out that the accelerated aging condition is the most sensitive measure of the radiation effects on thermoplastics in lieu of the results of real time aging. Therefore data is presented on parts aged at 60°C for 21 days after irradiation.

LDPE (density-0.917) has a high level of stability to sterilizing doses of gamma radiation. Gardner impact strength rose with increasing radiation, suggesting the occurrence of crosslinking. Failure in the Gardner test was ductile.

Tensile elongation at break increased with irradiation and aging. This suggests no chain scission. If chain scission had occurred to a significant degree, elongation as small as 10-15% might have signaled embrittlement. The observed increases are not consistent with chain scission.

The deflection at peak flexural load offered final proof that degradation was not a significant process operating in any of these polyothylenes after irradiation. Significant embrittlement due to chain scission is normally accompanied in polypropylene by reductions in deflection at peak flexural load of 25% or more, the values recorded for these polyethylenes were essentially constant over all radiation doses up to 75 kGy.

Color changes in irradiated polyolefins result from the effects of the radiation upon the phenolic antioxidants rather than from interaction between the radiation and the plastic molecules. The lack of coloration in the antioxidant free LDPE supports this conclusion.

The study verified that sterilizing doses of high energy radiation have negligible effect on the physical mechanical properties and color of LDPE. It does not appear that hindered amine light stabilizers are needed to protect polyethylenes against sterilizing doses of high energy radiation.

*R.C. Portnoy and V.R. Cross, "Method for Evaluating the Gamma Radiation Tolerance of Polypropylene for Medical Device Applications", Proceedings Society of Plastics Engineers ANTEC, XXXVI, 1826, Montreal (1991).

Reference: Portnoy, R. C., *The Response Of Various Polyethylenes To High Energy Radiation,* ANTEC 1994, conference proceedings - Society of Plastics Engineers, 1994.

Exxon: (Additives: 175 ppm BHT (2,6 di-t-butyl-4-methylphenol), <50 ppm BHEB (2,6 di-t-butyl-4-ethylphenol; characteristics: density: 0.922; melt index: 33.0 grams/10 min.; features: antioxidant stabilizer)

Samples were injection molded into ASTM test parts under standard conditions. Each set of parts was then divided into lots representing the various aging and irradiation conditions. Parts were irradiated at 0, 25, 50, 75 kGy (0.0, 2.5, 5.0, 7.5 Mrads). One group of parts was then tested immediately after irradiation and another after 21 days aging at 60°C. A third group of parts is in ambient temperature storage for testing at a later date. The ASTM properties tested were the tensile elongation at break and Gardner impact strength at 23°C (for polyolefins a significant fall in Gardner impact strength indicated embrittlement). The Hunter b color of each sample using Gardner discs was also determined. The investigators used their proprietary flex-to-failure test, which they have found to be a sensitive measure of embrittlement due to irradiation. Using this test they determined the deflection at peak flexural load and the mode of flexural failure, either ductile or brittle. Past work pointed out that the accelerated aging condition is the most sensitive measure of the radiation effects on thermoplastics in lieu of the results of real time aging. Therefore data is presented on parts aged at 60°C for 21 days after irradiation.

LDPE (density-0.922) has a high level of stability to sterilizing doses of gamma radiation. Gardner impact strength rose with increasing radiation, suggesting the occurrence of crosslinking. Failure in the Gardner test was ductile.

Tensile elongation at break increased with irradiation and aging. This suggests no chain scission. If chain scission had occurred to a significant degree, elongation as small as 10-15% might have signaled embrittlement. The observed increases are not consistent with chain scission.

The deflection at peak flexural load offered final proof that degradation was not a significant process operating in any of these polyothylenes after irradiation. Significant embrittlement due to chain scission is normally accompanied in polypropylene by reductions in deflection at peak flexural load of 25% or more, the values recorded for these polyethylenes were essentially constant over all radiation doses up to 75 kGy (7.5 Mrads).

Color changes in irradiated polyolefins result from the effects of the radiation upon the phenolic antioxidants rather than from interaction between the radiation and the plastic molecules. LDPE demonstrated a relatively large increase in color indicating a particular sensitivity to coloration after irradiation for the particular additive package used.

The study verified that sterilizing doses of high energy radiation have negligible effect on the physical mechanical properties of LDPE. The changes in color that occurred were most likely due to the choice and quantity of phenolic antioxidant used in the additive formulation. Phenolic additives can be used that do not cause coloration.

The study verified that sterilizing doses of high energy radiation have negligible effect on the physical mechanical properties and color of LDPE. It does not appear that hindered amine light stabilizers are needed to protect polyethylenes against sterilizing doses of high energy radiation.

*R.C. Portnoy and V.R. Cross, "Method for Evaluating the Gamma Radiation Tolerance of Polypropylene for Medical Device Applications", Proceedings Society of Plastics Engineers ANTEC, XXXVI, 1826, Montreal (1991).

Reference: Portnoy, R. C., *The Response Of Various Polyethylenes To High Energy Radiation,* ANTEC 1994, conference proceedings - Society of Plastics Engineers, 1994.

TABLE 80: Effect of Gamma Radiation Sterilization on Low Density Polyethylene

Material Family	LOW DENSITY POLYETHYLENE					
Material Supplier/Name	EXXON					
Material Note	density: 0.917; melt index: 12.0 g/10 min.	density: 0.917; melt index: 12.0 g/10 min.	density: 0.917; melt index: 12.0 g/10 min.	Additives: 175 ppm BHT (2,6 di-t-butyl-4-methylphenol), <50 ppm BHEB (2,6 di-t-butyl-4-ethylphenol; density: 0.922; melt index: 33.0 grams/10 min.; antioxidant stabilizer	Additives: 175 ppm BHT (2,6 di-t-butyl-4-methylphenol), <50 ppm BHEB (2,6 di-t-butyl-4-ethylphenol; density: 0.922; melt index: 33.0 grams/10 min.; antioxidant stabilizer	Additives: 175 ppm BHT (2,6 di-t-butyl-4-methylphenol), <50 ppm BHEB (2,6 di-t-butyl-4-ethylphenol; density: 0.922; melt index: 33.0 grams/10 min.; antioxidant stabilizer
Reference No.	23	23	23	23	23	23

EXPOSURE CONDITIONS

Type	Gamma Radiation	Gamma Radiation	Gamma Radiation	Gamma Radiation	Gamma Radiation	Gamma Radiation
Radiation Dose (Mrads)	2.5	5	7.5	2.5	5	7.5

POST EXPOSURE CONDITIONING III

Temperature (°C)	60	60	60	60	60	60
Time (hours)	504	504	504	504	504	504

PROPERTIES RETAINED (%)

Elongation	106.3 (ak)	112.2 (ak)	102.1 (ak)	107.3 (ak)	116.4 (ak)	107.9 (ak)
Gardner Impact	113.1 (cq)	111.1 (cq)	113.7 (cq)	103 (cq)	104.8 (cq)	97 (cq)
Deflection @ Failure	98.2 (i)	98.2 (i)	100 (i)	100 (i)	100 (i)	100 (i)

SURFACE and APPEARANCE

Δb Color	0.2 (z)	0.4 (z)	0.7 (z)	2.4 (z)	2.9 (z)	2.6 (z)

Gamma Radiation Resistance

Dow Chemical: Dowlex 2535

The physical properties of Linear Low Density Polyethylene (LLDPE) do not deteriorate upon exposure to gamma radiation. LLDPE showed little discoloration when exposed to 10.0 Mrads. Its ΔE increased only 0.5 between 2.5 and 10.0 Mrads exposure. LLDPE regains its initial optical appearance after 10.0 Mrads upon storage in a fluorescent light environment. Data collected two weeks after exposure show no significant loss in tensile yield strength. Impact strength, measured by notched Izod, is also unaffected.

Reference: Sturdevant, Marianne F., *Sterilization Compatibility of Rigid Thermoplastic Materials,* supplier technical report (301-1548) - Dow Chemical Company, 1988.

Dow Chemical: (features: natural resin)

Exposure to gamma radiation doses up to 10 Mrads does not alter the tensile and impact properties. However, the flexural property of the resin decreases upon irradiation. The flexural strength of the polymer drops from 9762 psi for un-irradiated control to 1192 psi and 1376 psi after exposure to 2.5 Mrads and 10 Mrads, respectively. This change in flexural strength is attributed to possible crosslinking and increased crystallization of the polymer.

Reference: Sturdevant, Marianne F., *The Long-term Effects of Ethylene Oxide and Gamma Radiation Sterilization on the Properties of Rigid Thermoplastic Materials,* ANTEC 1990, conference proceedings - Society of Plastics Engineers, 1990.

Exxon: (Additives: 200 ppm Irganox 1076, 850 ppm Zinc Stearate; characteristics: density: 0.926; melt index: 52.0 grams/10 min.; features: antioxidant stabilizer)

Samples were injection molded into ASTM test parts under standard conditions. Each set of parts was then divided into lots representing the various aging and irradiation conditions. Parts were irradiated at 0, 25, 50, 75 kGy (0.0, 2.5, 5.0, 7.5 Mrads). One group of parts was then tested immediately after irradiation and another after 21 days aging at 60°C. A third group of parts is in ambient temperature storage for testing at a later date. The ASTM properties tested were the tensile elongation at break and Gardner impact strength at 23°C (for polyolefins a significant fall in Gardner impact strength indicated embrittlement). The Hunter b color of each sample using Gardner discs was also determined. The investigators used their proprietary flex-to-failure test*, which they have found to be a sensitive measure of embrittlement due to irradiation. Using this test they determined the deflection at peak flexural load and the mode of flexural failure, either ductile or brittle. Past work pointed out that the accelerated aging condition is the most sensitive measure of the radiation effects on thermoplastics in lieu of the results of real time aging. Therefore data is presented on parts aged at 60°C for 21 days after irradiation.

LLDPE (desity-0.926) has a high level of stability to sterilizing doses of gamma radiation. Gardner impact strength rose with increasing radiation, suggesting the occurrence of crosslinking. Failure in the Gardner test was ductile.

Tensile elongation at break increased with irradiation and aging. This suggests no chain scission. If chain scission had occurred to a significant degree, elongation as small as 10-15% might have signaled embrittlement. The observed increases are not consistent with chain scission.

The deflection at peak flexural load offered final proof that degradation was not a significant process operating in any of these polyothylenes after irradiation. Significant embrittlement due to chain scission is normally accompanied in polypropylene by reductions in deflection at peak flexural load of 25% or more, the values recorded for these polyethylenes were essentially constant over all radiation doses up to 75 kGy (7.5 Mrads).

Color changes in irradiated polyolefins result from the effects of the radiation upon the phenolic antioxidants rather than from interaction between the radiation and the plastic molecules. LLDPE demonstrated a lack of severe coloration indicating that certain antioxidant additive packages will not color significantly with moderate doses of radiation.

The study verified that sterilizing doses of high energy radiation have negligible effect on the physical mechanical properties and color of LLDPE. It does not appear that hindered amine light stabilizers are needed to protect polyethylenes against sterilizing doses of high energy radiation.

*R.C. Portnoy and V.R. Cross, "Method for Evaluating the Gamma Radiation Tolerance of Polypropylene for Medical Device Applications", Proceedings Society of Plastics Engineers ANTEC, XXXVI, 1826, Montreal (1991).

Reference: Portnoy, R. C., *The Response Of Various Polyethylenes To High Energy Radiation,* ANTEC 1994, conference proceedings - Society of Plastics Engineers, 1994.

Ethylene Oxide (EtO) Resistance

Dow Chemical: Dowlex 2535

Notched Izod impact strength of Linear Low Density Polyethylene (LLDP) was not affected by up to five cycles of ethylene oxide gas. The first day following exposure residual EtO measured less than 300 ppm. After day seven only 3 ppm of residual EtO were detected. LLDP is well suited to this sterilization method.

Reference: Sturdevant, Marianne F., *Sterilization Compatibility of Rigid Thermoplastic Materials,* supplier technical report (301-1548) - Dow Chemical Company, 1988.

Dow Chemical: (features: natural resin)

LLDPE is compatible with multiple cycles of ethylene oxide sterilization due primarily to its excellent chemical resistance characteristics. Samples were tested up to five cycles.

Reference: Sturdevant, Marianne F., *The Long-term Effects of Ethylene Oxide and Gamma Radiation Sterilization on the Properties of Rigid Thermoplastic Materials,* ANTEC 1990, conference proceedings - Society of Plastics Engineers, 1990.

TABLE 81: Effect of Gamma Radiation Sterilization on Linear Low Density Polyethylene

Material Family	LINEAR LOW DENSITY POLYETHYLENE					
Material Supplier/Name	DOW					
Material Note	natural resin	natural resin	natural resin	natural resin	natural resin	natural resin
Reference No.	5	5	5	5	5	5
EXPOSURE CONDITIONS						
Type	Gamma Radiation	Gamma Radiation	Gamma Radiation	Gamma Radiation	Gamma Radiation	Gamma Radiation
Details	source: Cobalt 60	source: Cobalt 60	source: Cobalt 60	source: Cobalt 60	source: Cobalt 60	source: Cobalt 60
Radiation Dose (Mrads)	2.5	2.5	2.5	10	10	10
Note	test lab: Radiations Sterilizers Inc.	test lab: Radiations Sterilizers Inc.	test lab: Radiations Sterilizers Inc.	test lab: Radiations Sterilizers Inc.	test lab: Radiations Sterilizers Inc.	test lab: Radiations Sterilizers Inc.
POST EXPOSURE CONDITIONING						
Note	type: storage in dark	type: storage in dark	type: storage in dark	type: storage in dark	type: storage in dark	type: storage in dark
Temperature (°C)	21	21	21	21	21	21
Time (hours)	336	4368	8760	336	4368	8760
PROPERTIES RETAINED (%)						
Tensile Strength	72.9 (ht)	97.5 (ht)	96 (ht)	96.2 (ht)	101.4 (ht)	102 (ht)
Tensile Strength @ Yield	101.7 (in)	104.1 (in)	102.1 (in)	102.8 (in)	107.7 (in)	105.4 (in)
Elongation	>80.1 (av)	84.6 (av)	68.8 (av)	90.4 (av)	89.9 (av)	87.8 (av)
Dart Impact (total energy)	104.1 (fb)	95.9 (fb)	95.9 (fb)	112.3 (fb)	94 (fb)	113.1 (fb)
Notched Izod Impact	100 (fq)	100 (fq)	100 (fq)	100 (fq)	100 (fq)	100 (fq)

TABLE 82: Effect of Gamma Radiation Sterilization on Linear Low Density Polyethylene

Material Family	LINEAR LOW DENSITY POLYETHYLENE			
Material Supplier/Name	DOW DOWLEX 2535			
Reference No.	3	3	3	3
EXPOSURE CONDITIONS				
Type	Gamma Radiation	Gamma Radiation	Gamma Radiation	Gamma Radiation
Details	source: Cobalt 60	source: Cobalt 60	source: Cobalt 60	source: Cobalt 60
Radiation Dose (Mrads)	2.5	10	2.5	10
POST EXPOSURE CONDITIONING				
Note	type: storage under fluorescent light	type: storage under fluorescent light	type: storage in dark	type: storage in dark
Temperature (°C)	21	21	21	21
Time (hours)	336	336	336	336
PROPERTIES RETAINED (%)				
Tensile Strength @ Yield	100 (ii)	100 (ii)	100 (ii)	100 (ii)
Notched Izod Impact	100 (fp)	100 (fp)	100 (fp)	100 (fp)

TABLE 83: Effect of Gamma Radiation Sterilization on Linear Low Density Polyethylene

Material Family	LINEAR LOW DENSITY POLYETHYLENE		
Material Supplier/Name	EXXON		
Material Note	Additives: 200 ppm Irganox 1076, 850 ppm Zinc Stearate; density: 0.926; melt index: 52.0 grams/10 min.; antioxidant stabilizer	Additives: 200 ppm Irganox 1076, 850 ppm Zinc Stearate; density: 0.926; melt index: 52.0 grams/10 min.; antioxidant stabilizer	Additives: 200 ppm Irganox 1076, 850 ppm Zinc Stearate; density: 0.926; melt index: 52.0 grams/10 min.; antioxidant stabilizer
Reference No.	23	23	23
EXPOSURE CONDITIONS			
Type	Gamma Radiation	Gamma Radiation	Gamma Radiation
Radiation Dose (Mrads)	2.5	5	7.5
POST EXPOSURE CONDITIONING			
Temperature (°C)	60	60	60
Time (hours)	504	504	504
PROPERTIES RETAINED (%)			
Elongation	102.4 (ak)	120.7 (ak)	163.4 (ak)
Gardner Impact	101.6 (cq)	106.2 (cq)	109.3 (cq)
Deflection @ Failure	101.9 (i)	106.5 (i)	109.3 (i)
SURFACE and APPEARANCE			
Δb Color	1.4 (z)	1.8 (z)	2.2 (z)

TABLE 84: Effect of Ethylene Oxide Sterilization on Linear Low Density Polyethylene

Material Family	LINEAR LOW DENSITY POLYETHYLENE					
Material Supplier/Name	DOW					
Material Note	natural resin	natural resin	natural resin	natural resin	natural resin	natural resin
Reference No.	5	5	5	5	5	5

EXPOSURE CONDITIONS

Type	Ethylene Oxide	Ethylene Oxide	Ethylene Oxide	Ethylene Oxide	Ethylene Oxide	Ethylene Oxide
Details	12% EtO and 88% Freon	12% EtO and 88% Freon	12% EtO and 88% Freon	12% EtO and 88% Freon	12% EtO and 88% Freon	12% EtO and 88% Freon
Concentration	660 mg/l	660 mg/l	660 mg/l	660 mg/l	660 mg/l	660 mg/l
Number of Cycles	1	1	1	5	5	5
Note	RH: 60%; test lab: Ethox Corp.	RH: 60%; test lab: Ethox Corp.	RH: 60%; test lab: Ethox Corp.	RH: 60%; test lab: Ethox Corp.	RH: 60%; test lab: Ethox Corp.	RH: 60%; test lab: Ethox Corp.
Temperature (°C)	49	49	49	49	49	49
Time (hours)	≥6	≥6	≥6	≥6	≥6	≥6

PRE EXPOSURE CONDITIONING

Preconditioning Note	time: 8 hours; temperature: 37.8°C; RH: 60%	time: 8 hours; temperature: 37.8°C; RH: 60%	time: 8 hours; temperature: 37.8°C; RH: 60%	time: 8 hours; temperature: 37.8°C; RH: 60%	time: 8 hours; temperature: 37.8°C; RH: 60%	time: 8 hours; temperature: 37.8°C; RH: 60%

POST EXPOSURE CONDITIONING

Note	type: evacuation; pressure: 127 mm Hg	type: evacuation; pressure: 127 mm Hg	type: evacuation; pressure: 127 mm Hg	type: evacuation; pressure: 127 mm Hg	type: evacuation; pressure: 127 mm Hg	type: evacuation; pressure: 127 mm Hg

POST EXPOSURE CONDITIONING II

Note	type: aeration	type: aeration	type: aeration	type: aeration	type: aeration	type: aeration
Temperature (°C)	32.2	32.2	32.2	32.2	32.2	32.2
Time (hours)	≥16	≥16	≥16	≥16	≥16	≥16

POST EXPOSURE CONDITIONING III

Note	type: storage in dark	type: storage in dark	type: storage in dark	type: storage in dark	type: storage in dark	type: storage in dark
Temperature (°C)	21	21	21	21	21	21
Time (hours)	336	4368	8760	336	4368	8760

PROPERTIES RETAINED (%)

Tensile Strength		88.1 (ht)	82.3 (ht)	94.3 (ht)	104.3 (ht)	84.6 (ht)
Tensile Strength @ Yield		102.6 (in)	105.1 (in)	108.4 (in)	104.2 (in)	104.1 (in)
Elongation		84.9 (au)	84.6 (au)	92.9 (au)	107.7 (au)	88.3 (au)
Dart Impact (total energy)	100.4 (ff)	98.1 (ff)	89.6 (ff)	97 (ff)	89.2 (ff)	89.6 (ff)
Notched Izod Impact	100 (fq)	100 (fq)	100 (fq)	100 (fq)	100 (fq)	100 (fq)

TABLE 85: Effect of Ethylene Oxide Sterilization on Linear Low Density Polyethylene

Material Family	LINEAR LOW DENSITY POLYETHYLENE		
Material Supplier/Name	DOW DOWLEX 2535		
Reference No.	3	3	3
EXPOSURE CONDITIONS			
Type	Ethylene Oxide	Ethylene Oxide	Ethylene Oxide
Details	12% EtO and 88% Freon	12% EtO and 88% Freon	12% EtO and 88% Freon
Concentration	660 mg/l	660 mg/l	660 mg/l
Number of Cycles	1	5	1
Note	RH: 60%	RH: 60%	RH: 60%
Temperature (°C)	49	49	49
Time (hours)	≥6	≥6	≥6
PRE EXPOSURE CONDITIONING			
Preconditioning Note	time: 8 hours; temperature: 37.8°C; RH: 60%	time: 8 hours; temperature: 37.8°C; RH: 60%	time: 8 hours; temperature: 37.8°C; RH: 60%
POST EXPOSURE CONDITIONING			
Note	type: evacuation; pressure: 127 mm Hg	type: evacuation; pressure: 127 mm Hg	type: evacuation; pressure: 127 mm Hg
POST EXPOSURE CONDITIONING II			
Note	type: aeration	type: aeration	type: aeration
Temperature (°C)	32	32	32
Time (hours)	≥16	≥16	
POST EXPOSURE CONDITIONING III			
Note	type: storage in dark; RH: 50%	type: storage in dark; RH: 50%	
Temperature (°C)	21	21	
Time (hours)	336	336	
RESIDUALS (ppm)			
Residuals Determined			ethylene oxide
24 hour Aeration			244
48 hour Aeration			106
72 hour Aeration			39
96 hour Aeration			18
168 hour Aeration			3
PROPERTIES RETAINED (%)			
Tensile Strength @ Yield	100 (ii)	100 (ii)	
Notched Izod Impact	100 (fp)	100 (fp)	

GRAPH 115: Post Gamma Radiation Exposure Time vs. Delta E Color Change of Linear Low Density Polyethylene

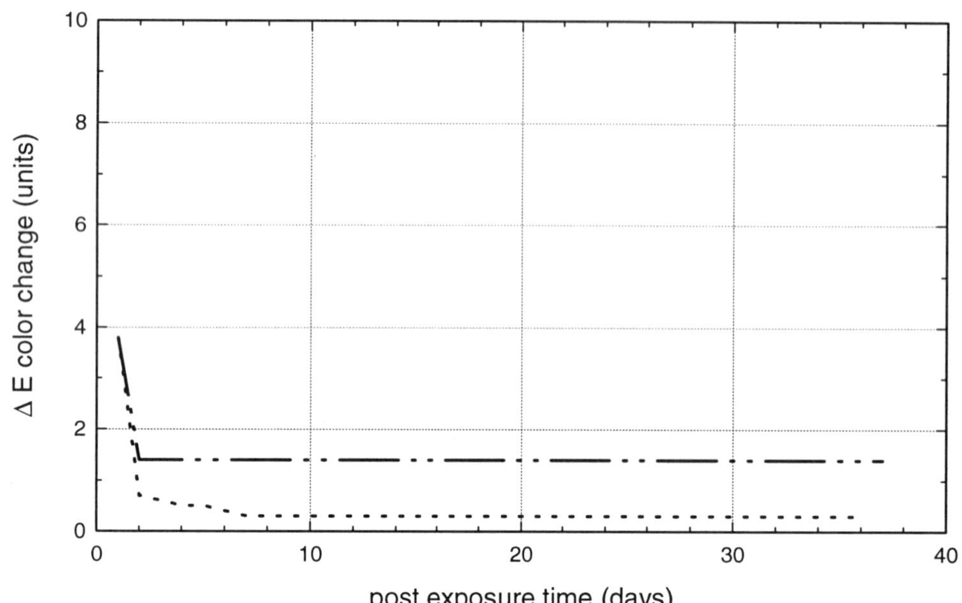

GRAPH 116: Post Gamma Radiation Exposure Time vs. Delta E Color Change of Linear Low Density Polyethylene

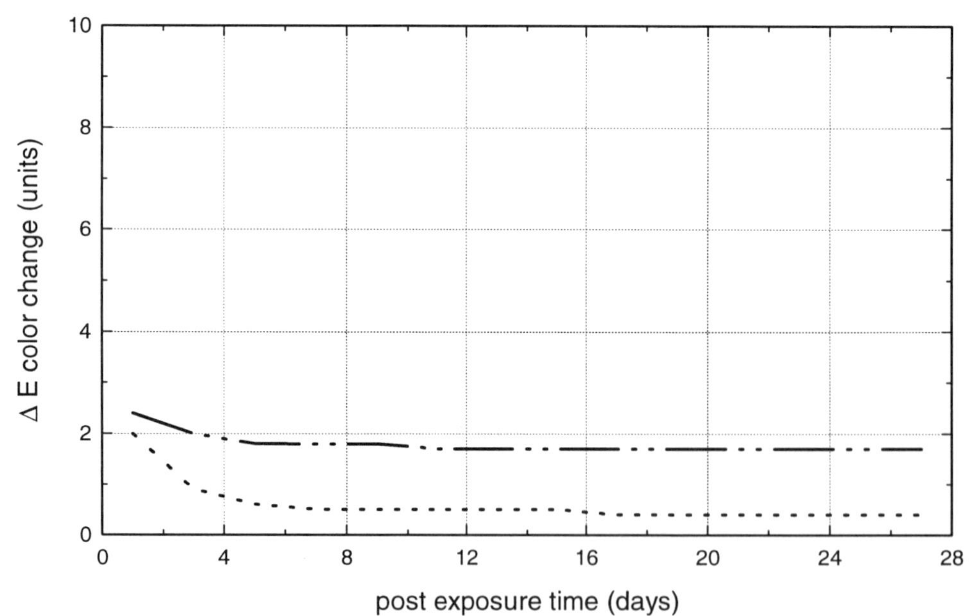

Radiation Resistance

Hoechst AG: Hostalen

If oxygen is excluded, polyethylene is crosslinked by the action of high energy, ionizing radiation. The degree of crosslinking is solely dependent on the radiation dose. Gamma rays, X-rays and electrons have the same effect. Crosslinking improves resistance to stress cracking but at sufficiently high dose levels can reduce elongation at break and toughness. Measurable changes result at radiation doses above 0.1 MJ/kg (10.0 Mrads); 1 MJ/kg (100 Mrads) lowers elongation at break to a few percent of the initial value; tensile stress at yield remains virtually unchanged up to this level.

When oxygen is present, for example in air, the effect of high energy radiation depends in a complex way on the dose rate (=dose/time), dose level and wall thickness of the polyethylene. In the case of parts with large wall thickness, short term exposure to high doses has much the same effect as exposure with oxygen excluded. Long term exposure (months to years) on the other hand causes a permanent reduction in toughness even at low doses as a result of oxygen diffusion. For example, a dose of 10 kJ/kg (1.0 Mrads) up to 20.0 kJ/kg (2.0 Mrads) at 1mm wall thickness lowers elongation at break to 10% of the initial value. With greater wall thickness, the material is less at risk but hairline cracks on the damaged outer surface can lead to fracture even at low stress.

Reference: *Hostalen Polymer Materials,* supplier design guide (HDKR 101 E 9050/022) - Hoechst AG.

Radiation Resistance

Dow Chemical: HDPE 12350

The samples had a density of 0.950 and a melt index of 12. They showed an increase in Izod impact after exposure to 10 Mrads of both gamma and E-beam radiation. The tensile modulus decreased by 18% after 2.5 Mrads of radiation and by 27% after 10 Mrads of radiation. The samples showed some color shift after radiation. The lower dosed samples returned to their original colors while the 10 Mrad samples maintained some visible yellow discoloration. ΔE ranged from 1.5 to 3.0.

Reference: Hermanson, Nancy J., Steffens, John F., *The Physical and Visual Property Changes in Thermoplastic Resins After Exposure to High Energy Sterilization - Gamma Versus Electron Beam,* ANTEC 1993, conference proceedings - Society of Plastics Engineers, 1993.

Dow Chemical: HDPE 25355

The samples had a density of 0.955 and a melt index of 25. They showed an increase in Izod impact after exposure to 10 Mrads of both gamma and E-beam radiation. The tensile modulus decreased by 22% after 2.5 Mrads of radiation and by 28% after 10 Mrads of radiation. The samples showed some color shift after radiation. The lower dosed samples returned to their original colors while the 10 Mrad samples maintained some visible yellow discoloration. ΔE ranged from 1.5 to 3.0.

Reference: Hermanson, Nancy J., Steffens, John F., *The Physical and Visual Property Changes in Thermoplastic Resins After Exposure to High Energy Sterilization - Gamma Versus Electron Beam,* ANTEC 1993, conference proceedings - Society of Plastics Engineers, 1993.

Hoechst AG: Hostalen

If oxygen is excluded, polyethylene is crosslinked by the action of high energy, ionizing radiation. The degree of crosslinking is solely dependent on the radiation dose. Gamma rays, X-rays and electrons have the same effect. Crosslinking improves resistance to stress cracking but at sufficiently high dose levels can reduce elongation at break and toughness. Measurable changes result at radiation doses above 0.1 MJ/kg (10.0 Mrads); 1 MJ/kg (100 Mrads) lowers elongation at break to a few percent of the initial value; tensile stress at yield remains virtually unchanged up to this level.

When oxygen is present, for example in air, the effect of high energy radiation depends in a complex way on the dose rate (=dose/time), dose level and wall thickness of the polyethylene. In the case of parts with large wall thicknesses, short term exposure to high doses has much the same effect as exposure with oxygen excluded. Long term exposure (months to years) on the other hand causes a permanent reduction in toughness even at low doses as a result of oxygen diffusion. For example, a dose of 10 kJ/kg (1.0 Mrads) up to 20.0 kJ/kg (2.0 Mrads) at 1mm wall thickness lowers elongation at break to 10% of the initial value. With greater wall thicknesses, the material is less at risk but hairline cracks on the damaged outer surface can lead to fracture even at low stress.

Reference: *Hostalen Polymer Materials,* supplier design guide (HDKR 101 E 9050/022) - Hoechst AG.

Phillips: Marlex

Data indicate that polymer crosslinking occurs with gamma irradiation accompanied by an increase in density, tensile strength and hardness and by a decrease in solubility. Irradiation of Marlex high density polyethylene also increases resistance to environmental stress cracking.

Reference: *Engineering Properties Of Marlex Resins,* supplier design guide (TSM-243) - Phillips 66 Company, 1983.

Gamma Radiation Resistance

Exxon: (Additives: 300 ppm Irganox 1076, 500 ppm Zinc Stearate; characteristics: density: 0.963; melt index: 8.2 grams/10 min.; features: antioxidant stabilizer)

Samples were injection molded into ASTM test parts under standard conditions. Each set of parts were then divided into lots representing the various aging and irradiation conditions. Parts were irradiated at 0, 25, 50, 75 kGy (0.0, 2.5, 5.0, 7.5 Mrads). One group of parts was then tested immediately after irradiation and another after 21 days aging at 60°C. A third group of parts is in ambient temperature storage for testing at a later date. The ASTM properties tested were the tensile elongation at break and Gardner impact strength at 23°C (for polyolefins a significant fall in Gardner impact strength indicated embrittlement). The Hunter b color of each sample using Gardner discs was also determined. The investigators used their proprietary flex-to-failure test*, which they have found to be a sensitive measure of embrittlement due to irradiation. Using this test they determined the deflection at peak flexural load and the mode of flexural failure, either ductile or brittle. Past work pointed out that the accelerated aging condition is the most sensitive measure of the radiation effects on thermoplastics in lieu of the results of real time aging. Therefore data is presented on parts aged at 60°C for 21 days after irradiation.

HDPE (density-0.963) has a high level of stability to sterilizing doses of gamma radiation. Gardner impact strength rose with increasing radiation, suggesting the occurrence of crosslinking. Failure in the Gardner test was ductile.

Tensile elongation at break increased with irradiation and aging. This suggests no chain scission. If chain scission had occurred to a significant degree, elongation as small as 10-15% might have signaled embrittlement. The observed increases are not consistent with chain scission.

The deflection at peak flexural load offered final proof that degradation was not a significant process operating in any of these polyothylenes after irradiation. Significant embrittlement due to chain scission is normally accompanied in polypropylene by reductions in deflection at peak flexural load of 25% or more, the values recorded for these polyethylenes were essentially constant over all radiation doses up to 75 kGy (7.5 Mrads).

Color changes in irradiated polyolefins result from the effects of the radiation upon the phenolic antioxidants rather than from interaction between the radiation and the plastic molecules. HDPE demonstrated a relatively large increase in color indicating a particular sensitivity to coloration after irradiation for the particular additive package used.

The study verified that sterilizing doses of high energy radiation have negligible effect on the physical and mechanical properties of HDPE. The changes in color that occurred were most likely due to the choice and quantity of phenolic antioxidant used in the additive formulation. Phenolic additives can be used that do not cause coloration. It does not appear that hindered amine light stabilizers are needed to protect polyethylenes against sterilizing doses of high energy radiation.

*R.C. Portnoy and V.R. Cross, "Method for Evaluating the Gamma Radiation Tolerance of Polypropylene for Medical Device Applications", Proceedings Society of Plastics Engineers ANTEC, XXXVI, 1826, Montreal (1991).

Reference: Portnoy, R. C., *The Response Of Various Polyethylenes To High Energy Radiation,* ANTEC 1994, conference proceedings - Society of Plastics Engineers, 1994.

DuPont: Tyvek 1059B (composition: spunbonded olefin; form: sheet); **Tyvek 1073B** (composition: spunbonded olefin; form: sheet)

The physical and bacterial barrier properties of Tyvek are not compromised at 5.0 Mrads exposure.

Reference: *Sterile Packaging Of Tyvek,* supplier marketing literature (H-19226) - Du Pont Company.

Ethylene Oxide (EtO) Resistance

Dow Chemical: HDPE 10062

Repeated cycles of EtO sterilization reduce the tensile modulus by 50%. A slight decrease in tensile yield eight weeks after exposure to 2 EtO cycles was also seen. The ultimate tensile elongation after exposure to CFC-12/EtO gas mixture increased by 2 fold or more. The samples that were sterilized using HCFC-12/EtO initially showed the same increase in tensile elongation, but after eight weeks the ultimate elongation was similar to the original elongation. All samples exhibited a 30% decrease in instrumented dart impact results after exposure to EtO. The changes in physical properties are thought to be crosslinking and/or chain scission of the polymer chains brought on by exposure to EtO. While physical properties are affected color does not change.

Reference: Hermanson, Nancy J., *Effects Of Alternate Carriers Of Ethylene Oxide Sterilant On Thermoplastics,* supplier technical report (301-02018) - Dow Chemical Company.

DuPont: **Tyvek 1059B** (composition: spunbonded olefin; form: sheet); **Tyvek 1073B** (composition: spunbonded olefin; form: sheet)

Tyvek absorbs significantly less EtO than ordinary medical paper during commercial sterilization cycles - and at the end of a week Tyvek contains no measurable amounts of EtO.

Reference: *Sterile Packaging Of Tyvek,* supplier marketing literature (H-19226) - Du Pont Company.

Steam Resistance

DuPont: **Tyvek 1059B** (composition: spunbonded olefin; form: sheet); **Tyvek 1073B** (composition: spunbonded olefin; form: sheet)

Tyvek is a high density polyethylene and will soften dramatically at temperatures around 135°C. For that reason it is not recommended for use in hospital steam sterilization where temperatures exceed 138°C. However, under carefully controlled conditions of 121°C to 127°C, it will perform well.

Reference: *Sterile Packaging Of Tyvek,* supplier marketing literature (H-19226) - Du Pont Company.

TABLE 86: Effect of Gamma Radiation Sterilization on High Density Polyethylene

Material Family	HIGH DENSITY POLYETHYLENE						
Material Supplier/Name	DUPONT TYVEK 1073B		DUPONT TYVEK 1059B		PHILLIPS MARLEX		
Material Note	spunbonded olefin; sheet	spunbonded olefin; sheet	spunbonded olefin; sheet	spunbonded olefin; sheet			
Reference No.	45	45	45	45	101	101	101
EXPOSURE CONDITIONS							
Type	Gamma Radiation	Gamma Radiation	Gamma Radiation	Gamma Radiation	Gamma Radiation	Gamma Radiation	Gamma Radiation
Radiation Dose (Mrads)	2.5	5	2.5	5	1	10	100
PROPERTIES RETAINED (%)							
Tensile Strength	91.1 (he)	69.6 (he)	84.8 (he)	69.5 (he)	128.6 (iw)	122.1 (iw)	143.3 (iw)
Elongation	101.6 (ai)	83.4 (ai)	98.2 (ai)	83.5 (ai)			
Elongation @ Yield					115.4 (bh)	115.4 (bh)	7.7 (bh)
SURFACE and APPEARANCE							
Hardness Units Change					D4	D6	D6
Yellowness Index note	no color change	very slight color change	no color change	very slight color change			

TABLE 87: Effect of Gamma Radiation Sterilization on High Density Polyethylene

Material Family	HIGH DENSITY POLYETHYLENE							
Material Supplier/Name	DOW HDPE 12350				DOW HDPE 25355			
Reference No.	1	1	1	1	1	1	1	1
EXPOSURE CONDITIONS								
Type	Gamma Radiation	Gamma Radiation	Gamma Radiation	Gamma Radiation	Gamma Radiation	Gamma Radiation	Gamma Radiation	Gamma Radiation
Radiation Dose (Mrads)	2.5	2.5	10	10	2.5	2.5	10	10
Note	test lab: SteriGenics	test lab: SteriGenics	test lab: SteriGenics	test lab: SteriGenics	test lab: SteriGenics	test lab: SteriGenics	test lab: SteriGenics	test lab: SteriGenics
POST EXPOSURE CONDITIONING								
Note	type: aging	type: aging	type: aging	type: aging	type: aging	type: aging	type: aging	type: aging
Time (hours)	168	1344	168	1344	168	1344	168	1344
PROPERTIES RETAINED (%)								
Tensile Strength @ Yield	100 (iq)	100 (iq)	100 (iq)	100 (iq)	100 (iq)	95.7 (iq)	100 (iq)	100 (iq)
Elongation	100 (bo)	100 (bo)	100 (bo)	100 (bo)	100 (bo)	100 (bo)	100 (bo)	100 (bo)
Flexural Strength	110.5 (cn)	110.5 (cn)	121.1 (cn)	121.1 (cn)	109.5 (cn)	109.5 (cn)	114.3 (cn)	119 (cn)
Modulus	112.2 (gy)	77.8 (gy)	92.2 (gy)	82.2 (gy)	87.1 (gx)	75 (gx)	79.3 (gx)	77.6 (gx)
Flexural Modulus	106.8 (cd)	104.1 (cd)	109.5 (cd)	105.4 (cd)	117.8 (cc)	108.2 (cc)	131.5 (cc)	117.8 (cc)
Dart Impact (total energy)	103.2 (ek)	100 (ek)	109.7 (ek)	103.2 (ek)	106.7 (ep)	100 (ep)	100 (ep)	103.3 (ep)
Dart Impact (peak energy)	100 (dk)	100 (dk)	105.9 (dk)	100 (dk)	106.3 (dn)	106.3 (dn)	106.3 (dn)	112.5 (dn)
Notched Izod Impact	120.8 (gf)	130.2 (gf)	181.1 (gf)	181.1 (gf)	100 (gd)	100 (gd)	116.2 (gd)	116.2 (gd)
Heat Deflection Temperature	102.2 (m)	87 (m)	87 (m)	87 (m)	102.4 (m)	117.1 (m)	131.7 (m)	122 (m)
Vicat Softening Point	100 (ku)	100 (ku)	100 (ku)	100 (ku)	100 (ku)	100 (ku)	100 (ku)	100 (ku)
SURFACE and APPEARANCE								
ΔL Color	-1.4 (ac)	-1.2 (ac)	-2.4 (ac)	-2.4 (ac)	-1.3 (ac)	-1.4 (ac)	-2.8 (ac)	-3.2 (ac)
Δa Color	-0.6 (o)	-1 (o)	0.4 (o)	-1.3 (o)	-0.5 (o)	-0.9 (o)	0.5 (o)	-1 (o)
Δb Color	4.5 (t)	5 (t)	6.7 (t)	8.6 (t)	4.1 (t)	5.1 (t)	7 (t)	9.4 (t)

TABLE 88: Effect of Gamma Radiation Sterilization on High Density Polyethylene

Material Family	HIGH DENSITY POLYETHYLENE		
Material Supplier/Name	EXXON		
Material Note	Additives: 300 ppm Irganox 1076, 500 ppm Zinc Stearate; density: 0.963; melt index: 8.2 grams/10 min.; antioxidant stabilizer	Additives: 300 ppm Irganox 1076, 500 ppm Zinc Stearate; density: 0.963; melt index: 8.2 grams/10 min.; antioxidant stabilizer	Additives: 300 ppm Irganox 1076, 500 ppm Zinc Stearate; density: 0.963; melt index: 8.2 grams/10 min.; antioxidant stabilizer
Reference No.	23	23	23
EXPOSURE CONDITIONS			
Type	Gamma Radiation	Gamma Radiation	Gamma Radiation
Radiation Dose (Mrads)	2.5	5	7.5
POST EXPOSURE CONDITIONING			
Temperature (°C)	60	60	60
Time (hours)	504	504	504
PROPERTIES RETAINED (%)			
Elongation	108.2 (ak)	134.2 (ak)	257.5 (ak)
Gardner Impact	103.6 (cq)	107.2 (cq)	106 (cq)
Deflection @ Failure	103.2 (i)	103.2 (i)	100 (i)
SURFACE and APPEARANCE			
Δb Color	3 (z)	4.7 (z)	5.4 (z)

TABLE 89: Effect of Electron Beam Radiation Sterilization on High Density Polyethylene

Material Family	HIGH DENSITY POLYETHYLENE											
Material Supplier/Name	DOW HDPE 12350				DOW HDPE 25355				DUPONT TYVEK 1073B		DUPONT TYVEK 1059B	
Material Note									spunbonded olefin; sheet		spunbonded olefin; sheet	
Reference No.	1	1	1	1	1	1	1	1	45	45	45	45

EXPOSURE CONDITIONS

Type	Electron Beam Radiation	Electron Beam Radiation	Electron Beam Radiation	Electron Beam Radiation	Electron Beam Radiation	Electron Beam Radiation	Electron Beam Radiation	Electron Beam Radiation	Electron Beam Radiation	Electron Beam Radiation	Electron Beam Radiation	Electron Beam Radiation
Radiation Dose (Mrads)	2.5	2.5	10	10	2.5	2.5	10	10	2.5	5	2.5	5
Note	test lab: E-Beam Services, Inc.	test lab: E-Beam Services, Inc.	test lab: E-Beam Services, Inc.	test lab: E-Beam Services, Inc.	test lab: E-Beam Services, Inc.	test lab: E-Beam Services, Inc.	test lab: E-Beam Services, Inc.	test lab: E-Beam Services, Inc.				

POST EXPOSURE CONDITIONING

Note	type: aging	type: aging	type: aging	type: aging	type: aging	type: aging	type: aging	type: aging				
Time (hours)	168	1344	168	1344	168	1344	168	1344				

PROPERTIES RETAINED (%)

Tensile Strength									101 (he)	78.3 (he)	76.9 (he)	71.6 (he)
Tensile Strength @ Yield	100 (iq)	95.2 (iq)	100 (iq)	100 (iq)	95.7 (iq)	95.7 (iq)	100 (iq)	100 (iq)				
Elongation	100 (bo)	100 (bo)	100 (bo)	100 (bo)	100 (bo)	100 (bo)	100 (bo)	100 (bo)	102.4 (ai)	85 (ai)	95.3 (ai)	85.9 (ai)
Flexural Strength	110.5 (cn)	110.5 (cn)	110.5 (cn)	115.8 (cn)	109.5 (cn)	109.5 (cn)	109.5 (cn)	100 (cn)				
Modulus	91.1 (gy)	75.6 (gy)	84.4 (gy)	73.3 (gy)	73.3 (gx)	71.6 (gx)	69 (gx)	71.6 (gx)				
Flexural Modulus	94.6 (cd)	102.7 (cd)	100 (cd)	112.2 (cd)	116.4 (cc)	106.8 (cc)	119.2 (cc)	109.6 (cc)				
Dart Impact (total energy)	103.2 (ek)	100 (ek)	116.1 (ek)	109.7 (ek)	103.3 (ep)	96.7 (ep)	110 (ep)	110 (ep)				
Dart Impact (peak energy)	100 (dk)	100 (dk)	100 (dk)	105.9 (dk)	106.3 (dn)	106.3 (dn)	112.5 (dn)	112.5 (dn)				
Notched Izod Impact	111.3 (gf)	130.2 (gf)	201.9 (gf)	201.9 (gf)	100 (gd)	100 (gd)	129.7 (gd)	116.2 (gd)				
Tear Resistance									124.4 (ae)	97.7 (ae)	70.8 (ae)	84.1 (ae)
Heat Deflection Temperature	89.1 (m)	108.7 (m)	84.8 (m)	121.7 (m)	119.5 (m)	131.7 (m)	131.7 (m)	141.5 (m)				
Vicat Softening Point	100 (ku)	100 (ku)	100 (ku)	100 (ku)	100 (ku)	100 (ku)	100.8 (ku)	100.8 (ku)				

SURFACE and APPEARANCE

ΔL Color	-0.4 (ac)	-0.3 (ac)	-0.9 (ac)	-1.4 (ac)	-0.5 (ac)	-0.5 (ac)	-1.1 (ac)	-1.9 (ac)				
Δa Color	-0.2 (o)	-0.3 (o)	-0.1 (o)	-0.7 (o)	-0.1 (o)	-0.1 (o)	0.1 (o)	-0.5 (o)				
Δb Color	1.2 (t)	1.5 (t)	3.6 (t)	5.3 (t)	1.2 (t)	1.6 (t)	3.2 (t)	5.6 (t)				

TABLE 90: Effect of Beta Radiation Sterilization on High Density Polyethylene

Material Family	HIGH DENSITY POLYETHYLENE							
Material Supplier/Name	PHILLIPS MARLEX							
Reference No.	101	101	101	101	101	101	101	101
EXPOSURE CONDITIONS								
Type	Beta Radiation	Beta Radiation	Beta Radiation	Beta Radiation	Beta Radiation	Beta Radiation	Beta Radiation	Beta Radiation
Radiation Dose (Mrads)	5	10	15	50	5	10	15	50
PROPERTIES RETAINED (%)								
Property Note	test temperature: 28°C	test temperature: 28°C	test temperature: 28°C	test temperature: 28°C	test temperature: 93°C	test temperature: 93°C	test temperature: 93°C	test temperature: 93°C
Tensile Strength	102.8 (ix)	106 (ix)	107.1 (ix)	111 (ix)	120.3 (iy)	125.8 (iy)	85.7 (iy)	120.3 (iy)
Elongation @ Yield	90 (bh)	110 (bh)	100 (bh)	100 (bh)	224.6 (bi)	311.4 (bi)	302.4 (bi)	79.6 (bi)
SURFACE and APPEARANCE								
Hardness Units Change	D3	D3	D4	D6				
Yellowness Index note	white	ivory	ivory	tan				

TABLE 91: Effect of Ethylene Oxide Sterilization on High Density Polyethylene

Material Family	HIGH DENSITY POLYETHYLENE							
Material Supplier/Name	DOW HDPE 10062							
Reference No.	6	6	6	6	6	6	6	6
EXPOSURE CONDITIONS								
Type	Ethylene Oxide	Ethylene Oxide	Ethylene Oxide	Ethylene Oxide	Ethylene Oxide	Ethylene Oxide	Ethylene Oxide	Ethylene Oxide
Details	12% EtO and 88% Freon	12% EtO and 88% Freon	12% EtO and 88% Freon	12% EtO and 88% Freon	8.6% EtO and 91.4% HCFC-124	8.6% EtO and 91.4% HCFC-124	8.6% EtO and 91.4% HCFC-124	8.6% EtO and 91.4% HCFC-124
Concentration	600 mg/l	600 mg/l	600 mg/l	600 mg/l				
Number of Cycles	1	1	2	2	1	1	2	2
Note	RH: 60%; test lab: Ethox Corp.	RH: 60%; test lab: Ethox Corp.	RH: 60%; test lab: Ethox Corp.	RH: 60%; test lab: Ethox Corp.	RH: 60%; test lab: Ethox Corp.	RH: 60%; test lab: Ethox Corp.	RH: 60%; test lab: Ethox Corp.	RH: 60%; test lab: Ethox Corp.
Temperature (°C)	48.9	48.9	48.9	48.9	48.9	48.9	48.9	48.9
Time (hours)	6	6	6	6	6	6	6	6
PRE EXPOSURE CONDITIONING								
Preconditioning Note	time: 18 hours; temperature: 37.8°C; RH: 60%	time: 18 hours; temperature: 37.8°C; RH: 60%	time: 18 hours; temperature: 37.8°C; RH: 60%	time: 18 hours; temperature: 37.8°C; RH: 60%	time: 18 hours; temperature: 37.8°C; RH: 60%	time: 18 hours; temperature: 37.8°C; RH: 60%	time: 18 hours; temperature: 37.8°C; RH: 60%	time: 18 hours; temperature: 37.8°C; RH: 60%
POST EXPOSURE CONDITIONING								
Note	type: aeration; pressure: 127 mm Hg	type: aeration; pressure: 127 mm Hg	type: aeration; pressure: 127 mm Hg	type: aeration; pressure: 127 mm Hg	type: aeration; pressure: 127 mm Hg	type: aeration; pressure: 127 mm Hg	type: aeration; pressure: 127 mm Hg	type: aeration; pressure: 127 mm Hg
Temperature (°C)	32.2	32.2	32.2	32.2	32.2	32.2	32.2	32.2
POST EXPOSURE CONDITIONING II								
Note	type: ambient conditions	type: ambient conditions	type: ambient conditions	type: ambient conditions	type: ambient conditions	type: ambient conditions	type: ambient conditions	type: ambient conditions
Time (hours)	168	1344	168	1344	168	1344	168	1344
PROPERTIES RETAINED (%)								
Tensile Strength @ Yield	99.2 (il)	99 (il)	99.7 (il)	94.5 (il)	101 (il)	94.8 (il)	101 (il)	92.1 (il)
Elongation	235.8 (bl)	259.3 (bl)	262.4 (bl)	178.8 (bl)	269 (bl)	128.8 (bl)	164.6 (bl)	90.7 (bl)
Modulus	67.9 (gv)	117 (gv)	81.3 (gv)	46.4 (gv)	53.6 (gv)	67.9 (gv)	51.8 (gv)	56.3 (gv)
Dart Impact (total energy)	96.2 (dv)	88.5 (du)	96.2 (dv)	103.8 (dv)	92.3 (dv)	100 (dv)	100 (dv)	96.2 (dv)
Dart Impact (peak energy)	70 (cz)	70 (cv)	70 (cz)	75 (cz)	70 (cz)	75 (cz)	70 (cz)	70 (cz)
SURFACE and APPEARANCE								
ΔE Color	0.66	0.49	0.31	0.71	0.19	0.16	0.33	0.6

TABLE 92: Effect of Ethylene Oxide Sterilization on High Density Polyethylene

Material Family	HIGH DENSITY POLYETHYLENE				
Material Supplier/Name	DOW HDPE 10062				DUPONT TYVEK 1073B
Material Note					spunbonded olefin; sheet
Reference No.	6	6	6	6	45

EXPOSURE CONDITIONS

Type	Ethylene Oxide	Ethylene Oxide	Ethylene Oxide	Ethylene Oxide	Ethylene Oxide
Details	12% EtO and 88% Freon	12% EtO and 88% Freon	8.6% EtO and 91.4% HCFC-124	8.6% EtO and 91.4% HCFC-124	
Concentration	600 mg/l	600 mg/l			
Number of Cycles	1	1	1	1	1
Note	RH: 60%; test lab: Ethox Corp.	RH: 60%; test lab: Ethox Corp.	RH: 60%; test lab: Ethox Corp.	RH: 60%; test lab: Ethox Corp.	
Temperature (°C)	48.9	48.9	48.9	48.9	
Time (hours)	6	6	6	6	

PRE EXPOSURE CONDITIONING

Preconditioning Note	time: 18 hours; temperature: 37.8°C; RH: 60%	time: 18 hours; temperature: 37.8°C; RH: 60%	time: 18 hours; temperature: 37.8°C; RH: 60%	time: 18 hours; temperature: 37.8°C; RH: 60%	

POST EXPOSURE CONDITIONING

Note	type: aeration; note: 10 air changes per hour	type: aeration; note: 30 air changes per hour	type: aeration; note: 10 air changes per hour	type: aeration; note: 30 air changes per hour	
Temperature (°C)	32.2	54.4	32.2	54.4	

RESIDUALS (ppm)

Residuals Determined	ethylene oxide	ethylene oxide	ethylene oxide	ethylene oxide	ethylene oxide
Little or No Aeration	388	388	377	377	3
17 hour Aeration		39		39	
24 hour Aeration	72	25	79	35	0.9
48 hour Aeration	26	4	54	20	0.7
72 hour Aeration	11		22		
168 hour Aeration					0

TABLE 93: Effect of Steam Sterilization on High Density Polyethylene

Material Family	HIGH DENSITY POLYETHYLENE		
Material Supplier/Name	DUPONT TYVEK 1073B		
Material Note	spunbonded olefin; sheet	spunbonded olefin; sheet	spunbonded olefin; sheet
Reference No.	45	45	45

EXPOSURE CONDITIONS

Type	Steam	Steam	Steam
Number of Cycles	1	1	1
Temperature (°C)	121.1	123.9	126.7

PROPERTIES RETAINED (%)

Tensile Strength	104.8 (hf)	115.5 (hf)	115 (hf)
Tensile Strength (transverse)	46.7 (hi)	46.7 (hi)	49.9 (hi)
Tear Resistance	109.3 (af)	116.3 (ae)	173.3 (ae)
Tear Resistance (transverse)	108.6 (ag)	109.7 (ag)	161.3 (ag)

Gamma Radiation Resistance

Dow Chemical: Attane

Attane copolymers can accept relatively high levels of radiation with little or no adverse effect on properties.

Reference: *Attane Ultra Low Density Ethylene-Octene Copolymers: Performance Plus Compared To LLDPE And EVA Resins In Flexible Packaging,* supplier marketing literature (305-1596-790) - Dow Chemical Company, 1989.

Gamma Radiation Resistance

Exxon: (composition: 19.3% EVA; Additives: <50 ppm BHT (2,6 di-t-butyl-4-methylphenol), 350 ppm BHEB (2,6 di-t-butyl-4-ethylphenol; characteristics: melt index: 22.5 grams/10 min.; features: antioxidant stabilizer)

Samples were injection molded into ASTM test parts under standard conditions. Each set of parts were then divided into lots representing the various aging and irradiation conditions. Parts were irradiated at 0, 25, 50, 75 kGy (0.0, 2.5, 5.0, 7.5 Mrads). One group of parts was then tested immediately after irradiation and another after 21 days aging at 60°C. A third group of parts is in ambient temperature storage for testing at a later date. The ASTM properties tested were the tensile elongation at break and Gardner impact strength at 23°C (for polyolefins a significant fall in Gardner impact strength indicated embrittlement). The Hunter b color of each sample using Gardner discs was also determined. The investigators used their proprietary flex-to-failure test*, which they have found to be a sensitive measure of embrittlement due to irradiation. Using this test they determined the deflection at peak flexural load and the mode of flexural failure, either ductile or brittle. Past work pointed out that the accelerated aging condition is the most sensitive measure of the radiation effects on thermoplastics in lieu of the results of real time aging. Therefore data is presented on parts aged at 60°C for 21 days after irradiation.

EVA (19.3%) has a high level of stability to sterilizing doses of gamma radiation. Gardner impact strength rose with increasing radiation, suggesting the occurrence of crosslinking. Failure in the Gardner test was ductile. EVA demonstrated a meaningful increase in impact strength, following expectations that a high EVA polyethylene is expected to be susceptible to crosslinking.

Tensile elongation at break increased with irradiation and aging. This suggests no chain scission. If chain scission had occurred to a significant degree, elongation as small as 10-15% might have signaled embrittlement. The observed increases are not consistent with chain scission.

The deflection at peak flexural load offered final proof that degradation was not a significant process operating in any of these polyothylenes after irradiation. Significant embrittlement due to chain scission is normally accompanied in polypropylene by reductions in deflection at peak flexural load of 25% or more, the values recorded for these polyethylenes were essentially constant over all radiation doses up to 75 kGy (7.5 Mrads).

Color changes in irradiated polyolefins result from the effects of the radiation upon the phenolic antioxidants rather than from interaction between the radiation and the plastic molecules. EVA was relatively highly colored and irradiation had no significant effect; the color remained essentially constant at all levels of irradiation and aging.

The study verified that sterilizing doses of high energy radiation have negligible effect on the physical mechanical properties and color of EVA. It does not appear that hindered amine light stabilizers are needed to protect polyethylenes against sterilizing doses of high energy radiation.

*R.C. Portnoy and V.R. Cross, "Method for Evaluating the Gamma Radiation Tolerance of Polypropylene for Medical Device Applications", Proceedings Society of Plastics Engineers ANTEC, XXXVI, 1826, Montreal (1991).

Reference: Portnoy, R. C., *The Response Of Various Polyethylenes To High Energy Radiation,* ANTEC 1994, conference proceedings - Society of Plastics Engineers, 1994.

TABLE 93a: Effect of Gamma Radiation Sterilization on Ethylene Vinal Acetate Polyethylene Copolymer

Material Family	ETHYLENE VINYL ACETATE POLYETHYLENE COPOLYMER		
Material Supplier/Name	EXXON		
Material Note	19.3% EVA; Additives :<50 ppm BHT (2,6 di-t-butyl-4-methylphenol), 350 ppm BHEB (2,6 di-t-butyl-4-ethylphenol; melt index: 22.5 g/10 min.; antioxidant stabilizer	19.3% EVA; Additives : <50 ppm BHT (2,6 di-t-butyl-4-methylphenol), 350 ppm BHEB (2,6 di-t-butyl-4-ethylphenol; melt index: 22.5 g/10 min.; antioxidant stabilizer	19.3% EVA; Additives : <50 ppm BHT (2,6 di-t-butyl-4-methylphenol), 350 ppm BHEB (2,6 di-t-butyl-4-ethylphenol; melt index: 22.5 g/10 min.; antioxidant stabilizer
Reference No.	23	23	23
EXPOSURE CONDITIONS			
Type	Gamma Radiation	Gamma Radiation	Gamma Radiation
Radiation Dose (Mrads)	2.5	5	7.5
POST EXPOSURE CONDITIONING			
Temperature (°C)	60	60	60
Time (hours)	504	504	504
PROPERTIES RETAINED (%)			
Elongation	100 (ak)	100 (ak)	100 (ak)
Gardner Impact	112.4 (cq)	117.5 (cq)	133 (cq)
Deflection @ Failure	96.4 (i)	100 (i)	98.5 (i)
SURFACE and APPEARANCE			
Δb Color	1 (z)	0.7 (z)	1.1 (z)

Radiation Resistance

Quantum: Petrothene

Waste containers and bags extruded from polypropylene can be sterilized by radiation.

Reference: *Performance Polyolefins For The Healthcare Market,* supplier marketing literature (6615/591) - Quantum Chemical Corporation, 1991.

Gamma Radiation Resistance

Himont: Profax PF511 (features: gamma radiation stabilized, high flow)

Pro-fax PF511 resists embrittlement after 5 Mrads of gamma radiation.

Reference: *Pro-fax PF511 High-Flow, Radiation-Resistant Polypropylene,* supplier marketing literature - Himont, 1989.

Eastman: Tenite P7673-960A (melt flow rate: 18 g/10 min.; features: gamma radiation stabilized); **Tenite P7673-984A** (melt flow rate: 18 g/10 min.; features: gamma radiation stabilized); **Tenite** (features: general purpose grade)

When a normally stabilized polypropylene is subjected to gamma radiation for sterilization purposes, two major changes may take place. A yellow color may develop due primarily to a chemical change in certain stabilizers that is initiated by radiation. Radiation will alter the chemical structure in such a way that it becomes more susceptible to oxidation. Therefore, the time required for an irradiated part to discolor or embrittle may be very short. Based on correlations between room temperature aging and accelerated aging at 43°C, the estimated shelf life of these materials after sterilization with gamma rays is three to five years. It appears from these correlations that aging for 30 days at 43°C is approximately equivalent to aging one year at 23°C. The materials studied were irradiated to 2.5 Mrads at an average dose rate of 0.1 Mrad/hour. Since a low rate is more harmful than a high dose rate, the data were generated under the worst conditions likely to be encountered.

The two most significant tests to monitor the effect of radiation on polypropylene are elongation at break and color. Samples were tested prior to irradiation (2.5 Mrads), immediately after irradiation but prior to oven aging, and at 30 day intervals (up to 180 days) after oven aging at 43°C. Elogation decreases to a certain level upon irradiation and apparently holds close to that level with time. Polypropylene with radiation stabilizers has high elongation after irradiation, indicating physical properties are maintained at a desirable level. Elongation of most regular polypropylenes drops to an unacceptable level after irradiation. The resulting embrittlement would make regular polypropylene unacceptable for producing medical devices that require radiation sterilization.

A color change due to radiation sterilization will be noticeable immediately after sterilization. This yellowness may decrease slightly during the early stages of aging. However, unless the material is properly stabilized, color will increase again with continued aging. Tensile yield strength and notched Izod do not appear to change significantly after irradiation.

Reference: *Tenite Polypropylene And Tenite Polyallomer Materials For Radiation Sterilization,* supplier technical report (MB-93) - Eastman Plastics, 1985.

Exxon: Polypropylene (composition: novel polypropylene copolymer with a hindered amine light stabilizer, no nucleator; features: gamma radiation stabilized, transparent, developmental material); **Polypropylene** (composition: polypropylene homopolymer with a hindered amine light stabilizer, no nucleator; features: gamma radiation stabilized, transparent, developmental material); **Polypropylene** (composition: normal ethylene random copolymer with a hindered amine light stabilizer, with nucleator; features: gamma radiation stabilized, transparent, developmental material)

Laboratory testing of this material has demonstrated that it retains color, elongation at break, and resistance to dart drop impact and penetration extremely well after irradiation and aging. In these respects it is very suitable for packaging medical devices.

Four principal techniques were used to achieve improvements in clarity and radiation tolerance of polypropylene: ethylene copolymerization, additives, processing improvements, and novel molecular characteristics. While any of these can be used to obtain some improvement in resin properties, the enhancement of properties such as clarity and radiation tolerance in polypropylene is usually best affected by a combination of these methods.

These techniques were used to design a polypropylene sheet that draws radiation tolerance and clarity from a combination of ethylene content, molecular structure, antioxidant additives, and forming technique. The material was produced from a novel type of copolymer of propylene and containing a hindered amine light stabilizer, but no nucleator. The sheet was blown with minimal orientation in an apparatus that provided exceptionally rapid chilling of the bubble. The result was a sheet containing a minimum of the hazy and radiation sensitive alpha crystalline form of polypropylene and a maximum of very clear, radiation tolerant disordered amorphous and slightly ordered smectic polypropylenes.

This material is far superior in terms of toughness and resistance to radiation as compared to two other test materials: 1) a sheet produced from a propylene homopolymer containing a similar radiation stabilization package and 2) a sheet prepared from a normal ethylene random copolymer containing a similar radiation stabilization package and a powerful polypropylene nucleating agent.

The superior performance of the material is also demonstrated by its retention of elongation at break and resistance to dart impact and puncture. The material's radiation tolerance is directly attributable to a design that incorporates contributions from all four of the significant factors in improving polypropylene performance. The use of random ethylene comonomer, unique molecular characteristics related to the incorporation of the random ethylene, a special process involving extremely rapid chilling of the blown melt, and a proper selection of additives all contribute synergistically to the result. The four important design factors combined have produced a highly radiation tolerant, very clear polypropylene sheet for use in medical device packaging.

Reference: Portnoy, R. C., Gulla, C. T., Kozimor, R. A., *Extremely Clear, Radiation Tolerant Polypropylene Sheet For Medical Device Packaging,* supplier technical report - Exxon Chemical Company.

Rexene: 13R9A (composition: random ethylene copolymers; features: gamma radiation stabilized)

Samples were exposed to 3.0 and 5.0 Mrads. The yellowness index, both light and dark, changes very little. Specimens exposed to nominal doses of gamma radiation undergo an increase in melt flow rate. The increase from a nominal 7.0 MFR to about 40 is a result of the chain scission effect gamma radiation has on polypropylene. Despite the scission effect, specimens maintain tensile strength and ultimate elongation essentially identical to the non irradiated control samples up through a nominal 3.0 Mrad dosage. At the higher dosage of nominal 5.0 Mrads, the ultimate elongation declines without any significant increase in tensile strength. The approximate 600% ultimate elongation obtained with the highest dosage still indicates a ductile, flexible, nonbrittle specimen. The yellowness index increases with increasing radiation dosages (range is from -1.7 for the control to about 5.0 for the highest dosage). Plaques exposed to fluorescent light decreased in yellowness to approach a neutral blue yellow color, while those kept in sealed dark containers increased slightly in yellowness.

Accelerated testing was performed on pellets, tensile bars and plaques exposed to nominal 1.5, 3.0 and 5.0 Mrads. The specimens were then separated into two groups, each with the varying Cobalt 60 dosages. One group was maintained at 40°C and the other at 60°C in hot air circulating ovens. Specimens at 40°C showed essentially no change over the test period regardless of dosage. Specimens at 60°C retained ductility and tensile strength over the test period. At all test levels, both ultimate elongation and tensile strength remain essentially constant.

Reference: *From Rexene Technology: Radiation Resistant Polypropylenes,* supplier technical report (EPP 7037 10/87 4M JP) - jRexene Products Company, 1987.

Rexene: 13R9A (composition: random ethylene copolymers; features: gamma radiation stabilized)

Rexene proprietary additive packages for polypropylene allow the modified resin to be exposed to sterilizing doses (5 Mrads) of gamma radiation and maintain useful physical properties for at least four years (at room temperature) after exposure.

Tensile bars of radiation resistant polypropylene (PP13R15A Lot 24669) were also tested after accelerated aging at both 80°C and 120°C for 3 and 6 weeks; bars were suspended in a forced draft oven. The samples were irradiated with a nominal 5.0 Mrads of gamma radiation from a Cobalt 60 source. After irradiation and ASTM conditioning, physical properties were determined. After 3 weeks at 80°C tensile strength at yield increases from 25.3 MPa immediately after irradiation to 28.6 MPa. Elongation at break decreases to 300% from 680%. Yellowness index increases from 5.0 to 6.9, an amount barely detectable to the human eye in side-by side comparisons. After 6 weeks at 80°C, samples maintain their physical properties. A group of tensile bars from the same batch were also stored in a controlled environment of 23°C and 50% relative humidity. After one year, physical property values lay between the values obtained for 0 and 3 weeks at 80°C. These results support the contention that three weeks at 80°C is equivalent to a year at ambient conditions.

Accelerated aging at 120°C can induce embrittlement and failure. Tensile bars were molded from Lot 24669 and then exposed to doses of 3.1 and 3.0 Mrads on consecutive days. After one day at 120°C embrittlement is evident in the substantial reduction of elongation at break. Oven aging was stopped after 7 days when embrittlement was so severe the tensile bars broke before a yield point was encountered. Also the yellowness index increased to nearly 20. Results from other experiments indicate that the radiation resistant polypropylene can withstand oven aging at 120°C for a week, with a significant increase in yellowness index. Other experiments show that a single dose of 6 Mrads does not cause failure. It appears as if two equal doses a day apart are more damaging than a single dose of double intensity.

The nature of the radiation source may influence post irradiation performance of polypropylene. To test this concern, samples of tensile bars made from PP13R9A Lot 24043 random copolymer were sent to four different locations and irradiated at a nominal 3.0 Mrads. Three locations used Cobalt 60 and one used Cesium 137 as primary sources of gamma radiation. Locations with Cobalt 60 had source intensities ranging from 1.6 megacuries to 3.5 megacuries. The Cesium source had an intensity of 1.5 megacuries. The yellowness index is slightly higher in samples exposed to the 2.0 megacurie source of Cobalt 60 compared to samples exposed to the 1.6 or 3.5 megacurie source of the same isotope. Compared to Cobalt 60 sources, radiation from Cesium 137 appears to cause significantly more yellow color in the radiation resistant polypropylene. Cobalt 60 emits two gammas at 1.17 Mev and 1.33 Mev and a beta with 0.31 Mev. Cesium 137 emits a 1.18 Mev gamma and a 0.51 Mev beta. Perhaps the more energetic beta causes more yellow color, or perhaps the decay of other isotopes causes the difference.

Reference: Lucas, Ben M., Paton, Samuel J., Thakker, M. T., *Radiation Resistant Polypropylene,* Medical Plastics '90 International Conference, Malmo, Sweden, conference proceedings - Rexene Products Company, 1990.

Ethylene Oxide (EtO) Resistance

Himont: Profax 6323

The tensile yield after 1 cycle of HCFC-124/EtO shows a decrease of 20%. After 2 cycles of the same carrier gas mixture there is no such decrease in tensile yield which suggests that a decrease in tensile strength is not typical. The tensile modulus is affected by EtO sterilization and is seen 8 weeks after EtO exposure. The tensile modulus after 2 cycles of EtO decreases by 78%. The instrumented impact data after sterilization appears to show a slight decrease.

Reference: Hermanson, Nancy J., *Effects Of Alternate Carriers Of Ethylene Oxide Sterilant On Thermoplastics,* supplier technical report (301-02018) - Dow Chemical Company.

Rexene: 13R9A (composition: random ethylene copolymers; features: gamma radiation stabilized)

Polypropylene can be ethylene oxide sterilized as the polymer is chemically very inert.

Reference: Lucas, Ben M., Paton, Samuel J., Thakker, M. T., *Radiation Resistant Polypropylene,* Medical Plastics '90 International Conference, Malmo, Sweden, conference proceedings - Rexene Products Company, 1990.

Quantum: Petrothene

Waste containers and bags extruded from polypropylene can be sterilized by ethylene oxide.

Reference: *Performance Polyolefins For The Healthcare Market,* supplier marketing literature (6615/591) - Quantum Chemical Corporation, 1991.

Steam Resistance

Rexene: 13R9A (composition: random ethylene copolymers; features: gamma radiation stabilized)

Polypropylene items can be sterilized in autoclaves as polypropylene can withstand temperatures up to 125°C.

Lucas, Ben M., Paton, Samuel J., Thakker, M. T., *Radiation Resistant Polypropylene,* Medical Plastics '90 International Conference, Malmo, Sweden, conference proceedings - Rexene Products Company, 1990.

Quantum: Petrothene

Waste containers and bags extruded from polypropylene can be autoclaved.

Reference: *Performance Polyolefins For The Healthcare Market,* supplier marketing literature (6615/591) - Quantum Chemical Corporation, 1991.

TABLE 94: Effect of Gamma Radiation Sterilization on Polypropylene

Material Family	**POLYPROPYLENE**
Material Supplier/Name	**HIMONT PROFAX PF511**
Material Note	gamma radiation stabilized, high flow
Reference No.	69

EXPOSURE CONDITIONS

Type	Gamma Radiation
Radiation Dose (Mrads)	5

PROPERTIES RETAINED (%)

Tensile Strength	100 (jb)

SURFACE and APPEARANCE

Δ Yellowness Index	2 (kv)

TABLE 95: Effect of Gamma Radiation Sterilization on Polypropylene

Material Family	POLYPROPYLENE						
Material Supplier/Name	REXENE 13R9A						
Material Note	random ethylene copolymer; gamma radiation stabilized	random ethylene copolymer; gamma radiation stabilized	random ethylene copolymer; gamma radiation stabilized	random ethylene copolymer; gamma radiation stabilized	random ethylene copolymer; gamma radiation stabilized	random ethylene copolymer; gamma radiation stabilized	random ethylene copolymer; gamma radiation stabilized
Reference No.	104	104	104	104	104	104	104

EXPOSURE CONDITIONS

Type	Gamma Radiation	Gamma Radiation	Gamma Radiation	Gamma Radiation	Gamma Radiation	Gamma Radiation	Gamma Radiation
Details	source: Cobalt 60	source: Cobalt 60	source: Cobalt 60	source: Cobalt 60	source: Cobalt 60	source: Cobalt 60	source: Cobalt 60
Radiation Dose (Mrads)	6.1	6.1	6.1	5	5	5	5
Note	doses of 3.1 Mrads and 3.0 Mrads on consecutive days	doses of 3.1 Mrads and 3.0 Mrads on consecutive days	doses of 3.1 Mrads and 3.0 Mrads on consecutive days				

POST EXPOSURE CONDITIONING

Note	type: ASTM conditioning	type: ASTM conditioning	type: ASTM conditioning	type: ASTM conditioning	type: ASTM conditioning	type: ASTM conditioning	type: ASTM conditioning

POST EXPOSURE CONDITIONING II

Note		type: forced draft air oven	type: forced draft air oven		type: forced draft air oven	type: forced draft air oven	type: ambient conditions
Temperature (°C)		120	120		80	80	23
Time (hours)	0	24	168	0	504	1008	8760

PROPERTIES RETAINED (%)

Tensile Strength @ Yield	106.7 (ih)	113.1 (ih)	0 (ih)	94.1 (ih)	106.3 (ih)	106.3 (ih)	104.8 (ih)
Elongation	100 (as)	6.5 (as)	3.2 (as)	100 (as)	50 (as)	60 (as)	53.3 (as)

SURFACE and APPEARANCE

Δ Yellowness Index	5.9 (kv)	9.7 (kv)	18.7 (kv)	1 (kv)	2.9 (kv)	3.2 (kv)	2.3 (kv)

TABLE 96: Effect of Gamma Radiation Sterilization on Polypropylene

Material Family	POLYPROPYLENE					
Material Supplier/Name	REXENE 13R9A					
Material Note	random ethylene copolymer; gamma radiation stabilized	random ethylene copolymer; gamma radiation stabilized	random ethylene copolymer; gamma radiation stabilized	random ethylene copolymer; gamma radiation stabilized	random ethylene copolymer; gamma radiation stabilized	random ethylene copolymer; gamma radiation stabilized
Reference No.	104	104	104	104	104	104
EXPOSURE CONDITIONS						
Type	Gamma Radiation	Gamma Radiation	Gamma Radiation	Gamma Radiation	Gamma Radiation	Gamma Radiation
Details	source: Cobalt 60	source: Cobalt 60	source: Cobalt 60	source: Cobalt 60	source: Cobalt 60	source: Cesium 137
Radiation Dose (Mrads)	5	5	3	3	3	3
Note			intensity: 1.6 megacuries	intensity: 2.0 megacuries	intensity: 3.5 megacuries	intensity: 1.5 megacuries
PRE EXPOSURE CONDITIONING						
Preconditioning Note	type: reprocessing	type: reprocessing				
Times Reprocessed	0	7				
POST EXPOSURE CONDITIONING						
Note	type: ASTM conditioning	type: ASTM conditioning				
PROPERTIES RETAINED (%)						
Tensile Strength @ Yield	103.4 (ih)	97.8 (ih)	98.9 (ih)	99.3 (ih)	96.7 (ih)	95.9 (ih)
Elongation	91.7 (as)	63.3 (as)	93.3 (as)	86.7 (as)	73.3 (as)	83.3 (as)
SURFACE and APPEARANCE						
Δ Yellowness Index	2.1 (kv)	1.4 (kv)	0.9 (kv)	3 (kv)	1.4 (kv)	10 (kv)

TABLE 97: Effect of Gamma Radiation Sterilization on Polypropylene

Material Family	POLYPROPYLENE					
Material Supplier/Name	REXENE 13R9A					
Material Note	random ethylene copolymer; gamma radiation stabilized	random ethylene copolymer; gamma radiation stabilized	random ethylene copolymer; gamma radiation stabilized	random ethylene copolymer; gamma radiation stabilized	random ethylene copolymer; gamma radiation stabilized	random ethylene copolymer; gamma radiation stabilized
Reference No.	99	99	99	98	98	98

EXPOSURE CONDITIONS

Type	Gamma Radiation	Gamma Radiation	Gamma Radiation	Gamma Radiation	Gamma Radiation	Gamma Radiation
Radiation Dose (Mrads)	1.5	3	5	1.5	3	5

POST EXPOSURE CONDITIONING

Note	type: storage under fluorescent light	type: storage under fluorescent light	type: storage under fluorescent light	type: storage under fluorescent light	type: storage under fluorescent light	type: storage under fluorescent light
Time (hours)	336	336	336	2190	2190	2190

PROPERTIES RETAINED (%)

Tensile Strength @ Yield	100.8 (ih)	101.8 (ih)	102.3 (ih)	100.7 (ih)	100.5 (ih)	101 (ih)
Elongation	100 (as)	100 (as)	82.9 (as)	100 (as)	100 (as)	82.9 (as)
Melt Flow Rate	430.6 (fk)	736.1 (fk)	625 (fk)	578.6 (fk)	857.1 (fk)	652.9 (fk)

SURFACE and APPEARANCE

Δ Yellowness Index	3.4 (kx)	4.9 (kx)	6.7 (kx)	2.4 (kx)	2.9 (kx)	3 (kx)

TABLE 98: Effect of Gamma Radiation Sterilization on Polypropylene

Material Family	POLYPROPYLENE					
Material Supplier/Name	REXENE 13R9A					
Material Note	random ethylene copolymer; gamma radiation stabilized	random ethylene copolymer; gamma radiation stabilized	random ethylene copolymer; gamma radiation stabilized	random ethylene copolymer; gamma radiation stabilized	random ethylene copolymer; gamma radiation stabilized	random ethylene copolymer; gamma radiation stabilized
Reference No.	98	98	98	98	98	98
EXPOSURE CONDITIONS						
Type	Gamma Radiation	Gamma Radiation	Gamma Radiation	Gamma Radiation	Gamma Radiation	Gamma Radiation
Radiation Dose (Mrads)	1.5	3	5	1.5	3	5
POST EXPOSURE CONDITIONING						
Note	type: storage under fluorescent light	type: storage under fluorescent light	type: storage under fluorescent light	type: storage under fluorescent light	type: storage under fluorescent light	type: storage under fluorescent light
Time (hours)	4380	4380	4380	8766	8766	8766
PROPERTIES RETAINED (%)						
Tensile Strength @ Yield	100.7 (ih)	100.2 (ih)	101 (ih)	99.5 (ih)	98.3 (ih)	97.8 (ih)
Elongation	100 (as)	100 (as)	75.7 (as)	100 (as)	100 (as)	100 (as)
Melt Flow Rate	567.6 (fk)	689.2 (fk)	689.2 (fk)	691.7 (fk)	805.6 (fk)	534.7 (fk)
SURFACE and APPEARANCE						
Δ Yellowness Index	2.4 (kx)	2.4 (kx)	2.8 (kx)	2.2 (kx)	2.2 (kx)	2.3 (kx)

TABLE 99: Effect of Gamma Radiation Sterilization on Polypropylene

Material Family	POLYPROPYLENE					
Material Supplier/Name	REXENE 13R9A					
Material Note	random ethylene copolymer; gamma radiation stabilized	random ethylene copolymer; gamma radiation stabilized	random ethylene copolymer; gamma radiation stabilized	random ethylene copolymer; gamma radiation stabilized	random ethylene copolymer; gamma radiation stabilized	random ethylene copolymer; gamma radiation stabilized
Reference No.	98	98	98	99	99	99
EXPOSURE CONDITIONS						
Type	Gamma Radiation	Gamma Radiation	Gamma Radiation	Gamma Radiation	Gamma Radiation	Gamma Radiation
Radiation Dose (Mrads)	1.5	3	5	15	3	5
POST EXPOSURE CONDITIONING						
Note	type: storage under fluorescent light	type: storage under fluorescent light	type: storage under fluorescent light	type: storage under fluorescent light	type: storage under fluorescent light	type: storage under fluorescent light
Time (hours)	13140	13140	13140	17531	17531	17531
PROPERTIES RETAINED (%)						
Tensile Strength @ Yield	99.3 (ih)	99.5 (ih)	99 (ih)	100 (ih)	100.2 (ih)	101 (ih)
Elongation	100 (as)	100 (as)	100 (as)	100 (as)	100 (as)	100 (as)
Melt Flow Rate	691.7 (fk)	805.6 (fk)	534.7 (fk)	554.1 (fk)	770.3 (fk)	752.7 (fk)
SURFACE and APPEARANCE						
Δ Yellowness Index	1.8 (kx)	2 (kx)	2 (kx)	0.9 (kx)	0.7 (kx)	0.7 (kx)

TABLE 100: Effect of Gamma Radiation Sterilization on Polypropylene

Material Family	POLYPROPYLENE					
Material Supplier/Name	REXENE 13R9A					
Material Note	random ethylene copolymer; gamma radiation stabilized	random ethylene copolymer; gamma radiation stabilized	random ethylene copolymer; gamma radiation stabilized	random ethylene copolymer; gamma radiation stabilized	random ethylene copolymer; gamma radiation stabilized	random ethylene copolymer; gamma radiation stabilized
Reference No.	99	99	99	98	98	98
EXPOSURE CONDITIONS						
Type	Gamma Radiation	Gamma Radiation	Gamma Radiation	Gamma Radiation	Gamma Radiation	Gamma Radiation
Radiation Dose (Mrads)	1.5	3	5	1.5	3	5
POST EXPOSURE CONDITIONING						
Note	type: storage in dark	type: storage in dark	type: storage in dark	type: storage in dark	type: storage in dark	type: storage in dark
Time (hours)	336	336	336	2190	2190	2190
SURFACE and APPEARANCE						
Δ Yellowness Index	3.5 (kx)	4.9 (kx)	6.7 (kx)	4.5 (kx)	5.7 (kx)	6.9 (kx)

TABLE 101: Effect of Gamma Radiation Sterilization on Polypropylene

Material Family	POLYPROPYLENE					
Material Supplier/Name	REXENE 13R9A					
Material Note	random ethylene copolymer; gamma radiation stabilized	random ethylene copolymer; gamma radiation stabilized	random ethylene copolymer; gamma radiation stabilized	random ethylene copolymer; gamma radiation stabilized	random ethylene copolymer; gamma radiation stabilized	random ethylene copolymer; gamma radiation stabilized
Reference No.	98	98	98	99	99	99

EXPOSURE CONDITIONS

Type	Gamma Radiation	Gamma Radiation	Gamma Radiation	Gamma Radiation	Gamma Radiation	Gamma Radiation
Radiation Dose (Mrads)	1.5	3	5	1.5	3	5

POST EXPOSURE CONDITIONING

Note	type: storage in dark	type: storage in dark	type: storage in dark	type: storage in dark	type: storage in dark	type: storage in dark
Time (hours)	4380	4380	4380	8765	8765	8765

SURFACE and APPEARANCE

Δ Yellowness Index	3.4 (kx)	5.2 (kx)	7.1 (kx)	5.1 (kx)	7.2 (kx)	7.1 (kx)

TABLE 102: Effect of Gamma Radiation Sterilization on Polypropylene

Material Family	POLYPROPYLENE					
Material Supplier/Name	REXENE 13R9A					
Material Note	random ethylene copolymer; gamma radiation stabilized	random ethylene copolymer; gamma radiation stabilized	random ethylene copolymer; gamma radiation stabilized	random ethylene copolymer; gamma radiation stabilized	random ethylene copolymer; gamma radiation stabilized	random ethylene copolymer; gamma radiation stabilized
Reference No.	98	98	98	99	99	99

EXPOSURE CONDITIONS

Type	Gamma Radiation	Gamma Radiation	Gamma Radiation	Gamma Radiation	Gamma Radiation	Gamma Radiation
Radiation Dose (Mrads)	1.5	3	5	1.5	3	5

POST EXPOSURE CONDITIONING

Note	type: storage in dark	type: storage in dark	type: storage in dark	type: storage in dark	type: storage in dark	type: storage in dark
Time (hours)	13140	13140	13140	17531	17531	17531

SURFACE and APPEARANCE

Δ Yellowness Index	5.5 (kx)	6.6 (kx)	7.4 (kx)	5.6 (kx)	6.6 (kx)	7.3 (kx)

TABLE 103: Effect of Gamma Radiation Sterilization on Polypropylene

Material Family	POLYPROPYLENE					
Material Supplier/Name	EXXON					
Material Note	polypropylene homopolymer with a hindered amine light stabilizer, no nucleator; gamma radiation stabilized, transparent, developmental material		novel polypropylene copolymer with a hindered amine light stabilizer, no nucleator; gamma radiation stabilized, transparent, developmental material		normal ethylene random copolymer with a hindered amine light stabilizer, with nucleator; gamma radiation stabilized, transparent, developmental material	
Reference No.	67	67	67	67	67	67
EXPOSURE CONDITIONS						
Type	Gamma Radiation	Gamma Radiation	Gamma Radiation	Gamma Radiation	Gamma Radiation	Gamma Radiation
Details	source: Cobalt 60	source: Cobalt 60	source: Cobalt 60	source: Cobalt 60	source: Cobalt 60	source: Cobalt 60
Radiation Dose (Mrads)	2.5	5	2.5	5	2.5	5
Note	dose rate: 10 kGy/hr	dose rate: 10 kGy/hr	dose rate: 10 kGy/hr	dose rate: 10 kGy/hr	dose rate: 10 kGy/hr	dose rate: 10 kGy/hr
POST EXPOSURE CONDITIONING						
Time (hours)	0	0	0	0	0	0
PROPERTIES RETAINED (%)						
Elongation	77.1 (al)	25.7 (al)	92 (al)	79 (al)	91 (al)	78 (al)
Dart Impact (total energy)	47.6 (dq)	28.6 (dq)				

TABLE 104: Effect of Gamma Radiation Sterilization on Polypropylene

Material Family	POLYPROPYLENE			
Material Supplier/Name	EXXON			
Material Note	novel polypropylene copolymer with a hindered amine light stabilizer, no nucleator; gamma radiation stabilized, transparent, developmental material		normal ethylene random copolymer with a hindered amine light stabilizer, with nucleator; gamma radiation stabilized, transparent, developmental material	
Reference No.	67	67	67	67
EXPOSURE CONDITIONS				
Type	Gamma Radiation	Gamma Radiation	Gamma Radiation	Gamma Radiation
Details	source: Cobalt 60	source: Cobalt 60	source: Cobalt 60	source: Cobalt 60
Radiation Dose (Mrads)	2.5	5	2.5	5
Note	dose rate: 10 kGy/hr	dose rate: 10 kGy/hr	dose rate: 10 kGy/hr	dose rate: 10 kGy/hr
POST EXPOSURE CONDITIONING				
Note	type: aging	type: aging	type: aging	type: aging
Time (hours)	504	504	504	504
PROPERTIES RETAINED (%)				
Elongation	95 (al)	79 (al)	68 (al)	0 (al)

TABLE 105: Effect of Gamma Radiation Sterilization on Polypropylene

Material Family	POLYPROPYLENE				
Material Supplier/Name	EASTMAN TENITE				
Material Note	general purpose grade	general purpose grade	general purpose grade	general purpose grade	general purpose grade
Reference No.	55	55	55	55	55
EXPOSURE CONDITIONS					
Type	Gamma Radiation	Gamma Radiation	Gamma Radiation	Gamma Radiation	Gamma Radiation
Radiation Dose (Mrads)	2.5	2.5	2.5	2.5	2.5
Note	dose rate: 0.1 Mrads/hr.	dose rate: 0.1 Mrads/hr.	dose rate: 0.1 Mrads/hr.	dose rate: 0.1 Mrads/hr.	dose rate: 0.1 Mrads/hr.
POST EXPOSURE CONDITIONING					
Note		type: accelerated aging	type: accelerated aging	type: accelerated aging	type: accelerated aging
Temperature (°C)		43.3	43.3	43.3	43.3
Time (hours)	0	720	1440	2880	4320
PROPERTIES RETAINED (%)					
Elongation	16 (aq)	16 (aq)	16 (aq)	16 (aq)	16 (aq)
SURFACE and APPEARANCE					
Δb Color	6.3 (x)	7.1 (x)	6.9 (x)	6.8 (x)	7.1 (x)

TABLE 106: Effect of Gamma Radiation Sterilization on Polypropylene

Material Family	POLYPROPYLENE				
Material Supplier/Name	EASTMAN TENITE P7673-960A				
Material Note	melt flow rate: 18 g/10 min.; gamma radiation stabilized				
Reference No.	55	55	55	55	55

EXPOSURE CONDITIONS

Type	Gamma Radiation	Gamma Radiation	Gamma Radiation	Gamma Radiation	Gamma Radiation
Radiation Dose (Mrads)	2.5	2.5	2.5	2.5	2.5
Note	dose rate: 0.1 Mrads/hr.	dose rate: 0.1 Mrads/hr.	dose rate: 0.1 Mrads/hr.	dose rate: 0.1 Mrads/hr.	dose rate: 0.1 Mrads/hr.

POST EXPOSURE CONDITIONING

Note		type: accelerated aging	type: accelerated aging	type: accelerated aging	type: accelerated aging
Temperature (°C)		43.3	43.3	43.3	43.3
Time (hours)	0	720	1440	2880	4320

PROPERTIES RETAINED (%)

Elongation	71.4 (aq)	66.7 (aq)	83.3 (aq)	73.8 (aq)	73.8 (aq)

SURFACE and APPEARANCE

Δb Color	6.1 (x)	5.4 (x)	5.4 (x)	6 (x)	6 (x)

TABLE 107: Effect of Gamma Radiation Sterilization on Polypropylene

Material Family	POLYPROPYLENE				
Material Supplier/Name	EASTMAN TENITE P7673-984A				
Material Note	melt flow rate: 18 g/10 min.; gamma radiation stabilized	melt flow rate: 18 g/10 min.; gamma radiation stabilized	melt flow rate: 18 g/10 min.; gamma radiation stabilized	melt flow rate: 18 g/10 min.; gamma radiation stabilized	melt flow rate: 18 g/10 min.; gamma radiation stabilized
Reference No.	55	55	55	55	55
EXPOSURE CONDITIONS					
Type	Gamma Radiation	Gamma Radiation	Gamma Radiation	Gamma Radiation	Gamma Radiation
Radiation Dose (Mrads)	2.5	2.5	2.5	2.5	2.5
Note	dose rate: 0.1 Mrads/hr.	dose rate: 0.1 Mrads/hr.	dose rate: 0.1 Mrads/hr.	dose rate: 0.1 Mrads/hr.	dose rate: 0.1 Mrads/hr.
POST EXPOSURE CONDITIONING					
Note		type: accelerated aging	type: accelerated aging	type: accelerated aging	type: accelerated aging
Temperature (°C)		43.3	43.3	43.3	43.3
Time (hours)	0	720	1440	2880	4320
PROPERTIES RETAINED (%)					
Elongation	57.5 (aq)	68.5 (aq)	65.8 (aq)	49.3 (aq)	57.5 (aq)
SURFACE and APPEARANCE					
Δb Color	1.25 (x)	0.95 (x)	1.35 (x)	1.25 (x)	1.15 (x)

TABLE 108: Effect of Ethylene Oxide Sterilization on Polypropylene

Material Family	POLYPROPYLENE							
Material Supplier/Name	HIMONT PROFAX 6323							
Reference No.	6	6	6	6	6	6	6	6
EXPOSURE CONDITIONS								
Type	Ethylene Oxide	Ethylene Oxide	Ethylene Oxide	Ethylene Oxide	Ethylene Oxide	Ethylene Oxide	Ethylene Oxide	Ethylene Oxide
Details	12% EtO and 88% Freon	12% EtO and 88% Freon	12% EtO and 88% Freon	12% EtO and 88% Freon	8.6% EtO and 91.4% HCFC-124	8.6% EtO and 91.4% HCFC-124	8.6% EtO and 91.4% HCFC-124	8.6% EtO and 91.4% HCFC-124
Concentration	600 mg/l	600 mg/l	600 mg/l	600 mg/l				
Number of Cycles	1	1	2	2	1	1	2	2
Note	RH: 60%; test lab: Ethox Corp.	RH: 60%; test lab: Ethox Corp.	RH: 60%; test lab: Ethox Corp.	RH: 60%; test lab: Ethox Corp.	RH: 60%; test lab: Ethox Corp.	RH: 60%; test lab: Ethox Corp.	RH: 60%; test lab: Ethox Corp.	RH: 60%; test lab: Ethox Corp.
Temperature (°C)	48.9	48.9	48.9	48.9	48.9	48.9	48.9	48.9
Time (hours)	6	6	6	6	6	6	6	6
PRE EXPOSURE CONDITIONING								
Preconditioning Note	time: 18 hours; temperature: 37.8°C; RH: 60%	time: 18 hours; temperature: 37.8°C; RH: 60%	time: 18 hours; temperature: 37.8°C; RH: 60%	time: 18 hours; temperature: 37.8°C; RH: 60%	time: 18 hours; temperature: 37.8°C; RH: 60%	time: 18 hours; temperature: 37.8°C; RH: 60%	time: 18 hours; temperature: 37.8°C; RH: 60%	time: 18 hours; temperature: 37.8°C; RH: 60%
POST EXPOSURE CONDITIONING								
Note	type: aeration; pressure: 127 mm Hg	type: aeration; pressure: 127 mm Hg	type: aeration; pressure: 127 mm Hg	type: aeration; pressure: 127 mm Hg	type: aeration; pressure: 127 mm Hg	type: aeration; pressure: 127 mm Hg	type: aeration; pressure: 127 mm Hg	type: aeration; pressure: 127 mm Hg
Temperature (°C)	32.2	32.2	32.2	32.2	32.2	32.2	32.2	32.2
POST EXPOSURE CONDITIONING II								
Note	type: ambient conditions	type: ambient conditions	type: ambient conditions	type: ambient conditions	type: ambient conditions	type: ambient conditions	type: ambient conditions	type: ambient conditions
Time (hours)	168	1344	168	1344	168	1344	168	1344
PROPERTIES RETAINED (%)								
Tensile Strength @ Yield	99.4 (il)	100 (il)	98.6 (il)	96.1 (il)	94 (il)	80 (il)	97.9 (il)	98.8 (il)
Elongation	156.7 (bl)	143.6 (bl)	131 (bl)	83.6 (bl)	94 (bl)	159.4 (bl)	84.2 (bl)	121.8 (bl)
Modulus	69.6 (gu)	109.8 (gu)	54.3 (gu)	20.7 (gu)	23.4 (gu)	96.2 (gu)	83.2 (gu)	22.3 (gu)
Dart Impact (total energy)	128.6 (eg)	100 (ee)	164.3 (eb)	42.9 (eb)	114.3 (ef)	142.9 (eb)	178.6 (ec)	142.9 (eb)
Dart Impact (peak energy)	71.4 (de)	71.4 (db)	100 (cw)	35.7 (dc)	78.6 (da)	107.1 (cw)	92.9 (cw)	100 (cw)
SURFACE and APPEARANCE								
ΔE Color	0.24	0.05	0.05	0.13	0	0.1	0.11	0.18

TABLE 109: Effect of Ethylene Oxide Sterilization on Polypropylene

Material Family	POLYPROPYLENE			
Material Supplier/Name	HIMONT PROFAX 6323			
Reference No.	6	6	6	6
EXPOSURE CONDITIONS				
Type	Ethylene Oxide	Ethylene Oxide	Ethylene Oxide	Ethylene Oxide
Details	12% EtO and 88% Freon	12% EtO and 88% Freon	8.6% EtO and 91.4% HCFC-124	8.6% EtO and 91.4% HCFC-124
Concentration	600 mg/l	600 mg/l		
Number of Cycles	1	1	1	1
Note	RH: 60%; test lab: Ethox Corp.	RH: 60%; test lab: Ethox Corp.	RH: 60%; test lab: Ethox Corp.	RH: 60%; test lab: Ethox Corp.
Temperature (°C)	48.9	48.9	48.9	48.9
Time (hours)	6	6	6	6
PRE EXPOSURE CONDITIONING				
Preconditioning Note	time: 18 hours; temperature: 37.8°C; RH: 60%	time: 18 hours; temperature: 37.8°C; RH: 60%	time: 18 hours; temperature: 37.8°C; RH: 60%	time: 18 hours; temperature: 37.8°C; RH: 60%
POST EXPOSURE CONDITIONING				
Note	type: aeration; note: 10 air changes per hour	type: aeration; note: 30 air changes per hour	type: aeration; note: 10 air changes per hour	type: aeration; note: 30 air changes per hour
Temperature (°C)	32.2	54.4	32.2	54.4
RESIDUALS (ppm)				
Residuals Determined	ethylene oxide	ethylene oxide	ethylene oxide	ethylene oxide
Little or No Aeration	451	451	415	415
17 hour Aeration		34		40
24 hour Aeration	138	25	163	36
48 hour Aeration	49	14	140	26
72 hour Aeration	25		64	

Gamma Radiation Resistance

Eastman: **Tenite M7853-343A** (characteristics: melt flow rate: 2 g/10 min.; features: gamma radiation stabilized); **Tenite M7853-368A** (characteristics: melt flow rate: 12 g/10 min.; features: gamma radiation stabilized); **Tenite** (features: general purpose grade)

When a normally stabilized polyallomer is subjected to gamma radiation for sterilization purposes, two major changes may take place. A yellow color may develop due primarily to a chemical change in certain stabilizers that is initiated by radiation. Radiation will alter the chemical structure in such a way that it becomes more susceptible to oxidation. Therefore, the time required for an irradiated part to discolor or embrittle may be very short. Based on correlations between room temperature aging and accelerated aging at 43°C the estimated shelf life of these materials after sterilization with gamma rays is three to five years. It appears from these correlations that aging for 30 days at 43°C is approximately equivalent to aging one year at 23°C. The materials studied were irradiated to 2.5 Mrads at an average dose rate of 0.1 Mrad/hour. Since a low rate is more harmful than a high dose rate, the data were generated under the worst conditions likely to be encountered.

The two most significant tests to monitor the effect of radiation on polyallomer are elongation at break and color. Samples were tested prior to irradiation (2.5 Mrads), immediately after irradiation but prior to oven aging, and at 30 day intervals (up to 180 days) after oven aging at 43°C. A color change due to radiation sterilization will be noticeable immediately after sterilization. This yellowness may decrease slightly during the early stages of aging. However, unless the material is properly stabilized, color will increase again with continued aging. Tensile yield strength and notched Izod do not appear to change significantly after irradiation.

Reference: *Tenite Polypropylene And Tenite Polyallomer Materials For Radiation Sterilization,* supplier technical report (MB-93) - Eastman Plastics, 1985.

TABLE 110: Effect of Gamma Radiation Sterilization on Polyallomer

Material Family	POLYALLOMER									
Material Supplier/Name	EASTMAN TENITE M7853-368A					EASTMAN TENITE M7853-343A				
Material Note	melt flow rate: 12 g/10 min.; gamma radiation stabilized					melt flow rate: 2 g/10 min.; gamma radiation stabilized				
Reference No.	55	55	55	55	55	55	55	55	55	55
EXPOSURE CONDITIONS										
Type	Gamma Radiation	Gamma Radiation	Gamma Radiation	Gamma Radiation	Gamma Radiation	Gamma Radiation	Gamma Radiation	Gamma Radiation	Gamma Radiation	Gamma Radiation
Radiation Dose (Mrads)	2.5	2.5	2.5	2.5	2.5	2.5	2.5	2.5	2.5	2.5
Note	dose rate: 0.1 Mrads/hr.	dose rate: 0.1 Mrads/hr.	dose rate: 0.1 Mrads/hr.	dose rate: 0.1 Mrads/hr.	dose rate: 0.1 Mrads/hr.	dose rate: 0.1 Mrads/hr.	dose rate: 0.1 Mrads/hr.	dose rate: 0.1 Mrads/hr.	dose rate: 0.1 Mrads/hr.	dose rate: 0.1 Mrads/hr.
POST EXPOSURE CONDITIONING										
Note	type: accelerated aging	type: accelerated aging	type: accelerated aging	type: accelerated aging		type: accelerated aging	type: accelerated aging	type: accelerated aging	type: accelerated aging	
Temperature (°C)	43.3	43.3	43.3	43.3		43.3	43.3	43.3	43.3	
Time (hours)	4320	2880	1440	720	0	4320	2880	1440	720	0
PROPERTIES RETAINED (%)										
Elongation	61.8 (aq)	52.8 (aq)	99.2 (aq)	97.6 (aq)	97.6 (aq)	86.5 (aq)	106.3 (aq)	67.6 (aq)	76.6 (aq)	81.1 (aq)
SURFACE and APPEARANCE										
Δ a Color	2.9 (x)	3.2 (x)	3 (x)	2.9 (x)	3 (x)	3.8 (x)	3.9 (x)	3.15 (x)	2.8 (x)	3.1 (x)

TABLE 111: Effect of Gamma Radiation Sterilization on Polyallomer

Material Family	POLYALLOMER				
Material Supplier/Name	EASTMAN TENITE				
Material Note	general purpose grade	general purpose grade	general purpose grade	general purpose grade	general purpose grade
Reference No.	55	55	55	55	55
EXPOSURE CONDITIONS					
Type	Gamma Radiation	Gamma Radiation	Gamma Radiation	Gamma Radiation	Gamma Radiation
Radiation Dose (Mrads)	2.5	2.5	2.5	2.5	2.5
Note	dose rate: 0.1 Mrads/hr.	dose rate: 0.1 Mrads/hr.	dose rate: 0.1 Mrads/hr.	dose rate: 0.1 Mrads/hr.	dose rate: 0.1 Mrads/hr.
POST EXPOSURE CONDITIONING					
Note	type: accelerated aging	type: accelerated aging	type: accelerated aging	type: accelerated aging	
Temperature (°C)	43.3	43.3	43.3	43.3	
Time (hours)	4320	2880	1440	720	0
PROPERTIES RETAINED (%)					
Elongation	44.2 (aq)	47.4 (aq)	49.5 (aq)	52.6 (aq)	63.2 (aq)
SURFACE and APPEARANCE					
Δ a Color	8.8 (x)	8.8 (x)	8.4 (x)	7.4 (x)	6.35 (x)

Radiation Resistance

Phillips: Ryton R-10 5002C (composition: glass and mineral reinforced; features: black); **Ryton R-10 7006A** (composition: glass and mineral reinforced; features: natural resin); **Ryton R-4** (composition: glass reinforced)

Ryton R-4 exhibits no significant deterioration of mechanical properties with relatively high exposures to gamma radiation. By comparison, the effects of similarly intense exposures on the mechanical performance of other Ryton PPS compounds appear significant.

Reference: *Ryton Polyphenylene Sulfide Resins Engineering Properties Guide,* supplier design guide (1065(a)-89 A 02) - Phillips 66 Company, 1989.

TABLE 112: Effect of Gamma Radiation Sterilization on Polypropylene Sulfide

Material Family	POLYPHENYLENE SULFIDE					
Material Supplier/Name	**PHILLIPS RYTON R-10 5002C**	**PHILLIPS RYTON R-10 7006A**	**PHILLIPS RYTON R-4**			
Material Note	glass and mineral reinforced; black	glass and mineral reinforced; natural resin	glass reinforced	glass reinforced	glass reinforced	glass reinforced
Reference No.	102	102	102	102	102	102

EXPOSURE CONDITIONS

Type	Gamma Radiation	Gamma Radiation	Gamma Radiation	Gamma Radiation	Gamma Radiation	Gamma Radiation
Radiation Dose (Mrads)	300	300	300	500	1000	5000

PROPERTIES RETAINED (%)

Tensile Strength	97.3 (he)	102.2 (he)	100 (he)			
Elongation	101.1 (ai)	132.4 (ai)	100.9 (ai)			
Flexural Strength	98.7 (ck)	101.5 (ck)	93.3 (ck)	103.5 (ck)	105.2 (ck)	99.3 (ck)
Modulus	102.7 (bz)	101.5 (bz)	101.7 (bz)	95.3 (bz)	96.8 (bz)	95.8 (bz)
Notched Izod Impact	95.9 (fm)	96.5 (fm)	95.9 (fm)			

Radiation Resistance

Dow Chemical: Styron 666D

The physical properties did not change after exposure to any dose either gamma or E-beam radiation. The yellowness shifted slightly after high doses of radiation (10 Mrads), but this discoloration faded by eight weeks. Photo-bleaching had no effect on this resin.

Reference: Hermanson, Nancy J., Steffens, John F., *The Physical and Visual Property Changes in Thermoplastic Resins After Exposure to High Energy Sterilization - Gamma Versus Electron Beam,* ANTEC 1993, conference proceedings - Society of Plastics Engineers, 1993.

BASF AG: Polystyrol (features: transparent)

The effect of gamma radiation on Polystyrol depends on the dose and dose rate, the temperature, the geometry of the sample, and the ambient medium. If atmospheric oxygen is excluded, Polystyrol is one of the most radiation-resistant plastics available. If it is irradiated in air, the radiation dose that suffices to cause damage is much lower.

Reference: *Polystyrol Product Line, Properties, Processing,* supplier design guide (B 564 e/2.93) - BASF Aktiengesellschaft, 1993.

Gamma Radiation Resistance

Dow Chemical: Styron 666APR (features: transparent)

Gamma radiation sterilization of styrenics is acceptable. The physical properties of the styrenics are not harmed when exposed to dosage levels of up to 10.0 Mrads. Discoloration of the resins occurs immediately upon irradiation. After exposure to 2.5 Mrads, most of the discoloration is recoverable with time. Higher dosages of 10.0 Mrads cause some amount of permanent discoloration of the resin. Exposure to 2.5 Mrads of radiation caused GPPS to discolor the least of all transparent resins with a yellowness index increase of less than 1.0 from the control. GPPS had an almost five fold increase in yellowness index values between 2.5 and 10.0 Mrads. No difference is seen when comparing the physical properties of the irradiated samples stored in fluorescent light versus those stored in complete darkness. The polystyrenes are surpassed only by polyethylenes in the ability to achieve minimal color shift with exposure to gamma radiation.

Reference: Sturdevant, Marianne F., *Sterilization Compatibility of Rigid Thermoplastic Materials,* supplier technical report (301-1548) - Dow Chemical Company, 1988.

Dow Chemical: (features: transparent, natural resin)

Unmodified styrenics are among the most stable polymers to gamma radiation exposure. Polystyrene is especially compatible with gamma sterilization as it maintains nearly all its crystal clarity upon exposure. Color reversal will occur within one month after irradiation. At exposure levels up to 10 Mrads, the samples showed no significant changes over the twelve month span of this study.

Reference: Sturdevant, Marianne F., *The Long-term Effects of Ethylene Oxide and Gamma Radiation Sterilization on the Properties of Rigid Thermoplastic Materials,* ANTEC 1990, conference proceedings - Society of Plastics Engineers, 1990.

Ethylene Oxide (EtO) Resistance

Dow Chemical: Styron 666APR (features: transparent)

Upon exposure to EtO some loss in tensile properties occurs. To maintain optimum performance GPPS should have only minimal exposure to EtO. Exposure of up to five cycles did not affect the notched Izod impact strength. GPPS loses tensile properties with exposure to EtO. On initial exposure to EtO, GPPS loses 23% of its tensile strength and maintains that level through five cycles of exposure. The loss in tensile strength is attributed to poor chemical resistance to ethylene oxide. After 20 days the resin achieved a residual EtO level of less than 200 ppm.

Reference: Sturdevant, Marianne F., *Sterilization Compatibility of Rigid Thermoplastic Materials,* supplier technical report (301-1548) - Dow Chemical Company, 1988.

Dow Chemical: (features: transparent, natural resin)

Styrenics retain their properties upon exposure to one normal ethylene oxide sterilization cycle. However, care should be taken to minimize excessive or multiple exposure to ethylene oxide as it may cause embrittlement and chemical attack leading to stress cracking. Five repeated sterilization cycles caused some embrittlement. The embrittlement is seen as a loss of tensile elongation at break and a decrease in instrumented dart impact energy. After multiple EtO cycles the embrittlement appears to compound with time. The elongation and instrumented impact strengths at six months and one year were significantly less than what was observed at two weeks after sterilization. Silver streaks, or crazes along the flow lines, were noted after exposure to five EtO cycles. This is indicative of stress cracking of the polymer with excessive exposure to ethylene oxide.

Reference: Sturdevant, Marianne F., *The Long-term Effects of Ethylene Oxide and Gamma Radiation Sterilization on the Properties of Rigid Thermoplastic Materials,* ANTEC 1990, conference proceedings - Society of Plastics Engineers, 1990.

Dow Chemical: Styron 666D (features: transparent)

Tensile strength at yield has a 17-27% decrease after 2 sterilization cycles for both EtO gas using 88% CFC-12 as the carrier gas and EtO with HCFC-124. The decrease is attributed to the chemical attack of EtO on styrenic resin which may cause embrittlement and stress cracking. GPPS maintains its practical toughness as shown in the instrumented dart impact data . An increase in percent haze is seen after two cycles of EtO Sterilization; this is attributed to the additional handling of the samples which increases the surface imperfections and thereby, increases the haziness of the transparent parts. EtO is a viable sterilization technique; however, care should be taken to minimize dosage.

Reference: Hermanson, Nancy J., *Effects Of Alternate Carriers Of Ethylene Oxide Sterilant On Thermoplastics,* supplier technical report (301-02018) - Dow Chemical Company.

TABLE 113: Effect of Gamma Radiation Sterilization on General Purpose Polysryrene

Material Family	GENERAL PURPOSE POLYSTYRENE					
Material Supplier/Name	DOW					
Material Note	transparent, natural resin	transparent, natural resin	transparent, natural resin	transparent, natural resin	transparent, natural resin	transparent, natural resin
Reference No.	5	5	5	5	5	5
EXPOSURE CONDITIONS						
Type	Gamma Radiation	Gamma Radiation	Gamma Radiation	Gamma Radiation	Gamma Radiation	Gamma Radiation
Details	source: Cobalt 60	source: Cobalt 60	source: Cobalt 60	source: Cobalt 60	source: Cobalt 60	source: Cobalt 60
Radiation Dose (Mrads)	2.5	2.5	2.5	10	10	10
Note	test lab: Radiations Sterilizers Inc.	test lab: Radiations Sterilizers Inc.	test lab: Radiations Sterilizers Inc.	test lab: Radiations Sterilizers Inc.	test lab: Radiations Sterilizers Inc.	test lab: Radiations Sterilizers Inc.
POST EXPOSURE CONDITIONING						
Note	type: storage in dark	type: storage in dark	type: storage in dark	type: storage in dark	type: storage in dark	type: storage in dark
Temperature (°C)	21	21	21	21	21	21
Time (hours)	336	4368	8760	336	4368	8760
PROPERTIES RETAINED (%)						
Tensile Strength	99.7 (hv)	100.4 (hv)	98.1 (hv)	96 (hv)	97.4 (hv)	96 (hv)
Tensile Strength @ Yield	99.7 (io)	100.4 (io)	98.1 (io)	96 (io)	97.4 (io)	96 (io)
Elongation	100 (ay)	100 (ay)	100 (ay)	100 (ay)	100 (ay)	100 (ay)
Dart Impact (total energy)	150 (fd)	66.7 (fd)	77.8 (fd)	827.8 (fg)	72.2 (fd)	44.4 (fd)
Notched Izod Impact	100 (fv)	100 (fv)	125 (fv)	75 (fu)	75 (fu)	125 (fv)

TABLE 114: Effect of Gamma Radiation Sterilization on General Purpose Polysryrene

Material Family	GENERAL PURPOSE POLYSTYRENE			
Material Supplier/Name	DOW STYRON 666APR			
Material Note	transparent	transparent	transparent	transparent
Reference No.	3	3	3	3

EXPOSURE CONDITIONS

Type	Gamma Radiation	Gamma Radiation	Gamma Radiation	Gamma Radiation
Details	source: Cobalt 60	source: Cobalt 60	source: Cobalt 60	source: Cobalt 60
Radiation Dose (Mrads)	2.5	10	2.5	10

POST EXPOSURE CONDITIONING

Note	type: storage under fluorescent light	type: storage under fluorescent light	type: storage in dark	type: storage in dark
Temperature (°C)	21	21	21	21
Time (hours)	336	336	336	336

PROPERTIES RETAINED (%)

Tensile Strength @ Yield	100 (ii)	97.1 (ii)	100 (ii)	97.1 (ii)
Notched Izod Impact	100 (fp)	100 (fp)	100 (fp)	100 (fp)

TABLE 115: Effect of Gamma Radiation Sterilization on General Purpose Polysryrene

Material Family	GENERAL PURPOSE POLYSTYRENE			
Material Supplier/Name	DOW STYRON 666D			
Reference No.	1	1	1	1

EXPOSURE CONDITIONS

Type	Gamma Radiation	Gamma Radiation	Gamma Radiation	Gamma Radiation
Radiation Dose (Mrads)	2.5	2.5	10	10
Note	test lab: SteriGenics	test lab: SteriGenics	test lab: SteriGenics	test lab: SteriGenics

POST EXPOSURE CONDITIONING

Note	type: aging	type: aging	type: aging	type: aging
Time (hours)	168	1344	168	1344

PROPERTIES RETAINED (%)

Tensile Strength @ Yield	100 (is)	100 (is)	97.8 (is)	100 (is)
Elongation	100 (bu)	100 (bu)	100 (bu)	100 (bu)
Flexural Strength	102.2 (co)	102.2 (co)	100 (co)	101.1 (co)
Modulus	91.4 (gz)	104.3 (gz)	106.1 (gz)	94.5 (gz)
Flexural Modulus	100.3 (cf)	105.7 (cf)	103.3 (cf)	103.3 (cf)
Dart Impact (total energy)	100 (en)	100 (en)	100 (en)	100 (en)
Dart Impact (peak energy)	100 (do)	100 (do)	100 (do)	100 (do)
Notched Izod Impact	128.6 (gd)	100 (gd)	100 (gd)	100 (gd)
Heat Deflection Temperature	97.4 (m)	98.7 (m)	100 (m)	101.3 (m)
Vicat Softening Point	101 (ku)	100 (ku)	101 (ku)	100 (ku)

SURFACE and APPEARANCE

Δ Yellowness Index	0.9 (kw)	0.2 (kw)	3.4 (kw)	2.2 (kw)
Haze Retained (%)	4.1	2.08	2.77	0.74
Light Transmission Retained (%)	100	109.9	97.8	109.9

TABLE 116: Effect of Electron Beam Radiation Sterilization on General Purpose Polysryrene

Material Family	GENERAL PURPOSE POLYSTYRENE			
Material Supplier/Name	DOW STYRON 666D			
Reference No.	1	1	1	1
EXPOSURE CONDITIONS				
Type	Electron Beam Radiation	Electron Beam Radiation	Electron Beam Radiation	Electron Beam Radiation
Radiation Dose (Mrads)	2.5	2.5	10	10
Note	test lab: E-Beam Services, Inc.	test lab: E-Beam Services, Inc.	test lab: E-Beam Services, Inc.	test lab: E-Beam Services, Inc.
POST EXPOSURE CONDITIONING				
Note	type: aging	type: aging	type: aging	type: aging
Time (hours)	168	1344	168	1344
PROPERTIES RETAINED (%)				
Tensile Strength @ Yield	97.8 (is)	100 (is)	97.8 (is)	100 (is)
Elongation	100 (bu)	100 (bu)	100 (bu)	100 (bu)
Flexural Strength	95.7 (co)	102.2 (co)	96.7 (co)	102.2 (co)
Modulus	101.8 (gz)	96.6 (gz)	104 (gz)	90.2 (gz)
Flexural Modulus	101.8 (cf)	104.2 (cf)	101.8 (cf)	103.6 (cf)
Dart Impact (total energy)	100 (en)	200 (en)	100 (en)	100 (en)
Dart Impact (peak energy)	100 (do)	100 (do)	100 (do)	100 (do)
Notched Izod Impact	100 (gd)	100 (gd)	100 (gd)	100 (gd)
Heat Deflection Temperature	97.4 (m)	98.7 (m)	105.3 (m)	109.2 (m)
Vicat Softening Point	99 (ku)	100 (ku)	101 (ku)	100 (ku)
SURFACE and APPEARANCE				
Δ Yellowness Index	0 (kw)	-0.4 (kw)	1.5 (kw)	0.8 (kw)
Haze Retained (%)	2.19	1.99	3.54	1.95
Light Transmission Retained (%)	98.9	109.9	97.8	109.9

TABLE 117: Effect of Ethylene Oxide Sterilization on Dow General Purpose Polysryrene

Material Family	GENERAL PURPOSE POLYSTYRENE					
Material Supplier/Name	DOW					
Material Note	transparent, natural resin	transparent, natural resin	transparent, natural resin	transparent, natural resin	transparent, natural resin	transparent, natural resin
Reference No.	5	5	5	5	5	5

EXPOSURE CONDITIONS

Type	Ethylene Oxide	Ethylene Oxide	Ethylene Oxide	Ethylene Oxide	Ethylene Oxide	Ethylene Oxide
Details	12% EtO and 88% Freon	12% EtO and 88% Freon	12% EtO and 88% Freon	12% EtO and 88% Freon	12% EtO and 88% Freon	12% EtO and 88% Freon
Concentration	660 mg/l	660 mg/l	660 mg/l	660 mg/l	660 mg/l	660 mg/l
Number of Cycles	1	1	1	5	5	5
Note	RH: 60%; test lab: Ethox Corp.	RH: 60%; test lab: Ethox Corp.	RH: 60%; test lab: Ethox Corp.	RH: 60%; test lab: Ethox Corp.	RH: 60%; test lab: Ethox Corp.	RH: 60%; test lab: Ethox Corp.
Temperature (°C)	49	49	49	49	49	49
Time (hours)	≥ 6	≥ 6	≥ 6	≥ 6	≥ 6	≥ 6

PRE EXPOSURE CONDITIONING

Preconditioning Note	time: 8 hours; temperature: 37.8°C; RH: 60%	time: 8 hours; temperature: 37.8°C; RH: 60%	time: 8 hours; temperature: 37.8°C; RH: 60%	time: 8 hours; temperature: 37.8°C; RH: 60%	time: 8 hours; temperature: 37.8°C; RH: 60%	time: 8 hours; temperature: 37.8°C; RH: 60%

POST EXPOSURE CONDITIONING

Note	type: evacuation; pressure: 127 mm Hg	type: evacuation; pressure: 127 mm Hg	type: evacuation; pressure: 127 mm Hg	type: evacuation; pressure: 127 mm Hg	type: evacuation; pressure: 127 mm Hg	type: evacuation; pressure: 127 mm Hg

POST EXPOSURE CONDITIONING II

Note	type: aeration	type: aeration	type: aeration	type: aeration	type: aeration	type: aeration
Temperature (°C)	32.2	32.2	32.2	32.2	32.2	32.2
Time (hours)	≥ 16	≥ 16	≥ 16	≥ 16	≥ 16	≥ 16

POST EXPOSURE CONDITIONING III

Note	type: storage in dark	type: storage in dark	type: storage in dark	type: storage in dark	type: storage in dark	type: storage in dark
Temperature (°C)	21	21	21	21	21	21
Time (hours)	336	4368	8760	336	4368	8760

PROPERTIES RETAINED (%)

Tensile Strength	74.9 (hy)	94.5 (hy)	90.1 (hy)	76.3 (hy)	79.5 (hy)	76.2 (hy)
Tensile Strength @ Yield	74.9 (ip)	94.5 (ip)	90.1 (ip)	76.3 (ip)	79.5 (ip)	76.2 (ip)
Elongation	50 (av)	100 (av)	50 (av)	50 (av)	50 (av)	50 (av)
Dart Impact (total energy)	133.3 (fb)	83.3 (fb)	66.7 (fb)	116.7 (fb)	38.9 (fb)	50 (fb)
Notched Izod Impact	125 (fs)	100 (fs)	75 (fs)	100 (fs)	75 (fs)	75 (fs)

TABLE 118: Effect of Ethylene Oxide Sterilization on General Purpose Polysryrene

Material Family	GENERAL PURPOSE POLYSTYRENE							
Material Supplier/Name	DOW STYRON 666D							
Material Note	transparent	transparent	transparent	transparent	transparent	transparent	transparent	transparent
Reference No.	6	6	6	6	6	6	6	6
EXPOSURE CONDITIONS								
Type	Ethylene Oxide	Ethylene Oxide	Ethylene Oxide	Ethylene Oxide	Ethylene Oxide	Ethylene Oxide	Ethylene Oxide	Ethylene Oxide
Details	12% EtO and 88% Freon	12% EtO and 88% Freon	12% EtO and 88% Freon	12% EtO and 88% Freon	8.6% EtO and 91.4% HCFC-124	8.6% EtO and 91.4% HCFC-124	8.6% EtO and 91.4% HCFC-124	8.6% EtO and 91.4% HCFC-124
Concentration	600 mg/l	600 mg/l	600 mg/l	600 mg/l				
Number of Cycles	1	1	2	2	1	1	2	2
Note	RH: 60%; test lab: Ethox Corp.	RH: 60%; test lab: Ethox Corp.	RH: 60%; test lab: Ethox Corp.	RH: 60%; test lab: Ethox Corp.	RH: 60%; test lab: Ethox Corp.	RH: 60%; test lab: Ethox Corp.	RH: 60%; test lab: Ethox Corp.	RH: 60%; test lab: Ethox Corp.
Temperature (°C)	48.9	48.9	48.9	48.9	48.9	48.9	48.9	48.9
Time (hours)	6	6	6	6	6	6	6	6
PRE EXPOSURE CONDITIONING								
Preconditioning Note	time: 18 hours; temperature: 37.8°C; RH: 60%	time: 18 hours; temperature: 37.8°C; RH: 60%	time: 18 hours; temperature: 37.8°C; RH: 60%	time: 18 hours; temperature: 37.8°C; RH: 60%	time: 18 hours; temperature: 37.8°C; RH: 60%	time: 18 hours; temperature: 37.8°C; RH: 60%	time: 18 hours; temperature: 37.8°C; RH: 60%	time: 18 hours; temperature: 37.8°C; RH: 60%
POST EXPOSURE CONDITIONING								
Note	type: aeration; pressure: 127 mm Hg	type: aeration; pressure: 127 mm Hg	type: aeration; pressure: 127 mm Hg	type: aeration; pressure: 127 mm Hg	type: aeration; pressure: 127 mm Hg	type: aeration; pressure: 127 mm Hg	type: aeration; pressure: 127 mm Hg	type: aeration; pressure: 127 mm Hg
Temperature (°C)	32.2	32.2	32.2	32.2	32.2	32.2	32.2	32.2
POST EXPOSURE CONDITIONING II								
Note	type: ambient conditions	type: ambient conditions	type: ambient conditions	type: ambient conditions	type: ambient conditions	type: ambient conditions	type: ambient conditions	type: ambient conditions
Time (hours)	168	1344	168	1344	168	1344	168	1344
PROPERTIES RETAINED (%)								
Tensile Strength @ Yield	95.8 (io)	92.2 (io)	89.5 (io)	82.7 (io)	93.1 (io)	73 (io)	101.6 (io)	80.2 (io)
Elongation	100 (bk)	100 (bk)	100 (bk)	100 (bk)	100 (bk)	100 (bk)	200 (bk)	100 (bk)
Modulus	92.4 (gw)	98.1 (gw)	88 (gw)	93.8 (gw)	85.6 (gw)	85.4 (gw)	90.1 (gw)	85.8 (gw)
Dart Impact (total energy)	100 (dy)	90.9 (dy)	90.9 (dy)	90.9 (dy)	81.8 (dy)	127.3 (dy)	90.9 (dy)	127.3 (dy)
Dart Impact (peak energy)	77.8 (cx)	55.6 (cx)	55.6 (cx)	66.7 (cx)	66.7 (cx)	100 (ey)	77.8 (ew)	111.1 (cx)
SURFACE and APPEARANCE								
Yellowness Index	0.93	0.8	1.06	1.03	1.06	0.95	0.97	1.21
Haze Retained (%)	4.98	5.84	7.4	7	2.15	4.6	10.39	8.73
Transparency Retained (%)	90	91	90	90	91	90	89	90

TABLE 119: Effect of Ethylene Oxide Sterilization on Dow Styron 666D General Purpose Polysryrene

Material Family	GENERAL PURPOSE POLYSTYRENE			
Material Supplier/Name	DOW STYRON 666D			
Material Note	transparent	transparent	transparent	transparent
Reference No.	6	6	6	6
EXPOSURE CONDITIONS				
Type	Ethylene Oxide	Ethylene Oxide	Ethylene Oxide	Ethylene Oxide
Details	12% EtO and 88% Freon	12% EtO and 88% Freon	8.6% EtO and 91.4% HCFC-124	8.6% EtO and 91.4% HCFC-124
Concentration	600 mg/l	600 mg/l		
Number of Cycles	1	1	1	1
Note	RH: 60%; test lab: Ethox Corp.	RH: 60%; test lab: Ethox Corp.	RH: 60%; test lab: Ethox Corp.	RH: 60%; test lab: Ethox Corp.
Temperature (°C)	48.9	48.9	48.9	48.9
Time (hours)	6	6	6	6
PRE EXPOSURE CONDITIONING				
Preconditioning Note	time: 18 hours; temperature: 37.8°C; RH: 60%	time: 18 hours; temperature: 37.8°C; RH: 60%	time: 18 hours; temperature: 37.8°C; RH: 60%	time: 18 hours; temperature: 37.8°C; RH: 60%
POST EXPOSURE CONDITIONING				
Note	type: aeration; note: 10 air changes per hour	type: aeration; note: 30 air changes per hour	type: aeration; note: 10 air changes per hour	type: aeration; note: 30 air changes per hour
Temperature (°C)	32.2	54.4	32.2	54.4
RESIDUALS (ppm)				
Residuals Determined	ethylene oxide	ethylene oxide	ethylene oxide	ethylene oxide
Little or No Aeration	1002	1002	1006	1006
24 hour Aeration	429	171	356	312
48 hour Aeration		112		290
72 hour Aeration	129	105	176	119
168 hour Aeration	77		108	

TABLE 120: Effect of Ethylene Oxide Sterilization on General Purpose Polysryrene

Material Family	GENERAL PURPOSE POLYSTYRENE		
Material Supplier/Name	DOW STYRON 666APR		
Material Note	transparent	transparent	transparent
Reference No.	3	3	3

EXPOSURE CONDITIONS

Type	Ethylene Oxide	Ethylene Oxide	Ethylene Oxide
Details	12% EtO and 88% Freon	12% EtO and 88% Freon	12% EtO and 88% Freon
Concentration	660 mg/l	660 mg/l	660 mg/l
Number of Cycles	1	5	1
Note	RH: 60%	RH: 60%	RH: 60%
Temperature (°C)	49	49	49
Time (hours)	≥ 6	≥ 6	≥ 6

PRE EXPOSURE CONDITIONING

Preconditioning Note	time: 8 hours; temperature: 37.8°C; RH: 60%	time: 8 hours; temperature: 37.8°C; RH: 60%	time: 8 hours; temperature: 37.8°C; RH: 60%

POST EXPOSURE CONDITIONING

Note	type: evacuation; pressure: 127 mm Hg	type: evacuation; pressure: 127 mm Hg	type: evacuation; pressure: 127 mm Hg

POST EXPOSURE CONDITIONING II

Note	type: aeration	type: aeration	type: aeration
Temperature (°C)	32	32	32
Time (hours)	≥ 16	≥ 16	

POST EXPOSURE CONDITIONING III

Note	type: storage in dark; RH: 50%	type: storage in dark; RH: 50%	
Temperature (°C)	21	21	
Time (hours)	336	336	

RESIDUALS (ppm)

Residuals Determined			ethylene oxide
24 hour Aeration			511
72 hour Aeration			260
696 hour Aeration			152
720 hour Aeration			145
768 hour Aeration			113
840 hour Aeration			69
888 hour Aeration			56

PROPERTIES RETAINED (%)

Tensile Strength @ Yield	75.7 (ii)	77.1 (ii)	
Notched Izod Impact	100 (fp)	100 (fp)	

GRAPH 117: **Post Gamma Radiation Exposure Time vs. Yellowness Index of General Purpose Polystyrene**

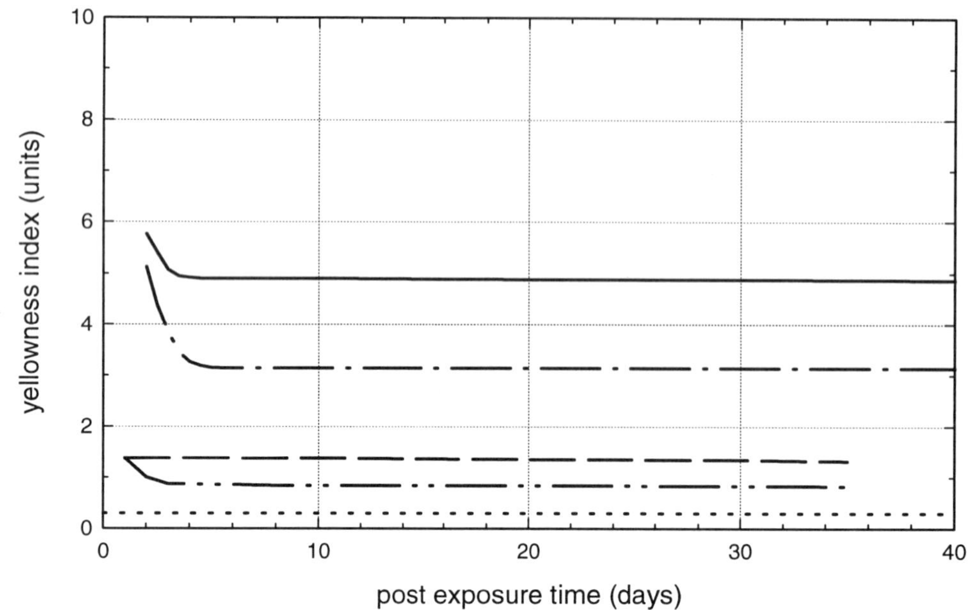

....	Dow Styron 666APR GPPS (transpar.); before exposure
—··—··	Dow Styron 666APR GPPS (transpar.); 2.5 Mrads; light storage
— — — —	Dow Styron 666APR GPPS (transpar.); 2.5 Mrads; dark storage
———	Dow Styron 666APR GPPS (transpar.); 10 Mrads; dark storage
—·—·—	Dow Styron 666APR GPPS (transpar.); 10 Mrads; light storage
Reference No.	3

GRAPH 118: **Post Gamma Radiation Exposure Time vs. Yellowness Index of General Purpose Polystyrene**

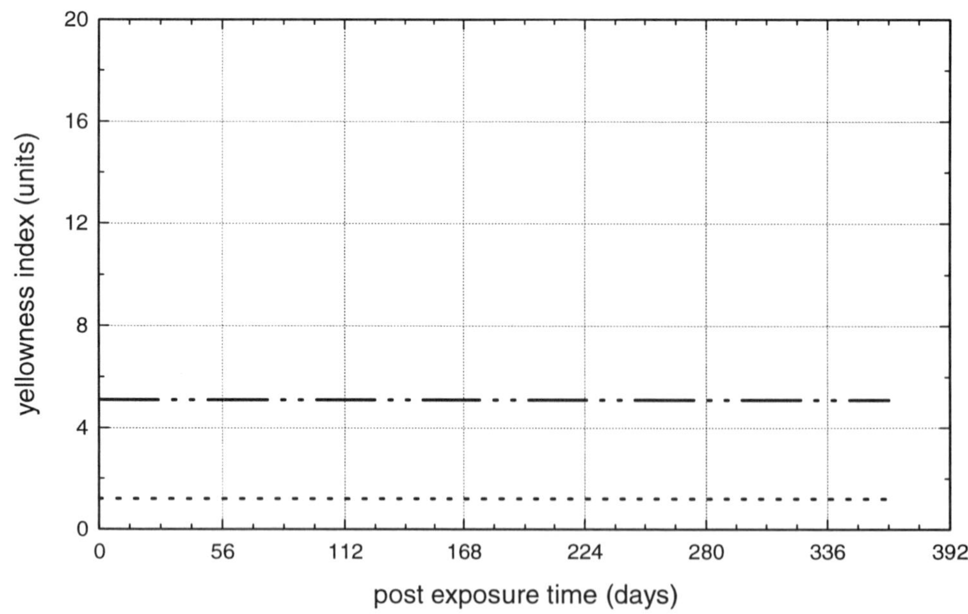

....	Dow GPPS; 2.5 Mrad
—··—··	Dow GPPS; 10 Mrad
Reference No.	5

GRAPH 119: Post Gamma Radiation Exposure Time vs. Percent Light Transmission of General Purpose Polystyrene

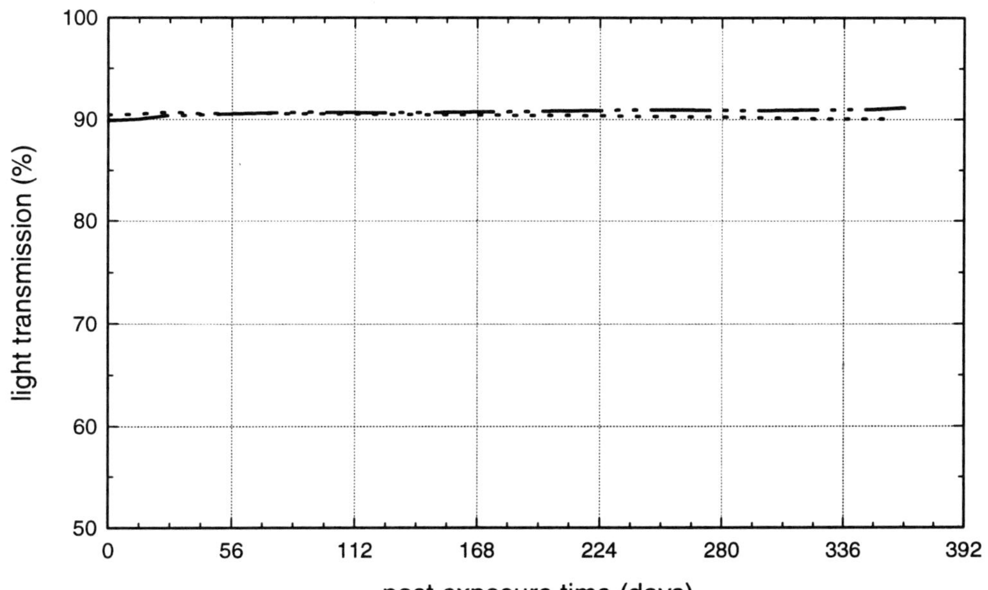

....	Dow GPPS; 2.5 Mrad
—··—··	Dow GPPS; 10 Mrad
Reference No.	5

Radiation Resistance

Dow Chemical: Styron 478 (features: transparent)

No significant changes in the physical properties of HIPS were seen after sterilization. However, the resin did yellow (b scale) after sterilization; this discoloration faded with time to nearly its original color. The 10 Mrad gamma sample had a ΔE of 2.3 which translates to slight yellow that is visible to the eye. Exposing HIPS to fluorescent light accelerated the color reversal.

Reference: Hermanson, Nancy J., Steffens, John F., *The Physical and Visual Property Changes in Thermoplastic Resins After Exposure to High Energy Sterilization - Gamma Versus Electron Beam,* ANTEC 1993, conference proceedings - Society of Plastics Engineers, 1993.

BASF AG: Polystyrol

The resistance of high-impact Polystyrol to gamma radiation is decidedly lower than that of general purpose products.

Reference: *Polystyrol Product Line, Properties, Processing,* supplier design guide (B 564 e/2.93) - BASF Aktiengesellschaft, 1993.

Gamma Radiation Resistance

Dow Chemical: Styron 484

Gamma radiation sterilization of styrenics is acceptable. The physical properties of the styrenics are not harmed when exposed to dosage levels of up to 10.0 Mrads. Discoloration of the resins occurs immediately upon irradiation. After exposure to 2.5 Mrads, most of the discoloration is recoverable with time. Exposure to 2.5 Mrads of radiation caused an increase in ΔE of less than 2. Higher dosages of 10.0 Mrads cause some amount of permanent discoloration of the resin. There was a slow increase in the tensile strength of HIPS with the increase in gamma dosage. The losses in impact properties and increase in tensile strength are thought to be attributed to breakdown and/or crosslinking occurring in the rubber phase. No difference is seen when comparing the physical properties of the irradiated samples stored in fluorescent light versus those stored in complete darkness. However, storage in light accelerates the bleach back process. The polystyrenes are surpassed only by polyethylenes in ability to achieve minimal color shift with exposure to gamma radiation.

Reference: Sturdevant, Marianne F., *Sterilization Compatibility of Rigid Thermoplastic Materials,* supplier technical report (301-1548) - Dow Chemical Company, 1988.

Dow Chemical: (features: natural resin)

The retention of impact strength of gamma sterilized rubber-modified styrenic polymers is dependent upon the degree of crosslinking that occurs in the butadiene rubber phase. The higher the radiation dosage, the greater the crosslinking, and the lower the ultimate impact strength. At the sterilization exposure level of 10 Mrads, HIPS showed losses in impact strength accompanied by a slight increase in tensile strength and a decrease in tensile elongation at break. This change is attributed to the crosslinking of the butadiene rubber matrix.

If the rubber content is high enough, crosslinking becomes the dominating factor in determining the physical property characteristics of the polymer upon irradiation. Crosslinked butadiene rubber losses its impact strength, thus at dosages sufficient to crosslink all the rubber, the enhanced impact properties originally provided by the rubber modifier are lost. The remaining impact strength of the material will be no better than that of unmodified polymer. Comparing the notched Izod impact strength at 2.5 Mrads and 10 Mrads, one can see the loss in properties with the increase in radiation dosage.

Reference: Sturdevant, Marianne F., *The Long-term Effects of Ethylene Oxide and Gamma Radiation Sterilization on the Properties of Rigid Thermoplastic Materials,* ANTEC 1990, conference proceedings - Society of Plastics Engineers, 1990.

Ethylene Oxide (EtO) Resistance

Dow Chemical: Styron 484

Upon exposure to EtO some loss in tensile properties occurs.

Reference: Sturdevant, Marianne F., *Sterilization Compatibility of Rigid Thermoplastic Materials,* supplier technical report (301-1548) - Dow Chemical Company, 1988.

Dow Chemical: (features: natural resin)

Physical property retention shows compatibility with ethylene oxide sterilization. Samples retained their physical integrity after one cycle of EtO sterilization. However, because of the styrenic components in these polymer matrices, over exposure to EtO sterilization should be avoided. Five repeated sterilization cycles caused some embrittlement. The embrittlement is seen as a loss of tensile elongation at break and a decrease in instrumented dart impact energy. After multiple EtO, cycles the embrittlement appears to compound with time. The elongation and instrumented impact strengths at six months and one year were significantly less than what was observed at two weeks after sterilization. Because of the styrenic component of the bulk polymer matrix, a slight decrease in the tensile elongation at break is also observed.

Reference: Sturdevant, Marianne F., *The Long-term Effects of Ethylene Oxide and Gamma Radiation Sterilization on the Properties of Rigid Thermoplastic Materials,* ANTEC 1990, conference proceedings - Society of Plastics Engineers, 1990.

TABLE 121: Effect of Gamma Radiation Sterilization on Impact Polystyrene

Material Family	IMPACT POLYSTYRENE					
Material Supplier/Name	DOW					
Material Note	natural resin	natural resin	natural resin	natural resin	natural resin	natural resin
Reference No.	5	5	5	5	5	5

EXPOSURE CONDITIONS

Type	Gamma Radiation	Gamma Radiation	Gamma Radiation	Gamma Radiation	Gamma Radiation	Gamma Radiation
Details	source: Cobalt 60	source: Cobalt 60	source: Cobalt 60	source: Cobalt 60	source: Cobalt 60	source: Cobalt 60
Radiation Dose (Mrads)	2.5	2.5	2.5	10	10	10
Note	test lab: Radiations Sterilizers Inc.	test lab: Radiations Sterilizers Inc.	test lab: Radiations Sterilizers Inc.	test lab: Radiations Sterilizers Inc.	test lab: Radiations Sterilizers Inc.	test lab: Radiations Sterilizers Inc.

POST EXPOSURE CONDITIONING

Note	type: storage in dark	type: storage in dark	type: storage in dark	type: storage in dark	type: storage in dark	type: storage in dark
Temperature (°C)	21	21	21	21	21	21
Time (hours)	336	4368	8760	336	4368	8760

PROPERTIES RETAINED (%)

Tensile Strength	99.3 (ht)	99.8 (ht)	101.9 (ht)	99.3 (ht)	100.6 (ht)	100.3 (ht)
Tensile Strength @ Yield	101.4 (ik)	101.6 (ik)	102.8 (ik)	105.5 (ik)	108.3 (ik)	108.8 (ik)
Elongation	85.5 (au)	90.9 (au)	89.1 (au)	67.3 (au)	63.6 (au)	61.8 (au)
Dart Impact (total energy)	102 (fc)	66.7 (fc)	61.4 (fc)	63.4 (fc)	61.4 (fc)	47.1 (fc)
Notched Izod Impact	93.3 (fw)	93.3 (fw)	93.3 (fw)	73.3 (fw)	73.3 (fw)	73.3 (fw)

TABLE 122: Effect of Gamma Radiation Sterilization on Impact Polystyrene

Material Family	IMPACT POLYSTYRENE							
Material Supplier/Name	DOW STYRON 478				DOW STYRON 484			
Material Note	transparent	transparent	transparent	transparent				
Reference No.	1	1	1	1	3	3	3	3

EXPOSURE CONDITIONS

Type	Gamma Radiation	Gamma Radiation	Gamma Radiation	Gamma Radiation	Gamma Radiation	Gamma Radiation	Gamma Radiation	Gamma Radiation
Radiation Dose (Mrads)	2.5	2.5	10	10	2.5	10	2.5	10
Note	test lab: SteriGenics	test lab: SteriGenics	test lab: SteriGenics	test lab: SteriGenics	source: Cobalt 60	source: Cobalt 60	source: Cobalt 60	source: Cobalt 60

POST EXPOSURE CONDITIONING

Note	type: aging	type: aging	type: aging	type: aging	type: storage under fluorescent light	type: storage under fluorescent light	type: storage in dark	type: storage in dark
Temperature (°C)					21	21	21	21
Time (hours)	168	1344	168	1344	336	336	336	336

PROPERTIES RETAINED (%)

Tensile Strength @ Yield	100 (ir)	104.3 (ir)	100 (ir)	104.3 (ir)	103.4 (ii)	110.3 (ii)	103.4 (ii)	106.9 (ii)
Elongation	78.4 (bq)	86.3 (bq)	84.3 (bq)	78.4 (bq)				
Flexural Strength	104.2 (cn)	97.9 (cn)	108.3 (cn)	100 (cn)				
Modulus	100.5 (ha)	101 (ha)	99 (ha)	98.5 (ha)				
Flexural Modulus	103.8 (cf)	108 (cf)	106.1 (cf)	105.7 (cf)				
Dart Impact (total energy)	81.3 (ej)	81.3 (ej)	87.5 (ej)	87.5 (ej)				
Dart Impact (peak energy)	81.3 (dm)	81.3 (dm)	87.5 (dm)	87.5 (dm)				
Notched Izod Impact	100 (fz)	96.1 (fz)	87.5 (fz)	87.5 (fz)	93.3 (fp)	73.3 (fp)	93.3 (fp)	93.3 (fp)
Heat Deflection Temperature	101.2 (m)	101.8 (m)	101.8 (m)	103.1 (m)				
Vicat Softening Point	100.5 (ku)	100.5 (ku)	100.9 (ku)	100.5 (ku)				

SURFACE and APPEARANCE

ΔL Color	84.16 (ac)	-0.88 (ac)	-0.74 (ac)	-3.25 (ac)				
Δa Color	-2.39 (o)	-2.08 (o)	-2.22 (o)	-5.19 (o)				
Δb Color	-6.82 (t)	3.69 (t)	4.95 (t)	8.46 (t)				

TABLE 123: Effect of Electron Beam Radiation Sterilization on Impact Polystyrene

Material Family	IMPACT POLYSTYRENE			
Material Supplier/Name	DOW STYRON 478			
Material Note	transparent	transparent	transparent	transparent
Reference No.	1	1	1	1

EXPOSURE CONDITIONS

Type	Electron Beam Radiation	Electron Beam Radiation	Electron Beam Radiation	Electron Beam Radiation
Radiation Dose (Mrads)	2.5	2.5	10	10
Note	test lab: E-Beam Services, Inc.	test lab: E-Beam Services, Inc.	test lab: E-Beam Services, Inc.	test lab: E-Beam Services, Inc.

POST EXPOSURE CONDITIONING

Note	type: aging	type: aging	type: aging	type: aging
Time (hours)	168	1344	168	1344

PROPERTIES RETAINED (%)

Tensile Strength @ Yield	100 (ir)	100 (ir)	104.3 (ir)	104.3 (ir)
Elongation	86.3 (bq)	74.5 (bq)	72.5 (bq)	68.6 (bq)
Flexural Strength	100 (cn)	97.9 (cn)	106.3 (cn)	102.1 (cn)
Modulus	100 (ha)	95.6 (ha)	99 (ha)	95.6 (ha)
Flexural Modulus	106.6 (cf)	106.1 (cf)	105.7 (cf)	103.8 (cf)
Dart Impact (total energy)	100 (ej)	87.5 (ej)	93.8 (ej)	68.8 (ej)
Dart Impact (peak energy)	100 (dm)	87.5 (dm)	93.8 (dm)	68.8 (dm)
Notched Izod Impact	91.4 (fz)	96.1 (fz)	83.6 (fz)	87.5 (fz)
Heat Deflection Temperature	103.1 (m)	102.5 (m)	110.4 (m)	108 (m)
Vicat Softening Point	100.9 (ku)	100.5 (ku)	100.9 (ku)	100.9 (ku)

SURFACE and APPEARANCE

ΔL Color	-2.33 (ac)	-1.28 (ac)	-0.6 (ac)	-1.05 (ac)
Δa Color	-3.97 (o)	-2.35 (o)	-1.9 (o)	-2.74 (o)
Δb Color	11.64 (t)	2.18 (t)	3.75 (t)	5.27 (t)

TABLE 124: Effect of Ethylene Oxide Sterilization on Impact Polystyrene

Material Family	IMPACT POLYSTYRENE							
Material Supplier/Name	DOW						DOW STYRON 484	
Material Note	natural resin	natural resin	natural resin	natural resin	natural resin	natural resin		
Reference No.	5	5	5	5	5	5	3	3
EXPOSURE CONDITIONS								
Type	Ethylene Oxide	Ethylene Oxide	Ethylene Oxide	Ethylene Oxide	Ethylene Oxide	Ethylene Oxide	Ethylene Oxide	Ethylene Oxide
Details	12% EtO and 88% Freon	12% EtO and 88% Freon	12% EtO and 88% Freon	12% EtO and 88% Freon	12% EtO and 88% Freon	12% EtO and 88% Freon	12% EtO and 88% Freon	12% EtO and 88% Freon
Concentration	660 mg/l	660 mg/l	660 mg/l	660 mg/l	660 mg/l	660 mg/l	660 mg/l	660 mg/l
Number of Cycles	1	1	1	5	5	5	1	5
Note	RH: 60%; test lab: Ethox Corp.	RH: 60%; test lab: Ethox Corp.	RH: 60%; test lab: Ethox Corp.	RH: 60%; test lab: Ethox Corp.	RH: 60%; test lab: Ethox Corp.	RH: 60%; test lab: Ethox Corp.	RH: 60%	RH: 60%
Temperature (°C)	49	49	49	49	49	49	49	49
Time (hours)	≥ 6	≥ 6	≥ 6	≥ 6	≥ 6	≥ 6	≥ 6	≥ 6
PRE EXPOSURE CONDITIONING								
Preconditioning Note	time: 8 hours; temperature: 37.8°C; RH: 60%	time: 8 hours; temperature: 37.8°C; RH: 60%	time: 8 hours; temperature: 37.8°C; RH: 60%	time: 8 hours; temperature: 37.8°C; RH: 60%	time: 8 hours; temperature: 37.8°C; RH: 60%	time: 8 hours; temperature: 37.8°C; RH: 60%	time: 8 hours; temperature: 37.8°C; RH: 60%	time: 8 hours; temperature: 37.8°C; RH: 60%
POST EXPOSURE CONDITIONING								
Note	type: evacuation; pressure: 127 mm Hg	type: evacuation; pressure: 127 mm Hg	type: evacuation; pressure: 127 mm Hg	type: evacuation; pressure: 127 mm Hg	type: evacuation; pressure: 127 mm Hg	type: evacuation; pressure: 127 mm Hg	type: evacuation; pressure: 127 mm Hg	type: evacuation; pressure: 127 mm Hg
POST EXPOSURE CONDITIONING II								
Note	type: aeration	type: aeration	type: aeration	type: aeration	type: aeration	type: aeration	type: aeration	type: aeration
Temperature (°C)	32.2	32.2	32.2	32.2	32.2	32.2	32	32
Time (hours)	≥ 16	≥ 16	≥ 16	≥ 16	≥ 16	≥ 16	≥ 16	≥ 16
POST EXPOSURE CONDITIONING III								
Note	type: storage in dark	type: storage in dark	type: storage in dark	type: storage in dark	type: storage in dark	type: storage in dark	type: storage in dark; RH: 50%	type: storage in dark; RH: 50%
Temperature (°C)	21	21	21	21	21	21	21	21
Time (hours)	336	4368	8760	336	4368	8760	336	336
PROPERTIES RETAINED (%)								
Tensile Strength	86.7 (hu)	85.9 (hu)	87.5 (hu)	81.8 (hu)	75.4 (hu)	76.5 (hu)		
Tensile Strength @ Yield	98.5 (in)	100.2 (in)	101.4 (in)	98.5 (in)	99.6 (in)	100.9 (in)	100 (ii)	100 (ii)
Elongation	60 (aw)	52.7 (aw)	52.7 (aw)	45.5 (aw)	30.9 (aw)	30.9 (aw)		
Dart Impact (total energy)	70.6 (fa)	62.1 (fa)	69.9 (fa)	80.4 (fa)	68.6 (fa)	68.6 (fa)		
Notched Izod Impact	100 (fx)	100 (fx)	93.3 (fx)	100 (fx)	86.7 (fx)	93.3 (fx)	100 (fp)	100 (fp)

GRAPH 120: Post Gamma Radiation Exposure Time vs. Delta E Color Change of Impact Polystyrene

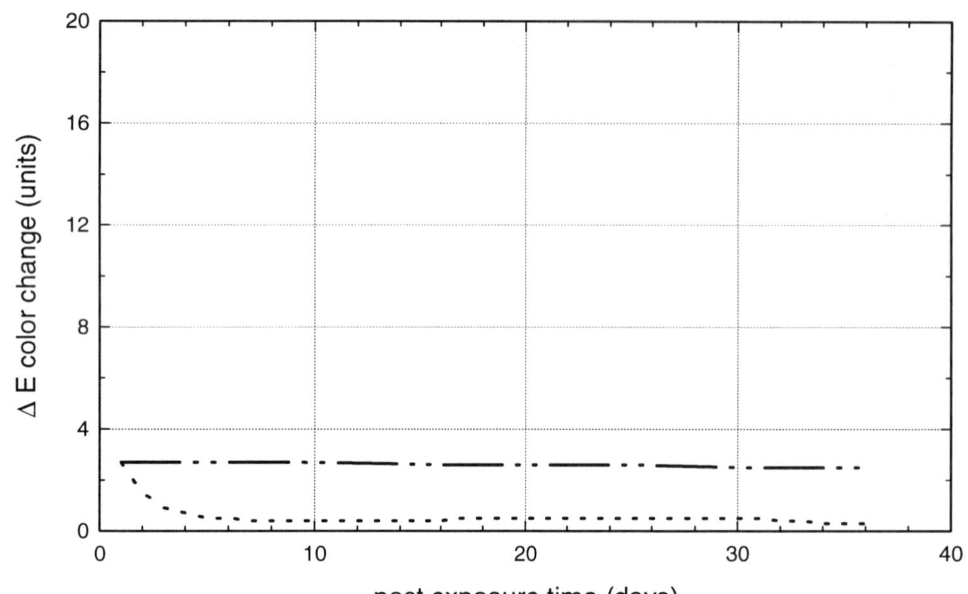

....	Dow Styron 484 IPS; 2.5 Mrads; light storage
—·—··	Dow Styron 484 IPS; 2.5 Mrads; dark storage
Reference No.	3

GRAPH 121: Post Gamma Radiation Exposure Time vs. Delta E Color Change of Impact Polystyrene

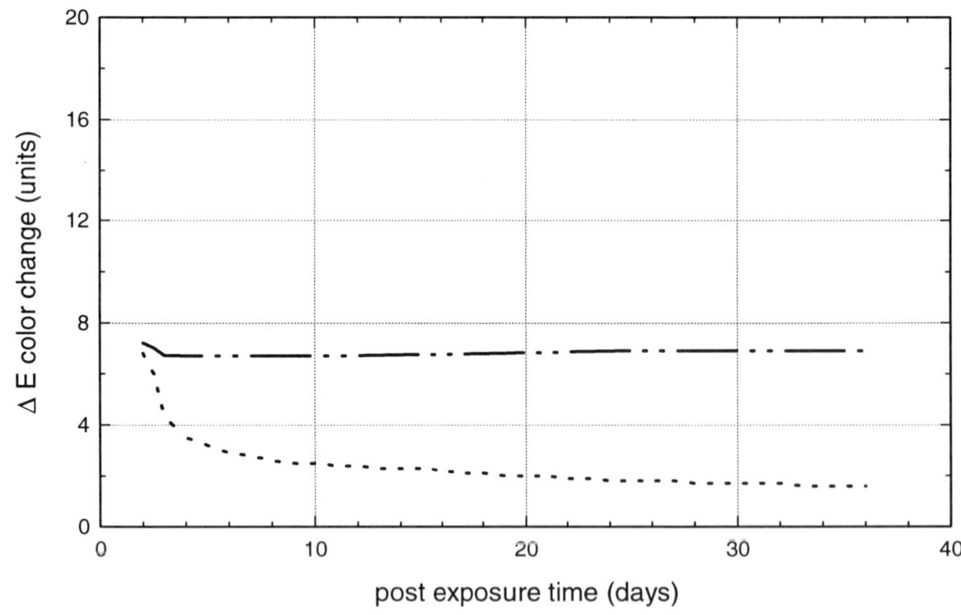

....	Dow Styron 484 IPS; 10 Mrads; light storage
—·—··	Dow Styron 484 IPS; 10 Mrads; dark storage
Reference No.	3

Gamma Radiation Resistance

BASF AG: Ultrason S (features: transparent, amber tint)

Ultrason S offers very high resistance to gamma radiation over the entire range of service temperatures. Gassing is very slight. The trasmittance for gamma rays is very high.

Reference: *Ultrason E, Ultrason S Product Line, Properties, Processing,* supplier design guide (B 602 e/10.92) - BASF Aktiengesellschaft, 1992.

Amoco Performance Products: Udel (features: transparent, amber tint)

In cobalt-60 gamma radiation exposures conducted by the Australian Atomic Energy Commission, the polymer was found to be completely soluble in chloroform following doses up to 500 Mrads. The University of Queensland reported polysulfone's resistance to such radiation to be the highest yet found for any organic polymer, being at least an order of magnitude greater than that of any olefin or vinyl polymer. Polycarbonate was found to undergo chain scission at the bisphenol A linkages.

Similar findings were made at Leichester University, where gamma radiolysis was found to damage polysulfone only slowly compared to other polymers.

Reference: *The Radiation Response of Udel Polysulfone,* supplier technical report (Number: 101) - Amoco Performance Products, Inc.

Amoco Performance Products: Udel (features: transparent, amber tint)

Udel Polysulfone is well suited for sterilization by gamma radiation (cobalt-60 source), without loss of impact resistance.

Reference: *Engineering Plastics For The Medical And Health Care Fields,* supplier marketing literature (G-F-50022) - Amoco Performance Products, Inc.

Amoco Performance Products: Udel P1700 (features: transparent, amber tint)

Polysulfone evidenced a modest change (downward) in melt flow when irradiated at 2.5 or 4 Mrads, indicating reasonably good resistance to molecular change at these dosages. At 6 Mrads, however, the melt flow dropped to about 1/15 of its original level, signifying the development of substantial crosslinking. Polysulfone discolors during irradiation. The level of discoloration increases with the increase in radiation dosage. Because of higher initial color, polysulfone parts are always of deeper color than those of polycarbonate. Polycarbonate parts show a greater rate of change after sterilization than do parts made from polysulfone. There is no evidence of change in mechanical properties at any of the three irradiation levels. Tensile strength, elongation, modulus, and notched Izod impact values remain essentially unchanged. This is surprising, in view of the crosslinking noted above, and indicates that such molecular changes are not sufficient to affect these properties. There is no significant change in chemical resistance following irradiation. After irradiation polysulfone demonstrated very slight alterations in specimen dimensions. The dimensional change did not exceed 0.2% and should not pose performance problems.

Reference: *Gamma Irradiaiton Studies On Polysulfone And Polycarbonate,* supplier technical report (project no.: 214M01; file no.: 7845-R) - Amoco Performance Products, Inc., 1982.

Electron Beam Radiation

Amoco Performance Products: Udel (features: transparent, amber tint)

Davis and Gleaves* report a high degree of polysulfone stability towards electron irradiation as measured by crosslinking, molecular weight change, and chemical as well as mechanical properties. Using a four million electron volt linear accelerator electron beam, no mechanical deterioration of the polymer occurred as a result of ionizing radiation up to at least 200 Mrads. Even at 400 Mrads, although some gradual darkening was observed, transparency was still maintained.

*Davis, A., Gleaves M.H., Golden J.H., Huglin M.B.. "The Electron Irradiation Resistance of Polysulfone", Makromolekulare Chemie 129, p. 63 (1969).

Reference: *The Radiation Response of Udel Polysulfone,* supplier technical report (Number: 101) - Amoco Performance Products, Inc.

Ethylene Oxide (EtO) Resistance

Amoco Performance Products: Udel (features: transparent, amber tint)

Tests run in a typical gas sterilization apparatus and hospital aeration system showed residuals in polysulfone to be well within the FDA-suggested limit of 25 ppm. Polysulfone's physical properties and appearance remained unchanged.

Reference: *Engineering Plastics For The Medical And Health Care Fields,* supplier marketing literature (G-F-50022) - Amoco Performance Products, Inc.

Steam Resistance

Amoco Performance Products: Mindel S1000

Mindel S-1000's resistance to steam is such that its mechanical and impact properties are retained throughout 100 autoclave cycles (30 minutes each at 132°C).

Engineering Plastics For The Medical And Health Care Fields, supplier marketing literature (G-F-50022) - Amoco Performance Products, Inc.

Amoco Performance Products: Udel (features: transparent, amber tint)

Polysulfone is not structurally attacked through hydrolysis and offers outstanding performance in water environments. Products from it have been steam sterilized repeatedly without loss of performance. In one experiment, unstressed samples were exposed to over 1,000 three minute cycles at 140°C in an autoclave with no significant change in mechanical properties.

Udel Polysulfone Design Engineering Handbook, supplier design guide (F-47178) - Amoco Performance Products, Inc., 1988.

Amoco Performance Products: Udel (features: transparent, amber tint)

Polysulfone has outstanding resistance to pressurized steam, the most frequently encountered hospital sterilization method. Transparent polysulfone will also retain its clarity during extended service life.

Engineering Plastics For The Medical And Health Care Fields, supplier marketing literature (G-F-50022) - Amoco Performance Products, Inc.

BASF AG: Ultrason S (features: transparent, amber tint)

Ultrason parts can be repeatedly sterilized in superheated steam. After more than 100 sterilization cycles, samples remain transparent and largely retain their high level of mechanical properties. In order to avoid environmental stress cracking, the level of molded-in stresses in parts should be as low as possible. Likewise, high mechanical loads should be avoided during sterilization.

Ultrason E, Ultrason S Product Line, Properties, Processing, supplier design guide (B 602 e/10.92) - BASF Aktiengesellschaft, 1992.

Polysulfone (features: transparent, natural resin, amber tint)

PSO has excellent retention of tensile strength after 1500 vacuum autoclave cycles. The addition of morpholine caused widespread crazing within 50 cycles. The area and degree of crazing was independent of induced stress. However, crazing was more evident in locations where spotting had occurred. In feed water without morpholine the rate and extent of crazing were significantly reduced. After 1500 cycles there was only minor crazing in localized areas. Because crazing was most evident around areas of spotting , the feed water additives appear to have become concentrated - and therefore more aggressive - during the drying phase of sterilization.

The process that caused crazing is not immediately obvious. One suggestion is when hot steam is introduced into the autoclave, the outer surface of the trays begin to absorb moisture. As this happens, the material gains weight and expands locally. Because equilibrium through the thickness of the material cannot be achieved quickly, stress builds up as the outer layers attempt to expand while being constrained by the dry inner core. Repeated cycling, with or without morpholine, causes fatigue in these high stress areas. This is supported by the fact that all crazes and cracks are found slightly below the surface. The presence of morpholine, particularly at condensation points, seems to accelerate fatigue craze growth.

Testing was conducted using a commercial vacuum autoclave programmed to provide a 9 minute prevacuum conditioning cycle, a 5 minute sterilization time, and 4 minutes of drying time. Tests were also conducted with and without 50 ppm of morpholine introduced into the feed water system of the vacuum sterilization system. Morpholine is one of the most prevalent boiler feed water additives. Thermoformed trays (representative of medical sterilization trays) were tested and loaded with aluminum bar stock to model the weight of medical instruments.

Reference: Bonifant, Benjamin C., *Designing With Amorphous Thermoplastics,* Medical Device & Diagnostic Industry, trade journal - Cannon Communications, Inc., 1988.

Dry Heat Sterilization

Amoco Performance Products: Udel (features: transparent, amber tint)

A dry heat technique has been found to be the most effective means to reduce the incidence of hospital related infections traced to inhalation therapy equipment. A study by Jet Propulsion Laboratories* recommends an exposure of six hours in dry 125°C heat. Products made from polysulfone can be dry heat sterilized at 140°C. Polysulfone may replace components made of PVC, polycarbonate, modified polyphenylene oxide, nylon and acetal, according to the same study.

*Technical Memorandum 33-670 by Jet Propulsion Laboratories.

Reference: *Engineering Plastics For The Medical And Health Care Fields,* supplier marketing literature (G-F-50022) - Amoco Performance Products, Inc.

Chemical Sterilants

Amoco Performance Products: Udel (features: transparent, amber tint)

Udel Polysulfone performs well, retaining its strength in a wide variety of aqueous disinfectants, including buffered glutaraldehyde, phenol, quaternay ammonium, iodophor and formaldehyde types, and detergent germicide.

Reference: *Engineering Plastics For The Medical And Health Care Fields,* supplier marketing literature (G-F-50022) - Amoco Performance Products, Inc.

Polysulfone (features: transparent, natural resin, amber tint)

The effects of several common disinfecting solutions and chemicals on PSO have been tested. Samples were immersed in the solutions and examined for cracks and crazes. Glutaraldehyde, a common ingredient of cold sterilants, is very soluble in water, and regenerates the dialdehyde on vacuum distillation so that it maintains/increases its presence through autoclave cycles. PSO was exposed at various stress levels at both room temperature and 150°C. The resin performed well at room temperature, even after prolonged exposure (simulating accidental storage in solution.) However, exposure at elevated temperature showed that PSO was attacked.

Reference: *Ultem Resin: Advanced Technology For Reusable Medical Devices,* supplier marketing literature (ULT-314A 1/91) RTB) - General Electric Company, 1991.

TABLE 125: Effect of Gamma Radiation Sterilization on Polysulfone

Material Family	POLYSULFONE		
Material Supplier/Name	AMOCO UDEL P1700		
Material Note	transparent, amber tint	transparent, amber tint	transparent, amber tint
Reference No.	43	43	43
EXPOSURE CONDITIONS			
Type	Gamma Radiation	Gamma Radiation	Gamma Radiation
Details	source: Cobalt 60	source: Cobalt 60	source: Cobalt 60
Radiation Dose (Mrads)	2.5	4	6
Note	test lab: Isomedix, Inc.	test lab: Isomedix, Inc.	test lab: Isomedix, Inc.
PROPERTIES RETAINED (%)			
Tensile Strength	98.1 (he)	98.1 (he)	103.9 (he)
Elongation	96.1 (ai)	102.3 (ai)	86.7 (ai)
Modulus	102.8 (gs)	102.8 (gs)	111.1 (gs)
Notched Izod Impact	100 (fm)	100 (fm)	107.7 (fm)
Melt Flow Rate	85.2 (a)	74.1 (a)	6.5 (a)

TABLE 126: Effect of Steam Sterilization on Polysulfone

Material Family	Polysulfone
Material Supplier/Name	Amoco Udel
Material Note	transparent, amber tint
Reference No.	17
EXPOSURE CONDITIONS	
Type	Steam
Number of Cycles	40
Temperature (°C)	132.2
Time (hours)	0.5
PROPERTIES RETAINED (%)	
Dart Impact (total energy)	87 (ct)

GRAPH 122: **Steam Autoclave Exposure Time vs. Tensile Strength of Polysulfone**

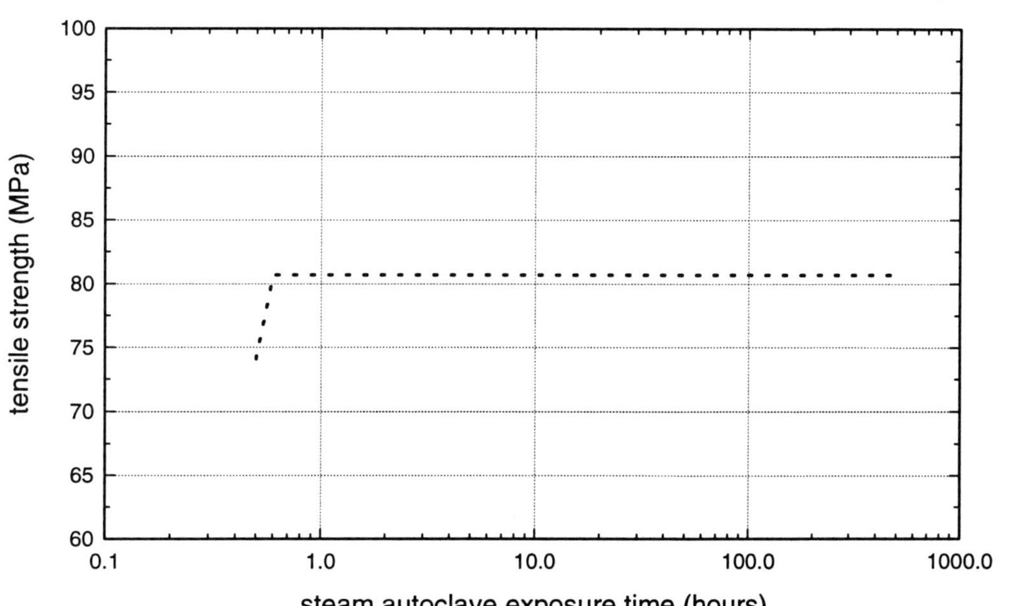

....	Amoco Udel Polysulfone
Reference No.	17

GRAPH 123: **Number of Steam Sterilization Cycles vs. Tensile Strength of Polysulfone**

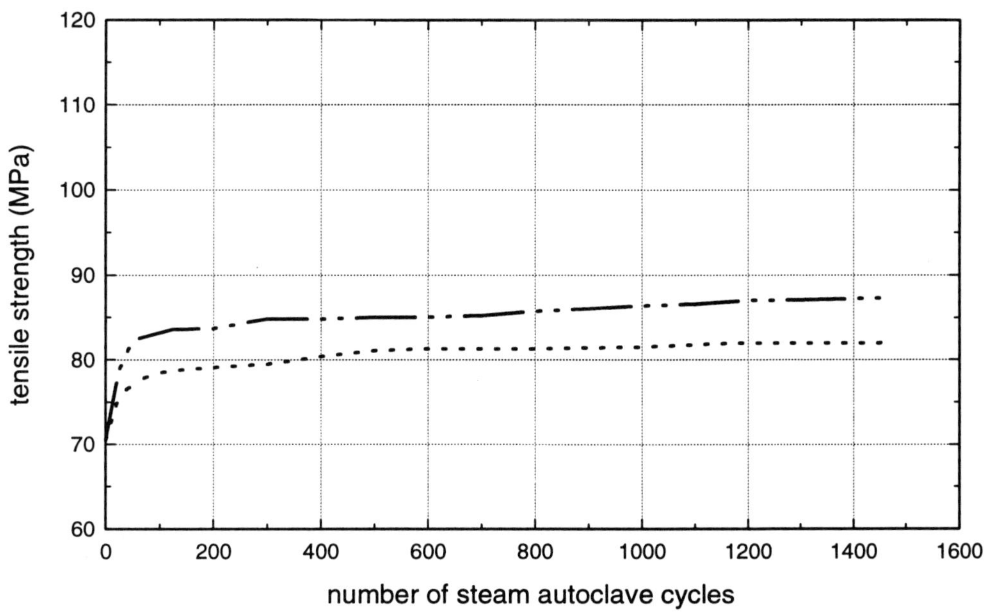

....	Polysulfone; no boiler additive
—··—··	Polysulfone; 50 ppm morpholine
Reference No.	52

GRAPH 124: Steam Autoclave Exposure Time vs. Elongation at Break of Polysulfone

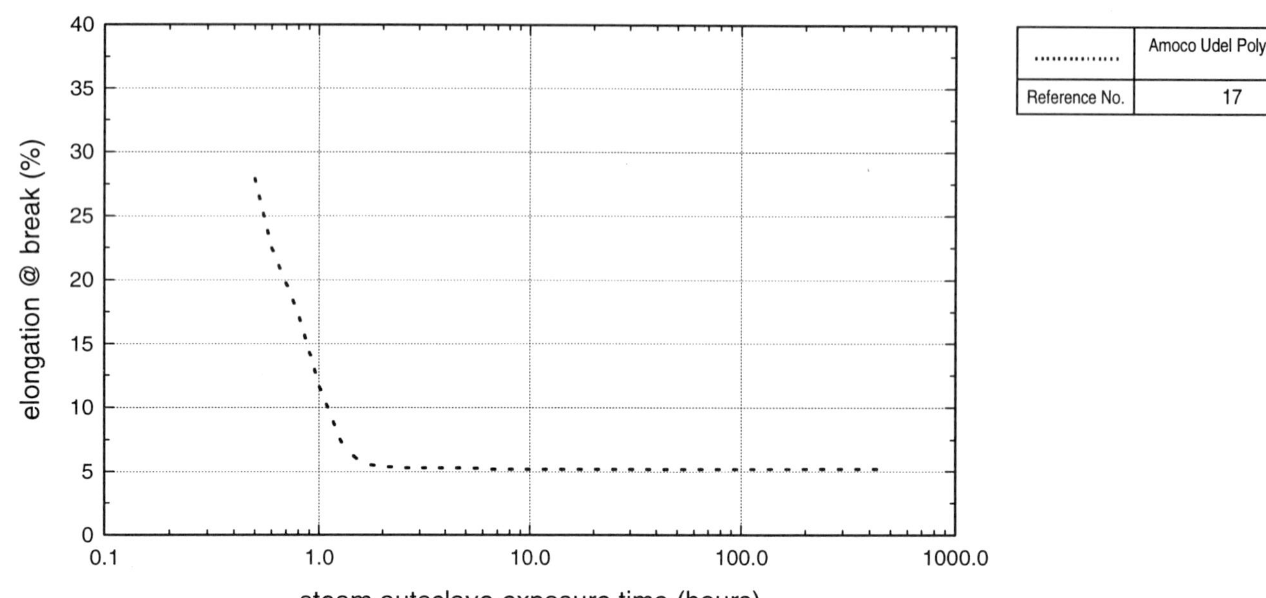

GRAPH 125: Steam Autoclave Exposure Time vs. Tensile Modulus of Polysulfone

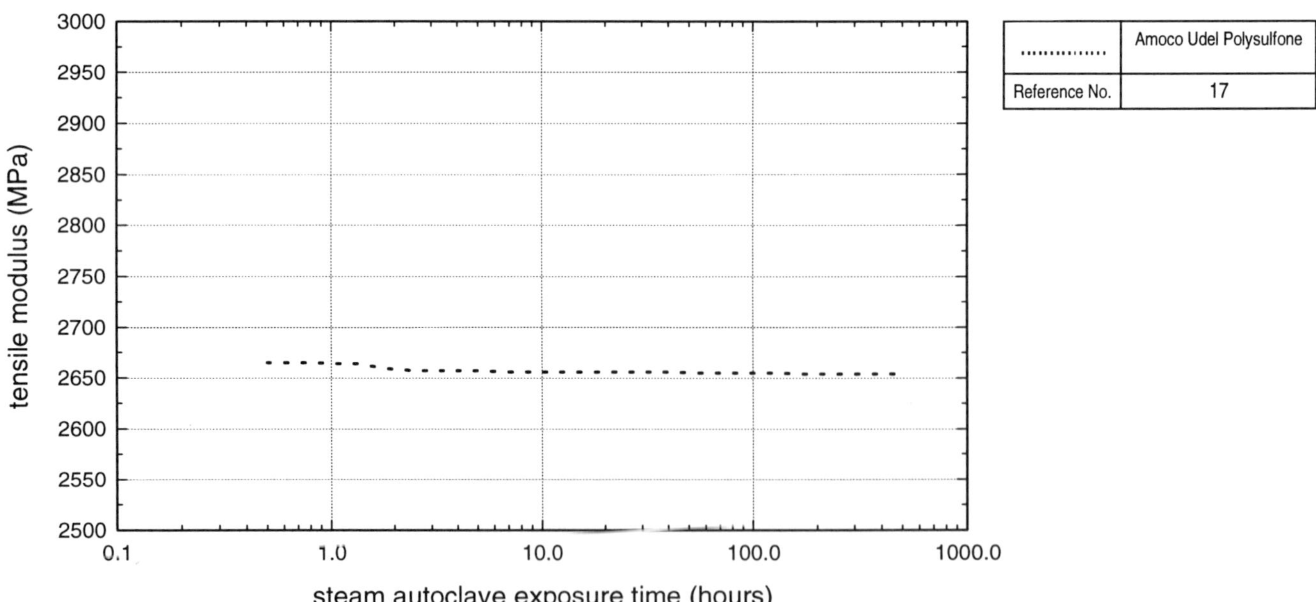

GRAPH 126: Steam Autoclave Exposure Time vs. Tensile Impact Strength of Polysulfone

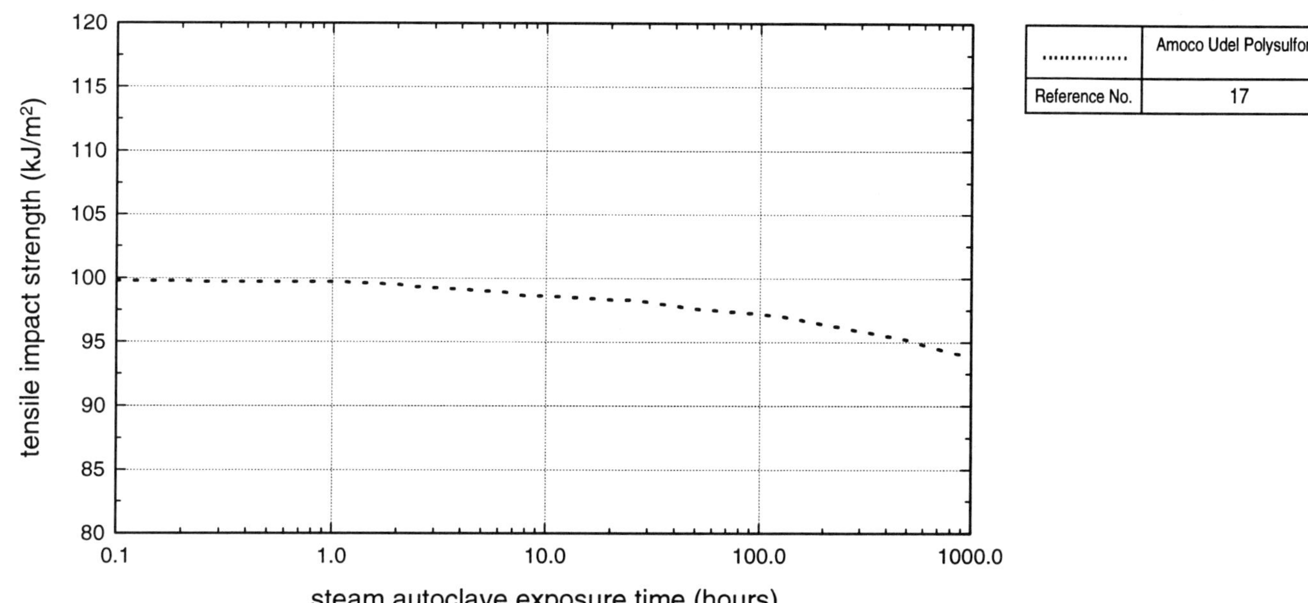

....	Amoco Udel Polysulfone
Reference No.	17

GRAPH 127: Number of Steam Sterilization Cycles vs. Drop Dart Impact Strength of Polysulfone

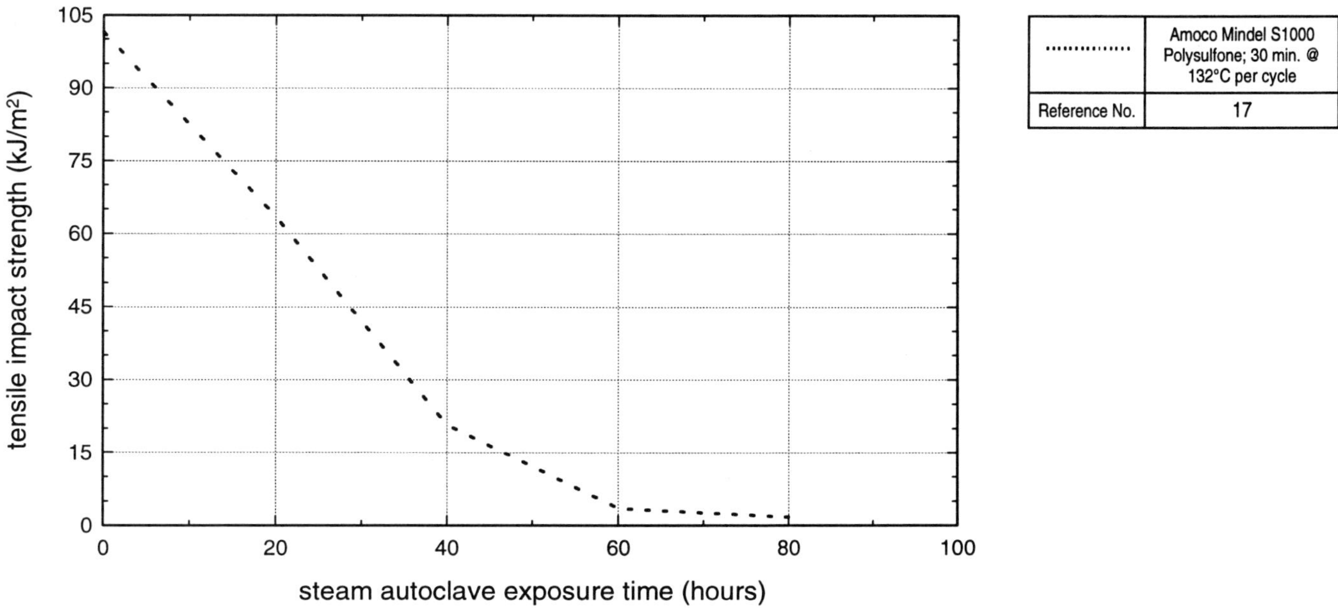

....	Amoco Mindel S1000 Polysulfone; 30 min. @ 132°C per cycle
Reference No.	17

Gamma Radiation Resistance

BASF AG: Ultrason E (features: transparent, amber tint)

Ultrason E offers very high resistance to gamma radiation over the entire range of service temperatures. Ultrason E products exposed to high-energy radiation suffer a noticeable decrease in tensile strength at yield and a significant decrease in ultimate elongation. Gassing is very slight. The trasmittance for gamma rays is very high.

Reference: *Ultrason E, Ultrason S Product Line, Properties, Processing,* supplier design guide (B 602 e/10.92) - BASF Aktiengesellschaft, 1992.

Victrex USA: Victrex PES (features: transparent, amber tint)

Gamma radiation can be applied to PES with no ill effects. One example of its use is a valve application where bacteria from milk and heavily sugared materials pass through the unit, which is then flushed with steam and then irradiated for a short period of time to kill any contaminating materials left behind.

Reference: *Victrex Polymers For Medical Applications,* supplier technical report - ICI Advanced Materials.

Steam Resistance

BASF AG: Ultrason E (features: transparent, amber tint)

Ultrason parts can be repeatedly sterilized in superheated steam. After more than 100 sterilization cycles, samples remain transparent and largely retain their high level of mechanical properties. Ultrason E differs from Ultrason S in that it absorbs more water and is thus suitable only for steam sterilization without a vacuum phase. In order to avoid environmental stress cracking, the level of molded-in stresses in parts should be as low as possible. Likewise, high mechanical loads should be avoided during sterilization.

Ultrason E, Ultrason S Product Line, Properties, Processing, supplier design guide (B 602 e/10.92) - BASF Aktiengesellschaft, 1992.

Polyehtersulfone (features: transparent, natural resin, amber tint)

PES has excellent retention of tensile strength after 1500 vacuum autoclave cycles. The addition of morpholine caused widespread crazing within 50 cycles. The area and degree of crazing was independent of induced stress. However, crazing was more evident in locations where spotting had occurred. In feed water without morpholine the rate and extent of crazing were significantly reduced. After 1500 cycles there was only minor crazing in localized areas. Because crazing was most evident around areas of spotting , the feed water additives appear to have become concentrated - and therefore more aggressive - during the drying phase of sterilization.

The process that caused crazing is not immediately obvious. One suggestion is when hot steam is introduced into the autoclave, the outer surface of the trays begin to absorb moisture. As this happens, the material gains weight and expands locally. Because equilibrium through the thickness of the material cannot be achieved quickly, stress builds up as the outer layers attempt to expand while being constrained by the dry inner core. Repeated cycling, with or without morpholine, causes fatigue in these high stress areas. This is supported by the fact that all crazes and cracks are found slightly below the surface. The presence of morpholine, particularly at condensation points, seems to accelerate fatigue craze growth.

Testing was conducted using a commercial vacuum autoclave programmed to provide a 9 minute prevacuum conditioning cycle, a 5 minute sterilization time, and 4 minutes of drying time. Tests were also conducted with and without 50 ppm of morpholine introduced into the feed water system of the vacuum sterilization system. Morpholine is one of the most prevalent boiler feed water additives. Thermoformed trays (representative of medical sterilization trays) were tested and loaded with aluminum bar stock to model the weight of medical instruments.

Bonifant, Benjamin C., *Designing With Amorphous Thermoplastics,* Medical Device & Diagnostic Industry, trade journal - Cannon Communications, Inc., 1988.

Victrex USA: Victrex PES (features: transparent, amber tint)

The product can resist steam sterilization up to 150°C, repeatedly. In the United States, the addition of boiler fluids can cause severe stress cracking when coupled with high temperature steam usage. In tests at hospitals throughout the United States, PES has shown to perform well and withstand many cycles under this system.

Reference: *Victrex Polymers For Medical Applications,* supplier technical report - ICI Advanced Materials.

Dry Heat Sterilization

Victrex USA: Victrex PES (features: transparent, amber tint)

PES's ability to withstand hot air sterilization up to 180°C has allowed PES into the dental industry where hot air sterilization is the normal mode of procedure.

Reference: *Victrex Polymers For Medical Applications,* supplier technical report - ICI Advanced Materials.

TABLE 127: Effect of Gamma Radiation Sterilization on Polyethersulfone

Material Family	POLYETHERSULFONE			
Material Supplier/Name	VICTREX PES			
Material Note	transparent, amber tint	transparent, amber tint	transparent, amber tint	transparent, amber tint
Reference No.	76	76	76	76

EXPOSURE CONDITIONS

Type	Gamma Radiation	Gamma Radiation	Gamma Radiation	Gamma Radiation
Radiation Dose (Mrads)	50	100	200	250

PROPERTIES RETAINED (%)

Tensile Strength @ Yield	101.3 (ib)	95 (ib)	105 (ib)	92.5 (ib)
Elongation	85.7 (ak)	21.4 (ak)	24.3 (ak)	12.9 (ak)

TABLE 128: Effect of Beta Radiation Sterilization on Polyethersulfone

Material Family	POLYETHERSULFONE					
Material Supplier/Name	VICTREX PES					
Material Note	transparent, amber tint	transparent, amber tint	transparent, amber tint	transparent, amber tint	transparent, amber tint	transparent, amber tint
Reference No.	76	76	76	76	76	76

EXPOSURE CONDITIONS

Type	Beta Radiation	Beta Radiation	Beta Radiation	Beta Radiation	Beta Radiation	Beta Radiation
Radiation Dose (Mrads)	30	60	30	25	25	50
Temperature (°C)	20	100	150	200	210	220

PROPERTIES RETAINED (%)

Tensile Strength @ Yield	102.4 (ib)	100 (ib)	98.8 (ib)	95.3 (ib)	96.5 (ib)	92.9 (ib)
Elongation	69.2 (ak)	28.5 (ak)	76.9 (ak)	72.3 (ak)	64.6 (ak)	38.5 (ak)

GRAPH 128: Number of Steam Sterilization Cycles vs. Tensile Strength of Polyethersulfone

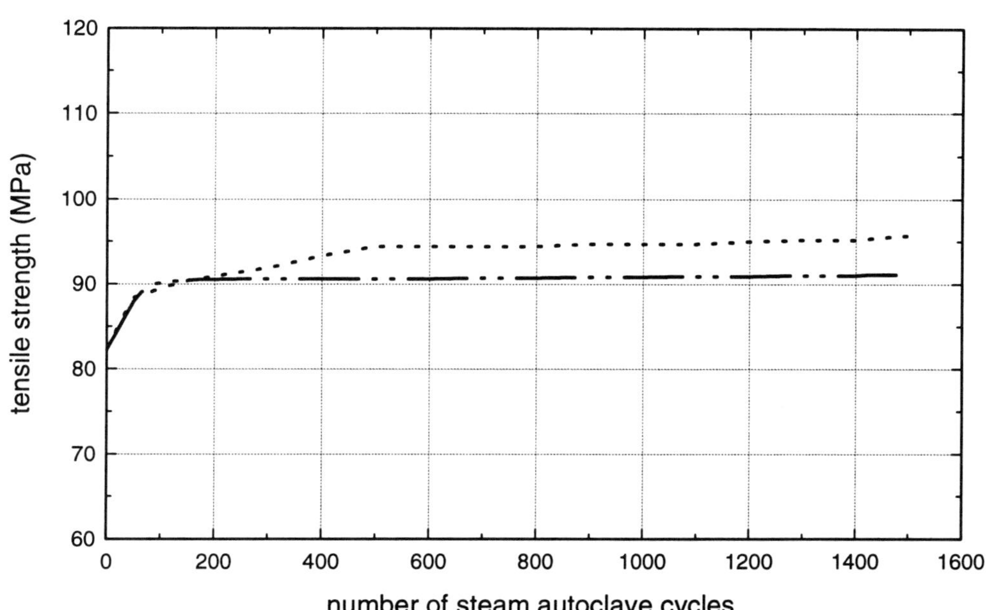

....	PES; no boiler additive
—··—··	PES; 50 ppm morpholine
Reference No.	52

GRAPH 129: Number of Steam Sterilization Cycles vs. Tensile Impact Strength of Polyethersulfone

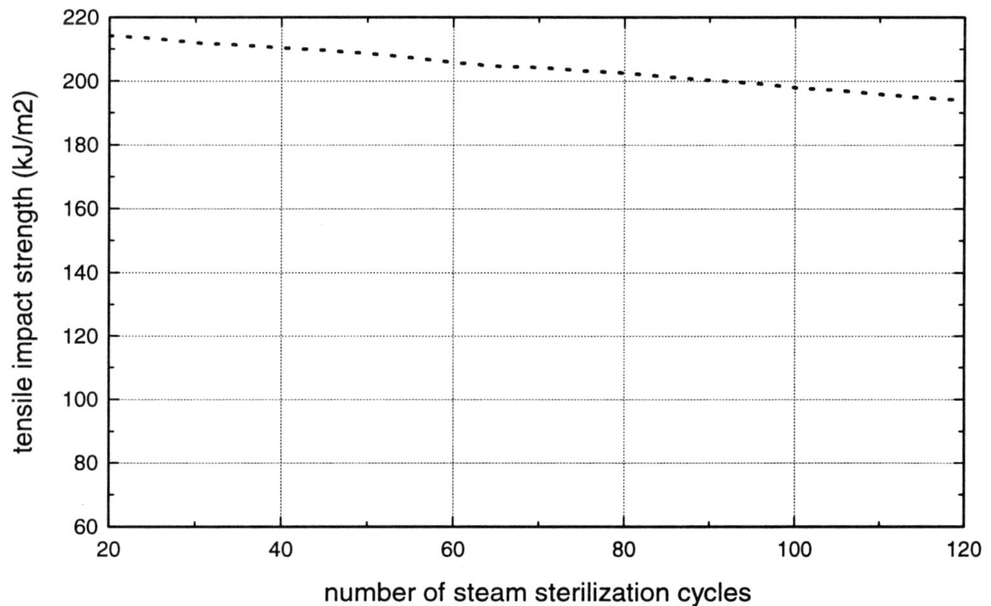

....	PES; 30 min. @ 132°C per cycle; test method: ASTM D1822
Reference No.	18

Steam Resistance

Amoco Performance Products: Radel

Radel resins have superior resistance to steam and boiling water. Hydrolytic stability of injection molded, 1/8 inch thick test bars was tested by exposure to 132°C steam containing 50 ppm of morpholine. (Morpholine or similar amine chemicals are typically present for corrosion resistance in steam coming from central supply sources. Such amines can be aggressive toward plastics.)

The molded bars were tested both unstressed and with 500 and 1000 psi constant applied stress. The stress was applied via weights attached to one end while fixing the other end horizontally as a cantilevered beam. Steam exposure was 30 mins/cycle returning to room temperature before starting another cycle.

Tensile impact (ASTM D1822) after steam exposure shows Radel A-200 maintaining good tensile impact through 1000+ cycles while polyetherimide deteriorated significantly.

Reference: *Radel Polyarylsulfone...A Family Of Thermoplastic Materials Engineered For High Performance In Harsh Environments,* supplier marketing literature (F-49896) - Amoco Performance Products, Inc.

GRAPH 130: Number of Steam Sterilization Cycles vs. Tensile Impact Strength of Polyarylsulfone

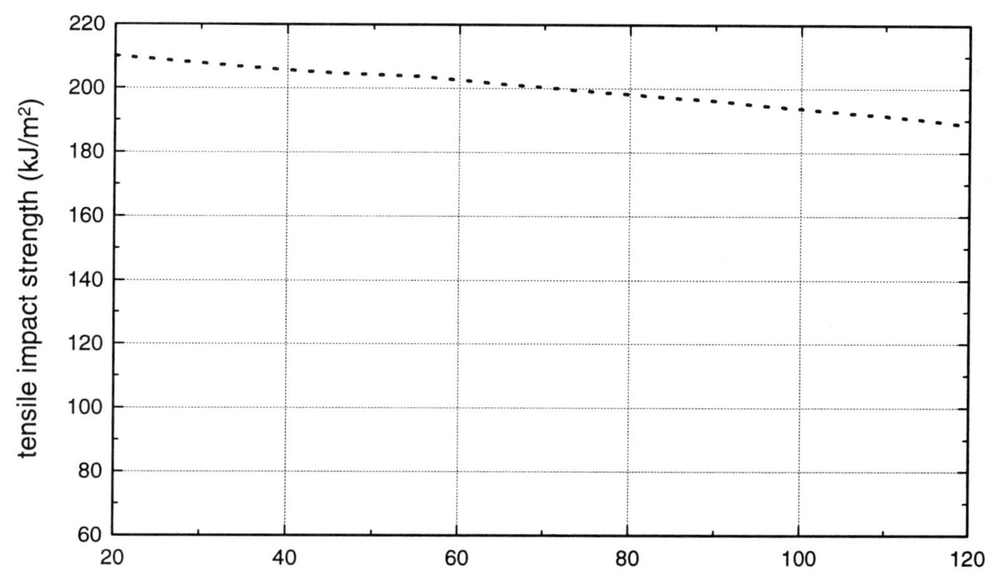

....	Amoco Radel A200 PAS; 30 min. @ 132 °C per cycle; test method: ASTM D1822
Reference No.	18

Radiation Resistance

Dow Chemical: Isoplast 2531 (features: transparent)

The physical properties were unaffected by high energy sterilization. The samples did show a 40-60% increase in ultimate tensile elongation; however a standard deviation of 32% reduces the significance of this result. Color decreased with time, but significant discoloration remained.

Reference: Hermanson, Nancy J., Steffens, John F., *The Physical and Visual Property Changes in Thermoplastic Resins After Exposure to High Energy Sterilization - Gamma Versus Electron Beam,* ANTEC 1993, conference proceedings - Society of Plastics Engineers, 1993.

Dow Chemical: Isoplast 2510

The physical properties were unaffected by high energy sterilization. Color decreased with time, but significant discoloration remained. This permanent color change can be masked with pigments, minimizing the visible color shift.

Reference: Hermanson, Nancy J., Steffens, John F., *The Physical and Visual Property Changes in Thermoplastic Resins After Exposure to High Energy Sterilization - Gamma Versus Electron Beam,* ANTEC 1993, conference proceedings - Society of Plastics Engineers, 1993.

Gamma Radiation Resistance

Dow Chemical: Isoplast 301 (features: transparent)

The pysical properties of Transparent Rigid Thermoplastic Polyurethanes are not significantly affected by exposure to gamma radiation. Discoloration is dramatic upon exposure to gamma radiation. Yellowness index increases from 6.3 before exposure to 77.6 after exposure. The discoloration appears to be permanent with bleach-back minimal. Storage in light does not affect the bleach back and optical properties exposed at both 2.5 and 10.0 Mrads.

Reference: Sturdevant, Marianne F., *Sterilization Compatibility of Rigid Thermoplastic Materials,* supplier technical report (301-1548) - Dow Chemical Company, 1988.

Dow Chemical: Isoplast 101

The pysical properties of Opague Rigid Thermoplastic Polyurethanes are not significantly affected by exposure to gamma radiation. Discoloration is dramatic upon exposure to gamma radiation. The discoloration appears to be permanent with bleach-back minimal. Storage in light does not affect the bleach back and optical properties exposed at both 2.5 and 10.0 Mrads.

Reference: Sturdevant, Marianne F., *Sterilization Compatibility of Rigid Thermoplastic Materials,* supplier technical report (301-1548) - Dow Chemical Company, 1988.

Dow Chemical: (features: transparent, natural resin)

Samples exhibit a substantial degree of discoloration upon irradiation. Transparent RTPU has a slow rate of color reversal with a large portion of its original discoloration still visible after one year. An increase in dosage, from 2.5 to 10 Mrads, increases the amount of initial and permanent discoloration. The physical properties are relatively unaffected up to a year after exposure to 10 Mrads of gamma radiation.

Reference: Sturdevant, Marianne F., *The Long-term Effects of Ethylene Oxide and Gamma Radiation Sterilization on the Properties of Rigid Thermoplastic Materials,* ANTEC 1990, conference proceedings - Society of Plastics Engineers, 1990.

Dow Chemical: (features: opaque, natural resin)

Samples exhibit a substantial degree of discoloration upon irradiation. An increase in dosage, from 2.5 to 10 Mrads, increases the amount of initial and permanent discoloration. The physical properties are relatively unaffected up to a year after exposure to 10 Mrads of gamma radiation.

Reference: Sturdevant, Marianne F., *The Long-term Effects of Ethylene Oxide and Gamma Radiation Sterilization on the Properties of Rigid Thermoplastic Materials,* ANTEC 1990, conference proceedings - Society of Plastics Engineers, 1990.

Ethylene Oxide (EtO) Resistance

Dow Chemical: Isoplast 301 (features: transparent)

Transparent Rigid Thermoplastic Polyurethanes exposed to five cycles of ethylene oxide gas retained their notched Izod impact strength. The tensile yield strength was also unaffected. Transparent RTPU measured less than 300 ppm of residual EtO on the first day following exposure and continued to de-gas rapidly.

Reference: Sturdevant, Marianne F., *Sterilization Compatibility of Rigid Thermoplastic Materials,* supplier technical report (301-1548) - Dow Chemical Company, 1988.

Dow Chemical: Isoplast 101

Opaque Rigid Thermoplastic Polyurethanes exposed to five cycles of ethylene oxide gas retained their notched Izod impact strength. The tensile yield strength was also unaffected. Opaque RTPU measured less than 300 ppm of residual EtO on the first day following exposure and continued to de-gas rapidly.

Reference: Sturdevant, Marianne F., *Sterilization Compatibility of Rigid Thermoplastic Materials,* supplier technical report (301-1548) - Dow Chemical Company, 1988.

Dow Chemical: (features: transparent, natural resin)

Transparent RTPU is compatible with multiple cycles of ethylene oxide sterilization due primarily to its excellent chemical resistance characteristics. Samples were tested up to five cycles.

Reference: Sturdevant, Marianne F., *The Long-term Effects of Ethylene Oxide and Gamma Radiation Sterilization on the Properties of Rigid Thermoplastic Materials,* ANTEC 1990, conference proceedings - Society of Plastics Engineers, 1990.

Dow Chemical: (features: opaque, natural resin)

Opaque RTPU is compatible with multiple cycles of ethylene oxide sterilization due primarily to its excellent chemical resistance characteristics. Samples were tested up to five cycles.

Reference: Sturdevant, Marianne F., *The Long-term Effects of Ethylene Oxide and Gamma Radiation Sterilization on the Properties of Rigid Thermoplastic Materials,* ANTEC 1990, conference proceedings - Society of Plastics Engineers, 1990.

TABLE 129: Effect of Gamma Radiation Sterilization on Rigid Thermoplastic Urethane

Material Family	RIGID THERMOPLASTIC URETHANE					
Material Supplier/Name	DOW					
Material Note	opaque, natural resin	opaque, natural resin	opaque, natural resin	opaque, natural resin	opaque, natural resin	opaque, natural resin
Reference No.	5	5	5	5	5	5
EXPOSURE CONDITIONS						
Type	Gamma Radiation	Gamma Radiation	Gamma Radiation	Gamma Radiation	Gamma Radiation	Gamma Radiation
Details	source: Cobalt 60	source: Cobalt 60	source: Cobalt 60	source: Cobalt 60	source: Cobalt 60	source: Cobalt 60
Radiation Dose (Mrads)	2.5	2.5	2.5	10	10	10
Note	test lab: Radiations Sterilizers Inc.	test lab: Radiations Sterilizers Inc.	test lab: Radiations Sterilizers Inc.	test lab: Radiations Sterilizers Inc.	test lab: Radiations Sterilizers Inc.	test lab: Radiations Sterilizers Inc.
POST EXPOSURE CONDITIONING						
Note	type: storage in dark	type: storage in dark	type: storage in dark	type: storage in dark	type: storage in dark	type: storage in dark
Temperature (°C)	21	21	21	21	21	21
Time (hours)	336	4368	8760	336	4368	8760
PROPERTIES RETAINED (%)						
Tensile Strength	116.1 (hy)	112.6 (hy)	109 (hy)	(hy)	107.6 (hy)	112.6 (hy)
Tensile Strength @ Yield	93.4 (il)	94.7 (il)	92.3 (il)	(il)	94.7 (il)	96.5 (il)
Elongation	113.9 (au)	109.2 (au)	101.2 (au)	(au)	106.4 (au)	112.7 (au)
Dart Impact (total energy)	101.6 (ff)	106.3 (ff)	101.9 (ff)	98.2 (ff)	93.7 (ff)	100.2 (ff)
Notched Izod Impact	102.4 (fx)	98.8 (fx)	97.6 (fx)	100.8 (fx)	100.4 (fx)	95.9 (fx)

TABLE 130: Effect of Gamma Radiation Sterilization on Rigid Thermoplastic Urethane

Material Family	RIGID THERMOPLASTIC URETHANE					
Material Supplier/Name	DOW					
Material Note	transparent, natural resin	transparent, natural resin	transparent, natural resin	transparent, natural resin	transparent, natural resin	transparent, natural resin
Reference No.	5	5	5	5	5	5
EXPOSURE CONDITIONS						
Type	Gamma Radiation	Gamma Radiation	Gamma Radiation	Gamma Radiation	Gamma Radiation	Gamma Radiation
Details	source: Cobalt 60	source: Cobalt 60	source: Cobalt 60	source: Cobalt 60	source: Cobalt 60	source: Cobalt 60
Radiation Dose (Mrads)	2.5	2.5	2.5	10	10	10
Note	test lab: Radiations Sterilizers Inc.	test lab: Radiations Sterilizers Inc.	test lab: Radiations Sterilizers Inc.	test lab: Radiations Sterilizers Inc.	test lab: Radiations Sterilizers Inc.	test lab: Radiations Sterilizers Inc.
POST EXPOSURE CONDITIONING						
Note	type: storage in dark	type: storage in dark	type: storage in dark	type: storage in dark	type: storage in dark	type: storage in dark
Temperature (°C)	21	21	21	21	21	21
Time (hours)	336	4368	8760	336	4368	8760
PROPERTIES RETAINED (%)						
Tensile Strength	109.5 (hy)	108.9 (hy)	102.6 (hy)	106 (hy)	99.2 (hy)	89.8 (hy)
Tensile Strength @ Yield	94.3 (il)	93.3 (il)	92.4 (il)	94.4 (il)	94.1 (il)	95.4 (il)
Elongation	109.5 (au)	116.8 (au)	104.4 (au)	112.4 (au)	91.2 (au)	92.7 (au)
Dart Impact (total energy)	93.8 (ff)	93.8 (ff)	89.9 (ff)	90.2 (ff)	87 (ff)	87.3 (ff)
Notched Izod Impact	85 (fx)	105 (fx)	75 (fx)	80 (fx)	95 (fx)	90 (fx)

TABLE 131: Effect of Gamma Radiation Sterilization on Rigid Thermoplastic Urethane

Material Family	RIGID THERMOPLASTIC URETHANE							
Material Supplier/Name	DOW ISOPLAST 301		DOW ISOPLAST 101		DOW ISOPLAST 301		DOW ISOPLAST 101	
Material Note	transparent	transparent			transparent	transparent		
Reference No.	3	3	3	3	3	3	3	3

EXPOSURE CONDITIONS

Type	Gamma Radiation	Gamma Radiation	Gamma Radiation	Gamma Radiation	Gamma Radiation	Gamma Radiation	Gamma Radiation	Gamma Radiation
Details	source: Cobalt 60	source: Cobalt 60	source: Cobalt 60	source: Cobalt 60	source: Cobalt 60	source: Cobalt 60	source: Cobalt 60	source: Cobalt 60
Radiation Dose (Mrads)	2.5	10	2.5	10	2.5	10	2.5	10

POST EXPOSURE CONDITIONING

Note	type: storage under fluorescent light	type: storage under fluorescent light	type: storage under fluorescent light	type: storage under fluorescent light	type: storage in dark	type: storage in dark	type: storage in dark	type: storage in dark
Temperature (°C)	21	21	21	21	21	21	21	21
Time (hours)	336	336	336	336	336	336	336	336

PROPERTIES RETAINED (%)

Tensile Strength @ Yield	94.4 (ii)	96.3 (ii)	94.8 (ii)	96.1 (ii)	94.4 (ii)	94.4 (ii)	94.8 (ii)	(ii)
Notched Izod Impact	100 (fp)	84.2 (fp)	101.2 (fp)	98.4 (fp)	100 (fp)	84.2 (fp)	101.2 (fp)	98.4 (fp)

TABLE 132: Effect of Gamma Radiation Sterilization on Rigid Thermoplastic Urethane

Material Family	RIGID THERMOPLASTIC URETHANE			
Material Supplier/Name	DOW ISOPLAST 2510			
Reference No.	1	1	1	1

EXPOSURE CONDITIONS

Type	Gamma Radiation	Gamma Radiation	Gamma Radiation	Gamma Radiation
Radiation Dose (Mrads)	2.5	2.5	10	10
Note	test lab: SteriGenics	test lab: SteriGenics	test lab: SteriGenics	test lab: SteriGenics

POST EXPOSURE CONDITIONING

Note	type: aging	type: aging	type: aging	type: aging
Time (hours)	168	1344	168	1344

PROPERTIES RETAINED (%)

Tensile Strength @ Yield	102 (ir)	118 (ir)	100 (ir)	116 (ir)
Elongation	103.1 (br)	96.4 (br)	96.9 (br)	100.5 (br)
Flexural Strength	100 (cn)	98.6 (cn)	98.6 (cn)	97.2 (cn)
Modulus	93 (hb)	94.1 (hb)	92.4 (hb)	90.8 (hb)
Flexural Modulus	100 (cf)	102.6 (cf)	105.3 (cf)	98.9 (cf)
Dart Impact (total energy)	82.5 (eh)	100 (eh)	98.4 (eh)	92.1 (eh)
Dart Impact (peak energy)	66.7 (dh)	102 (dh)	86.3 (dh)	82.4 (dh)
Notched Izod Impact	94.2 (gg)	98.3 (gg)	83.8 (gg)	102.2 (gg)
Heat Deflection Temperature	98.5 (m)	98.5 (m)	100 (m)	100 (m)
Vicat Softening Point	100 (ku)	97.8 (ku)	98.9 (ku)	97.8 (ku)

SURFACE and APPEARANCE

ΔL Color	-13.2 (ac)	-8 (ac)	-20.4 (ac)	-12.9 (ac)
Δa Color	8.8 (o)	5.2 (o)	12.8 (o)	7.2 (o)
Δb Color	8.2 (t)	5.3 (t)	8.7 (t)	9.4 (t)

TABLE 133: Effect of Gamma Radiation Sterilization on Rigid Thermoplastic Urethane

Material Family	RIGID THERMOPLASTIC URETHANE			
Material Supplier/Name	DOW ISOPLAST 2531			
Material Note	transparent	transparent	transparent	transparent
Reference No.	1	1	1	1
EXPOSURE CONDITIONS				
Type	Gamma Radiation	Gamma Radiation	Gamma Radiation	Gamma Radiation
Radiation Dose (Mrads)	2.5	2.5	10	10
Note	test lab: SteriGenics	test lab: SteriGenics	test lab: SteriGenics	test lab: SteriGenics
POST EXPOSURE CONDITIONING				
Note	type: aging	type: aging	type: aging	type: aging
Time (hours)	168	1344	168	1344
PROPERTIES RETAINED (%)				
Tensile Strength @ Yield	125 (iq)	107.8 (iq)	125 (iq)	125 (iq)
Elongation	87.9 (bs)	160.6 (bs)	160.6 (bs)	139.4 (bs)
Flexural Strength	103 (cn)	99 (cn)	100 (cn)	102 (cn)
Modulus	111.7 (gx)	100 (gx)	100.4 (gx)	102.2 (gx)
Flexural Modulus	103.8 (ce)	101.7 (ce)	100.4 (ce)	107.2 (ce)
Dart Impact (total energy)	98.7 (eo)	97.3 (eo)	89.3 (eo)	33.3 (eo)
Dart Impact (peak energy)	92.7 (dp)	105.5 (dp)	85.5 (dp)	34.5 (dp)
Notched Izod Impact	125 (gc)	107.8 (gc)	125 (gc)	125 (gc)
Heat Deflection Temperature	102.6 (m)	100 (m)	97.4 (m)	100 (m)
Vicat Softening Point	100 (ku)	100 (ku)	99.1 (ku)	97.2 (ku)
SURFACE and APPEARANCE				
Δ Yellowness Index	64.5 (kw)	48.3 (kw)	114.8 (kw)	85.8 (kw)
Haze (%)	2.23	0.89	1.35	3.89
Transparency Retained (%)	67	85.7	41.8	58.2

TABLE 134: Effect of Electron Beam Radiation Sterilization on Rigid Thermoplastic Urethane

Material Family	RIGID THERMOPLASTIC URETHANE			
Material Supplier/Name	DOW ISOPLAST 2531			
Material Note	transparent	transparent	transparent	transparent
Reference No.	1	1	1	1

EXPOSURE CONDITIONS

Type	Electron Beam Radiation	Electron Beam Radiation	Electron Beam Radiation	Electron Beam Radiation
Radiation Dose (Mrads)	2.5	2.5	10	10
Note	test lab: E-Beam Services, Inc.	test lab: E-Beam Services, Inc.	test lab: E-Beam Services, Inc.	test lab: E-Beam Services, Inc.

POST EXPOSURE CONDITIONING

Note	type: aging	type: aging	type: aging	type: aging
Time (hours)	168	1344	168	1344

PROPERTIES RETAINED (%)

Tensile Strength @ Yield	107.8 (iq)	107.8 (iq)	125 (iq)	117.2 (iq)
Elongation	160.6 (bs)	163.6 (bs)	140.4 (bs)	141.4 (bs)
Flexural Strength	103 (cn)	101 (cn)	103 (cn)	100 (cn)
Modulus	98.7 (gx)	97.4 (gx)	89.6 (gx)	94.4 (gx)
Flexural Modulus	102.1 (ce)	113.2 (ce)	97.4 (ce)	100 (ce)
Dart Impact (total energy)	100 (eo)	94.7 (eo)	84 (eo)	73.3 (eo)
Dart Impact (peak energy)	94.5 (dp)	92.7 (dp)	83.6 (dp)	72.7 (dp)
Notched Izod Impact	107.8 (gc)	107.8 (gc)	125 (gc)	117.2 (gc)
Heat Deflection Temperature	98.7 (m)	100 (m)	105.3 (m)	107.9 (m)
Vicat Softening Point	99.1 (ku)	99.1 (ku)	99.1 (ku)	97.2 (ku)

SURFACE and APPEARANCE

Δ Yellowness Index	64 (kw)	53.1 (kw)	136.2 (kw)	124 (kw)
Haze (%)	2.24	2.73	2.07	-1.02
Transparency Retained (%)	68.1	81.3	37.4	51.6

TABLE 135: Effect of Electron Beam Radiation Sterilization on Rigid Thermoplastic Urethane

Material Family	RIGID THERMOPLASTIC URETHANE			
Material Supplier/Name	DOW ISOPLAST 2510			
Reference No.	1	1	1	1

EXPOSURE CONDITIONS

Type	Electron Beam Radiation	Electron Beam Radiation	Electron Beam Radiation	Electron Beam Radiation
Radiation Dose (Mrads)	2.5	2.5	10	10
Note	test lab: E-Beam Services, Inc.	test lab: E-Beam Services, Inc.	test lab: E-Beam Services, Inc.	test lab: E-Beam Services, Inc.

POST EXPOSURE CONDITIONING

Note	type: aging	type: aging	type: aging	type: aging
Time (hours)	168	1344	168	1344

PROPERTIES RETAINED (%)

Tensile Strength @ Yield	100 (ir)	118 (ir)	102 (ir)	112 (ir)
Elongation	122.8 (br)	88.6 (br)	96.9 (br)	91.7 (br)
Flexural Strength	104.2 (cn)	97.2 (cn)	111.3 (cn)	95.8 (cn)
Modulus	100.5 (hb)	90.8 (hb)	89.2 (hb)	91.9 (hb)
Flexural Modulus	101.1 (cf)	97.9 (cf)	101.6 (cf)	101.1 (cf)
Dart Impact (total energy)	100 (eh)	93.7 (eh)	95.2 (eh)	93.7 (eh)
Dart Impact (peak energy)	92.2 (dh)	82.4 (dh)	84.3 (dh)	82.4 (dh)
Notched Izod Impact	103.5 (gg)	99.1 (gg)	99.1 (gg)	98.7 (gg)
Heat Deflection Temperature	98.5 (m)	98.5 (m)	107.6 (m)	100 (m)
Vicat Softening Point	100 (ku)	97.8 (ku)	97.8 (ku)	97.8 (ku)

SURFACE and APPEARANCE

ΔL Color	-13.1 (ac)	-7.7 (ac)	-21.1 (ac)	-7.9 (ac)
Δa Color	8.7 (o)	4.6 (o)	15.2 (o)	4.6 (o)
Δb Color	10.2 (t)	5.4 (t)	16.7 (t)	5.6 (t)

TABLE 136: Effect of Ethylene Oxide Sterilization on Rigid Thermoplastic Urethane

Material Family	RIGID THERMOPLASTIC URETHANE					
Material Supplier/Name	DOW					
Material Note	opaque, natural resin	opaque, natural resin	opaque, natural resin	opaque, natural resin	opaque, natural resin	opaque, natural resin
Reference No.	5	5	5	5	5	5

EXPOSURE CONDITIONS

Type	Ethylene Oxide	Ethylene Oxide	Ethylene Oxide	Ethylene Oxide	Ethylene Oxide	Ethylene Oxide
Details	12% EtO and 88% Freon	12% EtO and 88% Freon	12% EtO and 88% Freon	12% EtO and 88% Freon	12% EtO and 88% Freon	12% EtO and 88% Freon
Concentration	660 mg/l	660 mg/l	660 mg/l	660 mg/l	660 mg/l	660 mg/l
Number of Cycles	1	1	1	5	5	5
Note	RH: 60%; test lab: Ethox Corp.	RH: 60%; test lab: Ethox Corp.	RH: 60%; test lab: Ethox Corp.	RH: 60%; test lab: Ethox Corp.	RH: 60%; test lab: Ethox Corp.	RH: 60%; test lab: Ethox Corp.
Temperature (°C)	49	49	49	49	49	49
Time (hours)	≥ 6	≥ 6	≥ 6	≥ 6	≥ 6	≥ 6

PRE EXPOSURE CONDITIONING

Preconditioning Note	time: 8 hours; temperature: 37.8°C; RH: 60%	time: 8 hours; temperature: 37.8°C; RH: 60%	time: 8 hours; temperature: 37.8°C; RH: 60%	time: 8 hours; temperature: 37.8°C; RH: 60%	time: 8 hours; temperature: 37.8°C; RH: 60%	time: 8 hours; temperature: 37.8°C; RH: 60%

POST EXPOSURE CONDITIONING

Note	type: evacuation; pressure: 127 mm Hg	type: evacuation; pressure: 127 mm Hg	type: evacuation; pressure: 127 mm Hg	type: evacuation; pressure: 127 mm Hg	type: evacuation; pressure: 127 mm Hg	type: evacuation; pressure: 127 mm Hg

POST EXPOSURE CONDITIONING II

Note	type: aeration	type: aeration	type: aeration	type: aeration	type: aeration	type: aeration
Temperature (°C)	32.2	32.2	32.2	32.2	32.2	32.2
Time (hours)	≥ 16	≥ 16	≥ 16	≥ 16	≥ 16	≥ 16

POST EXPOSURE CONDITIONING III

Note	type: storage in dark	type: storage in dark	type: storage in dark	type: storage in dark	type: storage in dark	type: storage in dark
Temperature (°C)	21	21	21	21	21	21
Time (hours)	336	4368	8760	336	4368	8760

PROPERTIES RETAINED (%)

Tensile Strength	106.1 (hz)	93.2 (hz)	91.4 (hz)	101.6 (hz)	104.4 (hz)	87.6 (hz)
Tensile Strength @ Yield	98.6 (ij)	95.9 (ij)	93.3 (ij)	101.2 (ij)	99.3 (ij)	101.5 (ij)
Elongation	79.8 (av)	96.5 (av)	97.1 (av)	104.6 (av)	101.7 (av)	96.5 (av)
Dart Impact (total energy)	104.6 (fh)	96 (fh)	104.6 (fh)	105.8 (fh)	95.6 (fh)	97 (fh)
Notched Izod Impact	96.3 (fr)	103.3 (fr)	96.7 (fr)	96.3 (fr)	99.2 (fr)	89.8 (fr)

TABLE 137: Effect of Ethylene Oxide Sterilization on Rigid Thermoplastic Urethane

Material Family	RIGID THERMOPLASTIC URETHANE					
Material Supplier/Name	DOW					
Material Note	transparent, natural resin	transparent, natural resin	transparent, natural resin	transparent, natural resin	transparent, natural resin	transparent, natural resin
Reference No.	5	5	5	5	5	5

EXPOSURE CONDITIONS

Type	Ethylene Oxide	Ethylene Oxide	Ethylene Oxide	Ethylene Oxide	Ethylene Oxide	Ethylene Oxide
Details	12% EtO and 88% Freon	12% EtO and 88% Freon	12% EtO and 88% Freon	12% EtO and 88% Freon	12% EtO and 88% Freon	12% EtO and 88% Freon
Concentration	660 mg/l	660 mg/l	660 mg/l	660 mg/l	660 mg/l	660 mg/l
Number of Cycles	1	1	1	5	5	5
Note	RH: 60%; test lab: Ethox Corp.	RH: 60%; test lab: Ethox Corp.	RH: 60%; test lab: Ethox Corp.	RH: 60%; test lab: Ethox Corp.	RH: 60%; test lab: Ethox Corp.	RH: 60%; test lab: Ethox Corp.
Temperature (°C)	49	49	49	49	49	49
Time (hours)	≥ 6	≥ 6	≥ 6	≥ 6	≥ 6	≥ 6

PRE EXPOSURE CONDITIONING

Preconditioning Note	time: 8 hours; temperature: 37.8°C; RH: 60%	time: 8 hours; temperature: 37.8°C; RH: 60%	time: 8 hours; temperature: 37.8°C; RH: 60%	time: 8 hours; temperature: 37.8°C; RH: 60%	time: 8 hours; temperature: 37.8°C; RH: 60%	time: 8 hours; temperature: 37.8°C; RH: 60%

POST EXPOSURE CONDITIONING

Note	type: evacuation; pressure: 127 mm Hg	type: evacuation; pressure: 127 mm Hg	type: evacuation; pressure: 127 mm Hg	type: evacuation; pressure: 127 mm Hg	type: evacuation; pressure: 127 mm Hg	type: evacuation; pressure: 127 mm Hg

POST EXPOSURE CONDITIONING II

Note	type: aeration	type: aeration	type: aeration	type: aeration	type: aeration	type: aeration
Temperature (°C)	32.2	32.2	32.2	32.2	32.2	32.2
Time (hours)	≥ 16	≥ 16	≥ 16	≥ 16	≥ 16	≥ 16

POST EXPOSURE CONDITIONING III

Note	type: storage in dark	type: storage in dark	type: storage in dark	type: storage in dark	type: storage in dark	type: storage in dark
Temperature (°C)	21	21	21	21	21	21
Time (hours)	336	4368	8760	336	4368	8760

PROPERTIES RETAINED (%)

Tensile Strength	104.6 (hz)	102.5 (hz)	97.5 (hz)	100.9 (hz)	93.7 (hz)	101.1 (hz)
Tensile Strength @ Yield	96.6 (ij)	92.7 (ij)	92.4 (ij)	97.7 (ij)	92.5 (ij)	96.1 (ij)
Elongation	104.4 (av)	102.2 (av)	105.1 (av)	121.2 (av)	100.7 (av)	113.9 (av)
Dart Impact (total energy)	91.6 (fh)	83.2 (fh)	84.5 (fh)	91.5 (fh)	83.9 (fh)	81.2 (fh)
Notched Izod Impact	90 (fr)	95 (fr)	95 (fr)	100 (fr)	90 (fr)	95 (fr)

TABLE 138: Effect of Ethylene Oxide Sterilization on Rigid Thermoplastic Urethane

Material Family	RIGID THERMOPLASTIC URETHANE			
Material Supplier/Name	DOW ISOPLAST 301		DOW ISOPLAST 101	
Material Note	transparent	transparent		
Reference No.	3	3	3	3

EXPOSURE CONDITIONS

Type	Ethylene Oxide	Ethylene Oxide	Ethylene Oxide	Ethylene Oxide
Details	12% EtO and 88% Freon	12% EtO and 88% Freon	12% EtO and 88% Freon	12% EtO and 88% Freon
Concentration	660 mg/l	660 mg/l	660 mg/l	660 mg/l
Number of Cycles	1	5	1	5
Note	RH: 60%	RH: 60%	RH: 60%	RH: 60%
Temperature (°C)	49	49	49	49
Time (hours)	≥ 6	≥ 6	≥ 6	≥ 6

PRE EXPOSURE CONDITIONING

Preconditioning Note	time: 8 hours; temperature: 37.8°C; RH: 60%	time: 8 hours; temperature: 37.8°C; RH: 60%	time: 8 hours; temperature: 37.8°C; RH: 60%	time: 8 hours; temperature: 37.8°C; RH: 60%

POST EXPOSURE CONDITIONING

Note	type: evacuation; pressure: 127 mm Hg	type: evacuation; pressure: 127 mm Hg	type: evacuation; pressure: 127 mm Hg	type: evacuation; pressure: 127 mm Hg

POST EXPOSURE CONDITIONING II

Note	type: aeration	type: aeration	type: aeration	type: aeration
Temperature (°C)	32	32	32	32
Time (hours)	≥ 16	≥ 16	≥ 16	≥ 16

POST EXPOSURE CONDITIONING III

Note	type: storage in dark; RH: 50%	type: storage in dark; RH: 50%	type: storage in dark; RH: 50%	type: storage in dark; RH: 50%
Temperature (°C)	21	21	21	21
Time (hours)	336	336	336	336

PROPERTIES RETAINED (%)

Tensile Strength @ Yield	96.3 (ii)	97.2 (ii)	98.7 (ii)	102.6 (ii)
Notched Izod Impact	94.7 (fp)	100 (fp)	96.3 (fp)	96.3 (fp)

TABLE 139: Effect of Ethylene Oxide Sterilization on Rigid Thermoplastic Urethane

Material Family	**RIGID THERMOPLASTIC URETHANE**
Material Supplier/Name	**DOW ISOPLAST 101**
Material Note	transparent
Reference No.	3

EXPOSURE CONDITIONS

Type	Ethylene Oxide
Details	12% EtO and 88% Freon
Concentration	660 mg/l
Number of Cycles	1
Note	RH: 60%
Temperature (°C)	49
Time (hours)	≥ 6

PRE EXPOSURE CONDITIONING

Preconditioning Note	time: 8 hours; temperature: 37.8°C; RH: 60%

POST EXPOSURE CONDITIONING

Note	type: evacuation; pressure: 127 mm Hg

POST EXPOSURE CONDITIONING II

Note	type: aeration
Temperature (°C)	32

RESIDUALS (ppm)

Residuals Determined	ethylene oxide
24 hour Aeration	160
72 hour Aeration	118
168 hour Aeration	86
744 hour Aeration	70
816 hour Aeration	36
864 hour Aeration	23

GRAPH 131: **Post Gamma Radiation Exposure Time vs. Yellowness Index of Rigid Thermoplastic Urethane**

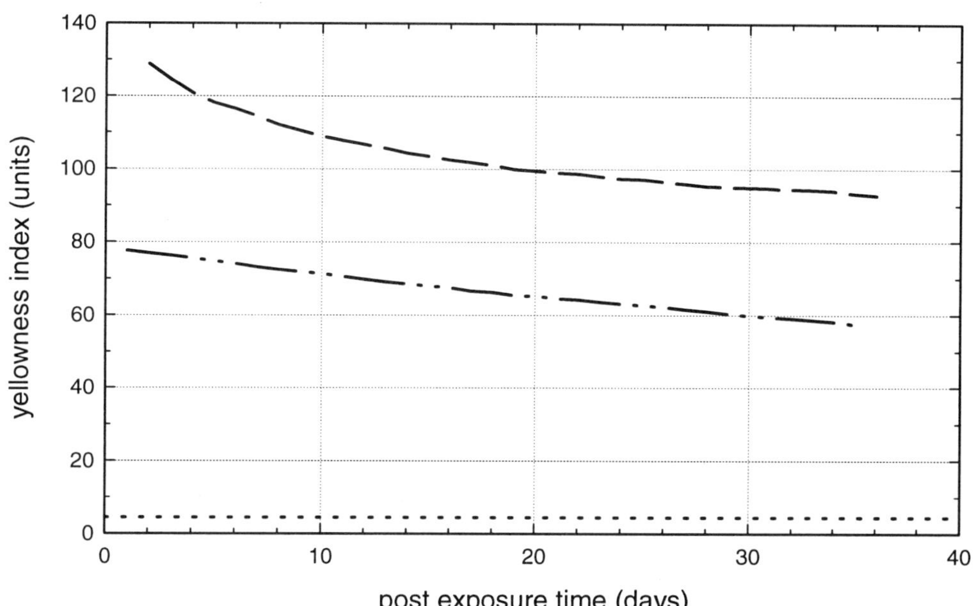

··············	Dow Isoplast 301 RTPU (transpar.); before exposure
—··—··	Dow Isoplast 301 RTPU (transpar.); 2.5 Mrads
— — — —	Dow Isoplast 301 RTPU (transpar.); 10 Mrads
Reference No.	3

GRAPH 132: **Post Gamma Radiation Exposure Time vs. Yellowness Index of Rigid Thermoplastic Urethane**

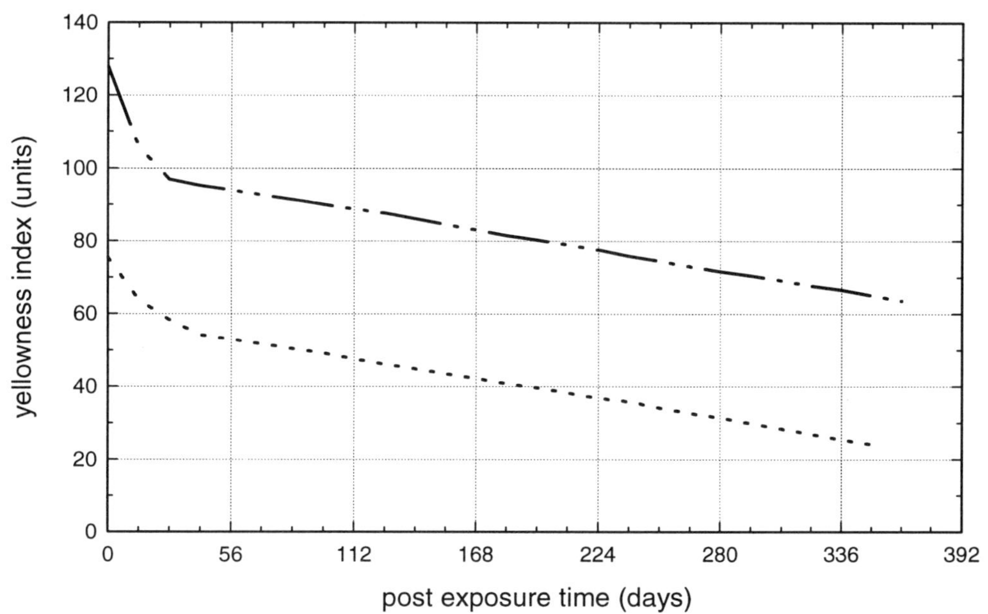

··············	Dow RTPU; 2.5 Mrad
—··—··	Dow RTPU; 10 Mrad
Reference No.	5

GRAPH 133: Post Gamma Radiation Exposure Time vs. Delta E Color Change of Rigid Thermoplastic Urethane

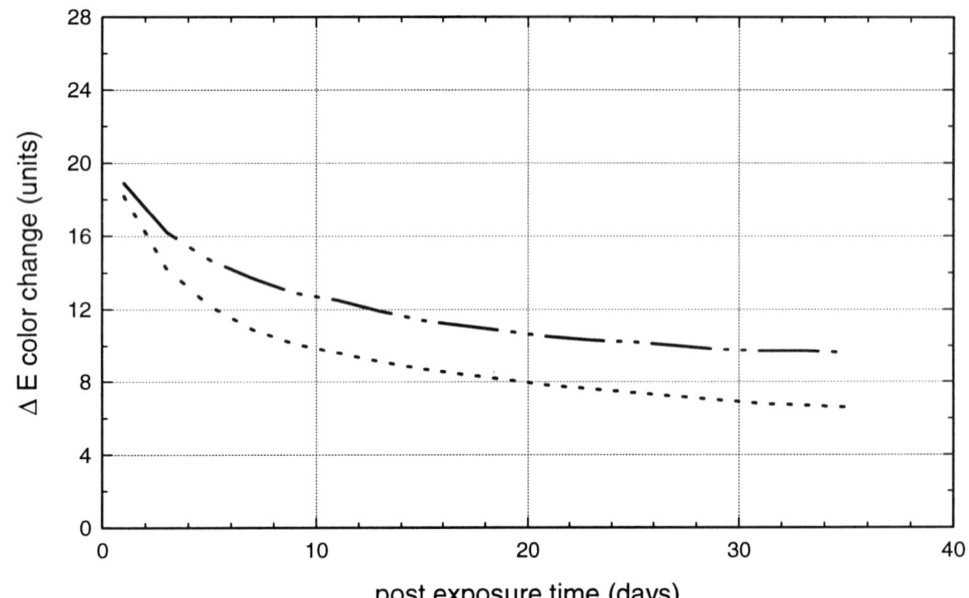

....	Dow Isoplast 101 RTPU; 2.5 Mrads; light storage
—··—··	Dow Isoplast 101 RTPU; 2.5 Mrads; dark storage
Reference No.	3

GRAPH 134: Post Gamma Radiation Exposure Time vs. Delta E Color Change of Rigid Thermoplastic Urethane

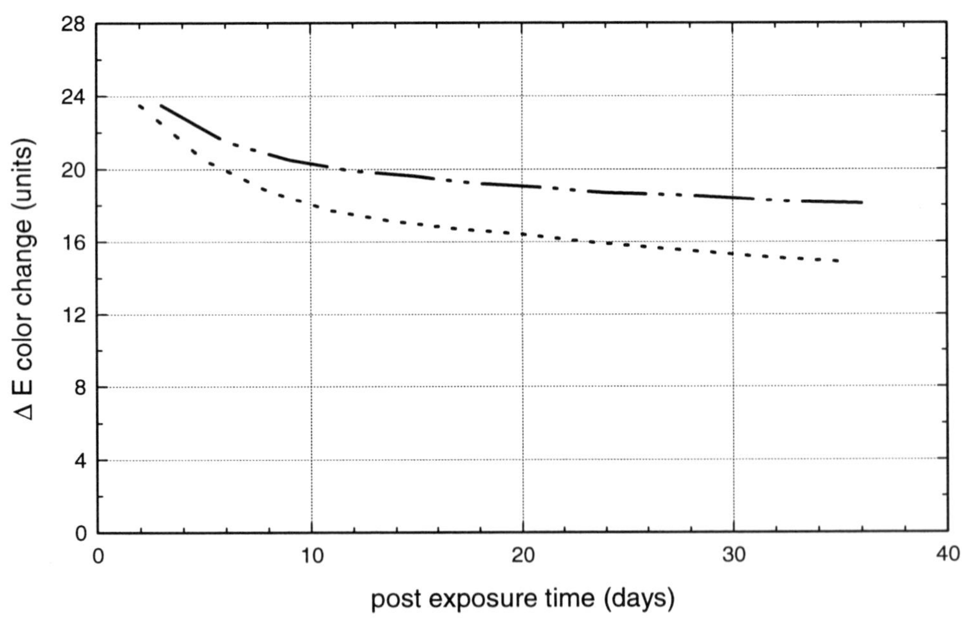

....	Dow Isoplast 101 RTPU; 10 Mrads; light storage
—··—··	Dow Isoplast 101 RTPU; 10 Mrads; dark storage
Reference No.	3

GRAPH 135: **Post Gamma Radiation Exposure Time vs. Percent Light Transmission of Rigid Thermoplastic Urethane**

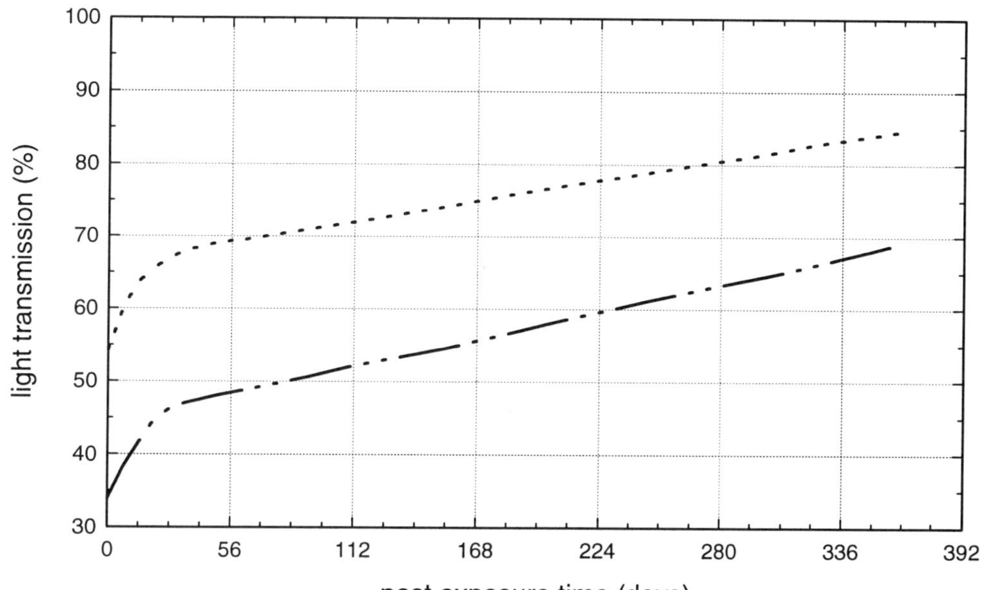

....	Dow RTPU; 2.5 Mrad
—··—··	Dow RTPU; 10 Mrad
Reference No.	5

Radiation Resistance

Dow Chemical: Tyril 1000B (features: transparent)

Neither gamma nor E-beam sterilization affected the physical properties. Initially the blue tint in SAN turned brown after irradiation; this color change decreased with time. After eight weeks, the SAN samples exposed to 2.5 Mrad of gamma or E-beam radiation had returned to their original color. No color difference was observed in samples stored in light versus dark; the samples were not sensitive to photo-bleaching.

Reference: Hermanson, Nancy J., Steffens, John F., *The Physical and Visual Property Changes in Thermoplastic Resins After Exposure to High Energy Sterilization - Gamma Versus Electron Beam,* ANTEC 1993, conference proceedings - Society of Plastics Engineers, 1993.

Gamma Radiation Resistance

Dow Chemical: Tyril 1000B (features: transparent)

Gamma radiation sterilization of styrenics is acceptable. The physical properties of the styrenics are not harmed when exposed to dosage levels of up to 10.0 Mrads. Discoloration of the resins occurs immediately upon irradiation. After exposure to 2.5 Mrads, most of the discoloration is recoverable with time. Higher dosages of 10.0 Mrads cause some amount of permanent discoloration of the resin. SAN had a two fold increase in yellowness index and ΔE values between 2.5 and 10.0 Mrads. No difference is seen when comparing the physical properties of the irradiated samples stored in fluorescent light versus those stored in complete darkness. Storage in light did not affect the bleach-back and optical properties of SAN

Reference: Sturdevant, Marianne F., *Sterilization Compatibility of Rigid Thermoplastic Materials,* supplier technical report (301-1548) - Dow Chemical Company, 1988.

Dow Chemical: (features: transparent, natural resin)

Unmodified styrenics are among the most stable polymers to gamma radiation exposure. SAN discolors more than polystyrene. At exposure levels up to 10 Mrads the samples showed no significant changes in physical properties over the twelve month span of this study.

Reference: Sturdevant, Marianne F., *The Long-term Effects of Ethylene Oxide and Gamma Radiation Sterilization on the Properties of Rigid Thermoplastic Materials,* ANTEC 1990, conference proceedings - Society of Plastics Engineers, 1990.

Monsanto: Lustran Sparkle 2090 (features: transparent, water clear)

Injection molded test specimens were irradiated at doses of 1.5, 2.5, 3.5 and 5.0 Mrads of gamma radiation. The Izod impact (ASTM D-256) value of transparent SAN (Lustran Sparkle) was statistically unaffected by radiation. There was no measurable change in tensile modulus (ASTM D-638). An increase in tensile stress (ASTM D-638) at yield of 5.0% was noted. There was no difference in tensile stress (ASTM D-638) at fail between the control and irradiated samples. Tensile elongation at yield (ASTM D-638) was unchanged. The results of tensile elongation at fail (ASTM D-638) were extremely variable. Virtually no change in flexural modulus (ASTM D-790) was noted. No obvious trends in flexural modulus (ASTM D-790) were observed, with values fluctuating plus or minus 5.0%. At 2.5 Mrads discoloration was minimal. At 10 Mrads the sample turned yellowish brown. During experimentation, a "fading" effect was observed. After time, the discoloration of all samples was barely perceptible.

Reference: *Effects Of Electron Beam And Gamma Radiation Sterilization Of Thermoplastics,* supplier technical report (7126A) - Monsanto Company, 1990.

Monsanto: Lustran SAN 31 (features: transparent, general purpose grade)

Injection molded test specimens were irradiated at doses of 1.5, 2.5, 3.5 and 5.0 Mrads of gamma radiation. The Izod impact (ASTM D-256) value of general purpose transparent SAN (Lustran 31) was statistically unaffected by radiation. There was no measurable change in tensile modulus (ASTM D-638). An increase in tensile stress (ASTM D-638) at yield of 5.0% was noted. There was no difference in tensile stress (ASTM D-638) at fail between the control and irradiated samples. Tensile elongation at yield (ASTM D-638) was unchanged. The results of tensile elongation at fail (ASTM D-638) were extremely variable. Virtually no change in flexural modulus (ASTM D-790) was noted. No obvious trends in flexural modulus (ASTM D-790) were observed, with values fluctuating plus or minus 5.0%. At 2.5 Mrads discoloration was minimal. At 10 Mrads the sample turned yellowish brown. During experimentation, a "fading" effect was observed. After time, the discoloration of all samples was barely perceptible.

Reference: *Effects Of Electron Beam And Gamma Radiation Sterilization Of Thermoplastics,* supplier technical report (7126A) - Monsanto Company, 1990.

Electron Beam Radiation

Monsanto: Lustran Sparkle 2090 (features: transparent, water clear)

Injection molded test specimens were irradiated at doses of 1.5, 2.5, 3.5 and 5.0 Mrads of electron beam radiation. The Izod impact (ASTM D-256) value of transparent SAN (Lustran Sparkle) was statistically unaffected by radiation. There was no measurable change in tensile modulus (ASTM D-638). An increase in tensile stress (ASTM D-638) at yield of 10% to 18% was noted. There was a 5% to 15% difference in tensile stress (ASTM D-638) at fail between the control and irradiated samples. Tensile elongation at yield (ASTM D-638) was unchanged. The results of tensile elongation at fail (ASTM D-638) were extremely variable. Virtually no change in flexural modulus (ASTM D-790) was noted. No obvious trends in flexural modulus (ASTM D-790) were observed, with values fluctuating plus or minus 5.0%. At 2.5 Mrads discoloration was minimal. At 10 Mrads the sample turned yellowish brown. During experimentation, a "fading" effect was observed. After time, the discoloration of all samples was barely perceptible.

Reference: *Effects Of Electron Beam And Gamma Radiation Sterilization Of Thermoplastics,* supplier technical report (7126A) - Monsanto Company, 1990.

Monsanto: Lustran SAN 31 (features: transparent, general purpose grade)

Injection molded test specimens were irradiated at doses of 1.5, 2.5, 3.5 and 5.0 Mrads of electron beam radiation. The Izod impact (ASTM D-256) value of general purpose transparent SAN (Lustran 31) was statistically unaffected by radiation. There was no measurable change in tensile modulus (ASTM D-638). An increase in tensile stress (ASTM D-638) at yield of 10% to 18% was noted. There was a 5% to 15% difference in tensile stress (ASTM D-638) at fail between the control and irradiated samples. Tensile elongation at yield (ASTM D-638) was unchanged. The results of tensile elongation at fail (ASTM D-638) were extremely variable. Virtually no change in flexural modulus (ASTM D-790) was noted. No obvious trends in flexural modulus (ASTM D-790) were observed, with values fluctuating plus or minus 5.0%. At 2.5 Mrads discoloration was minimal. At 10 Mrads the sample turned yellowish brown. During experimentation, a "fading" effect was observed. After time, the discoloration of all samples was barely perceptible.

Reference: *Effects Of Electron Beam And Gamma Radiation Sterilization Of Thermoplastics,* supplier technical report (7126A) - Monsanto Company, 1990.

Ethylene Oxide (EtO) Resistance

Dow Chemical: Tyril 1000B (features: transparent)

Upon exposure to EtO some loss in tensile properties occurs. To maintain optimum performance SAN should have only minimal exposure to EtO. Exposure of up to five cycles did not affect the notched Izod impact strength. At one cycle exposure, SAN does not lose a significant amount of its tensile strength. However, after five cycles of exposure, it loses 23% of its tensile yield strength. While the acrylonitrile content of SAN increases the copolymer's chemical resistance, caution must also be taken not to expose the resin to excessive amounts of EtO. The loss in tensile strength is attributed to poor chemical resistance to ethylene oxide. After 20 days the resin achieved a residual level of EtO of less than 200 ppm.

Reference: Sturdevant, Marianne F., *Sterilization Compatibility of Rigid Thermoplastic Materials,* supplier technical report (301-1548) - Dow Chemical Company, 1988.

Dow Chemical: (features: transparent, natural resin)

Styrenics retain their properties upon exposure to one normal ethylene oxide sterilization cycle. However, care should be taken to minimize excessive or multiple exposure to ethylene oxide as it may cause embrittlement and chemical attack leading to stress cracking. Five repeated sterilization cycles caused some embrittlement. The embrittlement is seen as a loss of tensile elongation at break and a decrease in instrumented dart impact energy. After multiple EtO, cycles the embrittlement appears to compound with time. The elongation and instrumented impact strengths at six months and one year were significantly less than what was observed at two weeks after sterilization. Silver streaks, or crazes along the flow lines, were noted after exposure to five EtO cycles. This is indicative of stress cracking of the polymer with excessive exposure to ethylene oxide.

Reference: Sturdevant, Marianne F., *The Long-term Effects of Ethylene Oxide and Gamma Radiation Sterilization on the Properties of Rigid Thermoplastic Materials,* ANTEC 1990, conference proceedings - Society of Plastics Engineers, 1990.

Dow Chemical: Tyril 1000B (features: transparent)

One cycle of EtO sterilization using CFC-12 or HCFC-124 as the carrier gas does not affect the tensile yield of SAN. However, 2 cycles of either sterilant gas mixture decreases the yield strength of SAN. For 2 cycles of the CFC-12 mixture, a 15% decrease is seen while the HCFC-124 mixture has a 21% decrease in tensile strength. The decrease is attributed to the chemical attack of EtO on the styrenic resin which may cause embrittlement and stress cracking. The instrumented impact is low for SAN (8-12 in-lb) and no significant change is found after sterilization. Yellowness index for SAN is about 3. The percent haze and light transmission are affected by handling and not significantly changed by EtO sterilization. EtO is a viable sterilization technique; however, care should be taken to minimize dosage.

Reference: Hermanson, Nancy J., *Effects Of Alternate Carriers Of Ethylene Oxide Sterilant On Thermoplastics,* supplier technical report (301-02018) - Dow Chemical Company.

Steam Resistance

BASF: Luran (features: transparent)

Steam sterilization is not recommended due to the material's low resistance to heat deformation.

Reference: Johnson, James A., supplier written correspondence - BASF Corporation, 1994.

TABLE 140: Effect of Gamma Radiation Sterilization on Styrene Acrylonitrile Copolymer

Material Family	STYRENE ACRYLONITRILE COPOLYMER					
Material Supplier/Name	DOW					
Material Note	transparent, natural resin	transparent, natural resin	transparent, natural resin	transparent, natural resin	transparent, natural resin	transparent, natural resin
Reference No.	5	5	5	5	5	5
EXPOSURE CONDITIONS						
Type	Gamma Radiation	Gamma Radiation	Gamma Radiation	Gamma Radiation	Gamma Radiation	Gamma Radiation
Details	source: Cobalt 60	source: Cobalt 60	source: Cobalt 60	source: Cobalt 60	source: Cobalt 60	source: Cobalt 60
Radiation Dose (Mrads)	2.5	2.5	2.5	10	10	10
Note	test lab: Radiations Sterilizers Inc.	test lab: Radiations Sterilizers Inc.	test lab: Radiations Sterilizers Inc.	test lab: Radiations Sterilizers Inc.	test lab: Radiations Sterilizers Inc.	test lab: Radiations Sterilizers Inc.
POST EXPOSURE CONDITIONING						
Note	type: storage in dark	type: storage in dark	type: storage in dark	type: storage in dark	type: storage in dark	type: storage in dark
Temperature (°C)	21	21	21	21	21	21
Time (hours)	336	4368	8760	336	4368	8760
PROPERTIES RETAINED (%)						
Tensile Strength	100 (hv)	98.9 (hv)	94.3 (hv)	98 (hv)	98.8 (hv)	95.1 (hv)
Tensile Strength @ Yield	100 (io)	98.9 (io)	94.3 (io)	98 (io)	98.8 (io)	95.1 (io)
Elongation	66.7 (ay)	66.7 (ay)	66.7 (ay)	66.7 (ay)	66.7 (ay)	66.7 (ay)
Dart Impact (total energy)	94.1 (fd)	76.5 (fd)	58.8 (fd)	123.5 (fd)	76.5 (fd)	58.8 (fd)
Notched Izod Impact	66.7 (fv)	166.7 (fv)	100 (fu)	166.7 (fv)	100 (fv)	166.7 (fv)

TABLE 141: Effect of Gamma Radiation Sterilization on Styrene Acrylonitrile Copolymer

Material Family	STYRENE ACRYLONITRILE COPOLYMER			
Material Supplier/Name	DOW TYRIL 1000B			
Material Note	transparent	transparent	transparent	transparent
Reference No.	3	3	3	3

EXPOSURE CONDITIONS

Type	Gamma Radiation	Gamma Radiation	Gamma Radiation	Gamma Radiation
Details	source: Cobalt 60	source: Cobalt 60	source: Cobalt 60	source: Cobalt 60
Radiation Dose (Mrads)	2.5	10	2.5	10

POST EXPOSURE CONDITIONING

Note	type: storage under fluorescent light	type: storage under fluorescent light	type: storage in dark	type: storage in dark
Temperature (°C)	21	21	21	21
Time (hours)	336	336	336	336

PROPERTIES RETAINED (%)

Tensile Strength @ Yield	100 (ii)	100 (ii)	101.9 (ii)	99 (ii)
Notched Izod Impact	100 (fp)	100 (fp)	75 (fp)	100 (fp)

TABLE 142: Effect of Gamma Radiation Sterilization on Styrene Acrylonitrile Copolymer

Material Family	STYRENE ACRYLONITRILE COPOLYMER			
Material Supplier/Name	DOW TYRIL 1000B			
Material Note	transparent	transparent	transparent	transparent
Reference No.	1	1	1	1
EXPOSURE CONDITIONS				
Type	Gamma Radiation	Gamma Radiation	Gamma Radiation	Gamma Radiation
Radiation Dose (Mrads)	2.5	2.5	10	10
Note	test lab: SteriGenics	test lab: SteriGenics	test lab: SteriGenics	test lab: SteriGenics
POST EXPOSURE CONDITIONING				
Note	type: aging	type: aging	type: aging	type: aging
Time (hours)	168	1344	168	1344
PROPERTIES RETAINED (%)				
Tensile Strength @ Yield	102.9 (ir)	100 (ir)	102.9 (ir)	102.9 (ir)
Elongation	100 (bq)	66.7 (bq)	100 (bq)	66.7 (bq)
Flexural Strength	102.4 (cn)	98.4 (cn)	100.8 (cn)	96.9 (cn)
Modulus	100.6 (hc)	101.1 (hc)	101.1 (hc)	94.3 (hc)
Flexural Modulus	103.3 (ch)	102.7 (ch)	104.9 (ch)	102.7 (ch)
Dart Impact (total energy)	100 (el)	200 (el)	100 (el)	100 (el)
Dart Impact (peak energy)	100 (dl)	100 (dl)	100 (dl)	100 (dl)
Notched Izod Impact	100 (ge)	131.3 (ge)	100 (ge)	100 (ge)
Heat Deflection Temperature	97.6 (m)	100 (m)	94 (m)	100 (m)
Vicat Softening Point	100 (ku)	99.1 (ku)	100 (ku)	99.1 (ku)
SURFACE and APPEARANCE				
Δ Yellowness Index	15.31 (kw)	-0.54 (kw)	42.84 (kw)	5.21 (kw)
Haze (%)	3.47	1.42	1.61	-0.01
Transperency Retained (%)	92.1	109	78.7	107.9

TABLE 143: Effect of Gamma Radiation Sterilization on Styrene Acrylonitrile Copolymer

Material Family	STYRENE ACRYLONITRILE COPOLYMER			
Material Supplier/Name	MONSANTO LUSTRAN SAN 31			
Material Note	transparent, general purpose grade	transparent, general purpose grade	transparent, general purpose grade	transparent, general purpose grade
Reference No.	88	88	88	88
EXPOSURE CONDITIONS				
Type	Gamma Radiation	Gamma Radiation	Gamma Radiation	Gamma Radiation
Radiation Dose (Mrads)	1.5	2.5	3.5	5
PROPERTIES RETAINED (%)				
Flexural Strength	99 (cl)	99 (cl)	92 (cl)	93 (cl)

TABLE 144: Effect of Electron Beam Radiation Sterilization on Styrene Acrylonitrile Copolymer

Material Family	STYRENE ACRYLONITRILE COPOLYMER			
Material Supplier/Name	DOW TYRIL 1000B			
Material Note	transparent	transparent	transparent	transparent
Reference No.	1	1	1	1

EXPOSURE CONDITIONS

Type	Electron Beam Radiation	Electron Beam Radiation	Electron Beam Radiation	Electron Beam Radiation
Radiation Dose (Mrads)	2.5	2.5	10	10
Note	test lab: E-Beam Services, Inc.	test lab: E-Beam Services, Inc.	test lab: E-Beam Services, Inc.	test lab: E-Beam Services, Inc.

POST EXPOSURE CONDITIONING

Note	type: aging	type: aging	type: aging	type: aging
Time (hours)	168	1344	168	1344

PROPERTIES RETAINED (%)

Tensile Strength @ Yield	101.4 (ir)	100 (ir)	102.9 (ir)	102.9 (ir)
Elongation	100 (bq)	66.7 (bq)	100 (bq)	100 (bq)
Flexural Strength	100 (cn)	98.4 (cn)	98.4 (cn)	96.9 (cn)
Modulus	117.3 (hc)	93.5 (hc)	111.6 (hc)	100.6 (hc)
Flexural Modulus	103.8 (ch)	103 (ch)	101.9 (ch)	101.1 (ch)
Dart Impact (total energy)	100 (el)	100 (el)	100 (el)	100 (el)
Dart Impact (peak energy)	100 (dl)	100 (dl)	100 (dl)	100 (dl)
Notched Izod Impact	131.3 (ge)	131.3 (ge)	131.3 (ge)	100 (ge)
Heat Deflection Temperature	100 (m)	95.2 (m)	102.4 (m)	103.6 (m)
Vicat Softening Point	100 (ku)	99.1 (ku)	100 (ku)	99.1 (ku)

SURFACE and APPEARANCE

Δ Yellowness Index	16.78 (kw)	-1.32 (kw)	39.69 (kw)	4.18 (kw)
Haze Retained (%)	4.16	-0.49	0.87	0.13
Haze (z-direction) Retained (%)	91	110.1	80.9	109

TABLE 145: Effect of Electron Beam Radiation Sterilization on Styrene Acrylonitrile Copolymer

Material Family	STYRENE ACRYLONITRILE COPOLYMER			
Material Supplier/Name	MONSANTO LUSTRAN SAN 31			
Material Note	transparent, general purpose grade	transparent, general purpose grade	transparent, general purpose grade	transparent, general purpose grade
Reference No.	88	88	88	88
EXPOSURE CONDITIONS				
Type	Electron Beam Radiation	Electron Beam Radiation	Electron Beam Radiation	Electron Beam Radiation
Radiation Dose (Mrads)	1.5	2.5	3.5	5
PROPERTIES RETAINED (%)				
Flexural Strength	102 (cl)	102 (cl)	99 (cl)	93 (cl)

TABLE 146: Effect of Ethylene Oxide Sterilization on Styrene Acrylonitrile Copolymer

Material Family	STYRENE ACRYLONITRILE COPOLYMER					
Material Supplier/Name	DOW					
Material Note	transparent, natural resin	transparent, natural resin	transparent, natural resin	transparent, natural resin	transparent, natural resin	transparent, natural resin
Reference No.	5	5	5	5	5	5

EXPOSURE CONDITIONS

Type	Ethylene Oxide	Ethylene Oxide	Ethylene Oxide	Ethylene Oxide	Ethylene Oxide	Ethylene Oxide
Details	12% EtO and 88% Freon	12% EtO and 88% Freon	12% EtO and 88% Freon	12% EtO and 88% Freon	12% EtO and 88% Freon	12% EtO and 88% Freon
Concentration	660 mg/l	660 mg/l	660 mg/l	660 mg/l	660 mg/l	660 mg/l
Number of Cycles	1	1	1	5	5	5
Note	RH: 60%; test lab: Ethox Corp.	RH: 60%; test lab: Ethox Corp.	RH: 60%; test lab: Ethox Corp.	RH: 60%; test lab: Ethox Corp.	RH: 60%; test lab: Ethox Corp.	RH: 60%; test lab: Ethox Corp.
Temperature (°C)	49	49	49	49	49	49
Time (hours)	≥ 6	≥ 6	≥ 6	≥ 6	≥ 6	≥ 6

PRE EXPOSURE CONDITIONING

Preconditioning Note	time: 8 hours; temperature: 37.8°C; RH: 60%	time: 8 hours; temperature: 37.8°C; RH: 60%	time: 8 hours; temperature: 37.8°C; RH: 60%	time: 8 hours; temperature: 37.8°C; RH: 60%	time: 8 hours; temperature: 37.8°C; RH: 60%	time: 8 hours; temperature: 37.8°C; RH: 60%

POST EXPOSURE CONDITIONING

Note	type: evacuation; pressure: 127 mm Hg	type: evacuation; pressure: 127 mm Hg	type: evacuation; pressure: 127 mm Hg	type: evacuation; pressure: 127 mm Hg	type: evacuation; pressure: 127 mm Hg	type: evacuation; pressure: 127 mm Hg

POST EXPOSURE CONDITIONING II

Note	type: aeration	type: aeration	type: aeration	type: aeration	type: aeration	type: aeration
Temperature (°C)	32.2	32.2	32.2	32.2	32.2	32.2
Time (hours)	≥ 16	≥ 16	≥ 16	≥ 16	≥ 16	≥ 16

POST EXPOSURE CONDITIONING III

Note	type: storage in dark	type: storage in dark	type: storage in dark	type: storage in dark	type: storage in dark	type: storage in dark
Temperature (°C)	21	21	21	21	21	21
Time (hours)	336	4368	8760	336	4368	8760

PROPERTIES RETAINED (%)

Tensile Strength	95 (hy)	78.8 (hy)	81.1 (hy)	77.7 (hy)	64.5 (hy)	79.1 (hy)
Tensile Strength @ Yield	95 (ip)	78.8 (ip)	81.1 (ip)	77.7 (ip)	64.5 (ip)	79.1 (ip)
Elongation	66.7 (av)	66.7 (av)	66.7 (av)	66.7 (av)	33.3 (av)	66.7 (av)
Dart Impact (total energy)	158.8 (fb)	158.8 (fb)	76.5 (fb)	94.1 (fb)	41.2 (fb)	52.9 (fb)
Notched Izod Impact	133.3 (fs)	100 (fs)	133.3 (fs)	66.7 (fs)	133.3 (fs)	100 (fs)

TABLE 147: Effect of Ethylene Oxide Sterilization on Styrene Acrylonitrile Copolymer

Material Family	STYRENE ACRYLONITRILE COPOLYMER							
Material Supplier/Name	DOW TYRIL 1000B							
Material Note	transparent	transparent	transparent	transparent	transparent	transparent	transparent	transparent
Reference No.	6	6	6	6	6	6	6	6

EXPOSURE CONDITIONS

Type	Ethylene Oxide	Ethylene Oxide	Ethylene Oxide	Ethylene Oxide	Ethylene Oxide	Ethylene Oxide	Ethylene Oxide	Ethylene Oxide
Details	12% EtO and 88% Freon	12% EtO and 88% Freon	12% EtO and 88% Freon	12% EtO and 88% Freon	8.6% EtO and 91.4% HCFC-124	8.6% EtO and 91.4% HCFC-124	8.6% EtO and 91.4% HCFC-124	8.6% EtO and 91.4% HCFC-124
Concentration	600 mg/l	600 mg/l	600 mg/l	600 mg/l				
Number of Cycles	1	1	2	2	1	1	2	2
Note	RH: 60%; test lab: Ethox Corp.	RH: 60%; test lab: Ethox Corp.	RH: 60%; test lab: Ethox Corp.	RH: 60%; test lab: Ethox Corp.	RH: 60%; test lab: Ethox Corp.	RH: 60%; test lab: Ethox Corp.	RH: 60%; test lab: Ethox Corp.	RH: 60%; test lab: Ethox Corp.
Temperature (°C)	48.9	48.9	48.9	48.9	48.9	48.9	48.9	48.9
Time (hours)	6	6	6	6	6	6	6	6

PRE EXPOSURE CONDITIONING

Preconditioning Note	time: 18 hours; temperature: 37.8°C; RH: 60%	time: 18 hours; temperature: 37.8°C; RH: 60%	time: 18 hours; temperature: 37.8°C; RH: 60%	time: 18 hours; temperature: 37.8°C; RH: 60%	time: 18 hours; temperature: 37.8°C; RH: 60%	time: 18 hours; temperature: 37.8°C; RH: 60%	time: 18 hours; temperature: 37.8°C; RH: 60%	time: 18 hours; temperature: 37.8°C; RH: 60%

POST EXPOSURE CONDITIONING

Note	type: aeration; pressure: 127 mm Hg	type: aeration; pressure: 127 mm Hg	type: aeration; pressure: 127 mm Hg	type: aeration; pressure: 127 mm Hg	type: aeration; pressure: 127 mm Hg	type: aeration; pressure: 127 mm Hg	type: aeration; pressure: 127 mm Hg	type: aeration; pressure: 127 mm Hg
Temperature (°C)	32.2	32.2	32.2	32.2	32.2	32.2	32.2	32.2

POST EXPOSURE CONDITIONING II

Note	type: ambient conditions	type: ambient conditions	type: ambient conditions	type: ambient conditions	type: ambient conditions	type: ambient conditions	type: ambient conditions	type: ambient conditions
Time (hours)	168	1344	168	1344	168	1344	168	1344

PROPERTIES RETAINED (%)

Tensile Strength @ Yield	98 (im)	93.5 (im)	84.2 (im)	85.7 (im)	103.1 (im)	100.5 (im)	79.1 (im)	78.9 (im)
Elongation	50 (bk)	100 (bk)	100 (bk)	100 (bk)	100 (bk)	100 (bk)	100 (bk)	100 (bk)
Modulus	94.3 (gw)	96.1 (gw)	95.3 (gw)	87.2 (gw)	91.1 (gw)	85.8 (gw)	92.1 (gw)	91.5 (gw)
Dart Impact (total energy)	80 (dt)	93.3 (dt)	66.7 (dt)	93.3 (dt)	80 (dt)	80 (dt)	66.7 (dt)	100 (dt)
Dart Impact (peak energy)	83.3 (es)	66.7 (ex)	58.3 (es)	75 (es)	75 (es)	75 (eu)	66.7 (es)	100 (eu)

SURFACE and APPEARANCE

Yellowness Index	2.9	2.61	2.8	2.71	3.12	2.89	2.83	2.7
Haze (%)	3.25	2.44	3.29	3.28	1.83	2.14	5.01	5.32
Transmittance (%)	91	91	90	91	91	91	91	91

TABLE 148: Effect of Ethylene Oxide Sterilization on Styrene Acrylonitrile Copolymer

Material Family	STYRENE ACRYLONITRILE COPOLYMER			
Material Supplier/Name	DOW TYRIL 1000B			
Material Note	transparent	transparent	transparent	transparent
Reference No.	6	6	6	6

EXPOSURE CONDITIONS

Type	Ethylene Oxide	Ethylene Oxide	Ethylene Oxide	Ethylene Oxide
Details	12% EtO and 88% Freon	12% EtO and 88% Freon	8.6% EtO and 91.4% HCFC-124	8.6% EtO and 91.4% HCFC-124
Concentration	600 mg/l	600 mg/l		
Number of Cycles	1	1	1	1
Note	RH: 60%; test lab: Ethox Corp.	RH: 60%; test lab: Ethox Corp.	RH: 60%; test lab: Ethox Corp.	RH: 60%; test lab: Ethox Corp.
Temperature (°C)	48.9	48.9	48.9	48.9
Time (hours)	6	6	6	6

PRE EXPOSURE CONDITIONING

Preconditioning Note	time: 18 hours; temperature: 37.8°C; RH: 60%	time: 18 hours; temperature: 37.8°C; RH: 60%	time: 18 hours; temperature: 37.8°C; RH: 60%	time: 18 hours; temperature: 37.8°C; RH: 60%

POST EXPOSURE CONDITIONING

Note	type: aeration; note: 10 air changes per hour	type: aeration; note: 30 air changes per hour	type: aeration; note: 10 air changes per hour	type: aeration; note: 30 air changes per hour
Temperature (°C)	32.2	54.4	32.2	54.4

RESIDUALS (ppm)

Residuals Determined	ethylene oxide	ethylene oxide	ethylene oxide	ethylene oxide
Little or No Aeration	586	586	573	573
24 hour Aeration	292	133	298	262
48 hour Aeration		109		132
72 hour Aeration	117	103	160	107
168 hour Aeration	68		100	

TABLE 149: Effect of Ethylene Oxide Sterilization on Styrene Acrylonitrile Copolymer

Material Family	STYRENE ACRYLONITRILE COPOLYMER		
Material Supplier/Name	DOW TYRIL 1000B		
Material Note	transparent	transparent	transparent
Reference No.	3	3	3

EXPOSURE CONDITIONS

Type	Ethylene Oxide	Ethylene Oxide	Ethylene Oxide
Details	12% EtO and 88% Freon	12% EtO and 88% Freon	12% EtO and 88% Freon
Concentration	660 mg/l	660 mg/l	660 mg/l
Number of Cycles	1	5	1
Note	RH: 60%	RH: 60%	RH: 60%
Temperature (°C)	49	49	49
Time (hours)	≥ 6	≥ 6	≥ 6

PRE EXPOSURE CONDITIONING

Preconditioning Note	time: 8 hours; temperature: 37.8°C; RH: 60%	time: 8 hours; temperature: 37.8°C; RH: 60%	time: 8 hours; temperature: 37.8°C; RH: 60%

POST EXPOSURE CONDITIONING

Note	type: evacuation; pressure: 127 mm Hg	type: evacuation; pressure: 127 mm Hg	type: evacuation; pressure: 127 mm Hg

POST EXPOSURE CONDITIONING II

Note	type: aeration	type: aeration	type: aeration
Temperature (°C)	32	32	32
Time (hours)	≥ 16	≥ 16	

POST EXPOSURE CONDITIONING III

Note	type: storage in dark; RH: 50%	type: storage in dark; RH: 50%	
Temperature (°C)	21	21	
Time (hours)	336	336	

RESIDUALS (ppm)

Residuals Determined			ethylene oxide
24 hour Aeration			373
72 hour Aeration			211
168 hour Aeration			119
720 hour Aeration			109
768 hour Aeration			96
840 hour Aeration			53
888 hour Aeration			39

PROPERTIES RETAINED (%)

Tensile Strength @ Yield	99 (ii)	76.7 (ii)	
Notched Izod Impact	100 (fp)	100 (fp)	

GRAPH 136: **Gamma Radiation Dose vs. Tensile Modulus Retained of Styrene Acrylonitrile Copolymer**

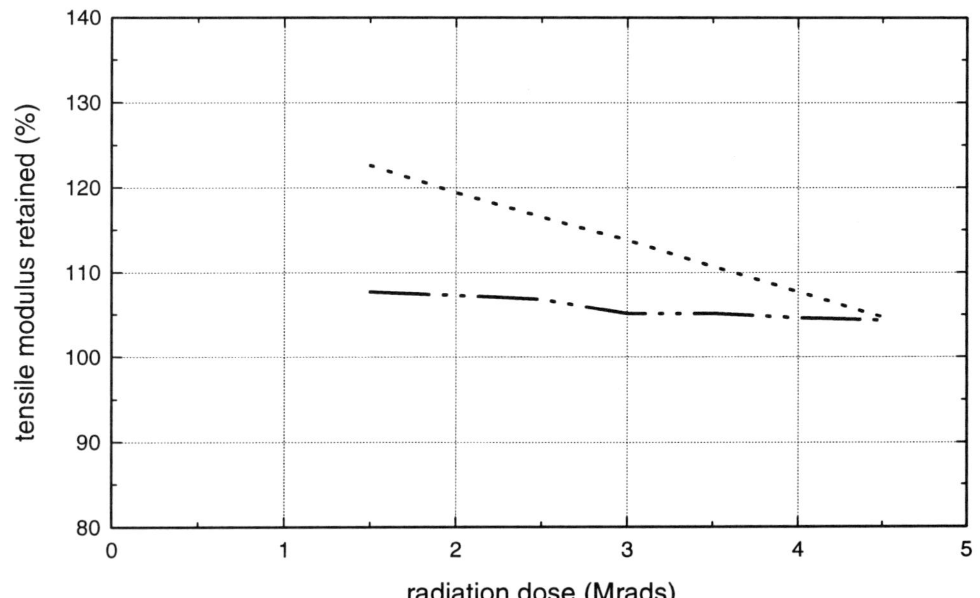

···············	Monsanto Lustran Sparkle 2090 SAN (transpar., water clear)
—··—··	Monsanto Lustran SAN 31 SAN
Reference No.	88

GRAPH 137: **Post Gamma Radiation Exposure Time vs. Yellowness Index of Styrene Acrylonitrile Copolymer**

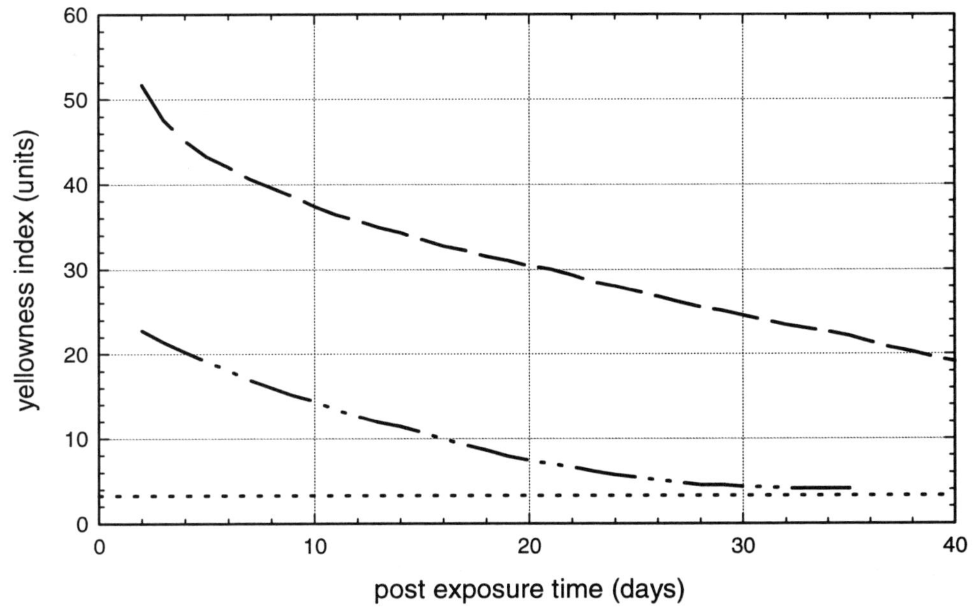

···············	Dow Tyril 1000B SAN (transpar.); before exposure
—··—··	Dow Tyril 1000B SAN (transpar.); 2.5 Mrads
— — — —	Dow Tyril 1000B SAN (transpar.); 10 Mrads
Reference No.	3

GRAPH 138: **Post Gamma Radiation Exposure Time vs. Yellowness Index of Styrene Acrylonitrile Copolymer**

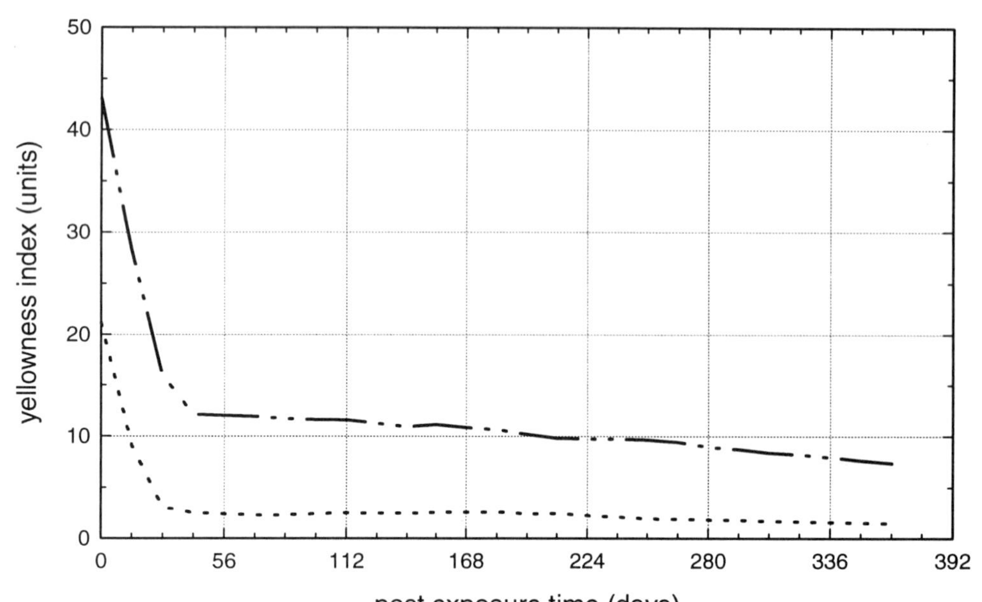

GRAPH 139: **Post Gamma Radiation Exposure Time vs. Percent Light Transmission of Styrene Acrylonitrile Copolymer**

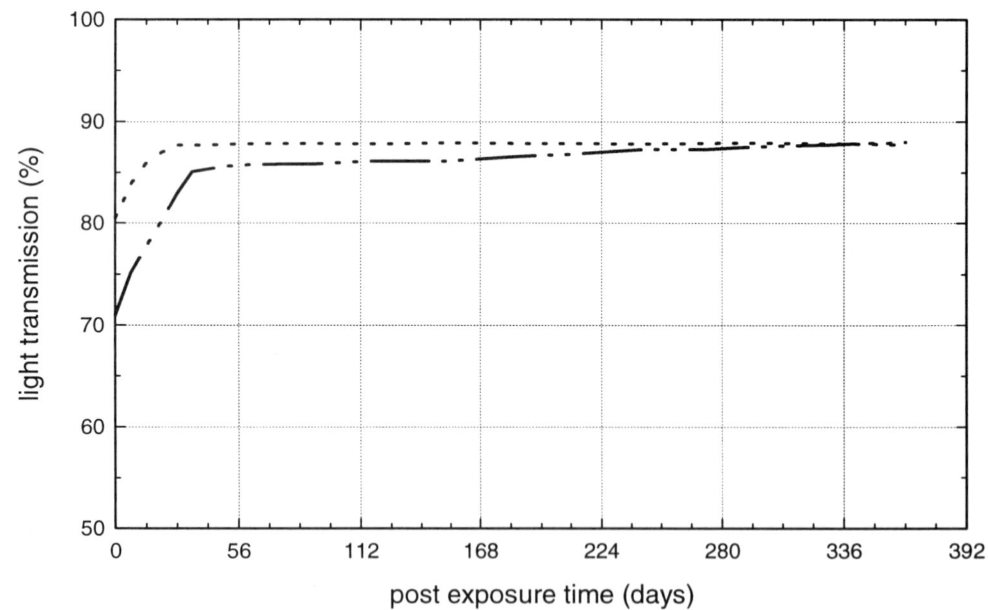

Gamma Radiation Resistance

Phillips: K Resin (features: transparent)

K-Resin exhibits good resistance to gamma radiation. Slight yellowing which tended to clear up in three weeks was observed. The benzene ring in these materials protect the structure from a breakdown. K-Resin, which also contains polybutadiene, is protected by the aid of the aromatic group, which absorbs the high energy radiation and dissipates them without destruction.

Reference: Atakent, A. I., *Gamma Radiation Study,* supplier technical report - Cyro Industries, 1990.

Phillips: K-Resin KR01 (features: transparent); **K-Resin KR03** (features: transparent, impact modified)

Although the level of gamma irradiation required for guaranteed sterilization is relatively low at 2.5 Mrads (recommended by USP), the influence of higher dosages as might occur on repeated exposure or a single higher exposure is important. Injection molded test specimens having thicknesses of 1/8 and 1/4 inch were produced from KR01 and KR03 polymers and subjected to gamma irradiation at dosage levels of 2.6, 5.1, 10.0, 20 and 30 Mrads. Physical properties were determined for exposed and non exposed specimens using standard ASTM methods. At an exposure of 2.6 Mrads, with the exception of a small decrease in flow rate, there was no loss in physical properties nor yellowing. With increasing dosage levels up through 30 Mrads, there was essentially no change in physical properties other than a continuing decrease in flow rate and a modest reduction of elongation of the KR03 sample. Although yellowing was only slight after 5.1 Mrads, yellow color development can occur at higher dosage levels.

Reference: *K-Resin A Clear Choice For Medical Packaging Devices,* supplier technical report (55-8801) - Phillips 66 Company, 1988.

Ethylene Oxide (EtO) Resistance

Phillips: K-Resin KR01 (features: transparent); **K-Resin KR03** (features: transparent, impact modified)

Injection molded K-Resin parts were ethylene oxide sterilized and the physical properties of both the sterilized and non sterilized parts were determined using ASTM methods. There was no property deterioration as a result of EtO sterilization.

Reference: *K-Resin A Clear Choice For Medical Packaging Devices,* supplier technical report (55-8801) - Phillips 66 Company, 1988.

TABLE 150: Effect of Gamma Radiation Sterilization on Styrene Butadiene Copolymer

Material Family	STYRENE BUTADIENE COPOLYMER									
Material Supplier/Name	PHILLIPS K-RESIN KR01		PHILLIPS K-RESIN KR03		PHILLIPS K-RESIN KR01			PHILLIPS K-RESIN KR03		
Material Note	transparent	transparent	transparent, impact modified	transparent, impact modified	transparent	transparent	transparent	transparent, impact modified	transparent, impact modified	transparent, impact modified
Reference No.	100	100	100	100	100	100	100	100	100	100
EXPOSURE CONDITIONS										
Type	Gamma Radiation	Gamma Radiation	Gamma Radiation	Gamma Radiation	Gamma Radiation	Gamma Radiation	Gamma Radiation	Gamma Radiation	Gamma Radiation	Gamma Radiation
Details	source: Cobalt 60	source: Cobalt 60	source: Cobalt 60	source: Cobalt 60	source: Cobalt 60	source: Cobalt 60	source: Cobalt 60	source: Cobalt 60	source: Cobalt 60	source: Cobalt 60
Radiation Dose (Mrads)	2.6	5.1	2.6	5.1	10	20	30	10	20	30
PROPERTIES RETAINED (%)										
Tensile Strength	102.5 (he)	102.5 (he)	97 (he)	97 (he)	103.7 (he)	104.9 (he)	109.8 (he)	100 (he)	101.4 (he)	100 (he)
Elongation	107.7 (ai)	107.7 (ai)	100 (ai)	91.7 (ai)	116.7 (ai)	133.3 (ai)	116.7 (ai)	86.6 (ai)	85.8 (ai)	66.1 (ai)
Modulus	97.2 (bz)	96.3 (bz)	100.4 (bz)	100 (bz)	100.4 (bz)	98.7 (bz)	99.6 (bz)	100 (bz)	101.2 (bz)	102 (bz)
Notched Izod Impact	100 (fm)	100 (fm)	120 (fm)	120 (fm)	100 (fm)	100 (fm)	100 (fm)	100 (fm)	100 (fm)	100 (fm)
Unnotched Izod Impact	93.1 (kq)	100 (kq)	100 (kq)	100 (kq)	100 (kq)	105.9 (kq)	123.5 (kq)	100 (kq)	96.9 (kq)	93.8 (kq)
Heat Deflection Temperature	101.8 (k)	103 (k)	100.6 (k)	100.6 (k)	106.2 (k)	104.9 (k)	108 (k)	106.3 (k)	102.5 (k)	107 (k)
Vicat Softening Point	101.5 (kt)	101.5 (kt)	99 (kt)	99.5 (kt)	104 (kt)	102.5 (kt)	101 (kt)	103 (kt)	101.5 (kt)	102 (kt)
Melt Flow Rate	89.4 (a)	71.2 (a)	85.7 (a)	74.6 (a)	25.3 (a)	2.5 (a)	0.5 (a)	31.9 (a)	8.3 (a)	0 (a)
SURFACE and APPEARANCE										
Hardness Units Change	D1 (gn)	D1 (gn)	D2 (gn)	D2 (gn)	D1 (gn)	D1 (gn)	D2 (gn)	D3 (gn)	D2 (gn)	D1 (gn)
Yellowness Index note	no yellowing	slight yellowing	no yellowing	slight yellowing	moderate to heavy yellowing	moderate to heavy yellowing	moderate to heavy yellowing	moderate to heavy yellowing	moderate to heavy yellowing	moderate to heavy yellowing

Gamma Radiation Resistance

BASF AG: Styrolux (features: transparent)

Samples were exposed to cobalt-60 at room temperature and demonstrated resistance to high energy radiation. A slight effect in the form of very faint yellowish tinge was observed, but this can disappear again.

Reference: *Styrolux Product Line, Properties, Processing,* supplier design guide (B 583 e/(950) 12.91) - BASF Aktiengesellschaft, 1992.

Gamma Radiation Resistance

Monsanto: Cadon 160 (features: general purpose grade)

Injection molded test specimens were irradiated at doses of 1.5, 2.5, 3.5 and 5.0 Mrads of gamma radiation. The Izod impact (ASTM D-256) value of general purpose ABS/SMA (Cadon 160) was statistically unaffected by radiation. There was no measurable change in tensile modulus (ASTM D-638). An increase in tensile stress (ASTM D-638) at yield of 5.0% was noted. There was no difference in tensile stress (ASTM D-638) at fail between the control and irradiated samples. Tensile elongation at yield (ASTM D-638) was unchanged. The results of tensile elongation at fail (ASTM D-638) were extremely variable. Virtually no change in flexural modulus (ASTM D-790) was noted. No obvious trends in flexural modulus (ASTM D-790) were observed, with values fluctuating plus or minus 5.0%. At 2.5 Mrads discoloration was minimal.

Reference: *Effects Of Electron Beam And Gamma Radiation Sterilization Of Thermoplastics*, supplier technical report (7126A) - Monsanto Company, 1990.

Monsanto: Cadon 112 (features: high impact)

Injection molded test specimens were irradiated at doses of 1.5, 2.5, 3.5 and 5.0 Mrads of gamma radiation. The Izod impact (ASTM D-256) value of high impact ABS/SMA (Cadon 112) fell by 10% to 20%. There was no measurable change in tensile modulus (ASTM D-638). An increase in tensile stress (ASTM D-638) at yield of 5.0% was noted. There was no difference in tensile stress (ASTM D-638) at fail between the control and irradiated samples. Tensile elongation at yield (ASTM D-638) was unchanged. The results of tensile elongation at fail (ASTM D-638) were extremely variable. Virtually no change in flexural modulus (ASTM D-790) was noted. No obvious trends in flexural modulus (ASTM D-790) were observed, with values fluctuating plus or minus 5.0%. At 2.5 Mrads discoloration was minimal.

Reference: *Effects Of Electron Beam And Gamma Radiation Sterilization Of Thermoplastics*, supplier technical report (7126A) - Monsanto Company, 1990.

Electron Beam Radiation

Monsanto: Cadon 160 (features: general purpose grade)

Injection molded test specimens were irradiated at doses of 1.5, 2.5, 3.5 and 5.0 Mrads of electron beam radiation. The Izod impact (ASTM D-256) value of general purpose ABS/SMA (Cadon 160) was statistically unaffected by radiation. There was no measurable change in tensile modulus (ASTM D-638). An increase in tensile stress (ASTM D-638) at yield of 10% to 18% was noted. There was a 5% to 15% difference in tensile stress (ASTM D-638) at fail between the control and irradiated samples. Tensile elongation at yield (ASTM D-638) was unchanged. The results of tensile elongation at fail (ASTM D-638) were extremely variable. Virtually no change in flexural modulus (ASTM D-790) was noted. No obvious trends in flexural modulus (ASTM D-790) were observed, with values fluctuating plus or minus 5.0%. At 2.5 Mrads discoloration was minimal.

Reference: *Effects Of Electron Beam And Gamma Radiation Sterilization Of Thermoplastics*, supplier technical report (7126A) - Monsanto Company, 1990.

Monsanto: Cadon 112 (features: high impact)

Injection molded test specimens were irradiated at doses of 1.5, 2.5, 3.5 and 5.0 Mrads of electron beam radiation. The Izod impact (ASTM D-256) value of high impact ABS/SMA (Cadon 112) fell be 10% to 20%. There was no measurable change in tensile modulus (ASTM D-638). An increase in tensile stress (ASTM D-638) at yield of 10% to 18% was noted. There was a 5% to 15% difference in tensile stress (ASTM D-638) at fail between the control and irradiated samples. Tensile elongation at yield (ASTM D-638) was unchanged. The results of tensile elongation at fail (ASTM D-638) were extremely variable. Virtually no change in flexural modulus (ASTM D-790) was noted. No obvious trends in flexural modulus (ASTM D-790) were observed, with values fluctuating plus or minus 5.0%. At 2.5 Mrads discoloration was minimal.

Reference: *Effects Of Electron Beam And Gamma Radiation Sterilization Of Thermoplastics*, supplier technical report (7126A) - Monsanto Company, 1990.

TABLE 151: Effect of Gamma Radiation Sterilization on Styrene Maleic Anhydride Copolymer

Material Family	STYRENE MALEIC ANHYDRIDE COPOLYMER			
Material Supplier/Name	MONSANTO CADON 112			
Material Note	high impact	high impact	high impact	high impact
Reference No.	88	88	88	88

EXPOSURE CONDITIONS

Type	Gamma Radiation	Gamma Radiation	Gamma Radiation	Gamma Radiation
Radiation Dose (Mrads)	1.5	2.5	3.5	5

PROPERTIES RETAINED (%)

Tensile Strength @ Yield	102 (ih)	103 (ih)	101 (ih)	102 (ih)

TABLE 152: Effect of Electron Beam Radiation Sterilization on Styrene Maleic Anhydride Copolymer

Material Family	STYRENE MALEIC ANHYDRIDE COPOLYMER			
Material Supplier/Name	MONSANTO CADON 112			
Material Note	high impact	high impact	high impact	high impact
Reference No.	88	88	88	88

EXPOSURE CONDITIONS

Type	Electron Beam Radiation	Electron Beam Radiation	Electron Beam Radiation	Electron Beam Radiation
Radiation Dose (Mrads)	1.5	2.5	3.5	5

PROPERTIES RETAINED (%)

Tensile Strength @ Yield	105 (ih)	106 (ih)	105 (ih)	108 (ih)

Gamma Radiation Resistance

Polyvinyl Chloride (features: transparent)

PVC turned amber most likely by crosslinking and formation of double bonds. The PVC demonstrated low transmission and high yellowness index values. Aromatic plasticizers used in PVC do not extend a good resistance to the system since they are blended in and not copolymerized.

Reference: Atakent, A. I., *Gamma Radiation Study,* supplier technical report - Cyro Industries, 1990.

Geon Company: Geon RX (features: transparent, medical grade)

Grades of rigid Geon Rx compounds are resistant to gamma radiation sterilization.

Reference: *BF Goodrich Geon RX Medical Grade Molding Compounds,* supplier marketing literature (CIM-014F) - BFGoodrich Company, 1988.

Geon Company: Geon 87614 (features: gamma radiation stabilized, transparent, medical grade); **Geon 87616** (features: gamma radiation stabilized, transparent, medical grade); **Geon 87619** (features: gamma radiation stabilized, transparent, medical grade)

These medical grade rigid vinyl compounds were developed to withstand gamma sterilization. The compounds meet United States Pharmacopia (USP) XXII Class VI biological certification before and after 5.0 Mrads of gamma irradiation. They also maintained good transparency with low discoloration after 5.0 Mrads of gamma irradiation.

Reference: *BF Goodrich Geon Rigid Vinyl Gamma Sterilizable Compounds,* supplier marketing literature (RX-020) - BFGoodrich Company, 1990.

Ethylene Oxide (EtO) Resistance

Dexter: Dural 752-C2H

Eight weeks after EtO sterilization a very slight decrease is seen in tensile yield. The instrumented impact properties decreased 25% after exposure to EtO. The decrease in impact is seen within one week of exposure.

Reference: Hermanson, Nancy J., *Effects Of Alternate Carriers Of Ethylene Oxide Sterilant On Thermoplastics,* supplier technical report (301-02018) - Dow Chemical Company.

Geon Company: Geon RX (features: transparent, medical grade)

Rigid Geon Rx compounds are ethylene oxide sterilizable while retaining undiminished properties and clarity.

Reference: *BF Goodrich Geon RX Medical Grade Molding Compounds,* supplier marketing literature (CIM-014F) - BFGoodrich Company, 1988.

TABLE 152a: Effect of Ethylene Oxide Sterilization on Polyvinyl Chloride

Material Family	POLYVINYL CHLORIDE							
Material Supplier/Name	DEXTER DURAL 752-C2H							
Reference No.	6	6	6	6	6	6	6	6

EXPOSURE CONDITIONS

Type	Ethylene Oxide	Ethylene Oxide	Ethylene Oxide	Ethylene Oxide	Ethylene Oxide	Ethylene Oxide	Ethylene Oxide	Ethylene Oxide
Details	12% EtO and 88% Freon	12% EtO and 88% Freon	12% EtO and 88% Freon	12% EtO and 88% Freon	8.6% EtO and 91.4% HCFC-124	8.6% EtO and 91.4% HCFC-124	8.6% EtO and 91.4% HCFC-124	8.6% EtO and 91.4% HCFC-124
Concentration	600 mg/l	600 mg/l	600 mg/l	600 mg/l				
Number of Cycles	1	1	2	2	1	1	2	2
Note	RH: 60%; test lab: Ethox Corp.	RH: 60%; test lab: Ethox Corp.	RH: 60%; test lab: Ethox Corp.	RH: 60%; test lab: Ethox Corp.	RH: 60%; test lab: Ethox Corp.	RH: 60%; test lab: Ethox Corp.	RH: 60%; test lab: Ethox Corp.	RH: 60%; test lab: Ethox Corp.
Temperature (°C)	48.9	48.9	48.9	48.9	48.9	48.9	48.9	48.9
Time (hours)	6	6	6	6	6	6	6	6

PRE EXPOSURE CONDITIONING

Preconditioning Note	time: 18 hours; temperature: 37.8 °C; RH: 60%	time: 18 hours; temperature: 37.8 °C; RH: 60%	time: 18 hours; temperature: 37.8 °C; RH: 60%	time: 18 hours; temperature: 37.8 °C; RH: 60%	time: 18 hours; temperature: 37.8 °C; RH: 60%	time: 18 hours; temperature: 37.8 °C; RH: 60%	time: 18 hours; temperature: 37.8 °C; RH: 60%	time: 18 hours; temperature: 37.8 °C; RH: 60%

POST EXPOSURE CONDITIONING

Note	type: aeration; pressure: 127 mm Hg	type: aeration; pressure: 127 mm Hg	type: aeration; pressure: 127 mm Hg	type: aeration; pressure: 127 mm Hg	type: aeration; pressure: 127 mm Hg	type: aeration; pressure: 127 mm Hg	type: aeration; pressure: 127 mm Hg	type: aeration; pressure: 127 mm Hg
Temperature (°C)	32.2	32.2	32.2	32.2	32.2	32.2	32.2	32.2

POST EXPOSURE CONDITIONING II

Note	type: ambient conditions	type: ambient conditions	type: ambient conditions	type: ambient conditions	type: ambient conditions	type: ambient conditions	type: ambient conditions	type: ambient conditions
Time (hours)	168	1344	168	1344	168	1344	168	1344

PROPERTIES RETAINED (%)

Tensile Strength @ Yield	100.1 (il)	96.7 (il)	99.6 (il)	96.4 (il)	95.6 (il)	94.5 (il)	99.6 (il)	97.3 (il)
Elongation	100 (bm)	92.3 (bm)	92.3 (bm)	84.6 (bm)	96.2 (bm)	88.5 (bm)	96.2 (bm)	80.8 (bm)
Modulus	114.4 (gu)	82.9 (gu)	100.7 (gu)	97.7 (gu)	87.2 (gu)	80.5 (gu)	110.1 (gu)	100 (gu)
Dart Impact (total energy)	95.8 (ea)	95.8 (ea)	97.9 (ea)	102.1 (ea)	97.9 (ea)	102.1 (ea)	100 (ea)	104.2 (ea)
Dart Impact (peak energy)	75 (dd)	72.5 (dd)	72.5 (dd)	80 (dd)	72.5 (dd)	75 (dd)	75 (dd)	80 (dd)

SURFACE and APPEARANCE

ΔE Color	0.1	0.13	0.15	0.15	0.11	0.17	0.18	0.14

TABLE 153: Effect of Ethylene Oxide Sterilization on Polyvinyl Chloride

Material Family	POLYVINYL CHLORIDE			
Material Supplier/Name	DEXTER DURAL 752-C2H			
Reference No.	6	6	6	6

EXPOSURE CONDITIONS

Type	Ethylene Oxide	Ethylene Oxide	Ethylene Oxide	Ethylene Oxide
Details	12% EtO and 88% Freon	12% EtO and 88% Freon	8.6% EtO and 91.4% HCFC-124	8.6% EtO and 91.4% HCFC-124
Concentration	600 mg/l	600 mg/l		
Number of Cycles	1	1	1	1
Note	RH: 60%; test lab: Ethox Corp.	RH: 60%; test lab: Ethox Corp.	RH: 60%; test lab: Ethox Corp.	RH: 60%; test lab: Ethox Corp.
Temperature (°C)	48.9	48.9	48.9	48.9
Time (hours)	6	6	6	6

PRE EXPOSURE CONDITIONING

Preconditioning Note	time: 18 hours; temperature: 37.8 °C; RH: 60%	time: 18 hours; temperature: 37.8 °C; RH: 60%	time: 18 hours; temperature: 37.8 °C; RH: 60%	time: 18 hours; temperature: 37.8 °C; RH: 60%

POST EXPOSURE CONDITIONING

Note	type: aeration; note: 10 air changes per hour	type: aeration; note: 30 air changes per hour	type: aeration; note: 10 air changes per hour	type: aeration; note: 10 air changes per hour
Temperature (°C)	32.2	54.4	32.2	54.4

RESIDUALS (ppm)

Residuals Determined	ethylene oxide	ethylene oxide	ethylene oxide	ethylene oxide
Little or No Aeration	399	399	307	307
17 hour Aeration		93		123
24 hour Aeration	203	78	178	119
48 hour Aeration	92	61	133	71
72 hour Aeration		36	118	

GRAPH 140: **Post Gamma Radiation Exposure Time vs. Delta Yellowness Index of Polyvinyl Chloride**

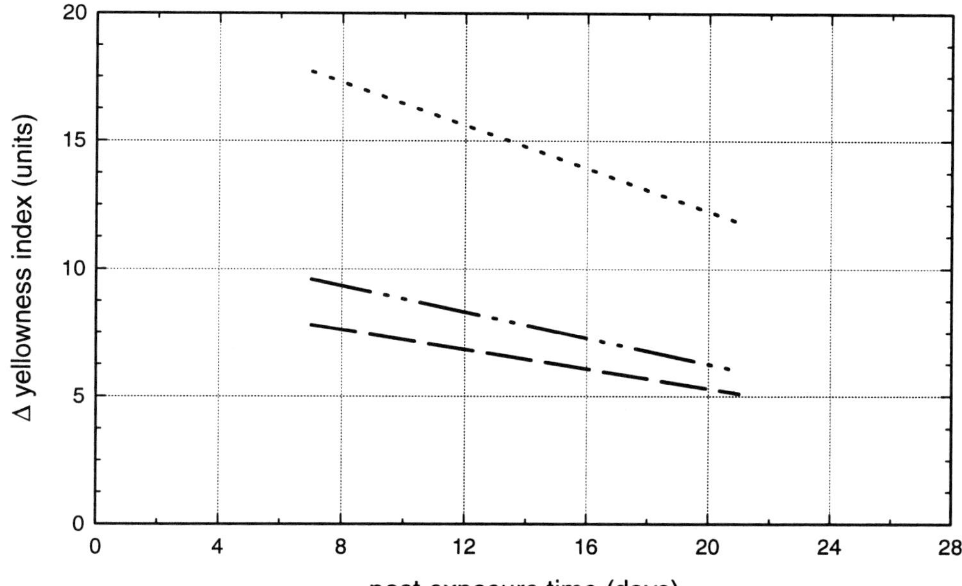

··············	Geon Co. Geon 87614 PVC
—··—··	Geon Co. Geon 87616 PVC
— — — —	Geon Co. Geon 87619 PVC
Reference No.	66

Gamma Radiation Resistance

Dow Chemical: Pulse 1370

PC/ABS loses physical properties linearly with increased dosages of gamma radiation. At 2.5 Mrads impact strength was retained, but the polymer lost 22% at 10.0 Mrads. The loss of impact properties is thought to be attributed to breakdown during the rubber phase. The optical properties bleach back in time and are accelerated by fluorescent light exposure.

Reference: Sturdevant, Marianne F., *Sterilization Compatibility of Rigid Thermoplastic Materials,* supplier technical report (301-1548) - Dow Chemical Company, 1988.

Ethylene Oxide (EtO) Resistance

Dow Chemical: Pulse 1370

The physical property performance is unaffected by EtO exposure. Exposure of up to five cycles of ethylene oxide gas did not affect the notched Izod impact strength. The tensile yield strength was also unaffected by multiple exposures. The first day following exposure, residual EtO levels measured over 500 ppm. After 20 days the residual level dropped to below 200ppm.

Reference: Sturdevant, Marianne F., *Sterilization Compatibility of Rigid Thermoplastic Materials,* supplier technical report (301-1548) - Dow Chemical Company, 1988.

TABLE 154: Effect of Gamma Radiation Sterilization on Polycarbonate ABS Alloy

Material Family	POLYCARBONATE ABS ALLOY			
Material Supplier/Name	DOW PULSE 1370			
Reference No.	3	3	3	3
EXPOSURE CONDITIONS				
Type	Gamma Radiation	Gamma Radiation	Gamma Radiation	Gamma Radiation
Details	source: Cobalt 60	source: Cobalt 60	source: Cobalt 60	source: Cobalt 60
Radiation Dose (Mrads)	2.5	10	2.5	10
POST EXPOSURE CONDITIONING				
Note	type: storage under fluorescent light	type: storage under fluorescent light	type: storage in dark	type: storage in dark
Temperature (°C)	21	21	21	21
Time (hours)	336	336	336	336
PROPERTIES RETAINED (%)				
Tensile Strength @ Yield	100 (ii)	101.4 (ii)	100 (ii)	101.4 (ii)
Notched Izod Impact	94.6 (fp)	77.5 (fp)	95.3 (fp)	77.5 (fp)

TABLE 155: Effect of Ethylene Oxide Sterilization on Polycarbonate ABS Alloy

Material Family	POLYCARBONATE ABS ALLOY		
Material Supplier/Name	DOW PULSE 1370		
Reference No.	3	3	3

EXPOSURE CONDITIONS

Type	Ethylene Oxide	Ethylene Oxide	Ethylene Oxide
Details	12% EtO and 88% Freon	12% EtO and 88% Freon	12% EtO and 88% Freon
Concentration	660 mg/l	660 mg/l	660 mg/l
Number of Cycles	1	5	1
Note	RH: 60%	RH: 60%	RH: 60%
Temperature (°C)	49	49	49
Time (hours)	≥ 6	≥ 6	≥ 6

PRE EXPOSURE CONDITIONING

Preconditioning Note	time: 8 hours; temperature: 37.8 °C; RH: 60%	time: 8 hours; temperature: 37.8 °C; RH: 60%	time: 8 hours; temperature: 37.8 °C; RH: 60%

POST EXPOSURE CONDITIONING

Note	type: evacuation; pressure: 127 mm Hg	type: evacuation; pressure: 127 mm Hg	type: evacuation; pressure: 127 mm Hg

POST EXPOSURE CONDITIONING II

Note	type: aeration	type: aeration	type: aeration
Temperature (°C)	32	32	32
Time (hours)	≥ 16	≥ 16	

POST EXPOSURE CONDITIONING III

Note	type: storage in dark; RH: 50%	type: storage in dark; RH: 50%	
Temperature (°C)	21	21	
Time (hours)	336	336	

RESIDUALS (ppm)

Residuals Determined			ethylene oxide
24 hour Aeration			502
72 hour Aeration			255
168 hour Aeration			182
744 hour Aeration			145
840 hour Aeration			93
888 hour Aeration			68

PROPERTIES RETAINED (%)

Tensile Strength @ Yield	101.4 (ii)	100 (ii)	
Notched Izod Impact	96.9 (fp)	100.8 (fp)	

GRAPH 141: Post Gamma Radiation Exposure Time vs. Delta E Color Change of Polycarbonate ABS Alloy

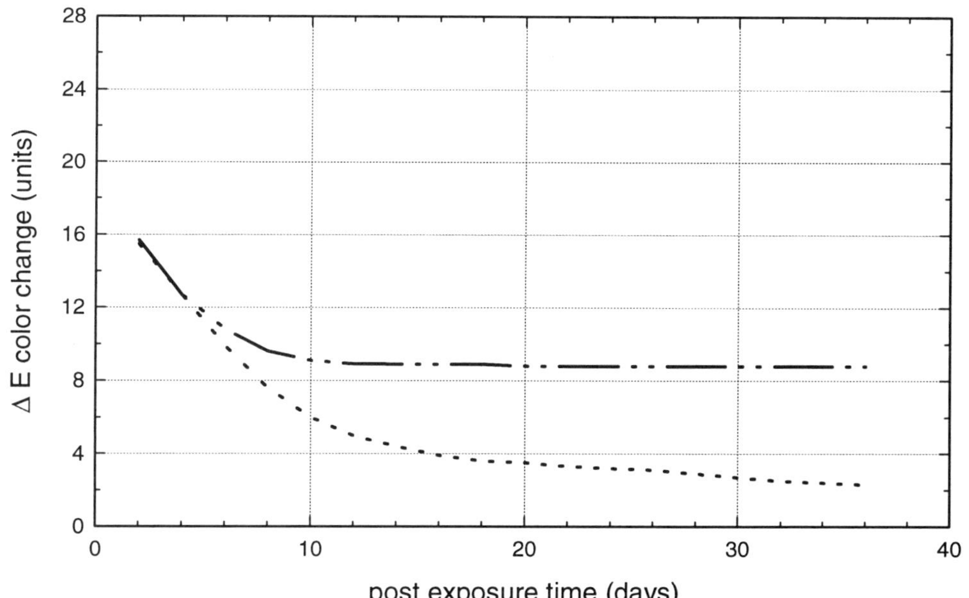

....	Dow Pulse 1370 PC ABS Alloy; 2.5 Mrads; light storage
—··—··	Dow Pulse 1370 PC ABS Alloy; 2.5 Mrads; dark storage
Reference No.	3

GRAPH 142: Post Gamma Radiation Exposure Time vs. Delta E Color Change of Polycarbonate ABS Alloy

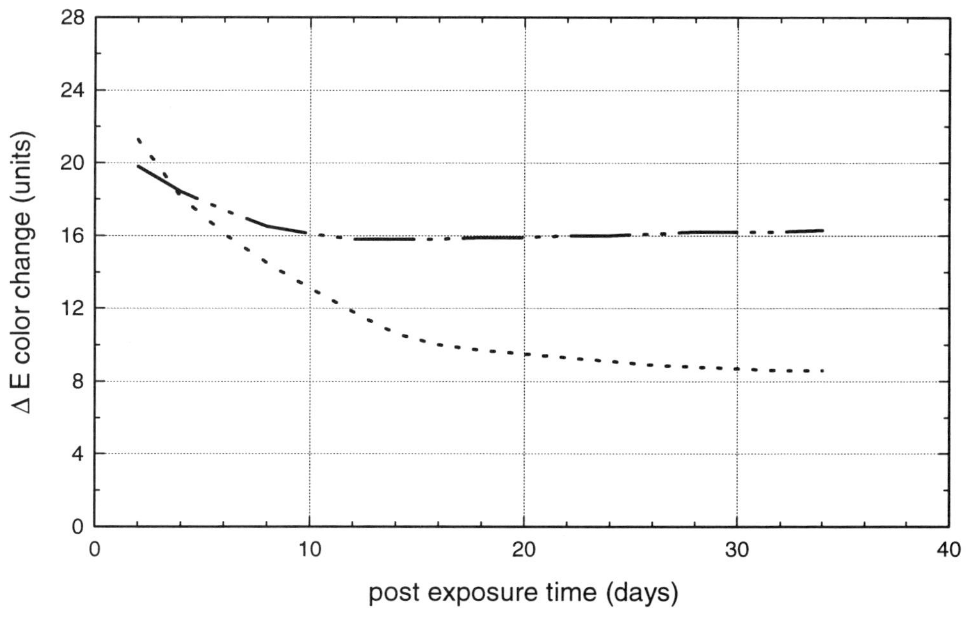

....	Dow Pulse 1370 PC ABS Alloy; 10 Mrads; light storage
—··—··	Dow Pulse 1370 PC ABS Alloy; 10 Mrads; dark storage
Reference No.	3

Gamma Radiation Resistance

Cyro: Cyrex 200-8005 (features: medical grade, white opaque)

At 2.5 Mrads there are no negative effects from gamma radiation. Cyrex 200-8005 alloy exhibits a high degree of whiteness prior to and after sterilization. Following exposure to gamma radiation, the whiteness associated with Cyrex 200-8005 shows a slight color shift immediately after exposure. There is no change in color after the initial color shift.

Reference: *CyRex 200 Acrylic-Polycarbonate Alloy,* supplier marketing literature (1767-194-5BP) - Cyro Industries, 1994.

GRAPH 143: **Beta Radiation Dose vs. Tensile Strength of Polycarbonate Polyester Alloy**

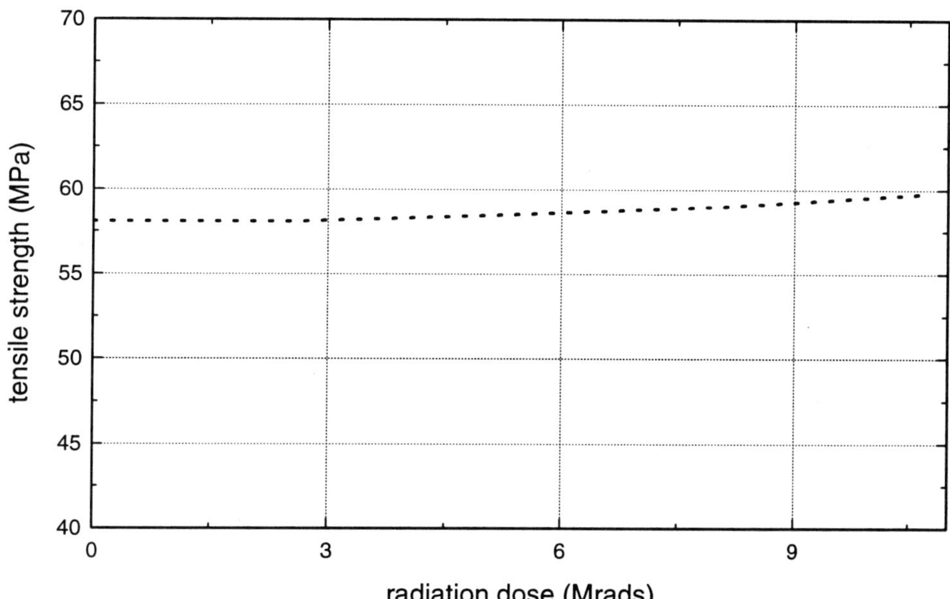

....	PC Polyester Alloy
Reference No.	105

GRAPH 144: **Beta Radiation Dose vs. Tensile Modulus of Polycarbonate Polyester Alloy**

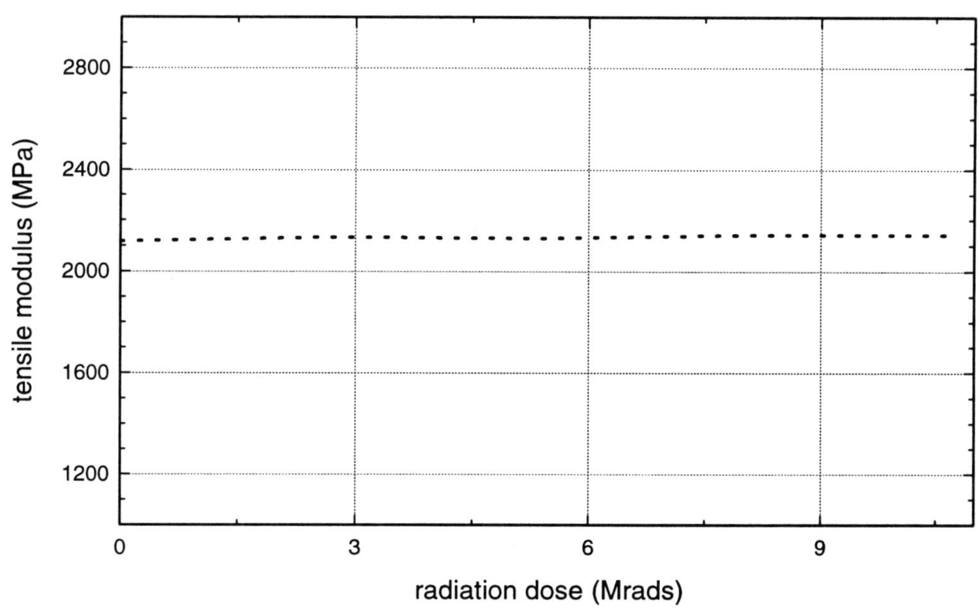

....	PC Polyeser Alloy
Reference No.	105

GRAPH 145: Beta Radiation Dose vs. Yellowness Index of Polycarbonate Polyester Alloy

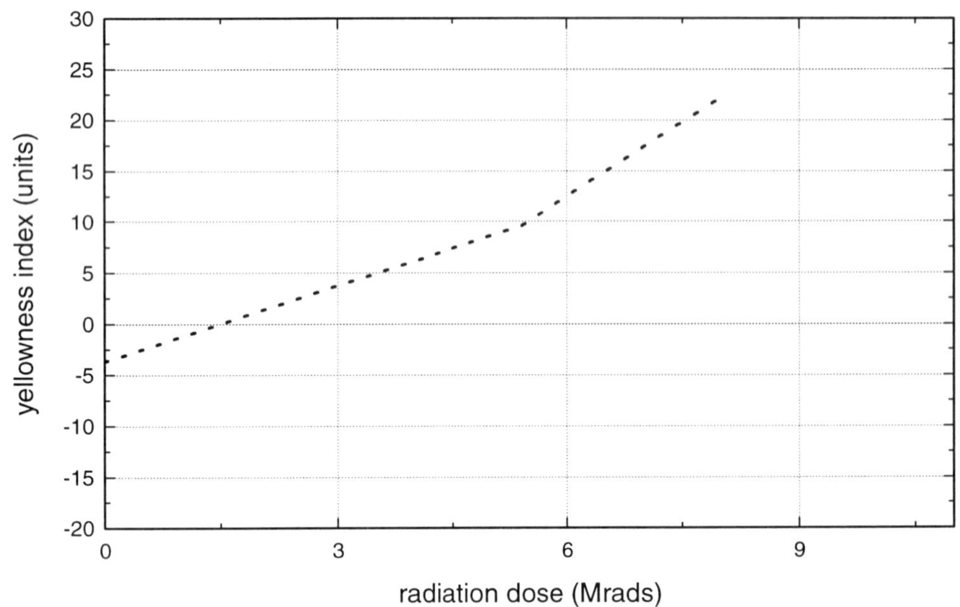

GRAPH 146: Post Beta Radiation Exposure Time vs. Yellowness Index of Polycarbonate Polyester Alloy

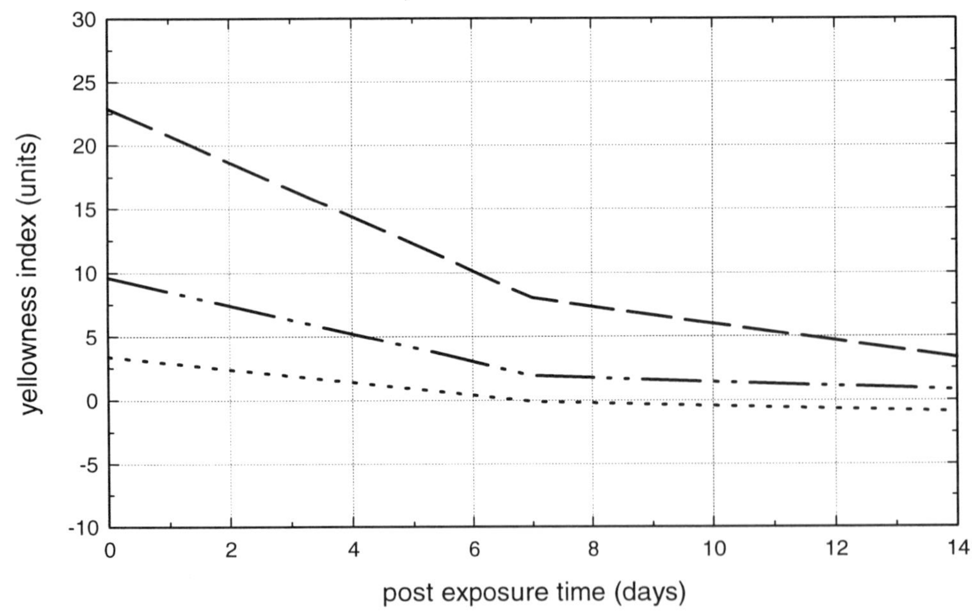

Gamma Radiation Resistance

Eastman Performance Plastics: Ektar MB DA003

Gamma irradiation with typical doses of 2.5 and 5.0 Mrads does not affect physical properties of the material.

Reference: *Ektar Polymers For Healthcare Products,* supplier marketing literature (P/MD-11) - Eastman Performance Plastics, 1991.

Eastman Performance Plastics: Ektar MB DA003 (features: transparent)

There is much less color shift in the alloy of copolyester and polycarbonate than in the polycarbonate alone. There are no significant changes in mechanical properties due to gamma radiation exposure through the 5 Mrad level.

Reference: Goulder, G. K., Seymour, R. W., *Polyesters And Copolyesters For Use In Medical Applications,* supplier technical report - Eastman Performace Plastics.

Ethylene Oxide (EtO) Resistance

Eastman Performance Plastics: Ektar MB DA003 (features: transparent)

No significant property changes were observed.

Reference: Goulder, G. K., Seymour, R. W., *Polyesters And Copolyesters For Use In Medical Applications,* supplier technical report - Eastman Performace Plastics.

Eastman Performance Plastics: Ektar MB DA003

Physical properties and color of Ektar polymers are virtually unaffected by ethylene oxide sterilization.

Reference: *Ektar Polymers For Healthcare Products,* supplier marketing literature (P/MD-11) - Eastman Performance Plastics, 1991.

Chemical Sterilants

Eastman Performance Plastics: Ektar MB DA003

IV solutions, betadine, bleach and other chemicals common to the healthcare environment have virtually no effect on Ektar MB DA003. At typical assembly strains, these materials are stress crack resistant to most solutions.

Reference: *Ektar Polymers For Healthcare Products,* supplier marketing literature (P/MD-11) - Eastman Performance Plastics, 1991.

Steam Resistance

Miles: Makroblend

Certain grades of Makrolon polycarbonate/PET blend can be used under standard autoclaving conditions (i.e., 121°C for 15 to 30 minutes).

Reference: *Medical Milestones,* supplier marketing literature (KU-F-2019(10F)/ 112/6/92) - Miles Inc., 1992.

TABLE 156: Effect of Ethylene Oxide Sterilization on Polyetherimide Polycarbonate Alloy

Material Family	POLYETHERIMIDE POLYCARBONATE ALLOY							
Material Supplier/Name	**GE ULTEM LTX100A**	**GE ULTEM LTX100B**	**GE ULTEM LTX100A**	**GE ULTEM LTX100B**	**GE ULTEM LTX100A**	**GE ULTEM LTX100B**	**GE ULTEM LTX100A**	**GE ULTEM LTX100B**
Material Note	opaque	opaque	opaque	opaque	opaque	opaque	opaque	opaque
Reference No.	107	107	107	107	107	107	107	107
EXPOSURE CONDITIONS								
Type	Ethylene Oxide	Ethylene Oxide	Ethylene Oxide	Ethylene Oxide	Ethylene Oxide	Ethylene Oxide	Ethylene Oxide	Ethylene Oxide
Details	100% EtO	100% EtO	100% EtO	100% EtO	100% EtO	100% EtO	100% EtO	100% EtO
Concentration	800 mg/l	800 mg/l	800 mg/l	800 mg/l	800 mg/l	800 mg/l	800 mg/l	800 mg/l
Temperature (°C)	51.7	51.7	51.7	51.7	51.7	51.7	51.7	51.7
Time (hours)	8	8	8	8	8	8	8	8
PRE EXPOSURE CONDITIONING								
Preconditioning Note	type: pre vacuum; pressure: 660-711 mm Hg; RH: 45-60%; dwell time: 15 minutes	type: pre vacuum; pressure: 660-711 mm Hg; RH: 45-60%; dwell time: 15 minutes	type: pre vacuum; pressure: 660-711 mm Hg; RH: 45-60%; dwell time: 15 minutes	type: pre vacuum; pressure: 660-711 mm Hg; RH: 45-60%; dwell time: 15 minutes	type: pre vacuum; pressure: 660-711 mm Hg; RH: 45-60%; dwell time: 15 minutes	type: pre vacuum; pressure: 660-711 mm Hg; RH: 45-60%; dwell time: 15 minutes	type: pre vacuum; pressure: 660-711 mm Hg; RH: 45-60%; dwell time: 15 minutes	type: pre vacuum; pressure: 660-711 mm Hg; RH: 45-60%; dwell time: 15 minutes
POST EXPOSURE CONDITIONING								
Note	type: aeration; note: ambient conditions	type: aeration; note: ambient conditions	type: aeration; note: ambient conditions	type: aeration; note: ambient conditions	type: aeration; note: mechanical with fan	type: aeration; note: mechanical with fan	type: aeration; note: mechanical with fan	type: aeration; note: mechanical with fan
Temperature (°C)	23	23	23	23	49	49	49	49
RESIDUALS (ppm)								
Residuals Determined	ethylene chlorohydrin	ethylene chlorohydrin	ethylene glycol	ethylene glycol	ethylene chlorohydrin	ethylene chlorohydrin	ethylene glycol	ethylene glycol
336 hour Aeration	<3	<3	<18	<23	<3	<3	<19	<21

Radiation Resistance

Advanced Elastomer Systems: Trefsin 3281-50 (features: medical grade, barrier properties); **Trefsin 3281-60** (features: medical grade, barrier properties)

Trefsin resins are resistant to 5 Mrads.

Reference: *Trefsin 3281 Food/Medical Grade Thermoplastic Elastomer,* supplier technical report - Advanced Elastomer Systems, 1991.

Gamma Radiation Resistance

Advanced Elastomer Systems: Santoprene 181-55 (features: medical grade); **Santoprene 281-45** (features: medical grade); **Santoprene 281-55** (features: medical grade); **Santoprene 281-64** (features: medical grade); **Santoprene 281-73** (features: medical grade); **Santoprene 281-87** (features: medical grade); **Santoprene 283-40** (features: medical grade); **Santoprene** (features: medical grade); **Santoprene 40D** (features: medical grade); **Santoprene 45A** (features: medical grade); **Santoprene 55A** (features: medical grade); **Santoprene 64A** (features: medical grade); **Santoprene 73A** (features: medical grade); **Santoprene 87A** (features: medical grade)

Medical grades can be exposed to up to 4 Mrads of cobalt 60 gamma radiation, with only minimal change in physical properties and no development of toxicity during subsequent storage of the irradiated rubber. The irradiation appears to cause significant but tolerable losses of tensile strength and ultimate elongation. The aging after irradiation results in no further loss, up to a period of one year.

The agarose overlay method was used to determine the cytotoxicity of the irradiated and subsequently stored specimens. No significant toxicity developed at exposures of 4.2 Mrads and lower, and only slight toxicity at 5.3 Mrads.

Reference: *Medical Grades Of Santoprene Thermoplastic Rubber,* supplier technical report (TCD00793) - Advanced Elastomer Systems, 1993.

Ethylene Oxide (EtO) Resistance

Advanced Elastomer Systems: Santoprene 181-55 (features: medical grade); **Santoprene 281-45** (features: medical grade); **Santoprene 281-55** (features: medical grade); **Santoprene 281-64** (features: medical grade); **Santoprene 281-73** (features: medical grade); **Santoprene 281-87** (features: medical grade); **Santoprene 283-40** (features: medical grade); **Santoprene** (features: medical grade); **Santoprene 40D** (features: medical grade); **Santoprene 45A** (features: medical grade); **Santoprene 55A** (features: medical grade); **Santoprene 64A** (features: medical grade); **Santoprene 73A** (features: medical grade); **Santoprene 87A** (features: medical grade)

Five standard hospital EtO sterilization cycles give no significant loss in hardness or tensile stress-strain properties for both 73 Shore A and 40 Shore D grades. Immediately after sterilization samples sterilized with EtO had residual amounts of 0.03% to 0.06% and samples sterilized with ethylene chlorohydrin and ethylene glycol had residual amounts of 0.02% to 0.03%. The residual amounts virtually disappeared after 31 days.

Reference: *Medical Grades Of Santoprene Thermoplastic Rubber,* supplier technical report (TCD00793) - Advanced Elastomer Systems, 1993.

Steam Resistance

Advanced Elastomer Systems: Santoprene 181-55 (features: medical grade); **Santoprene 281-45** (features: medical grade); **Santoprene 281-55** (features: medical grade); **Santoprene 281-64** (features: medical grade); **Santoprene 281-73** (features: medical grade); **Santoprene 281-87** (features: medical grade); **Santoprene 283-40** (features: medical grade); **Santoprene** (features: medical grade);

Santoprene 40D (features: medical grade); **Santoprene 45A** (features: medical grade); **Santoprene 55A** (features: medical grade); **Santoprene 64A** (features: medical grade); **Santoprene 73A** (features: medical grade); **Santoprene 87A** (features: medical grade)

Test specimens were evaluated for changes in harness, tensile stress-strain properties and swell after exposures of 10, 25, 50, 75 and 100 cycles. The changes in tensile properties are quite minor, insignificant and well within accepted ASTM D 2000 limits. The hardness changes are only marginally significant, but within accepted ASTM limits. The weight change in all cases is less than 2% and in most below 1%. These results indicate that 100 cycles of live steam autoclaving at 134°C does not significantly change properties.

Medical Grades Of Santoprene Thermoplastic Rubber, supplier technical report (TCD00793) - Advanced Elastomer Systems, 1993.

Advanced Elastomer Systems: Santoprene 40D (features: FDA grade); **Santoprene 73A** (features: FDA grade)

The effect of live steam on food contact grades at 126°C was determined for exposure times of 15, 30, 75 and 150 minutes for 73 Shore A and 40 Shore D hardness rubbers. Neither the hardness nor tensile stress strain properties changed significantly in any of the exposures. The steam exposure improved the compression set of both grades. The improvement is likely due primarily to annealing of the rubber rather than an intrinsic improvement in the compression set.

Reference: *Santoprene Thermoplastic Rubber For Food And Beverage Contact,* supplier technical report (TCD00693) - Advanced Elastomer Systems, 1993.

TABLE 157: Effect of Ethylene Oxide Sterilization on Olefinic Thermoplastic Elastomer

Material Family	OLEFINIC THERMOPLASTIC ELASTOMER					
Material Supplier/Name	ADVANCED ELASTOMER SYSTEMS SANTOPRENE 40D					
Material Note	medical grade	medical grade	medical grade	medical grade	medical grade	medical grade
Reference No.	9	9	9	9	9	9
EXPOSURE CONDITIONS						
Type	Ethylene Oxide	Ethylene Oxide	Ethylene Oxide	Ethylene Oxide	Ethylene Oxide	Ethylene Oxide
POST EXPOSURE CONDITIONING						
Note	type: aeration	type: aeration	type: aeration	type: aeration	type: aeration	type: aeration
RESIDUALS (ppm)						
Residuals Determined	ethylene oxide	ethylene oxide	ethylene glycol	ethylene glycol	ethylene chlorohydrin	ethylene chlorohydrin
Little or No Aeration	422		30		219	
744 hour Aeration		<5		9		<5

TABLE 158: Effect of Ethylene Oxide Sterilization on Olefinic Thermoplastic Elastomer

Material Family	OLEFINIC THERMOPLASTIC ELASTOMER					
Material Supplier/Name	ADVANCED ELASTOMER SYSTEMS SANTOPRENE 64A					
Material Note	medical grade	medical grade	medical grade	medical grade	medical grade	medical grade
Reference No.	9	9	9	9	9	9
EXPOSURE CONDITIONS						
Type	Ethylene Oxide	Ethylene Oxide	Ethylene Oxide	Ethylene Oxide	Ethylene Oxide	Ethylene Oxide
POST EXPOSURE CONDITIONING						
Note	type: aeration	type: aeration	type: aeration	type: aeration	type: aeration	type: aeration
RESIDUALS (ppm)						
Residuals Determined	ethylene oxide	ethylene oxide	ethylene glycol	ethylene glycol	ethylene chlorohydrin	ethylene chlorohydrin
Little or No Aeration	315		45		212	
744 hour Aeration		<5		20		8

TABLE 159: Effect of Ethylene Oxide Sterilization on Olefinic Thermoplastic Elastomer

Material Family	OLEFINIC THERMOPLASTIC ELASTOMER					
Material Supplier/Name	ADVANCED ELASTOMER SYSTEMS SANTOPRENE 73A					
Material Note	medical grade	medical grade	medical grade	medical grade	medical grade	medical grade
Reference No.	9	9	9	9	9	9
EXPOSURE CONDITIONS						
Type	Ethylene Oxide	Ethylene Oxide	Ethylene Oxide	Ethylene Oxide	Ethylene Oxide	Ethylene Oxide
POST EXPOSURE CONDITIONING						
Note	type: aeration	type: aeration	type: aeration	type: aeration	type: aeration	type: aeration
RESIDUALS (ppm)						
Residuals Determined	ethylene oxide	ethylene oxide	ethylene glycol	ethylene glycol	ethylene chlorohydrin	ethylene chlorohydrin
Little or No Aeration	468		68		296	
744 hour Aeration		8		13		<5

TABLE 160: Effect of Ethylene Oxide Sterilization on Olefinic Thermoplastic Elastomer

Material Family	OLEFINIC THERMOPLASTIC ELASTOMER					
Material Supplier/Name	ADVANCED ELASTOMER SYSTEMS SANTOPRENE 87A					
Material Note	medical grade	medical grade	medical grade	medical grade	medical grade	medical grade
Reference No.	9	9	9	9	9	9
EXPOSURE CONDITIONS						
Type	Ethylene Oxide	Ethylene Oxide	Ethylene Oxide	Ethylene Oxide	Ethylene Oxide	Ethylene Oxide
POST EXPOSURE CONDITIONING						
Note	type: aeration	type: aeration	type: aeration	type: aeration	type: aeration	type: aeration
RESIDUALS (ppm)						
Residuals Determined	ethylene oxide	ethylene oxide	ethylene glycol	ethylene glycol	ethylene chlorohydrin	ethylene chlorohydrin
Little or No Aeration	582		43		296	
744 hour Aeration		<5		17		<5

TABLE 161: Effect of Steam Sterilization on Olefinic Thermoplastic Elastomer

Material Family	OLEFINIC THERMOPLASTIC ELASTOMER							
Material Supplier/Name	ADVANCED ELASTOMER SYSTEMS SANTOPRENE 73A				ADVANCED ELASTOMER SYSTEMS SANTOPRENE 40D			
Material Note	FDA grade	FDA grade	FDA grade	FDA grade	FDA grade	FDA grade	FDA grade	FDA grade
Reference No.	10	10	10	10	10	10	10	10
EXPOSURE CONDITIONS								
Type	Steam	Steam	Steam	Steam	Steam	Steam	Steam	Steam
Note	live steam cycle; pressure: 0.14 MPa	live steam cycle; pressure: 0.14 MPa	live steam cycle; pressure: 0.14 MPa	live steam cycle; pressure: 0.14 MPa	live steam cycle; pressure: 0.14 MPa	live steam cycle; pressure: 0.14 MPa	live steam cycle; pressure: 0.14 MPa	live steam cycle; pressure: 0.14 MPa
Temperature (°C)	126	126	126	126	126	126	126	126
Time (hours)	0.25	0.5	1.25	2.5	0.25	0.5	1.25	2.5
PROPERTIES RETAINED (%)								
Tensile Strength	95 (ja)	91 (ja)	93 (ja)	90 (ja)	99 (ja)	99 (ja)	101 (ja)	101 (ja)
Elongation	91 (ar)	86 (ar)	90 (ar)	87 (ar)	93 (ar)	91 (ar)	93 (ar)	93 (ar)
Modulus	102 (d)	102 (d)	101 (d)	101 (d)	106 (d)	107 (d)	107 (d)	106 (d)
SURFACE and APPEARANCE								
Hardness Units Change	A0 (go)	A0 (go)	A0 (go)	A0 (go)	A0 (go)	A0 (go)	A0 (go)	A0 (go)

TABLE 162: Effect of Steam Sterilization on Olefinic Thermoplastic Elastomer

Material Family	OLEFINIC THERMOPLASTIC ELASTOMER									
Material Supplier/Name	ADVANCED ELASTOMER SYSTEMS SANTOPRENE 281-45					ADVANCED ELASTOMER SYSTEMS SANTOPRENE 281-55				
Material Note	medical grade	medical grade	medical grade	medical grade	medical grade	medical grade	medical grade	medical grade	medical grade	medical grade
Reference No.	9	9	9	9	9	9	9	9	9	9
EXPOSURE CONDITIONS										
Type	Steam Autoclave	Steam Autoclave	Steam Autoclave	Steam Autoclave	Steam Autoclave	Steam Autoclave	Steam Autoclave	Steam Autoclave	Steam Autoclave	Steam Autoclave
Number of Cycles	10	25	50	75	100	10	25	50	75	100
Note	live steam cycle, 5 min. rise, 5 min. @ 134°C, 5 min. cool	live steam cycle, 5 min. rise, 5 min. @ 134°C, 5 min. cool	live steam cycle, 5 min. rise, 5 min. @ 134°C, 5 min. cool	live steam cycle, 5 min. rise, 5 min. @ 134°C, 5 min. cool	live steam cycle, 5 min. rise, 5 min. @ 134°C, 5 min. cool	live steam cycle, 5 min. rise, 5 min. @ 134°C, 5 min. cool	live steam cycle, 5 min. rise, 5 min. @ 134°C, 5 min. cool	live steam cycle, 5 min. rise, 5 min. @ 134°C, 5 min. cool	live steam cycle, 5 min. rise, 5 min. @ 134°C, 5 min. cool	live steam cycle, 5 min. rise, 5 min. @ 134°C, 5 min. cool
Temperature (°C)	134	134	134	134	134	134	134	134	134	134
Time (hours)	0.083	0.083	0.083	0.083	0.083	0.083	0.083	0.083	0.083	0.083
PROPERTIES RETAINED (%)										
Tensile Strength	91 (jw)	68 (jw)	93 (jw)	95 (jw)	91 (jw)	86 (jw)	85 (jw)	84 (jw)	92 (jw)	91 (jw)
Elongation	103 (jd)	77 (jd)	103 (jd)	101 (jd)	96 (jd)	86 (jd)	81 (jd)	81 (jd)	78 (jd)	80 (jd)
Modulus	88 (b)	86 (b)	89 (b)	95 (b)	94 (b)	92 (b)	95 (b)	93 (b)	104 (b)	99 (b)
SURFACE and APPEARANCE										
Hardness Units Change	A1 (gn)	A1 (gn)	A1 (gn)	A2 (gn)	A2 (gn)	A2 (gn)	A1 (gn)	A4 (gn)	A4 (gn)	A4 (gn)

TABLE 163: Effect of Steam Sterilization on Olefinic Thermoplastic Elastomer

Material Family	OLEFINIC THERMOPLASTIC ELASTOMER				
Material Supplier/Name	ADVANCED ELASTOMER SYSTEMS SANTOPRENE 283-40				
Material Note	medical grade	medical grade	medical grade	medical grade	medical grade
Reference No.	9	9	9	9	9

EXPOSURE CONDITIONS

Type	Steam Autoclave	Steam Autoclave	Steam Autoclave	Steam Autoclave	Steam Autoclave
Number of Cycles	10	25	50	75	100
Note	live steam cycle, 5 min. rise, 5 min. @ 134°C, 5 min. cool	live steam cycle, 5 min. rise, 5 min. @ 134°C, 5 min. cool	live steam cycle, 5 min. rise, 5 min. @ 134°C, 5 min. cool	live steam cycle, 5 min. rise, 5 min. @ 134°C, 5 min. cool	live steam cycle, 5 min. rise, 5 min. @ 134°C, 5 min. cool
Temperature (degrees C)	134	134	134	134	134
Time (hours)	0.083	0.083	0.083	0.083	0.083

PROPERTIES RETAINED (%)

Tensile Strength	96 (jw)	96 (jw)	95 (jw)	98 (jw)	95 (jw)
Elongation	98 (jd)	96 (jd)	94 (jd)	86 (jd)	82 (jd)
Modulus	105 (b)	106 (b)	106 (b)	116 (b)	120 (b)

SURFACE and APPEARANCE

Hardness Units Change	D2 (gn)	D1 (gn)	D0 (gn)	D2 (gn)	D1 (gn)

GRAPH 147: Post Gamma Radiation Exposure Time vs. Tensile Strength Retained of Olefinic Thermoplastic Elastomer

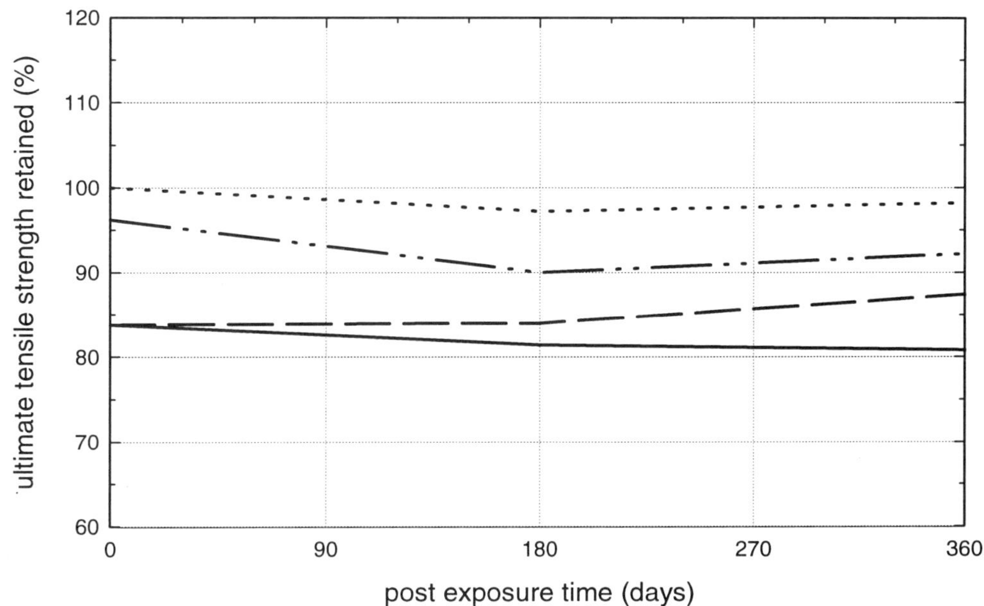

....	Adv. Elast. Santoprene 73A TPO (medical grade); 1.0 Mrad
—··—··	Adv. Elast. Santoprene 73A TPO (medical grade); 2.1 Mrad
– – – –	Adv. Elast. Santoprene 73A TPO (medical grade); 4.2 Mrad
———	Adv. Elast. Santoprene 73A TPO (medical grade); 5.3 Mrad
Reference No.	9

GRAPH 148: Post Gamma Radiation Exposure Time vs. Tensile Strength Retained of Olefinic Thermoplastic Elastomer

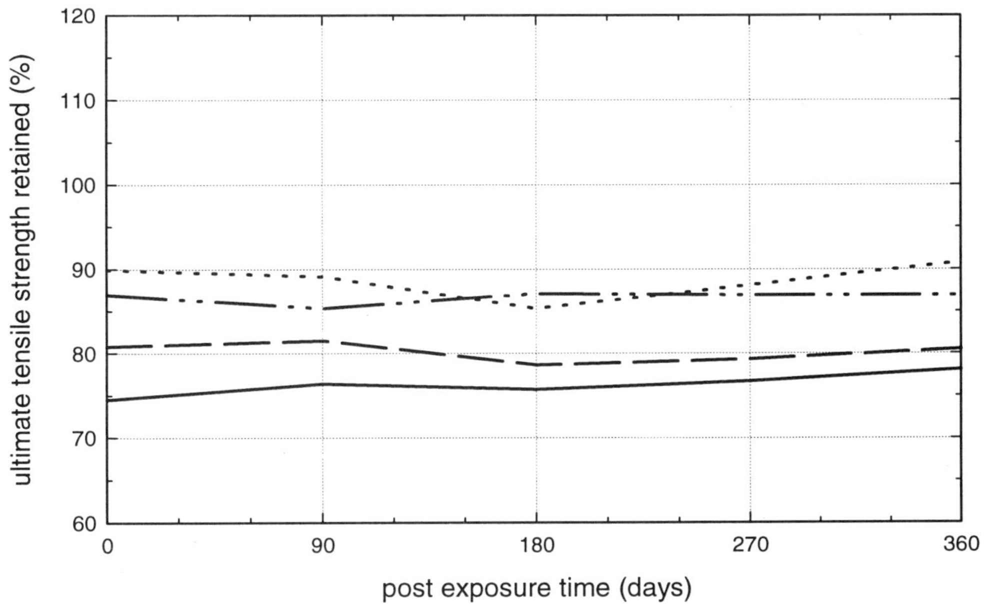

....	Adv. Elast. Santoprene 40D TPO (medical grade); 1.0 Mrad
—··—··	Adv. Elast. Santoprene 40D TPO (medical grade); 2.1 Mrad
– – – –	Adv. Elast. Santoprene 40D TPO (medical grade); 4.2 Mrad
———	Adv. Elast. Santoprene 40D TPO (medical grade); 5.3 Mrad
Reference No.	9

GRAPH 149: Post Gamma Radiation Exposure Time vs. Elongation Retained of Olefinic Thermoplastic Elastomer

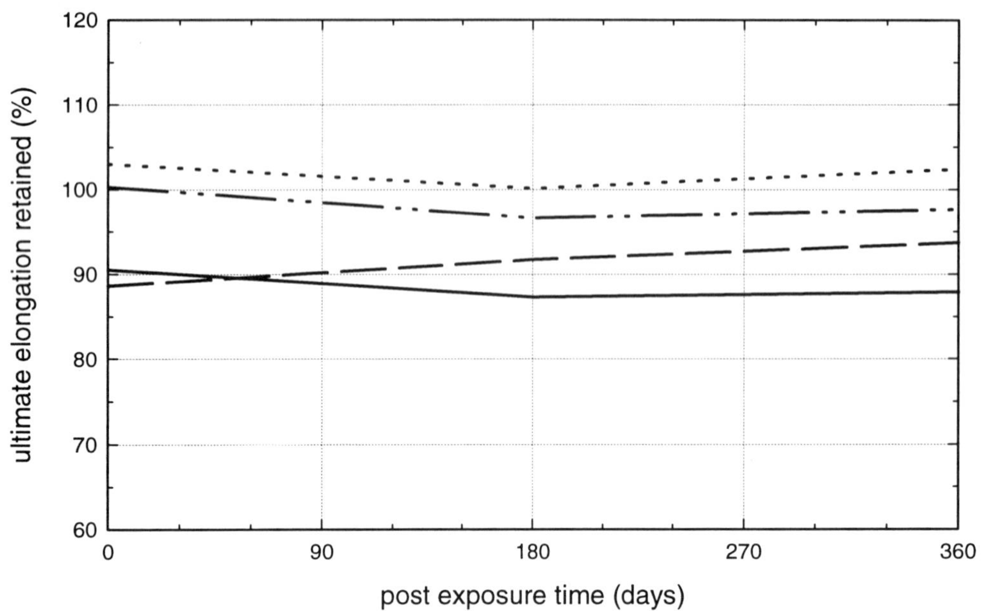

GRAPH 150: Post Gamma Radiation Exposure Time vs. Elongation Retained of Olefinic Thermoplastic Elastomer

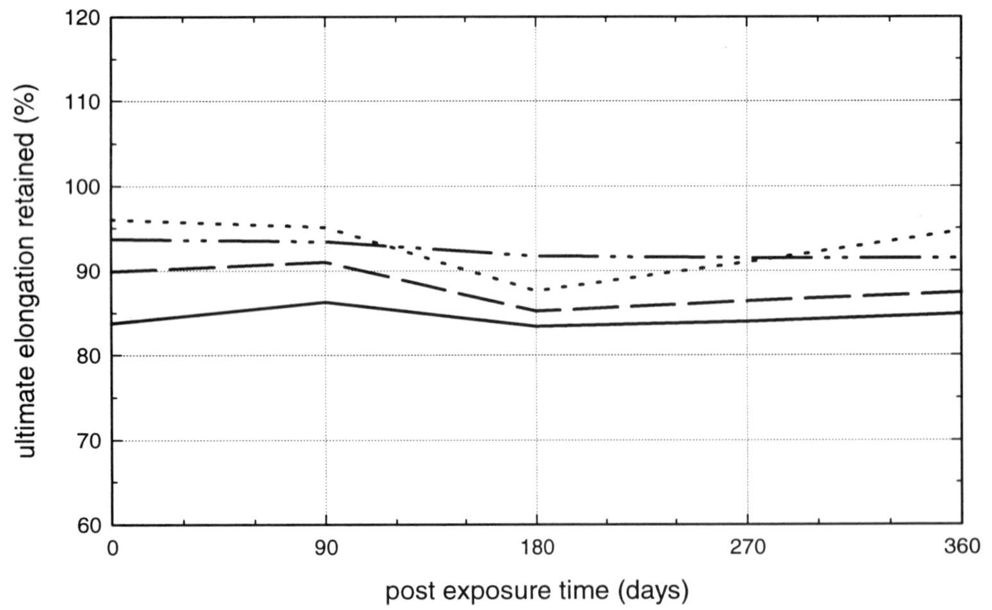

GRAPH 151: **Number of Ethylene Oxide Sterilization Cycles vs. Tensile Strength Retained of Olefinic Thermoplastic Elastomer**

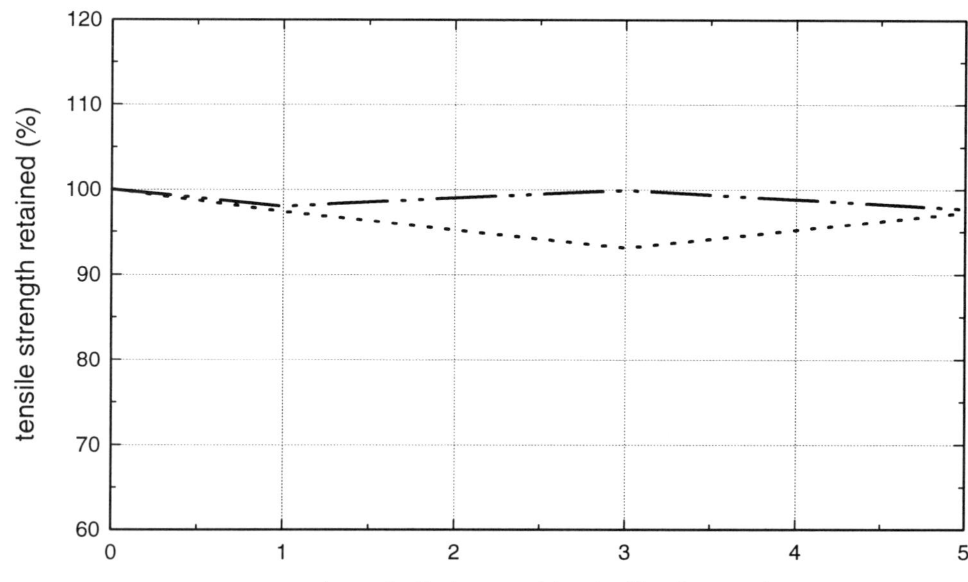

....	Adv. Elast. Santoprene 73A TPO (medical grade)
—·—··	Adv. Elast. Santoprene 40D TPO (medical grade)
Reference No.	9

GRAPH 152: **Number of Ethylene Oxide Sterilization Cycles vs. Elongation at Break Retained of Olefinic Thermoplastic Elastomer**

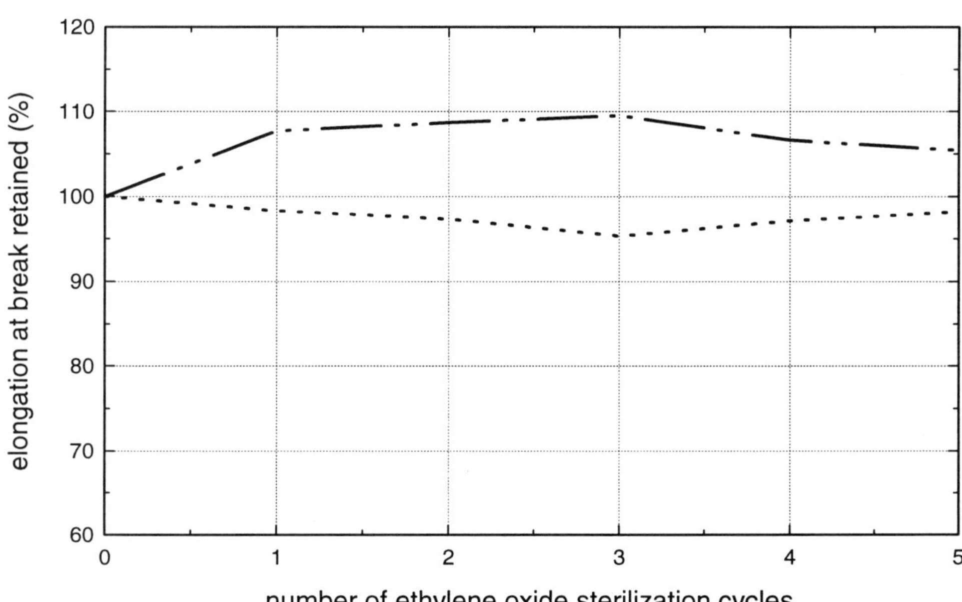

....	Adv. Elast. Santoprene 73A TPO (medical grade)
—·—··	Adv. Elast. Santoprene 40D TPO (medical grade)
Reference No.	9

GRAPH 153: **Number of Ethylene Oxide Sterilization Cycles vs. 100% Modulus Retained of Olefinic Thermoplastic Elastomer**

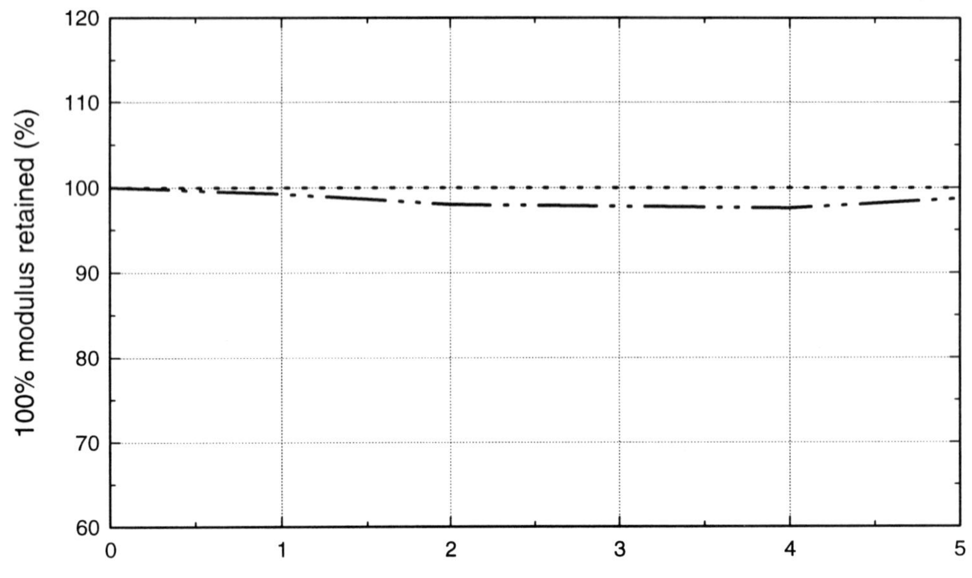

....	Adv. Elast. Santoprene 73A TPO (medical grade)
—··—··	Adv. Elast. Santoprene 40D TPO (medical grade)
Reference No.	9

Radiation Resistance

DuPont: Hytrel 4056 (characteristics: Shore D hardness: 40); **Hytrel 5556** (characteristics: Shore D hardness: 55); **Hytrel 7246** (characteristics: Shore D hardness: 72)

Injection molded slabs of Hytrel 4056, Hytrel 5556, and Hytrel 7246, 2mm (0.079 in) thick, were exposed to a 1.5 Mev electron beam in air at 23°C. The slabs were tested by ASTM methods. Exposure to 150 kJ/kg (15 Mrads) produces very little change in the properties of Hytrel.

Reference: *Hytrel Radiation Resistance,* supplier technical report (E-68116) - Du Pont Company, 1984.

Gamma Radiation Resistance

Eastman: Ecdel 9966 (composition: copolyester ether (COPE); features: transparent)

Ecdel elastomer film, when exposed to increasing levels of gamma radiation, experiences a decrease in inherent viscosity and a substantial reduction in tear strength.

Reference: *Ecdel Elastomers,* supplier design guide (MB-100A) - Eastman Plastics, 1990.

Ethylene Oxide (EtO) Resistance

Eastman: Ecdel 9966 (composition: copolyester ether (COPE); features: transparent)

The material may be sterilized by ethylene oxide.

Reference: *Ecdel Elastomers,* supplier design guide (MB-100A) - Eastman Plastics, 1990.

Steam Resistance

Eastman: Ecdel 9966 (composition: copolyester ether (COPE); features: transparent)

The material may be sterilized by steam autoclave.

Reference: *Ecdel Elastomers,* supplier design guide (MB-100A) - Eastman Plastics, 1990.

TABLE 164: Effect of Gamma Radiation Sterilization on Polyester Thermoplastic Elastomer

Material Family	POLYESTER THERMOPLASTIC ELASTOMER		
Material Supplier/Name	EASTMAN ECDEL 9966		
Material Note	copolyester ether (COPE); transparent	copolyester ether (COPE); transparent	copolyester ether (COPE); transparent
Reference No.	60	60	60
EXPOSURE CONDITIONS			
Type	Gamma Radiation	Gamma Radiation	Gamma Radiation
Radiation Dose (Mrads)	1.5	3	4.5
PROPERTIES RETAINED (%)			
Tensile Strength	84.7 (hp)	82.3 (hp)	80.1 (hp)
Tensile Strength @ Yield	105.1 (ie)	112.4 (ie)	124.8 (ie)
Elongation	111.5 (aj)	102.9 (aj)	97.6 (aj)
Modulus	94.4 (fl)	111.1 (fl)	116.7 (fl)
Tear Resistance	98.6 (gp)	74.5 (gp)	39 (gp)

TABLE 165: Effect of Electron Beam Radiation Sterilization on Polyester Thermoplastic Elastomer

Material Family	POLYESTER THERMOPLASTIC ELASTOMER								
Material Supplier/Name	DUPONT HYTREL 4056			DUPONT HYTREL 5556			DUPONT HYTREL 7246		
Material Note	Shore D hardness: 40	Shore D hardness: 40	Shore D hardness: 40	Shore D hardness: 55	Shore D hardness: 55	Shore D hardness: 55	Shore D hardness: 72	Shore D hardness: 72	Shore D hardness: 72
Reference No.	38	38	38	38	38	38	38	38	38
EXPOSURE CONDITIONS									
Type	Electron Beam Radiation	Electron Beam Radiation	Electron Beam Radiation	Electron Beam Radiation	Electron Beam Radiation	Electron Beam Radiation	Electron Beam Radiation	Electron Beam Radiation	Electron Beam Radiation
Radiation Dose (Mrads)	5	10	15	5	10	15	5	10	15
Note	1.5 Meg electron beam; RH: 70%	1.5 Meg electron beam; RH: 70%	1.5 Meg electron beam; RH: 70%	1.5 Meg electron beam; RH: 70%	1.5 Meg electron beam; RH: 70%	1.5 Meg electron beam; RH: 70%	1.5 Meg electron beam; RH: 70%	1.5 Meg electron beam; RH: 70%	1.5 Meg electron beam; RH: 70%
Temperature (°C)	23	23	23	23	23	23	23	23	23
PROPERTIES RETAINED (%)									
Tensile Strength	94.6 (jc)	94.6 (jc)	91.7 (jc)	104 (jc)	106.3 (jc)	111.4 (jc)	102.5 (jc)	104.8 (jc)	108.1 (jc)
Elongation	92.7 (at)	90.9 (at)	89.1 (at)	120.5 (at)	120.5 (at)	125.6 (at)	95.3 (at)	86 (at)	90.7 (at)
Modulus	107.4 (c)	91.2 (c)	89.7 (c)	100.7 (c)	100.7 (c)	98.6 (c)	107.3 (c)	108.6 (c)	111.8 (c)
SURFACE and APPEARANCE									
Hardness Units Change	D0 (gn)	D0 (gn)	D0 (gn)	D0 (gn)	D0 (gn)	D0 (gn)	D0 (gn)	D0 (gn)	D0 (gn)

Gamma Radiation Resistance

Consolidated Polymers: C-Flex R70-001 (composition: styrene ethylene butylene styrene block copolymer (SEBS); features: white opaque); **C-Flex R70-002** (composition: styrene ethylene butylene styrene block copolymer (SEBS); features: white opaque); **C-Flex R70-003** (composition: styrene ethylene butylene styrene block copolymer (SEBS); features: white opaque); **C-Flex R70-006** (composition: styrene ethylene butylene styrene block copolymer (SEBS); features: white opaque); **C-Flex R70-085** (composition: styrene ethylene butylene styrene block copolymer (SEBS); features: transparent, amber tint)

The following generally occurs after radiation sterilization: slight yellowing, slight decrease in tensile strength, and slight decrease in hardness.

Reference: *The Effects Of Cobalt 60 Radiation Upon The Physical Properties Of Selected C-Flex Elastomers,* Technical Information Document, supplier technical report - Consolidated Polymer Technologies, Inc.

Shell Chemical: Cariflex TR-1000; Cariflex TR-1101 (composition: styrene butadiene styrene block copolymer (SBS)); **Cariflex TR-1102** (composition: styrene butadiene styrene block copolymer (SBS); features: low molecular weight); **Cariflex TR-1107** (composition: styrene isoprene styrene block copolymer (SIS)); **Kraton 2000; Kraton 2104; Kraton 2109**

Sterilization, which is necessary in most pharmaceutical applications, should with thermoplastic rubbers occur at ambient temperatures. For this reason gamma irradiation has been found to be a very suitable method. These materials were given doses of 3, 9 and 15 Mrads from a cobalt 60 source in air at 23°C. The effect of irradiation on tensile strength, modulus, elongation at break and hardness was observed using 1 mm thick compression molded slabs. All materials investigated resisted the standard sterilization dosage of 2.5 Mrads, without being seriously affected. The level of 15 Mrads had a marked effect only on TR-1101.

Reference: Johnson, D., Draaijer, L. M., Schipper, G. P., *Sterilization Of "Cariflex" TR-1000 Series And "Kraton" 2000 Compounds,* supplier technical report - Shell Development Co., 1983.

Shell Chemical: Kraton G2705 (composition: styrene ethylene butylene styrene block copolymer (SEBS))

Kraton G rubbers can be sterilized with gamma radiation without suffering large losses in physical properties. Test specimens were subjected to Cobalt 60 radiation doses of 3.0, 6.0, and 12.0 Mrads. After 21 months of aging under ambient conditions Kraton G-2705 saw little reduction in properties.

Reference: *Kraton Thermoplastic Rubber Medial Products,* supplier technical report (SC:1032-88) - Shell Chemical Company, 1988.

Evode Plastics: Evoprene G (composition: styrene ethylene butylene styrene block copolymer (SEBS))

The general purpose filled and unfilled grades can all be sterilized by gamma radiation. Advice on gamma sterilizing other grades should be sought.

Reference: *Technical Information Evoprene G,* supplier technical report (RDS 028/9240) - Evode Plastics.

Ethylene Oxide (EtO) Resistance

Shell Chemical: Kraton (composition: styrene ethylene butylene styrene block copolymer (SEBS))

All SEBS rubbers can be sterilized by ethylene oxide without suffering large losses in physical properties.

Reference: *Applications Of Styrene-Butadiene-Styrene (SBS) And Styrene-Ethylene-Butylene-Styrene (SEBS) In The Medical Industry,* supplier technical report - Shell Chemical Company.

Consolidated Polymers: C-Flex (composition: styrene ethylene butylene styrene block copolymer (SEBS))

Use a standard hospital sterilization cycle. Ambient aeration time should be 5 -7 days depending on packaging. With a vacuum/heated aeration chamber any standard cycle should work.

Reference: *Autoclaving/Steam & Gas Sterilization,* Technical Information Document, supplier technical report - Consolidated Polymer Technologies, Inc.

Shell Chemical: Cariflex TR-1000; Kraton 2000; Kraton 2109

EtO is widely used in the pharmaceutical industry and is suitable for TR products. It is essential that any articles exposed to ethylene oxide be thoroughly aired after disinfection, as the material is very toxic and causes skin blistering. KRATON 2109 desorbed EtO much faster than vulcanized polyisoprene.

Reference: Johnson, D., Draaijer, L. M., Schipper, G. P., *Sterilization Of "Cariflex" TR-1000 Series And "Kraton" 2000 Compounds,* supplier technical report - Shell Development Co., 1983.

Evode Plastics: Evoprene G (composition: styrene ethylene butylene styrene block copolymer (SEBS))

The general purpose filled and unfilled grades can all be sterilized by ethylene oxide. Advice on ethylene oxide sterilizing other grades should be sought.

Reference: *Technical Information Evoprene G,* supplier technical report (RDS 028/9240) - Evode Plastics.

Steam Resistance

Shell Chemical: Kraton (composition: styrene ethylene butylene styrene block copolymer (SEBS))

All SEBS rubbers can be sterilized by steam without suffering large losses in physical properties.

Applications Of Styrene-Butadiene-Styrene (SBS) And Styrene-Ethylene-Butylene-Styrene (SEBS) In The Medical Industry, supplier technical report - Shell Chemical Company.

Consolidated Polymers: C-Flex (composition: styrene ethylene butylene styrene block copolymer (SEBS))

Not recommended for flash autoclaving (steam temperatures of greater than 135°C and exposure times of approximately 5 minutes). Recommended steam sterilization conditions are: steam temperature of 116 - 131°C and exposure times of 20 - 25 minutes.

Autoclaving/Steam & Gas Sterilization, Technical Information Document, supplier technical report - Consolidated Polymer Technologies, Inc.

Shell Chemical: Cariflex TR-1000; Kraton 2000

Not suitable - temperatures of the order of 100°C or higher will cause severe deformation.

Johnson, D., Draaijer, L. M., Schipper, G. P., *Sterilization Of "Cariflex" TR-1000 Series And "Kraton" 2000 Compounds,* supplier technical report - Shell Development Co., 1983.

Evode Plastics: Evoprene G (composition: styrene ethylene butylene styrene block copolymer (SEBS))

The general purpose filled and unfilled grades can all be sterilized by steam. Advice on steam sterilizing other grades should be sought.

Reference: *Technical Information Evoprene G,* supplier technical report (RDS 028/9240) - Evode Plastics.

Dry Heat Sterilization

Shell Chemical: Cariflex TR-1000; Kraton 2000

Not suitable - temperatures of the order of 100°C or higher will cause severe deformation.

Reference: Johnson, D., Draaijer, L. M., Schipper, G. P., *Sterilization Of "Cariflex" TR-1000 Series And "Kraton" 2000 Compounds,* supplier technical report - Shell Development Co., 1983.

TABLE 166: Effect of Gamma Radiation Sterilization on Styrenic Thermoplastic Elastomer

Material Family	STYRENIC THERMOPLASTIC ELASTOMER					
Material Supplier/Name	CONSOLIDATED POLYMERS C-FLEX R70-001			CONSOLIDATED POLYMERS C-FLEX R70-006		
Material Note	styrene ethylene butylene styrene block copolymer (SEBS); white opaque	styrene ethylene butylene styrene block copolymer (SEBS); white opaque	styrene ethylene butylene styrene block copolymer (SEBS); white opaque	styrene ethylene butylene styrene block copolymer (SEBS); white opaque	styrene ethylene butylene styrene block copolymer (SEBS); white opaque	styrene ethylene butylene styrene block copolymer (SEBS); white opaque
Reference No.	40	40	40	40	40	40

EXPOSURE CONDITIONS

Type	Gamma Radiation	Gamma Radiation	Gamma Radiation	Gamma Radiation	Gamma Radiation	Gamma Radiation
Details	source: Cobalt 60	source: Cobalt 60	source: Cobalt 60	source: Cobalt 60	source: Cobalt 60	source: Cobalt 60
Radiation Dose (Mrads)	3.02	4.64	7.66	3.02	4.64	7.66

POST EXPOSURE CONDITIONING

Note	type: ambient conditions	type: ambient conditions	type: ambient conditions	type: ambient conditions	type: ambient conditions	type: ambient conditions
Time (hours)	336	336	336	336	336	336

PROPERTIES RETAINED (%)

Tensile Strength	90.3 (ke)	80.6 (kd)	68.5 (jz)	83.7 (kb)	73.2 (kh)	62.6 (ki)
Elongation	97.4 (jl)	93.4 (jn)	92.1 (jg)	95.9 (jh)	101.4 (jh)	94.5 (jp)

SURFACE and APPEARANCE

Hardness Units Change	A-3 (gn)	A-3 (gn)	A-4 (gn)	A0 (gn)	A-4 (gn)	A-8 (gn)
Yellowness Index note	slight yellowing	slight yellowing	slight yellowing	slight yellowing	slight yellowing	slight yellowing

TABLE 167: Effect of Gamma Radiation Sterilization on Styrenic Thermoplastic Elastomer

Material Family	STYRENIC THERMOPLASTIC ELASTOMER					
Material Supplier/Name	CONSOLIDATED POLYMERS C-FLEX R70-002			CONSOLIDATED POLYMERS C-FLEX R70-003		
Material Note	styrene ethylene butylene styrene block copolymer (SEBS); white opaque	styrene ethylene butylene styrene block copolymer (SEBS); white opaque	styrene ethylene butylene styrene block copolymer (SEBS); white opaque	styrene ethylene butylene styrene block copolymer (SEBS); white opaque	styrene ethylene butylene styrene block copolymer (SEBS); white opaque	styrene ethylene butylene styrene block copolymer (SEBS); white opaque
Reference No.	40	40	40	40	40	40

EXPOSURE CONDITIONS

Type	Gamma Radiation	Gamma Radiation	Gamma Radiation	Gamma Radiation	Gamma Radiation	Gamma Radiation
Details	source: Cobalt 60	source: Cobalt 60	source: Cobalt 60	source: Cobalt 60	source: Cobalt 60	source: Cobalt 60
Radiation Dose (Mrads)	3.02	4.64	7.66	3.02	4.64	7.66

POST EXPOSURE CONDITIONING

Note	type: ambient conditions	type: ambient conditions	type: ambient conditions	type: ambient conditions	type: ambient conditions	type: ambient conditions
Time (hours)	336	336	336	336	336	336

PROPERTIES RETAINED (%)

Tensile Strength	78 (kf)	73.9 (jx)	59.1 (kg)	93.2 (kg)	79.6 (ka)	61.5 (kl)
Elongation	108.8 (je)	108.8 (jo)	105.9 (jj)	110.3 (ji)	97.1 (jm)	98.5 (jq)

SURFACE and APPEARANCE

Hardness Units Change	A-2 (gn)	A-3 (gn)	A-2 (gn)	A-2 (gn)	A-4 (gn)	A-5 (gn)
Yellowness Index note	slight yellowing	slight yellowing	slight yellowing	slight yellowing	slight yellowing	slight yellowing

TABLE 168: Effect of Gamma Radiation Sterilization on Styrenic Thermoplastic Elastomer

Material Family	STYRENIC THERMOPLASTIC ELASTOMER				
Material Supplier/Name	CONSOLIDATED POLYMERS C-FLEX R70-085		SHELL KRATON G2705		
Material Note	styrene ethylene butylene styrene block copolymer (SEBS); transparent, amber tint	styrene ethylene butylene styrene block copolymer (SEBS); transparent, amber tint	styrene ethylene butylene styrene block copolymer (SEBS)	styrene ethylene butylene styrene block copolymer (SEBS)	styrene ethylene butylene styrene block copolymer (SEBS)
Reference No.	40	40	87	87	87

EXPOSURE CONDITIONS

Type	Gamma Radiation	Gamma Radiation	Gamma Radiation	Gamma Radiation	Gamma Radiation
Details	source: Cobalt 60	source: Cobalt 60			
Radiation Dose (Mrads)	2	5	3	6	12

POST EXPOSURE CONDITIONING

Note	type: ambient conditions	type: ambient conditions	type: ambient conditions	type: ambient conditions	type: ambient conditions
Time (hours)	336	336	15330	15330	15330

PROPERTIES RETAINED (%)

Tensile Strength (transverse)	102.5 (kn)	107 (ko)	112 (he)	97 (he)	85 (he)
Elongation			110 (ai)	108 (ai)	112 (ai)
Elongation @ Yield (transverse)	95 (jt)	99.7 (jv)			
300% Modulus			96 (e)	90 (e)	85 (e)

SURFACE and APPEARANCE

Hardness Units Change	A-1 (gn)	A-1 (gn)	A-7 (gn)	A-7 (gn)	A-12 (gn)
Yellowness Index note	slight yellowing	slight yellowing			

Radiation Resistance

Dow Chemical: Pellethane 2363-75D (features: transparent)

There is permanent color change associated with TPU's and high energy radiation sterilization. Gamma radiation caused greater color shifts than electron beam. Initially the 2.5 Mrad samples had larger yellowness index values than the 10 Mrad samples, within 8 weeks this trend reversed for gamma, but not for E-beam. The 2.5 Mrad sample maintained a greater yellowness index than the 10 Mrad E-beam sample. These color results are not typical; increasing doses of high energy radiation sterilization generally cause greater color changes in thermoplastics than lower doses.

TPU with a shore hardness of 75D exposed to 10 Mrads of E-beam radiation showed that the tensile and flexural moduli decreased by approximately 50%. The same resin sterilized at 10 Mrads with gamma had a decrease of 20% in tensile modulus and no significant decrease in flexural modulus.

Reference: Hermanson, Nancy J., Steffens, John F., *The Physical and Visual Property Changes in Thermoplastic Resins After Exposure to High Energy Sterilization - Gamma Versus Electron Beam,* ANTEC 1993, conference proceedings - Society of Plastics Engineers, 1993.

Dow Chemical: Pellethane 2363-80A (features: transparent)

There is permanent color change associated with TPU's and high energy radiation sterilization. Gamma radiation caused greater color shifts than electron beam.

Reference: Hermanson, Nancy J., Steffens, John F., *The Physical and Visual Property Changes in Thermoplastic Resins After Exposure to High Energy Sterilization - Gamma Versus Electron Beam,* ANTEC 1993, conference proceedings - Society of Plastics Engineers, 1993.

Gamma Radiation Resistance

Dow Chemical: Pellethane 2363-55D (features: transparent); **Pellethane 2363-80A** (features: transparent)

Molded sheets of Pellethane 2363-80A and 2363-55D elastomers were both treated with 2.75 and 9.6 Mrads of radiation. Properties were measured soon after irradiation and after 33 months at room temperature. Although the samples were darker immediately after treatment, they became lighter on storage, and no significant changes in physical properties were observed.

Reference: *Pellethane Polyurethane Elastomers - 2363 Series For Medical Applications,* supplier marketing literature (306-180-792XSMG) - Dow Chemical Company, 1992.

Thermedics: Tecoflex EG60D (composition: aliphatic polyurethane; features: medical grade); **Tecoflex EG80A** (composition: aliphatic polyurethane; features: medical grade)

Tecoflex extruded tapes were sterilized with single (2.5 Mrads) and double (5.0 Mrads) doses of gamma radiation. Testing was done in compliance with ASTM D-638-72 for tensile and elongation properties. After 5.0 Mrads exposure and 15 months post irradiation the aliphatic polyurethanes continue to retain their physical properties. While Tecoflex retains its physical properties after exposure to both 2.5 and 5.0 Mrads, the manufacturer recommends against exposure to 5.0 Mrads due to excessive yellowing of the material.

Reference: *Tecoflex Medical Grade Aliphatic Polyurethanes,* supplier technical report - Thermedics Inc.

Ethylene Oxide (EtO) Resistance

Dow Chemical: Pellethane 2363-55D (features: transparent); **Pellethane 2363-80A** (features: transparent)

Ethylene oxide has no noticeable effect on material characteristics.

Reference: *Pellethane Polyurethane Elastomers - 2363 Series For Medical Applications,* supplier marketing literature (306-180-792XSMG) - Dow Chemical Company, 1992.

Miles: Texin (features: general purpose grade)

Parts molded or extruded from Texin resins may be sterilized by ethylene oxide. The number of sterilization cycles that parts made from Texin elastomers can withstand depends upon part design and processing parameters. Therefore, the manufacturer must evaluate each part to determine the number of sterilization cycles appropriate for the part's exact performance requirements.

Reference: *Texin Thermoplastic Polyurethane For Medical Applications,* supplier marketing literature (KU-E 1002(5)L/ 427-5M-5/92) - Miles Inc., 1992.

Steam Resistance

Miles: Texin (features: general purpose grade)

Steam autoclave and boiling water sterilization techniques are not recommended and should be avoided.

Texin Thermoplastic Polyurethane For Medical Applications, supplier marketing literature (KU-E 1002(5)L/ 427-5M-5/92) - Miles Inc., 1992.

Dow Chemical: Pellethane 2363-55D (features: transparent); Pellethane 2363-80A (features: transparent)

Steam sterilization is not recommended for parts manufactured from Pellethane 2363. When steam sterilization is used, hydrolysis of the solid polyurethane can give rise to aromatic amine impurities.

Reference: *Pellethane Polyurethane Elastomers - 2363 Series For Medical Applications,* supplier marketing literature (306-180-792XSMG) - Dow Chemical Company, 1992.

Dry Heat Sterilization

Dow Chemical: Pellethane 2363-55D (features: transparent); Pellethane 2363-80A (features: transparent)

Dry heat sterilization for 6 hours at 123°C has no significant effect on mechanical properties, but will result in a slight amount of discoloration. The softer elastomers may cloud up due to the surface migration of the processing aids.

Reference: *Pellethane Polyurethane Elastomers - 2363 Series For Medical Applications,* supplier marketing literature (306-180-792XSMG) - Dow Chemical Company, 1992.

Miles: Texin (features: general purpose grade)

Parts molded or extruded from Texin resins may be sterilized by dry heat. The number of sterilization cycles that parts made from Texin elastomers can withstand depends upon part design and processing parameters. Therefore, the manufacturer must evaluate each part to determine the number of sterilization cycles appropriate for the part's exact performance requirements.

Reference: *Texin Thermoplastic Polyurethane For Medical Applications,* supplier marketing literature (KU-E 1002(5)L/ 427-5M-5/92) - Miles Inc., 1992.

TABLE 169: Effect of Gamma Radiation Sterilization on Urethane Thermoplastic Elastomer

Material Family	URETHANE THERMOPLASTIC ELASTOMER							
Material Supplier/Name	DOW PELLETHANE 2363-80A				DOW PELLETHANE 2363-55D			
Material Note	transparent	transparent	transparent	transparent	transparent	transparent	transparent	transparent
Reference No.	8	8	8	8	8	8	8	8
EXPOSURE CONDITIONS								
Type	Gamma Radiation	Gamma Radiation	Gamma Radiation	Gamma Radiation	Gamma Radiation	Gamma Radiation	Gamma Radiation	Gamma Radiation
Details	source: Cobalt 60	source: Cobalt 60	source: Cobalt 60	source: Cobalt 60	source: Cobalt 60	source: Cobalt 60	source: Cobalt 60	source: Cobalt 60
Radiation Dose (Mrads)	2.75	2.75	9.6	9.6	2.75	2.75	9.6	9.6
POST EXPOSURE CONDITIONING								
Note		type: storage in dark		type: storage in dark		type: storage in dark		type: storage in dark
Temperature (degrees C)		23		23		23		23
Time (hours)	0	24090	0	24090	0	24090	0	24090
PROPERTIES RETAINED (%)								
Tensile Strength	101.6 (he)	106.7 (he)	104.1 (he)	109.7 (he)	93.3 (he)	103.4 (he)	94.5 (he)	106.4 (he)
Elongation	102.3 (ai)	100 (ai)	111.6 (ai)	109.3 (ai)	83.7 (ai)	86.1 (ai)	83.7 (ai)	90.7 (ai)
Modulus	103.4 (b)	123.9 (b)	102.3 (b)	118.2 (b)	99.6 (b)	110.8 (b)	98.4 (b)	112.4 (b)
300% Modulus	99 (e)	125 (e)	87 (e)	101.4 (e)	105.6 (e)	110.3 (e)	103.4 (e)	109.7 (e)

TABLE 170: Effect of Gamma Radiation Sterilization on Urethane Thermoplastic Elastomer

Material Family	URETHANE THERMOPLASTIC ELASTOMER					
Material Supplier/Name	THERMEDICS TECOFLEX EG80A					
Material Note	aliphatic polyurethane; medical grade	aliphatic polyurethane; medical grade	aliphatic polyurethane; medical grade	aliphatic polyurethane; medical grade	aliphatic polyurethane; medical grade	aliphatic polyurethane; medical grade
Reference No.	95	95	95	95	95	95

EXPOSURE CONDITIONS

Type	Gamma Radiation	Gamma Radiation	Gamma Radiation	Gamma Radiation	Gamma Radiation	Gamma Radiation
Details	source: Cobalt 60	source: Cobalt 60	source: Cobalt 60	source: Cobalt 60	source: Cobalt 60	source: Cobalt 60
Radiation Dose (Mrads)	2.5	2.5	2.5	5	5	5

POST EXPOSURE CONDITIONING

Note	type: aging	type: aging	type: aging	type: aging	type: aging	type: aging
Time (hours)	336	1176	10950	336	1176	10950

PROPERTIES RETAINED (%)

Tensile Strength	98 (he)	93 (he)	100 (he)	89 (he)	92 (he)	99 (he)
Elongation	112 (ai)	116 (ai)	107 (ai)	116 (ai)	117 (ai)	114 (ai)

SURFACE and APPEARANCE

Yellowness Index note				yellowing	yellowing	yellowing

TABLE 171: Effect of Gamma Radiation Sterilization on Urethane Thermoplastic Elastomer

Material Family	URETHANE THERMOPLASTIC ELASTOMER					
Material Supplier/Name	THERMEDICS TECOFLEX EG60D					
Material Note	aliphatic polyurethane; medical grade	aliphatic polyurethane; medical grade	aliphatic polyurethane; medical grade	aliphatic polyurethane; medical grade	aliphatic polyurethane; medical grade	aliphatic polyurethane; medical grade
Reference No.	95	95	95	95	95	95
EXPOSURE CONDITIONS						
Type	Gamma Radiation	Gamma Radiation	Gamma Radiation	Gamma Radiation	Gamma Radiation	Gamma Radiation
Details	source: Cobalt 60	source: Cobalt 60	source: Cobalt 60	source: Cobalt 60	source: Cobalt 60	source: Cobalt 60
Radiation Dose (Mrads)	2.5	2.5	2.5	5	5	5
POST EXPOSURE CONDITIONING						
Note	type: aging	type: aging	type: aging	type: aging	type: aging	type: aging
Time (hours)	336	1176	10950	336	1176	10950
PROPERTIES RETAINED (%)						
Tensile Strength	97 (he)	92 (he)	98 (he)	85 (he)	79 (he)	75 (he)
Elongation	97 (ai)	113 (ai)	100 (ai)	95 (ai)	105 (ai)	90 (ai)
SURFACE and APPEARANCE						
Yellowness Index				yellowing	yellowing	yellowing

TABLE 172: Effect of Gamma Radiation Sterilization on Urethane Thermoplastic Elastomer

Material Family	URETHANE THERMOPLASTIC ELASTOMER							
Material Supplier/Name	DOW PELLETHANE 2363-75D				DOW PELLETHANE 2363-80A			
Material Note	transparent	transparent	transparent	transparent	transparent	transparent	transparent	transparent
Reference No.	1	1	1	1	1	1	1	1
EXPOSURE CONDITIONS								
Type	Gamma Radiation	Gamma Radiation	Gamma Radiation	Gamma Radiation	Gamma Radiation	Gamma Radiation	Gamma Radiation	Gamma Radiation
Radiation Dose (Mrads)	2.5	2.5	10	10	2.5	2.5	10	10
Note	test lab: SteriGenics	test lab: SteriGenics	test lab: SteriGenics	test lab: SteriGenics	test lab: SteriGenics	test lab: SteriGenics	test lab: SteriGenics	test lab: SteriGenics
POST EXPOSURE CONDITIONING								
Note	type: aging	type: aging	type: aging	type: aging	type: aging	type: aging	type: aging	type: aging
Time (hours)	168	1344	168	1344	168	1344	168	1344
PROPERTIES RETAINED (%)								
Tensile Strength @ Yield	98.2 (is)	96.4 (is)	94.5 (is)	98.2 (is)				
Elongation	72.5 (bt)	67.7 (bt)	91.8 (bt)	74.4 (bt)				
Flexural Strength	98.5 (co)	91 (co)	106 (co)	92.5 (co)				
Modulus	85.1 (hb)	87.3 (hb)	76.9 (hb)	83.3 (hb)				
Flexural Modulus	93.4 (cg)	82.3 (cg)	90.9 (cg)	88.9 (cg)				
Dart Impact (total energy)	101.4 (em)	104.3 (em)	100 (em)	102.9 (em)				
Dart Impact (peak energy)	104.5 (dg)	106.8 (dg)	106.8 (dg)	115.9 (dg)				
Notched Izod Impact	75 (ga)	125 (ga)	100 (ga)	207.8 (ga)				
Heat Deflection Temperature	100 (m)	100 (m)	100 (m)	100 (m)	100 (m)	100 (m)	100 (m)	100 (m)
Vicat Softening Point	101.8 (ku)	103.6 (ku)	101.8 (ku)	101.8 (ku)	104.8 (ku)	100 (ku)	96.4 (ku)	97.6 (ku)
SURFACE and APPEARANCE								
Δ Yellowness Index	64.5 (kw)	20 (kw)	42.1 (kw)	28 (kw)	5.9 (kw)	3.57 (kw)	21.91 (kw)	20.75 (kw)
Haze (%)	2.97	4.24	1.85	-0.71	4.57	4	5.63	3.39
Transparency Retained (%)	76.5	114.8	87.7	58	101.3	124.4	107.7	120.5

TABLE 173: Effect of Electron Beam Radiation Sterilization on Urethane Thermoplastic Elastomer

Material Family	URETHANE THERMOPLASTIC ELASTOMER							
Material Supplier/Name	DOW PELLETHANE 2363-75D				DOW PELLETHANE 2363-80A			
Material Note	transparent	transparent	transparent	transparent	transparent	transparent	transparent	transparent
Reference No.	1	1	1	1	1	1	1	1
EXPOSURE CONDITIONS								
Type	Electron Beam Radiation	Electron Beam Radiation	Electron Beam Radiation	Electron Beam Radiation	Electron Beam Radiation	Electron Beam Radiation	Electron Beam Radiation	Electron Beam Radiation
Radiation Dose (Mrads)	2.5	2.5	10	10	2.5	2.5	10	10
Note	test lab: E-Beam Services, Inc.	test lab: E-Beam Services, Inc.	test lab: E-Beam Services, Inc.	test lab: E-Beam Services, Inc.	test lab: E-Beam Services, Inc.	test lab: E-Beam Services, Inc.	test lab: E-Beam Services, Inc.	test lab: E-Beam Services, Inc.
POST EXPOSURE CONDITIONING								
Note	type: aging	type: aging	type: aging	type: aging	type: aging	type: aging	type: aging	type: aging
Time (hours)	168	1344	168	1344	168	1344	168	1344
PROPERTIES RETAINED (%)								
Tensile Strength @ Yield	92.7 (is)	92.7 (is)	61.8 (is)	87.3 (is)				
Elongation	90.5 (bt)	69.3 (bt)	95.6 (bt)	69 (bt)				
Flexural Strength	101.5 (co)	97 (co)	68.7 (co)	58.2 (co)				
Modulus	88.2 (hb)	79.2 (hb)	53.8 (hb)	49.3 (hb)				
Flexural Modulus	88.9 (cg)	91.4 (cg)	55.6 (cg)	53 (cg)				
Dart Impact (total energy)	105.7 (em)	107.1 (em)	101.4 (em)	97.1 (em)				
Dart Impact (peak energy)	122.7 (dg)	134.1 (dg)	88.6 (dg)	86.4 (dg)				
Notched Izod Impact	82.8 (ga)	150 (ga)	142.2 (ga)	125 (ga)				
Heat Deflection Temperature	100 (m)	100 (m)	100 (m)	100 (m)	100 (m)	100 (m)	100 (m)	100 (m)
Vicat Softening Point	101.8 (ku)	107.3 (ku)	96.4 (ku)	94.5 (ku)	104.8 (ku)	104.8 (ku)	102.4 (ku)	104.8 (ku)
SURFACE and APPEARANCE								
Δ Yellowness Index	46 (kw)	21.9 (kw)	18.7 (kw)	8.6 (kw)	-1.52 (kw)	-1.09 (kw)	7.37 (kw)	9.04 (kw)
Haze (%)	4.48	0.64	2.99	3.21	3.75	4.02	5.83	2.75
Transparency Retained (%)	93.8	102.5	102.5	119.8	114.1	125.6	111.5	124.4

Gamma Radiation Resistance

Alpha Gary: Plasticized PVC (characteristics: Shore A hardness: 80; features: transparent); **Plasticized PVC** (characteristics: Shore A hardness: 84; features: transparent)

Consecutive reprocessing of the extrusion compound produces no change in tensile or elongation properties until the fifth pass. Presumably the increase in tensile and decrease in elongation is the result of crosslinking some of the PVC chains. Degradation by gamma radiation produces some chain scission, negating the effects of the crosslinking. This is seen by the drop in tensile and increase in elongation of the irradiated sample. An enhanced stabilizer package was added to the most processed sample and this seemed to reduce the effect on physicals by radiation.

The molding compound did not show any effects due to crosslinking. The molding compound is better lubricated and has a higher level of secondary stabilizer; consequently, it does not undergo the degree of crosslinking because of less shear in the processing and thus less damage to the molecular structure.

Sterilization by gamma radiation will cause a substantial shift in the color of vinyl compounds from blue to yellow. Vinyl compounds can be stabilized against this kind of damage. Stabilization normally prevents severe mechanical or physical deterioration. Some degree of color shift still occurs. This shift is proportional to the intensity of radiolytic exposure.

Blue to yellow shift occurs in the compounds after ambient storage for several months or years. This shift, caused by gradual depletion of the "reserve stability", will be accelerated by thermal processing. Since a judiciously selected regrind concentration minimizes the presence of "no reserve" materials in the blend, our reprocessing experiments show minimal, essentially unpredictable, color shift from the virgin material to the fifth pass. Radiation exposure causes an appearance change that appears fairly constant at each stage of the processing chain.

The superior stability of the molding compound can be seen by the smaller delta values after exposure to 2.5 Mrads radiation. More severe exposure at 3.5 Mrads eliminates this advantage. Above 2.5 Mrads, incremental radiation increases produce significant color changes; at 3.5 Mrads, delta values are 2 to 3 times those at 2.5 Mrads. Repeated processing causes no deterioration in the ability of the compounds to resist color changes upon exposure to gamma radiation.

Reference: Rakus, J., Brantley, T., Szabo, E., *Environmental Concerns With Vinyl Medical Devices,* ANTEC 1991, conference proceedings - Society of Plastics Engineers, 1991.

Geon Company: Geon RX (features: transparent, medical grade)

Grades of flexible Geon Rx compounds are resistant to gamma radiation sterilization. Geon Rx transparent flexible gamma resistant compounds exhibit no visually apparent color change at typical exposure levels

Reference: *BF Goodrich Geon RX Medical Grade Molding Compounds,* supplier marketing literature (CIM-014F) - BFGoodrich Company, 1988.

Ethylene Oxide (EtO) Resistance

Geon Company: Geon RX (features: transparent, medical grade)

Flexible Geon Rx compounds are ethylene oxide sterilizable while retaining undiminished properties and clarity.

Reference: *BF Goodrich Geon RX Medical Grade Molding Compounds,* supplier marketing literature (CIM-014F) - BFGoodrich Company, 1988.

TABLE 174: Effect of Gamma Radiation Sterilization on Polyvinyl Chloride Polyol

Material Family	POLYVINYL CHLORIDE POLYOL			
Material Supplier/Name	COLORITE			
Material Note	Shore A hardness: 65; transparent	Shore A hardness: 65; transparent	Shore A hardness: 75; transparent	Shore A hardness: 75; transparent
Reference No.	32	32	32	32
EXPOSURE CONDITIONS				
Type	Gamma Radiation	Gamma Radiation	Gamma Radiation	Gamma Radiation
Radiation Dose (Mrads)	2.5	4	2.5	4
PROPERTIES RETAINED (%)				
Tensile Strength	102.5 (he)	104.4 (he)	102 (he)	107 (he)
Elongation	104.7 (ai)	110.8 (ai)	90.9 (ai)	84.7 (ai)
SURFACE and APPEARANCE				
Δb Color	0.31 (z)	0.41 (z)	0.47 (z)	0.93 (z)

TABLE 175: Effect of Gamma Radiation Sterilization on Polyvinyl Chloride Polyol

Material Family	POLYVINYL CHLORIDE POLYOL						
Material Supplier/Name	ALPHA GARY						
Material Note	Shore A hardness: 80; transparent	Shore A hardness: 80; transparent	Shore A hardness: 80; transparent	Shore A hardness: 80; transparent	Shore A hardness: 80; transparent	Shore A hardness: 80; transparent	Shore A hardness: 80; transparent
Reference No.	34	34	34	34	34	34	34
EXPOSURE CONDITIONS							
Type	Gamma Radiation	Gamma Radiation	Gamma Radiation	Gamma Radiation	Gamma Radiation	Gamma Radiation	Gamma Radiation
Radiation Dose (Mrads)	2.5	2.5	2.5	2.5	2.5	2.5	2.5
Note							enhanced stabilizer system used
PRE EXPOSURE CONDITIONING							
Preconditioning Note	type: reprocessing; regrind level: 20%	type: reprocessing; regrind level: 20%	type: reprocessing; regrind level: 20%	type: reprocessing; regrind level: 20%	type: reprocessing; regrind level: 20%	type: reprocessing; regrind level: 20%	type: reprocessing; regrind level: 20%
Times Reprocessed	1	2	3	4	5	6	6
PROPERTIES RETAINED (%)							
Tensile Strength	95.8 (he)	99.5 (he)	104.2 (he)	103.7 (he)	101.9 (he)	89.8 (he)	96 (he)
Elongation	90.6 (ai)	103.4 (ai)	100 (ai)	93.3 (ai)	100 (ai)	116 (ai)	147.4 (ai)
SURFACE and APPEARANCE							
ΔE Color	3.7 (aa)	3.5 (aa)	3.2 (aa)	3.1 (aa)	2.9 (aa)	4 (aa)	3.8 (aa)
Δb Color	3.6 (v)	3.4 (v)	3.2 (v)	3 (v)	2.8 (v)	3.9 (v)	3.8 (v)

TABLE 176: Effect of Gamma Radiation Sterilization on Polyvinyl Chloride Polyol

Material Family	POLYVINYL CHLORIDE POLYOL						
Material Supplier/Name	ALPHA GARY						
Material Note	Shore A hardness: 80; transparent	Shore A hardness: 80; transparent	Shore A hardness: 80; transparent	Shore A hardness: 80; transparent	Shore A hardness: 80; transparent	Shore A hardness: 80; transparent	Shore A hardness: 80; transparent
Reference No.	34	34	34	34	34	34	34

EXPOSURE CONDITIONS

Type	Gamma Radiation	Gamma Radiation	Gamma Radiation	Gamma Radiation	Gamma Radiation	Gamma Radiation	Gamma Radiation
Radiation Dose (Mrads)	3.5	3.5	3.5	3.5	3.5	3.5	3.5
Note							specimen note: enhanced stabilizer system used

PRE EXPOSURE CONDITIONING

Preconditioning Note	type: reprocessing; regrind level: 20%	type: reprocessing; regrind level: 20%	type: reprocessing; regrind level: 20%	type: reprocessing; regrind level: 20%	type: reprocessing; regrind level: 20%	type: reprocessing; regrind level: 20%	type: reprocessing; regrind level: 20%
Times Reprocessed	1	2	3	4	5	6	6

PROPERTIES RETAINED (%)

Tensile Strength	102.3 (he)	96.7 (he)	100 (he)	100.5 (he)	102.8 (he)	88.1 (he)	91.2 (he)
Elongation	87.5 (ai)	103.4 (ai)	100 (ai)	93.3 (ai)	86.2 (ai)	108 (ai)	136.8 (ai)

SURFACE and APPEARANCE

ΔE Color	6.3 (aa)	5.7 (aa)	5.6 (aa)	5.7 (aa)	6 (aa)	5.9 (aa)	6.6 (aa)
Δb Color	6.2 (v)	5.5 (v)	5.5 (v)	5.6 (v)	5.8 (v)	5.8 (v)	6.5 (v)

TABLE 177: Effect of Gamma Radiation Sterilization on Polyvinyl Chloride Polyol

Material Family	POLYVINYL CHLORIDE POLYOL					
Material Supplier/Name	ALPHA GARY					
Material Note	Shore A hardness: 84; transparent	Shore A hardness: 84; transparent	Shore A hardness: 84; transparent	Shore A hardness: 84; transparent	Shore A hardness: 84; transparent	Shore A hardness: 84; transparent
Reference No.	34	34	34	34	34	34
EXPOSURE CONDITIONS						
Type	Gamma Radiation	Gamma Radiation	Gamma Radiation	Gamma Radiation	Gamma Radiation	Gamma Radiation
Radiation Dose (Mrads)	2.5	2.5	2.5	2.5	2.5	2.5
PRE EXPOSURE CONDITIONING						
Preconditioning Note	type: reprocessing; regrind level: 20%	type: reprocessing; regrind level: 20%	type: reprocessing; regrind level: 20%	type: reprocessing; regrind level: 20%	type: reprocessing; regrind level: 20%	type: reprocessing; regrind level: 20%
Times Reprocessed	1	2	3	4	5	6
PROPERTIES RETAINED (%)						
Tensile Strength	106.1 (he)	91 (he)	107.8 (he)	90.2 (he)	97.2 (he)	96.3 (he)
Elongation	88.5 (ai)	104.2 (ai)	92.6 (ai)	100 (ai)	108.3 (ai)	96.3 (ai)
SURFACE and APPEARANCE						
ΔE Color	3.3 (aa)	2.1 (aa)	1.5 (aa)	1.5 (aa)	1.5 (aa)	1.6 (aa)
Δb Color	3.1 (v)	2 (v)	1.3 (v)	1.4 (v)	1.5 (v)	1.5 (v)

TABLE 178: Effect of Gamma Radiation Sterilization on Polyvinyl Chloride Polyol

Material Family	POLYVINYL CHLORIDE POLYOL					
Material Supplier/Name	ALPHA GARY					
Material Note	Shore A hardness: 84; transparent	Shore A hardness: 84; transparent	Shore A hardness: 84; transparent	Shore A hardness: 84; transparent	Shore A hardness: 84; transparent	Shore A hardness: 84; transparent
Reference No.	34	34	34	34	34	34

EXPOSURE CONDITIONS

Type	Gamma Radiation	Gamma Radiation	Gamma Radiation	Gamma Radiation	Gamma Radiation	Gamma Radiation
Radiation Dose (Mrads)	3.5	3.5	3.5	3.5	3.5	3.5

PRE EXPOSURE CONDITIONING

Preconditioning Note	type: reprocessing; regrind level: 20%	type: reprocessing; regrind level: 20%	type: reprocessing; regrind level: 20%	type: reprocessing; regrind level: 20%	type: reprocessing; regrind level: 20%	type: reprocessing; regrind level: 20%
Times Reprocessed	1	2	3	4	5	6

PROPERTIES RETAINED (%)

Tensile Strength	92.5 (he)	89.1 (he)	102.4 (he)	89.8 (he)	92.1 (he)	96.3 (he)
Elongation	96.2 (ai)	104.2 (ai)	92.6 (ai)	96.2 (ai)	108.3 (ai)	92.6 (ai)

SURFACE and APPEARANCE

ΔE Color	6.3 (aa)	5.4 (aa)	4.9 (aa)	5.2 (aa)	5.1 (aa)	5.1 (aa)
Δb Color	6.2 (v)	5.3 (v)	5.3 (v)	6.2 (v)	6.3 (v)	5.9 (v)

Electron Beam Radiation

Atochem: Kynar Flex 2800

The E-beam radiation resistance of fluoropolymers is reported to be highly dependent on the crosslinking versus scission ratio of the particular resin. Short term radiation exposure testing data for Kynar Flex 2800 shows very little change in mechanical properties.

Reference: Seller, David A., Robinson, Donald N., *Potential Applications For Kynar Flex PVDF In The Nuclear Industry,* supplier technical report - Elf Atochem North America.

Gamma Radiation Resistance

Dow Corning: SILASTIC Rx (form: tubing; features: medical grade)

Samples were exposed to 2.5 Mrads of cobalt 60 gamma radiation. Tensile strength was essentially unchanged. Minimal but measurable increases were seen in durometer and tensile modulus at 200% elongation. Tear resistance (Die B) and elongation decreased. These physical property changes are similar to those seen when the tubing materials are subjected to oven post cures.

Reference: *Applied Innovations In Silicone Tubing,* supplier marketing literature (51-769C-91) - Dow Corning Medical.

Ethylene Oxide (EtO) Resistance

Dow Corning: SILASTIC Rx (form: tubing; features: medical grade)

Ethylene oxide will not deteriorate Silastic RX medical grade tubing. However, sterilization by EtO is not recommended unless sufficient data is available regarding the time required to complete outgassing of residual EtO.

Reference: *Applied Innovations In Silicone Tubing,* supplier marketing literature (51-769C-91) - Dow Corning Medical.

Dow Corning: SILASTIC Rx (form: tubing; features: medical grade)

Some device makers want to avoid ethylene oxide sterilization because of the combination of residuals and tighter EPA requirements. If EtO sterilization is required, the permeability of silicone is an advantage: the sterilization process is simple and tends to leave lower levels of residuals than in other tubing materials.

Reference: *Silicone Tubing Meets Biocompatibility and Sterilization Demands in Devices,* Biomaterials News, supplier newsletter (Form # 51-849; Volume 5, Issue 4) - Dow Corning Medical, 1991.

Steam Resistance

Dow Corning: SILASTIC Rx (form: tubing; features: medical grade)

Samples were tested using the following 3 autoclave methods: 1) High speed instrument (flash) autoclave - Place on nonlinting cloth or sterilizing paper in a clean open tray. Sterilize 10 minutes at 132°C at 30 psi (2kg/cm^2). 2) Standard gravity autoclave - Wrap in nonlinting cloth or sterilizing paper and place in a clean, open tray. Sterilize 30 minutes at 121°C at 15 psi (1 kg/cm^2). 3) Prevacuum high-temperature autoclave - Wrap in nonlinting cloth or sterilizing paper and place in a clean, open tray. Sterilize at normal cycle (30 to m35 minutes) at 121°C. Samples did not deteriorate after being repeatedly steam sterilized in an autoclave 25 times. No change in physical properties was noted.

Reference: *Applied Innovations In Silicone Tubing,* supplier marketing literature (51-769C-91) - Dow Corning Medical.

Reagent	Reagent Note	Conc. (%)	Temp. (°C)	Time (days)	Load	PDL Rating	Resistance Note	Material Note
ABS								
Ammonia			23	7		8	resistant	Monsanto Lustran
		10	23	7		9	no noticeable chg. in appear.	Toray Toyolac 900; transp.
		28	23	7		8	slightly yellowed	"
		100	23	28		7	sl. clouding/discoloration, expected life: mths. to yrs.	Dow Magnum
		100	52	28		7	"	"
Benzaldehyde			23	7		2	plastic severely crazed, softened or dissolved	Monsanto Lustran
		100	23	28		2	severe attack, softened in few hrs.	Dow Magnum
		100	52	28		2	"	"
Benzoic Acid			23	7		8	resistant	Monsanto Lustran
			23	28		9	plastic was unaffected for duration of test	Dow Magnum
			52	28		9	"	"
Benzyl Alcohol			23	7		2	plastic severely crazed, softened or dissolved	Monsanto Lustran
		100	23	28		2	severe attack, softened in few hrs.	Dow Magnum
		100	52	28		2	"	"
Betadine			23		strained	8	resistant	Lustran ABS 248
Boric Acid			23	7		8	"	Monsanto Lustran
		100	23	28		9	plastic was unaffected for duration of test	Dow Magnum
		100	52	28		9	"	"
Butyl Alcohol			23	7		8	resistant	Monsanto Lustran
		100	23	28		9	plastic was unaffected for duration of test	Dow Magnum
		100	52	28		2	severe attack, softened in few hrs.	"
Carbolic Acid			23	7		2	plastic severely crazed, softened or dissolved	Monsanto Lustran
Chlorine		10	23	28		5	moder. effect; expected life: wks. to mths.	Dow Magnum
		10	23	28		2	severe attack, softened in few hrs.	"
		10	52	28		5	moder. effect; expected life: wks. to mths.	"
		10	52	28		2	severe attack, softened in few hrs.	"
		100	23	28		2	"	"
		100	52	28		2	"	"
Cresol			23	7		2	plastic severely crazed, softened or dissolved	Monsanto Lustran
		100	23	28		2	severe attack, softened in few hrs.	Dow Magnum
		100	52	28		2	"	"
Ethyl Acetate			23	7	<6.9 MPa	2	exp. stress is level where part stress cracks	GE Cycolac
			23	7		0	dissolved	Toray Toyolac 900; transp.
			23	7		2	partially dissolves	Toray Toyolac Parel TP10; antistatic
		85-88	23	28		0	solvent	Dow Magnum
		85-88	52	28		0	"	"
		90	23	7		2	plastic severely crazed, softened or dissolved	Monsanto Lustran
		100	23	28		2	severe attack, softened in few hrs.	Dow Magnum
		100	52	28		2	"	"
Ethyl Alcohol			23	7	<6.9 MPa	2	exp. stress is level where part stress cracks	GE Cycolac
			23	7		2	plastic severely crazed, softened or dissolved	Monsanto Lustran
			23	7		2	"	"
			23	7		7	slightly whitens	Toray Toyolac Parel TP10; antistatic
			23	7		3	softens, whitens	Toray Toyolac 900; transp.
			23	28		7	sl. clouding/discoloration, expected life: mths. to yrs.	Dow Magnum
			52	28		2	severe attack, softened in few hrs.	"
		40	23	28		9	plastic was unaffected for duration of test	"
		40	52	28		7	sl. clouding/discoloration, expected life: mths. to yrs.	"
		50	22	1	0.55%	4	exp. strain is level where part stress cracks	Monsanto Lustran; 6.4 mm samples
		50	23	7	6.9-13.8 MPa	4	exp. stress is level where part stress cracks	GE Cycolac
		50	23	7		9	no noticeable chg. in appear.	Toray Toyolac 900; transp.
		95	23	7	strained	8	appear. unchanged	GE Cycolac KJU

Reagent	Reagent Note	Conc. (%)	Temp. (°C)	Time (days)	Load	PDL Rating	Resistance Note	Material Note
ABS (continued)								
Ethyl Alcohol		95	23	7		2		GE Cycolac KJU
		100	22	1	0.4%	2	exp. strain is level where part stress cracks	Monsanto Lustran; 6.4 mm samples
Ethylene Chlorohydrin		100	23	28		5	moder. effect; expected life: wks. to mths.	Dow Magnum
		100	52	28		2	severe attack, softened in few hrs.	"
Ethylene Glycol			23		0.25%	8		LNP; 30% gl.fib.
			23			9		"
			23	7		9	no noticeable chg. in appear.	Toray Toyolac 900; transp.
			23	7		8	resistant	Monsanto Lustran
			23	7		6	slightly whitens	Toray Toyolac Parel TP10; antistatic
			81			6		LNP; 30% gl.fib.
		100	23	28		9	plastic was unaffected for duration of test	Dow Magnum
		100	52	28		7	sl. clouding/discoloration, expected life: mths. to yrs.	"
Ethylene Oxide		100	23	28		2	severe attack, softened in few hrs.	"
		100	52	28		2	"	"
Fluorine		100	23	28		2	"	"
		100	52	28		2	"	"
Formaldehyde		30	23	7		8	resistant	Monsanto Lustran
		30	23	28		9	plastic was unaffected for duration of test	Dow Magnum
		30	52	28		7	sl. clouding/discoloration, expected life: mths. to yrs.	"
		37	23	7		9	no noticeable chg. in appear.	Toray Toyolac 900; transp.
		37	23	28		3	consid. chg.; expected life: days	Dow Magnum
		37	52	28		2	severe attack, softened in few hrs.	"
Hydrogen Peroxide		1	23	28		9	plastic was unaffected for duration of test	"
		1	52	28		9	"	"
		3	23	7		9	no noticeable chg. in appear.	Toray Toyolac 900; transp.
		3	23	7		8	resistant	Monsanto Lustran
		3	23	28		7	sl. clouding/discoloration, expected life: mths. to yrs.	Dow Magnum
		3	52	28		5	moder. effect; expected life: wks. to mths.	"
		30	23	28		5	"	"
		30	52	28		5	"	"
		90	23	28		5	"	"
		90	52	28		5	"	"
Iodine			23	7		2	plastic severely crazed, softened or dissolved	Monsanto Lustran
		100	23	28		2	severe attack, softened in few hrs.	Dow Magnum
		100	52	28		2	"	"
Isopropyl Alcohol			23	7	<6.9 MPa	2	exp. stress is level where part stress cracks	GE Cycolac
			23	7		9	no noticeable chg. in appear.	Toray Toyolac 900; transp.
			23	7		8	resistant	Monsanto Lustran
		50	23	7	6.9-13.8 MPa	4	exp. stress is level where part stress cracks	GE Cycolac
		95		0.042		1		transp.
		100	23	28		9	plastic was unaffected for duration of test	Dow Magnum
		100	52	28		2	severe attack, softened in few hrs.	"
Mercuric Chloride		100	23	28		7	sl. clouding/discoloration, expected life: mths. to yrs.	"
		100	52	28		7	"	"
Methyl Isobutyl Ketone			23	7	<6.9 MPa	2	exp. stress is level where part stress cracks	GE Cycolac
			23	7		2	plastic severely crazed, softened or dissolved	Monsanto Lustran
		100	23	28		0	solvent	Dow Magnum
		100	52	28		0	"	"
Morpholine		100	23	28		2	severe attack, softened in few hrs.	"
		100	52	28		2	"	"
Phenol			23			2	not acceptable	
			23	7	<6.9 MPa	2	exp. stress is level where part stress cracks	GE Cycolac
			65			2	not acceptable	
		5	23	7		2	plastic severely crazed, softened or dissolved	Monsanto Lustran
		5	23	7		2	softens, whitens	Toray Toyolac 900; transp.
		5	23	7		2	slightly blackened, partially dissolved	Toray Toyolac Parel TP10; antistatic

ABS (continued)

Reagent	Reagent Note	Conc. (%)	Temp. (°C)	Time (days)	Load	PDL Rating	Resistance Note	Material Note
Phenol		5	23	28		2	severe attack, softened in few hrs.	Dow Magnum
		5	52	28		2	"	"
Potassium Permanganate			23	7		8	resistant	Monsanto Lustran
		100	23	28		7	sl. clouding/discoloration, expected life: mths. to yrs.	Dow Magnum
		100	52	28		5	moder. effect; expected life: wks. to mths.	"
Propylene Oxide		100	23	28		2	severe attack, softened in few hrs.	"
		100	52	28		2	"	"
Salicylaldehyde			23	7		2	plastic severely crazed, softened or dissolved	Monsanto Lustran
		100	23	28		0	solvent	Dow Magnum
		100	52	28		0	"	"
Salicylic Acid		100	23	28		9	plastic was unaffected for duration of test	"
		100	23	28		7	sl. clouding/discoloration, expected life: mths. to yrs.	"
		100	52	28		9	plastic was unaffected for duration of test	"
		100	52	28		7	sl. clouding/discoloration, expected life: mths. to yrs.	"
Saline Solutions		10	23	7		9	no noticeable chg. in appear.	Toray Toyolac 900; transp.
Silver Nitrate			23	28		7	sl. clouding/discoloration, expected life: mths. to yrs.	Dow Magnum
			52	28		5	moder. effect; expected life: wks. to mths.	"
Sodium Hypochlorite			23	28		7	sl. clouding/discoloration, expected life: mths. to yrs.	"
			52	28		5	moder. effect; expected life: wks. to mths.	"
		15	23	28		7	sl. clouding/discoloration, expected life: mths. to yrs.	"
		15	52	28		7	"	"
Sulfuric Acid		1	23	7		9	no noticeable chg. in appear.	Toray Toyolac 900; transp.
		10	23		0.25%	5		LNP; 30% gl.fib.
		10	23			8		"
		10	23	7	strained	8	appear. unchanged	GE Cycolac KJU
		10	23	7		7		"
		10	23	7		9	no noticeable chg. in appear.	Toray Toyolac 900; transp.
		10	23	28		7	sl. clouding/discoloration, expected life: mths. to yrs.	Dow Magnum
		10	52	28		7	"	"
Toothpaste			23	7	<6.9 MPa	2	exp. stress is level where part stress cracks	GE Cycolac

ABS Nylon Alloy

Reagent	Reagent Note	Conc. (%)	Temp. (°C)	Time (days)	Load	PDL Rating	Resistance Note	Material Note
Ethyl Alcohol			23	7	0.5%	7		GE Elemid RM1
			23	7	1%	6		"
			23	7	1.5%	5		"
			23	7		4		"

ABS PU Alloy

Reagent	Reagent Note	Conc. (%)	Temp. (°C)	Time (days)	Load	PDL Rating	Resistance Note	Material Note
Ethylene Glycol			23	1		9	exc. resistance	Dow Prevail 3150
Sulfuric Acid		5	23	1		9	"	"

Acetal Resin

Reagent	Reagent Note	Conc. (%)	Temp. (°C)	Time (days)	Load	PDL Rating	Resistance Note	Material Note
Ammonia			20	<360		8	no irreversible chg.	BASF Ultraform
			70	<360		8	"	"
		20	23	<360		8	"	"
Benzaldehyde			23	<360		8	"	"
Benzoic Acid		20	23	<360		6	changes in mass; dimen.; props.	"
Benzyl Alcohol			23	<360		8	no irreversible chg.	"
Bleach			23	<360		2	rapid attack, stress cracks, irreversible damage	"
			23	<360		2	"	"
Boric Acid			23	<360		8	no irreversible chg.	"
Butyl Alcohol			23	<360		8	"	"
Chlorine						1	unsatisf. resistance	DuPont Delrin
			23	<360		2	rapid attack, stress cracks, irreversible damage	BASF Ultraform
Ethyl Acetate			23	180		6	no visible chg.	Hoechst Cel. Celcon M90
			23	180		6	"	Hoechst Cel. Celcon M270
			23	180		6	"	Hoechst Cel. Celcon M25
			23	<360		8	no irreversible chg.	BASF Ultraform
			23	365		4	no visible chg.	Hoechst Cel. Celcon M270

Acetal Resin (continued)

Reagent	Reagent Note	Conc. (%)	Temp. (°C)	Time (days)	Load	PDL Rating	Resistance Note	Material Note
Ethyl Acetate			23	365		4	no visible chg.	Hoechst Cel. Celcon M90
			23	365		4	"	Hoechst Cel. Celcon M25
			49	180		4	"	Hoechst Cel. Celcon M90
			49	180		4	"	Hoechst Cel. Celcon M270
			49	180		4	"	Hoechst Cel. Celcon M25
		100	23	7	strained	8	no stress cracks observed	DuPont Delrin 500
		100	23	365		4		DuPont Delrin
		100	50	365		4		"
Ethyl Alcohol			23	<360		8	no irreversible chg.	BASF Ultraform
			23	365		6	completely suitable	Mitsubishi Iupital F20-02
		40	23	<360		8	no irreversible chg.	BASF Ultraform
		50	23	180		7	no visible chg.	Hoechst Cel. Celcon M25
		50	23	180		7	"	Hoechst Cel. Celcon M270
		50	23	180		7	"	Hoechst Cel. Celcon M90
		50	23	365		6	"	"
		50	23	365		6	"	Hoechst Cel. Celcon M25
		50	23	365		6	"	Hoechst Cel. Celcon M270
		50	49	180		5	"	Hoechst Cel. Celcon M90
		50	49	180		5	"	Hoechst Cel. Celcon M25
		50	49	180		5	"	Hoechst Cel. Celcon M270
		95	23	180		7	"	"
		95	23	180		7	"	Hoechst Cel. Celcon M25
		95	23	180		7	"	Hoechst Cel. Celcon M90
		95	23	365		6	"	"
		95	23	365		6	"	Hoechst Cel. Celcon M270
		95	23	365		6	"	Hoechst Cel. Celcon M25
		95	49	180		5	"	Hoechst Cel. Celcon M90
		95	49	180		5	"	Hoechst Cel. Celcon M270
		95	49	180		5	"	Hoechst Cel. Celcon M25
		99.5	23	90		8		Mitsubishi Iupital F20-02
		100	23	365		5		DuPont Delrin
		100	50	365		6		"
		100	60	10		6		"
Ethylene Glycol			23	<360		8	no irreversible chg.	BASF Ultraform
		50	82	180		7	sligtly discolored	Hoechst Cel. Celcon M25
		50	82	180		7	"	Hoechst Cel. Celcon M90
		50	82	180		7	"	Hoechst Cel. Celcon M270
Ethylene Oxide			23	<360		8	no irreversible chg.	BASF Ultraform
Fluorine			23	<360		2	rapid attack, stress cracks, irreversible damage	"
Formaldehyde		37	23	7	strained	8	no stress cracks observed	DuPont Delrin 500
Hydrogen Peroxide		0.5	23	<360		8	no irreversible chg.	BASF Ultraform
		0.5	23	365		9		DuPont Delrin
		3	23	90		9		Mitsubishi Iupital F20-02
		3	23	180		8	no visible chg.	Hoechst Cel. Celcon M25
		3	23	180		8	"	Hoechst Cel. Celcon M270
		3	23	180		8	"	Hoechst Cel. Celcon M90
		3	23	365		8	"	"
		3	23	365		8	"	Hoechst Cel. Celcon M270
		3	23	365		8	"	Hoechst Cel. Celcon M25
		30	23	<360		4	not resistant to prolonged contact	BASF Ultraform
		90	29	28		1	unsatisf. resistance	DuPont Delrin
		90	66	7		1	"	"
Isopropyl Alcohol			23	<360		8	no irreversible chg.	BASF Ultraform
			60	<360		8	"	"
Phenol			>43	<360		2	rapid attack, stress cracks, irreversible damage	"
		5	23	180		4	no visible chg.	Hoechst Cel. Celcon M25
		5	23	180		4	"	Hoechst Cel. Celcon M90
		5	23	180		4	"	Hoechst Cel. Celcon M270
		5	23	365		4	discolored	Hoechst Cel. Celcon M25
		5	23	365		4	"	Hoechst Cel. Celcon M90
		5	23	365		4	"	Hoechst Cel. Celcon M270
		70	23	<360		2	rapid attack, stress cracks, irreversible damage	BASF Ultraform
		88	23	<360		2	"	"
		100	60	90		1	unsatisf. resistance	DuPont Delrin
Potassium Permanganate		1	23	<360		8	no irreversible chg.	BASF Ultraform
		10	60	275		6		DuPont Delrin
			23	<360		2	"	"

Acetal Resin (continued)

Reagent	Reagent Note	Conc. (%)	Temp. (°C)	Time (days)	Load	PDL Rating	Resistance Note	Material Note
Sodium Hypochlorite			23	180		7	usable under some conditions	Mitsubishi Iupital F20-02
		4.6	23	180		1	not recommended for use	Hoechst Cel. Celcon M90
		4.6	23	180		1	"	Hoechst Cel. Celcon M25
		4.6	23	180		1	"	Hoechst Cel. Celcon M270
		5	23	30		1	unsatisf. resistance	DuPont Delrin
		10	23	<360		2	rapid attack, stress cracks, irreversible damage	BASF Ultraform
Sulfuric Acid		1	23	180		7	usable under some conditions	Mitsubishi Iupital F20-02
		1	35	316		1	unsatisf. resistance	DuPont Delrin
		2	23	<360		2	rapid attack, stress cracks, irreversible damage	BASF Ultraform
		3	23	90		9		Mitsubishi Iupital F20-02
		3	23	180		8	no visible chg.	Hoechst Cel. Celcon M90
		3	23	180		8	"	Hoechst Cel. Celcon M270
		3	23	180		8	"	Hoechst Cel. Celcon M25
		3	23	365		8	"	"
		3	23	365		8	"	Hoechst Cel. Celcon M90
		3	23	365		8	"	Hoechst Cel. Celcon M270

ASA

Reagent	Reagent Note	Conc. (%)	Temp. (°C)	Time (days)	Load	PDL Rating	Resistance Note	Material Note
Ammonia		25	20	<=365		9	resistant for yrs.	BASF Luran S 757 R
		25	50	<=365		7	resistant for mths.	"
Benzaldehyde			20	<=365		0	dissolved	"
			50	<=365		0	"	"
Benzoic Acid			20	<=365		9	resistant for yrs.	"
			50	<=365		9	"	"
Benzyl Alcohol			20	<=365		0	dissolved	"
			50	<=365		0	"	"
Bleach		12	20	<=365		9	resistant for yrs.	"
		12	50	<=365		9	"	"
Boric Acid			20	<=365		9	"	"
			50	<=365		9	"	"
Butyl Alcohol			20	<=365		7	resistant for mths.	"
			25	7	0.5%	2		GE Geloy 1120
			25	7	1%	2		"
			25	7		2		"
			50	<=365		5	resistant for wks.	BASF Luran S 757 R
Chlorine			20	<=365		3	resistant for hrs.	"
			20	<=365		3	"	"
			20	<=365		5	resistant for wks.	"
			50	<=365		3	resistant for hrs.	"
			50	<=365		3	"	"
Ethyl Acetate			20	<=365		0	dissolved	"
			50	<=365		0	"	"
Ethyl Alcohol		40	20	<=365		9	resistant for yrs.	"
		40	50	<=365		7	resistant for mths.	"
		95	20	<=365		7	"	"
		95	50	<=365		4	resistant for days	"
Ethylene Glycol			20	<=365		9	resistant for yrs.	"
			50	<=365		9	"	"
Formaldehyde		30	20	<=365		7	resistant for mths.	"
		30	50	<=365		5	resistant for wks.	"
Hydrogen Peroxide		3	20	<=365		9	resistant for yrs.	"
		3	50	<=365		9	"	"
Iodine			20	<=365		5	resistant for wks.	"
Isopropyl Alcohol			25	7	0.5%	5		GE Geloy 1120
			25	7	1%	4		"
			25	7		4		"
Mercuric Chloride			20	<=365		9	resistant for yrs.	BASF Luran S 757 R
			50	<=365		9	"	"
Methyl Isobutyl Ketone			20	<=365		0	dissolved	"
			50	<=365		0	"	"
Phenol			20	<=365		3	resistant for hrs.	"
			50	<=365		3	"	"
		10	20	<=365		4	resistant for days	"
		10	50	<=365		4	"	"
Potassium Permanganate			20	<=365		7	resistant for mths.	"
			50	<=365		4	resistant for days	"

ASA (continued)

Reagent	Reagent Note	Conc. (%)	Temp. (°C)	Time (days)	Load	PDL Rating	Resistance Note	Material Note
Propylene Oxide			20	<=365		3	resistant for hrs.	"
			50	<=365		3	"	"
Salicylic Acid			20	<=365		9	resistant for yrs.	"
			50	<=365		7	resistant for mths.	"
Silver Nitrate			20	<=365		7	"	"
			50	<=365		7	"	"
Sodium Hypochlorite			20	<=365		9	resistant for yrs.	"
			50	<=365		7	resistant for mths.	"
Sulfuric Acid		10	20	<=365		9	resistant for yrs.	"
		10	50	<=365		9	"	"
Thymol			20	<=365		3	resistant for hrs.	"
			50	<=365		3	"	"

Acrylic Resin

Reagent	Reagent Note	Conc. (%)	Temp. (°C)	Time (days)	Load	PDL Rating	Resistance Note	Material Note
Ammonia			20			8	resistant	Cyro Acrylite
			20			8	"	Cyro XT
			20			8	"	Cyro Cyrolite G20
			23			8	"	ICI Diakon
			23			2	unsatisf. resistance	"
		10				8	no staining tendency	Aristech Aristech I-300
		10				8	"	Aristech Aristech S-300
		10				8	"	Aristech Aristech S-300; covered samples
		10				8	"	Aristech Aristech I-300; covered samples
		10				8	unaffected per ANSI Z124.1-1971	Aristech Aristech I-300
Benzaldehyde			20			2	susceptible to attack	Cyro Acrylite
			20			3	"	Cyro XT
			20			3	"	Cyro Cyrolite G20
			23			2	unsatisf. resistance	ICI Diakon
Benzoic Acid			23			8	exc. resistance	
			65			8	"	
Benzyl Alcohol			23			2	unsatisf. resistance	ICI Diakon
Bleach						6	superficial stain only	Aristech Aristech I-300
			20			8	resistant	Cyro Cyrolite G20
			20			8	"	Cyro XT
			20			8	"	Cyro Acrylite
			25	0.6		6	cleaned w/ mild abrasive	Aristech Aristech I-300
		<=2	20			8	resistant	Cyro Acrylite
		<=2	20			8	"	Cyro Cyrolite G20
		<=2	20			8	"	Cyro XT
Boric Acid			23			8	exc. resistance	
			65			8	"	
Butyl Alcohol			20			5	limited resistance	Cyro Acrylite
			20			3	susceptible to attack	Cyro XT
			20			3	"	Cyro Cyrolite G20
			23			2	not recommended for use	
			23			2	unsatisf. resistance	ICI Diakon
			65			2	not recommended for use	
Carbolic Acid			20			3	susceptible to attack	Cyro XT
			20			3	"	Cyro Cyrolite G20
			20			2	"	Cyro Acrylite
Chlorine			20			5	limited resistance	"
			20			3	susceptible to attack	Cyro XT
			20			3	"	Cyro Cyrolite G20
			23			5	fair - limited use	
			23			6	good - minor attack	
			65			5	fair - limited use	
			65			2	not recommended for use	
		2	23			5	some attack, sl. reduction in mech. props.	ICI Diakon
Cresol			20			2	susceptible to attack	Cyro Acrylite
			20			3	"	Cyro XT
			20			3	"	Cyro Cyrolite G20
Ethyl Acetate						2	do not use as a cleaner	Aristech Acrysteel IGP
						4	stained	Aristech Aristech I-300
						5	"	Aristech Aristech S-300
						4	"	Aristech Aristech I-300; covered samples
						6	superficial stain only	Aristech Aristech I-300

Acrylic Resin (continued)

Reagent	Reagent Note	Conc. (%)	Temp. (°C)	Time (days)	Load	PDL Rating	Resistance Note	Material Note
Ethyl Acetate						5	stained	Aristech Aristech S-300; covered samples
			20			3	susceptible to attack	Cyro XT
			20			3	"	Cyro Cyrolite G20
			20			2	"	Cyro Acrylite
			23			2	unsatisf. resistance	ICI Diakon
			23	7		2	attacked	Rohm & Haas Kamax T-170
			23	7		2	"	Rohm & Haas Kamax T-240
			23	7		2	"	Rohm & Haas Kamax T-260
			23	7		2	"	Rohm & Haas Kamax T-150
			25			2	not recommended for use	Rohm & Haas Plexiglas V
			25	0.6		6	cleaned w/ mild abrasive	Aristech Aristech I-300
			25	7		0	dissolved	Aristech Aristech GPA
			25	7		4	rubbery, swollen	Aristech Aristech I-300
			25	7		0	dissolved	Rohm & Haas Plexiglas VMHF
			25	7		0	"	Rohm & Haas Plexiglas V052
			25	7		0	"	Rohm & Haas Plexiglas V920
			25	7		0	"	Rohm & Haas Plexiglas V825
			25	7		0	"	Rohm & Haas Plexiglas V044
			25	7		0	"	Rohm & Haas Plexiglas VS
			25	7		0	"	Rohm & Haas Plexiglas V045FH
			25	7		0	"	Rohm & Haas Plexiglas V045
			25	7		0	"	Rohm & Haas Plexiglas VM
			25	7		0	"	Rohm & Haas Plexiglas V811
			65			2	not recommended for use	
		100	23	7		0	completely dissolved	Novacor NAS 10
Ethyl Alcohol						8	no staining tendency	Aristech Aristech S-300; covered samples
						8	"	Aristech Aristech I-300
						8	"	Aristech Aristech I-300; covered samples
						8	"	Aristech Aristech S-300
						8	unaffected per ANSI Z124.1-1971	Aristech Aristech I-300
			20			2	susceptible to attack	Cyro Acrylite
			23			2	not recommended for use	
			23			2	unsatisf. resistance	ICI Diakon
			23	7		7	no chg. in appear.	Rohm & Haas Kamax T-170
			23	7		8	"	Rohm & Haas Kamax T-260
			23	7		8	"	Rohm & Haas Kamax T-240
			23	7		8	"	Rohm & Haas Kamax T-150
			65			2	not recommended for use	
		10	23			6	not recommended for prolonged contact	ICI Diakon
		<=15	20			8	resistant	Cyro Acrylite
		<=15	20			8	"	Cyro XT
		<=15	20			8	"	Cyro Cyrolite G20
		>15	20			3	susceptible to attack	Cyro XT
		>15	20			3	"	Cyro Cyrolite G20
		50	23			5	some attack, sl. reduction in mech. props.	ICI Diakon
		50	23	7		4		Cyro Acrylite H-12
		50	23	7		3		Cyro Acrylite M-30
		50	23	7		4		Cyro Acrylite H-15
		50	23	7		9	no visible chg.	Novacor NAS 10
		50	23	28		9	"	"
		50	25			9	exc. resistance	Rohm & Haas Plexiglas V
		50	25	7		5		Rohm & Haas Plexiglas V045
		50	25	7		2		Rohm & Haas Plexiglas VM
		50	25	7		5		Rohm & Haas Plexiglas V825
		50	25	7		5		Rohm & Haas Plexiglas V044
		50	25	7		4		Rohm & Haas Plexiglas V920
		50	25	7		2		Rohm & Haas Plexiglas VMHF
		50	25	7		5		Rohm & Haas Plexiglas V052
		50	25	7		1		Rohm & Haas Plexiglas VS
		50	25	7		5		Rohm & Haas Plexiglas V045FH
		50	25	7		5		Rohm & Haas Plexiglas V811
		50	25	7		7		Aristech Aristech GPA
		50	25	7		5		Aristech Aristech I-300
		50	>25			9	exc. resistance	Rohm & Haas Plexiglas V
		95	23	7		9	no visible chg.	Novacor NAS 10
		95	23	28		8	"	"
		95	25			7	good resistance	Rohm & Haas Plexiglas V

Reagent	Reagent Note	Conc. (%)	Temp. (°C)	Time (days)	Load	PDL Rating	Resistance Note	Material Note
Acrylic Resin (continued)								
Ethyl Alcohol		95	25	7		2	swelled	Rohm & Haas Plexiglas VS
		95	25	7		2	“	Rohm & Haas Plexiglas VM
		95	25	7		2	“	Rohm & Haas Plexiglas VMHF
		95	25	7		3		Rohm & Haas Plexiglas V811
		95	25	7		3		Rohm & Haas Plexiglas V052
		95	25	7		3		Rohm & Haas Plexiglas V045FH
		95	25	7		3		Rohm & Haas Plexiglas V825
		95	25	7		1		Rohm & Haas Plexiglas V920
		95	25	7		3		Rohm & Haas Plexiglas V045
		95	25	7		3		Rohm & Haas Plexiglas V044
		95	25	7		6		Aristech Aristech I-300
		95	25	7		6		Aristech Aristech GPA
		95	>25			5	fair resistance	Rohm & Haas Plexiglas V
Ethylene Glycol			20			6	limited resistance	Cyro Cyrolite G20
			20			6	“	Cyro XT
			23			8	resistant	ICI Diakon
			23	0.0035	0.5%	8	no effect	Rohm & Haas Kamax T-170
			23	0.0035	0.5%	8	“	Rohm & Haas Kamax T-150
			23	0.0035	0.5%	8	“	Rohm & Haas Kamax T-260
			23	0.0035	0.5%	8	“	Rohm & Haas Kamax T-240
Ethylene Oxide			23			8	resistant	ICI Diakon
			23			5	some attack, sl. reduction in mech. props.	“
		100	54	.17-.25		9		gamma grade
		100	54	.17-.25		9		toughened
		100	54	.17-.25		9		gen. purp. grade
Formaldehyde		37	23			8	exc. resistance	
		37	65			6	good - minor attack	
		40	23			8	resistant	ICI Diakon
Hydrogen Peroxide						6	superficial stain only	Aristech Aristech I-300
			23			8	resistant	ICI Diakon
			25	0.6		6	cleaned w/ mild abrasive	Aristech Aristech I-300
		3				8	no staining tendency	Aristech Aristech I-300; covered samples
		3				8	“	Aristech Aristech S-300; covered samples
		3				8	“	Aristech Aristech S-300
		3				8	“	Aristech Aristech I-300
		3	20			8	suitable for packaging	Cyro XT; bottles
		3	23	7		9	no visible chg.	Novacor NAS 10
		3	23	28		9	“	“
		3	25			9	exc. resistance	Rohm & Haas Plexiglas V
		3	25	7		8		Rohm & Haas Plexiglas VS
		3	25	7		8		Rohm & Haas Plexiglas VMHF
		3	25	7		8		Rohm & Haas Plexiglas V045
		3	25	7		8		Rohm & Haas Plexiglas V044
		3	25	7		8		Rohm & Haas Plexiglas V045FH
		3	25	7		8		Rohm & Haas Plexiglas V825
		3	25	7		8		Rohm & Haas Plexiglas V811
		3	25	7		8		Rohm & Haas Plexiglas V920
		3	25	7		8		Rohm & Haas Plexiglas V052
		3	25	7		8		Rohm & Haas Plexiglas VM
		3	25	7		9		Aristech Aristech GPA
		3	25	7		8		Aristech Aristech I-300
		3	>25			9	exc. resistance	Rohm & Haas Plexiglas V
		28	23	7		9	no visible chg.	Novacor NAS 10
		28	23	28		9	“	“
		28	25			9	exc. resistance	Rohm & Haas Plexiglas V
		28	25	7		7		Rohm & Haas Plexiglas V811
		28	25	7		7		Rohm & Haas Plexiglas VS
		28	25	7		7		Rohm & Haas Plexiglas V825
		28	25	7		7		Rohm & Haas Plexiglas V920
		28	25	7		7		Rohm & Haas Plexiglas V045
		28	25	7		7		Rohm & Haas Plexiglas V044
		28	25	7		7		Rohm & Haas Plexiglas V052
		28	25	7		7		Rohm & Haas Plexiglas VMHF
		28	25	7		7		Rohm & Haas Plexiglas V045FH
		28	25	7		7		Rohm & Haas Plexiglas VM
		28	>25			9	exc. resistance	Rohm & Haas Plexiglas V
		<=40	20			6	limited resistance	Cyro Cyrolite G20
		<=40	20			6	“	Cyro XT

Acrylic Resin (continued)

Reagent	Reagent Note	Conc. (%)	Temp. (°C)	Time (days)	Load	PDL Rating	Resistance Note	Material Note
Hydrogen Peroxide		<=40	20			8	resistant	Cyro Acrylite
		90	23			2	unsatisf. resistance	ICI Diakon
Iodine						6	superficial stain only	Aristech Aristech I-300
		1	25	0.6		6	cleaned w/ mild abrasive	"
		5	20			3	susceptible to attack	Cyro XT
		5	20			3	"	Cyro Cyrolite G20
		5	20			2	"	Cyro Acrylite
Isopropyl Alcohol						8	acceptable as a cleaner	Aristech Acrysteel IGP
						8	unaffected per ANSI Z124.1-1971	Aristech Aristech I-300
			20			5	limited resistance	Cyro Acrylite
			20			3	susceptible to attack	Cyro Cyrolite G20
			20			3	"	Cyro XT
			23			5	some attack, sl. reduction in mech. props.	ICI Diakon
			23	0.0035	0.5%	8	no effect	Rohm & Haas Kamax T-150
			23	0.0035	0.5%	8	"	Rohm & Haas Kamax T-260
			23	0.0035	0.5%	8	"	Rohm & Haas Kamax T-240
			23	0.0035	0.5%	8	"	Rohm & Haas Kamax T-170
		10	23			5	some attack, sl. reduction in mech. props.	ICI Diakon
		50	23			5	"	"
		95		0.01		0	sample broke after 15 minutes	impact modified
		99	25	7		9		Aristech Aristech I-300
		99	25	7		9		Aristech Aristech GPA
		100	23	7		9	no visible chg.	Novacor NAS 10
		100	23	28		9	"	"
Phenol			20			3	susceptible to attack	Cyro XT
			20			3	"	Cyro Cyrolite G20
			20			2	"	Cyro Acrylite
			23			2	unsatisf. resistance	ICI Diakon
			65			2	not recommended for use	
		3				6	superficial stain only	Aristech Aristech I-300
		3	25	0.6		6	cleaned w/ mild abrasive	"
		5				4	stained	Aristech Aristech I-300; covered samples
		5				5	"	Aristech Aristech S-300; covered samples
		5				5	"	Aristech Aristech S-300
		5				4	"	Aristech Aristech I-300
		5	23	7		9	no visible chg.	Novacor NAS 10
		5	23	28		9	"	"
		5	25			2	not recommended for use	Rohm & Haas Plexiglas V
		5	25	7		2	attacked	Rohm & Haas Plexiglas V044
		5	25	7		2	"	Rohm & Haas Plexiglas V825
		5	25	7		2	"	Rohm & Haas Plexiglas V920
		5	25	7		2	"	Rohm & Haas Plexiglas VS
		5	25	7		2	"	Rohm & Haas Plexiglas VMHF
		5	25	7		2	"	Rohm & Haas Plexiglas V811
		5	25	7		2	"	Rohm & Haas Plexiglas V045
		5	25	7		2	"	Rohm & Haas Plexiglas V052
		5	25	7		2	"	Rohm & Haas Plexiglas V045FH
		5	25	7		2	"	Rohm & Haas Plexiglas VM
		5	25	7		2	attacked, colored	Aristech Aristech I-300
		5	25	7		2	"	Aristech Aristech GPA
		5	>25			2	not recommended for use	Rohm & Haas Plexiglas V
Potassium Permanganate			23			6	good - minor attack	
			23			8	resistant	ICI Diakon
			65			6	good - minor attack	
Sodium Hypochlorite						8	no staining tendency	Aristech Aristech I-300; covered samples
						8	"	Aristech Aristech S-300; covered samples
						8	"	Aristech Aristech S-300
						8	"	Aristech Aristech I-300
			23			8	resistant	ICI Diakon
		5	23	7		9	no visible chg.	Novacor NAS 10
		5	23	28		9	"	"
		5	25			9	exc. resistance	Rohm & Haas Plexiglas V
		5	25	7		9		Rohm & Haas Plexiglas V045FH
		5	25	7		9		Rohm & Haas Plexiglas V811
		5	25	7		9		Rohm & Haas Plexiglas V045
		5	25	7		9		Rohm & Haas Plexiglas V920

Reagent	Reagent Note	Conc. (%)	Temp. (°C)	Time (days)	Load	PDL Rating	Resistance Note	Material Note

Acrylic Resin (continued)

Reagent	Reagent Note	Conc. (%)	Temp. (°C)	Time (days)	Load	PDL Rating	Resistance Note	Material Note
Sodium Hypochlorite		5	25	7		9		Rohm & Haas Plexiglas VS
		5	25	7		9		Rohm & Haas Plexiglas V825
		5	25	7		9		Rohm & Haas Plexiglas VM
		5	25	7		9		Rohm & Haas Plexiglas VMHF
		5	25	7		9		Rohm & Haas Plexiglas V052
		5	25	7		9		Rohm & Haas Plexiglas V044
		5	>25			9	exc. resistance	Rohm & Haas Plexiglas V
		6	20			8	suitable for packaging	Cyro XT; bottles
Sulfuric Acid		3	23	7		9	no visible chg.	Novacor NAS 10
		3	23	7		8		Cyro Acrylite H-12
		3	23	7		8		Cyro Acrylite H-15
		3	23	7		8		Cyro Acrylite M-30
		3	23	28		9	no visible chg.	Novacor NAS 10
		3	25			9	exc. resistance	Rohm & Haas Plexiglas V
		3	25	7		9		Rohm & Haas Plexiglas V044
		3	25	7		8		Rohm & Haas Plexiglas VMHF
		3	25	7		8		Rohm & Haas Plexiglas VS
		3	25	7		9		Rohm & Haas Plexiglas V045FH
		3	25	7		8		Rohm & Haas Plexiglas VM
		3	25	7		9		Rohm & Haas Plexiglas V045
		3	25	7		9		Rohm & Haas Plexiglas V811
		3	25	7		8		Rohm & Haas Plexiglas V920
		3	25	7		9		Rohm & Haas Plexiglas V825
		3	25	7		9		Rohm & Haas Plexiglas V052
		3	25	7		9		Aristech Aristech GPA
		3	25	7		9		Aristech Aristech I-300
		3	>25			9	exc. resistance	Rohm & Haas Plexiglas V
		10	23			8	"	
		10	23			8	resistant	ICI Diakon
		10	65			6	good - minor attack	
		<=30	20			8	resistant	Cyro Acrylite
		<=30	20			8	"	Cyro XT
		<=30	20			8	"	Cyro Cyrolite G20

Acrylic Copolymer / Terpolymer

Reagent	Reagent Note	Conc. (%)	Temp. (°C)	Time (days)	Load	PDL Rating	Resistance Note	Material Note
Ethylene Oxide		100	54	.17-.25		7		Novacor NAS 30
		100	54	.17-.25		9		Novacor Zylar 90
Isopropyl Alcohol		95		0.042		9		Zylar 94-568
		95		0.042		9		Zylar ST 94-560
		95		0.042		9		Zylar ST 94-562
		95		0.042		9		Zylar ST 94-561
Toothpaste						8		Zylar ST 94-560
						6		Zylar 93-546

Acrylonitrile Copolymer

Reagent	Reagent Note	Conc. (%)	Temp. (°C)	Time (days)	Load	PDL Rating	Resistance Note	Material Note
Benzaldehyde			23	<112		2	test failed in < 16 weeks; softened; top concave	BP Chem. Barex 210
			38	<112		2	test failed in < 16 weeks; frost; softened	"
Bleach		5.25	23	365		5	satisf. at 1 yr.; good appear.	"
		5.25	38	112-140		3	satisf. for 16-20 wks.; cloudy; round bottom	"
Butyl Alcohol			23	365		8	satisf. at 1 yr.; good appear.	"
			38	365		6	"	"
Cresol			23	56		2	test failed; softened; cracking	"
			38	56		2	"	"
Ethyl Acetate			23	365		7	satisf. at 1 yr.; good appear.	"
			38	112-140		4	satisf. for 16-20 wks.; frost; softened	"
Ethyl Alcohol			23	365		6	satisf. at 1 yr.; good appear.	"
			23	365		8	satisf. at 1 yr.; frost	"
			38	365		4	satisf. at 1 yr.; very sl. frost; round bottom	"
			38	365		5	satisf. at 1 yr.; good appear.	"
Formaldehyde		37	23	365		7	"	"
		37	38	112-140		4	satisf. for 16-20 wks.; frost	"
Hydrogen Peroxide		6.3	23	<112		2	test failed in < 16 weeks; frost; rounded	"
		6.3	38	<112		2	"	"
Iodine			38	<112		3	test failed in < 16 weeks; discolored; opaque; round bottom	"

Reagent	Reagent Note	Conc. (%)	Temp. (°C)	Time (days)	Load	PDL Rating	Resistance Note	Material Note

Acrylonitrile Copolymer (continued)

Reagent	Reagent Note	Conc. (%)	Temp. (°C)	Time (days)	Load	PDL Rating	Resistance Note	Material Note
Isopropyl Alcohol			23	365		9	satisf. at 1 yr.	BP Chem. Barex 210
			38	365		3	satisf. at 1 yr.; good appear.; round bottom	"
Methyl Isobutyl Ketone			23	365		7	satisf. at 1 yr.; good appear.	"
			38	365		7	"	"
Phenol			23	21		3	top concave; softened	"
Sodium Hypochlorite		5.25	23	365		5	satisf. at 1 yr.; good appear.	"
		5.25	38	112-140		3	satisf. for 16-20 wks.; cloudy; round bottom	"
Sulfuric Acid		10	23	365		5	satisf. at 1 yr.; good appear.	"
		10	38	365		4	"	"

Aromatic Polyamide

Reagent	Reagent Note	Conc. (%)	Temp. (°C)	Time (days)	Load	PDL Rating	Resistance Note	Material Note
Ethyl Alcohol			21	41.7		9		DuPont Kevlar 29
Sulfuric Acid		10	21	4.2		8		"
		10	21	41.7		1		"

Cellulose Acetate

Reagent	Reagent Note	Conc. (%)	Temp. (°C)	Time (days)	Load	PDL Rating	Resistance Note	Material Note
Blood			23	7		9	appear. unchanged	Eastman Tenite Acetate
Butyl Alcohol			23	2		7	"	"
Carbolic Acid		5	23	7		0	decomposed, generally unsatisf.	"
Chlorine			23	7		3	crazed, brittle, unsatisf.	"
			23	7		4	"	"
			23	7		2	soft, swollen, unsatisf.	"
Ethyl Acetate			23	2		0	dissolved, generally unsatisf.	"
Ethyl Alcohol			23	2		1	soft, swollen, unsatisf.	"
		50	23	7		2	"	"
Ethylene Glycol			23	365		5	slightly softened	"
Ethylene Oxide			23	1		1	soft, swollen, unsatisf.	"
Formaldehyde		35	23	7		3	"	"
Hydrogen Peroxide		3	23	365		3	soft, bleached, unsatisf.	"
		5	23	2		5	slightly bleached	"
Iodine			23	2		5	stained brown, generally unsatisf.	"
Isopropyl Alcohol			23	2		1	swelled, unsatisf.	"
Mercuric Chloride		5	23	2		5	appear. unchanged	"
Methyl Isobutyl Ketone			23	2		1	swelled, surface attack, unsatisf.	"
Phenol		5	23	7		0	decomposed, generally unsatisf.	"
Potassium Permanganate			23	2		5	stained black	"
Silver Nitrate		3	23	2		6	slightly softened	"
Sodium Hypochlorite		5	23	2		5	soft, surface attack, unsatisf.	"
Sulfuric Acid		3	23	30		3	soft, unsatisf.	"
		10	23	365		0	decomposed, generally unsatisf.	"
Toothpaste			38	4		7	appear. unchanged	"

Cellulose Acetate Butyrate

Reagent	Reagent Note	Conc. (%)	Temp. (°C)	Time (days)	Load	PDL Rating	Resistance Note	Material Note
Benzaldehyde			23			0	dissolved, generally unsatisf.	Eastman Tenite Butyrate
Blood			23	7		9	appear. unchanged	"
Boric Acid		5	23	2		7	"	"
Butyl Alcohol			23	2		1	swelled, unsatisf.	"
Carbolic Acid		5	23	7		0	decomposed, generally unsatisf.	"
Chlorine			23	7		2	crazed, brittle, unsatisf.	"
			23	7		4	"	"
			23	7		2	soft, swollen, unsatisf.	"
Ethyl Acetate			23			0	dissolved, generally unsatisf.	"
Ethyl Alcohol			23	2		2	soft, unsatisf.	"
		50	23	7		2	"	"
Ethylene Glycol			23	365		5	appear. unchanged	"
Ethylene Oxide			23	1		1	swelled, soft, unsatisf.	"
			41	0.007		9	appear. unchanged	"
Formaldehyde		4	23	0.035		9	"	Eastman Tenite Butyrate; exposed 5 min./day for 5 days
		35	23	60		1	swelled, soft, unsatisf.	Eastman Tenite Butyrate
Hydrogen Peroxide		3	23	365		6	appear. unchanged	"
		5	23	2		7	"	"
Isopropyl Alcohol			23	2		1	soft, tacky, unsatisf.	"
Methyl Isobutyl Ketone			23			0	dissolved, generally unsatisf.	"
Phenol		5	23	7		0	decomposed, generally unsatisf.	"
Silver Nitrate		2.5	23	2		7	appear. unchanged	"

Reagent	Reagent Note	Conc. (%)	Temp. (°C)	Time (days)	Load	PDL Rating	Resistance Note	Material Note
Cellulose Acetate Butyrate (continued)								
Sodium Hypochlorite		30	23	13		6	appear. unchanged	Eastman Tenite Butyrate
Sulfuric Acid		3	23	365		6	sl. discolor.	"
		10	23	365		7	"	"
Cellulose Propionate								
Ethyl Acetate			23	365		0	dissolved, generally unsatisf.	Eastman Tenite Propionate
Ethyl Alcohol		50	23	365		2	soft, swollen, unsatisf.	"
		95	23	365		2	"	"
Ethylene Glycol			23	365		9	appear. unchanged	"
Formaldehyde		35	23	365		2	soft, swollen, unsatisf.	"
Hydrogen Peroxide		3	23	365		9	appear. unchanged	"
Phenol		5	23	365		0	disintegrated, generally unsatisf.	"
Sulfuric Acid		3	23	365		9	appear. unchanged	"
Ethylene Vinyl Alcohol Copolymer								
Ammonia		10				6	no chg. in appear.	Nippon Goh. Soarnol
Benzaldehyde			20	30		9		Eval Eval F-Series
			20	30		9		Eval Eval E-Series
			20	365		9		Eval Eval F-Series
			20	365		9		Eval Eval E-Series
Benzyl Alcohol			30			9	insoluble in reagent	Eval Eval; 0, 62.9, 73.4, 100 mol% VA contents
			30			1	soluble in reagent	Eval Eval; 40 mol% VA content
			100			9	insoluble in reagent	Eval Eval; 0, 73.4, 100 mol% VA contents
			100			5	partially soluble in reagent	Eval Eval; 62.9 mol% VA content
			100			1	soluble in reagent	Eval Eval; 40 mol% VA content
			150			0	insoluble in reagent	Eval Eval; 0, 73.4, 100 mol% VA contents
			150			5	partially soluble in reagent	Eval Eval; 62.9 mol% VA content
			150			1	soluble in reagent	Eval Eval; 40 mol% VA content
Butyl Alcohol			30			9	insoluble in reagent	Eval Eval; 0, 62.9, 73.4, 100 mol% VA contents
			30			5	partially soluble in reagent	Eval Eval; 40 mol% VA content
			100			9	insoluble in reagent	Eval Eval; 0, 62.9, 73.4, 100 mol% VA contents
			100			5	partially soluble in reagent	Eval Eval; 40 mol% VA content
			150			9	insoluble in reagent	Eval Eval; 0, 62.9, 73.4, 100 mol% VA contents
			150			1	soluble in reagent	Eval Eval; 40 mol% VA content
Carbolic Acid		5				3	becomes transpar.	Nippon Goh. Soarnol
Ethyl Acetate						9	no chg. in appear.	"
			20	30		9		Eval Eval F-Series
			20	30		9		Eval Eval E-Series
			20	365		9		Eval Eval F-Series
			20	365		9		Eval Eval E-Series
			30			9	insoluble in reagent	Eval Eval; 0, 40, 62.9, 73.4, 100 mol% VA contents
			100			9	"	Eval Eval; 40, 62.9, 73.4, 100 mol% VA contents
			100			5	partially soluble in reagent	Eval Eval; 0 mol% VA content
			150			9	insoluble in reagent	Eval Eval; 40, 62.9, 73.4, 100 mol% VA contents
			150			5	partially soluble in reagent	Eval Eval; 0 mol% VA content
Ethyl Alcohol						8	no chg. in appear.	Nippon Goh. Soarnol
			20	30		6		Eval Eval F-Series
			20	30		3		Eval Eval E-Series
			20	365		1		"
			20	365		4		Eval Eval F-Series
			30			9	insoluble in reagent	Eval Eval; 0, 40, 62.9, 73.4, 100 mol% VA contents
			100			9	"	Eval Eval; 0, 73.4, 100 mol% VA contents
			100			5	partially soluble in reagent	Eval Eval; 40, 62.9 mol% VA contents
			150			9	insoluble in reagent	Eval Eval; 0, 100 mol% VA contents
			150			5	partially soluble in reagent	Eval Eval; 40, 62.9, 73.4 mol% VA contents
Ethylene Chlorohydrin			30			9	insoluble in reagent	Eval Eval; 0, 100 mol% VA contents
			30			1	soluble in reagent	Eval Eval; 40, 62.9, 73.4 mol% VA contents
			100			9	insoluble in reagent	Eval Eval; 0, 100 mol% VA contents
			100			1	soluble in reagent	Eval Eval; 40, 62.9, 73.4 mol% VA contents

Reagent	Reagent Note	Conc. (%)	Temp. (°C)	Time (days)	Load	PDL Rating	Resistance Note	Material Note
Ethylene Vinyl Alcohol Copolymer (continued)								
Ethylene Chlorohydrin			150			9	insoluble in reagent	Eval Eval; 0, 100 mol% VA contents
			150			1	soluble in reagent	Eval Eval; 40, 62.9, 73.4 mol% VA contents
Ethylene Glycol			20	30		6		Eval Eval E-Series
			20	30		7		Eval Eval F-Series
			20	365		4		Eval Eval E-Series
			20	365		6		Eval Eval F-Series
			30			9	insoluble in reagent	Eval Eval; 0, 40, 62.9, 73.4, 100 mol% VA contents
			100			9	"	Eval Eval; 0 mol% VA content
			100			5	partially soluble in reagent	Eval Eval; 40, 73.4, 100 mol% VA contents
			100			1	soluble in reagent	Eval Eval; 62.9 mol% VA content
			150			9	insoluble in reagent	Eval Eval; 0 mol% VA content
			150			5	partially soluble in reagent	Eval Eval; 100 mol% VA content
			150			1	soluble in reagent	Eval Eval; 40, 62.9, 73.4 mol% VA contents
Hydrogen Peroxide		3				6	no chg. in appear.	Nippon Goh. Soarnol
Isopropyl Alcohol						9	"	"
			30			9	insoluble in reagent	Eval Eval; 0, 40, 62.9, 73.4, 100 mol% VA contents
			100			9	"	Eval Eval; 0, 62.9, 73.4, 100 mol% VA contents
			100			5	partially soluble in reagent	Eval Eval; 40 mol% VA content
			150			9	insoluble in reagent	Eval Eval; 0, 62.9, 73.4, 100 mol% VA contents
			150			1	soluble in reagent	Eval Eval; 40 mol% VA content
Methyl Isobutyl Ketone			30			9	insoluble in reagent	Eval Eval; 0, 40, 62.9, 73.4, 100 mol% VA contents
			100			9	"	Eval Eval; 62.9, 73.4, 100 mol% VA contents
			100			5	partially soluble in reagent	Eval Eval; 0, 40 mol% VA contents
			150			9	insoluble in reagent	Eval Eval; 62.9, 73.4, 100 mol% VA contents
			150			5	partially soluble in reagent	Eval Eval; 0, 40 mol% VA contents
Phenol			30			9	insoluble in reagent	Eval Eval; 0 mol% VA content
			30			1	soluble in reagent	Eval Eval; 40, 62.9, 73.4, 100 mol% VA contents
			100			9	insoluble in reagent	Eval Eval; 0 mol% VA content
			100			1	soluble in reagent	Eval Eval; 40, 62.9, 73.4, 100 mol% VA contents
			150			5	partially soluble in reagent	Eval Eval; 0 mol% VA content
			150			1	soluble in reagent	Eval Eval; 40, 62.9, 73.4, 100 mol% VA contents
Sulfuric Acid		3				7	no chg. in appear.	Nippon Goh. Soarnol
Ethylene Chlorotrifluoroethylene Copolymer								
Ammonia			23			8	recommended for use	Ausimont Halar
			149			8	"	"
		10	23			8	"	"
		10	121			8	"	"
Benzaldehyde			23	11		8	no stress cracks observed	"
			121	11		3	"	"
		10	23			8	recommended for use	"
		10	66			8	"	"
		10	121			2	not recommended for use	"
		>10	23			8	recommended for use	"
		>10	66			2	not recommended for use	"
Benzoic Acid			23			8	recommended for use	"
			121			8	"	"
Benzyl Alcohol			23			8	"	"
			149			8	"	"
Bleach			23			8	"	"
			23			8	"	"
			149			8	"	"
			149			8	"	"
Boric Acid			23			8	"	"
			149			8	"	"
Butyl Alcohol			23			8	"	"
			23			8	"	"
			23			8	"	"
			23	11		8	no stress cracks observed	"
			118	11		5	"	"

Ethylene Chlorotrifluoroethylene Copolymer (continued)

Reagent	Reagent Note	Conc. (%)	Temp. (°C)	Time (days)	Load	PDL Rating	Resistance Note	Material Note
Butyl Alcohol			149			8	recommended for use	Ausimont Halar
			149			8	"	"
			149			8	"	"
Chlorine			23			8	"	"
			23			8	"	"
			23			8	"	"
			23			8	"	"
			66			8	"	"
			121			2	not recommended for use	"
			121			8	recommended for use	"
			121			8	"	"
			121			8	"	"
Cresol			23			8	"	"
			66			8	"	"
			121			2	not recommended for use	"
Ethyl Acetate			23			8	recommended for use	"
			23	11		7	no stress cracks observed	"
			66			8	recommended for use	"
			71	11		3	no stress cracks observed	"
Ethyl Alcohol			23			8	recommended for use	"
			149			8	"	"
Ethylene Chlorohydrin			23			8	"	"
			66			2	not recommended for use	"
Ethylene Glycol			23			8	recommended for use	"
			23	7	0.25%	9		LNP; 20% gl.fib.
			23	7	0.25%	1		LNP; 20% carb.fib.
			23	7		9		"
			23	7		9		LNP; 20% gl.fib.
			82	3	0.25%	8		"
			82	3	0.25%	1		LNP; 20% carb.fib.
			82	3		7		"
			82	3		9		LNP; 20% gl.fib.
			149			8	recommended for use	Ausimont Halar
			149	1	0.25%	8		LNP; 20% gl.fib.
			149	1	0.25%	1		LNP; 20% carb.fib.
			149	1		8		LNP; 20% gl.fib.
			149	1		5		LNP; 20% carb.fib.
Ethylene Oxide			23			8	recommended for use	Ausimont Halar
			149			8	"	"
Fluorine			23			8	"	"
Formaldehyde		35	23			8	"	"
		35	66			8	"	"
		37	23			8	"	"
		37	66			8	"	"
		50	23			8	"	"
Hydrogen Peroxide			23			8	"	"
		50	23			8	"	"
		50	66			8	"	"
		90	23			8	"	"
		90	66			8	"	"
Iodine			23			8	"	"
			121			8	"	"
		10	23			8	"	"
		10	121			8	"	"
Isopropyl Alcohol			23			8	"	"
			149			8	"	"
Mercuric Chloride			23			8	"	"
			121			8	"	"
Methyl Isobutyl Ketone			23			8	"	"
			66			8	"	"
			121			2	not recommended for use	"
Phenol			23			8	recommended for use	"
			66			8	"	"
			121			2	not recommended for use	"
Potassium Permanganate		10	23			8	recommended for use	"
		10	149			8	"	"
		25	23			8	"	"
		25	149			8	"	"

Ethylene Chlorotrifluoroethylene Copolymer (continued)

Reagent	Reagent Note	Conc. (%)	Temp. (°C)	Time (days)	Load	PDL Rating	Resistance Note	Material Note
Propylene Oxide			23			2	not recommended for use	Ausimont Halar
			23	11		3	no stress cracks observed	"
Salicylaldehyde			23			8	recommended for use	"
			66			2	not recommended for use	"
Salicylic Acid			23			8	recommended for use	"
			66			8	"	"
Silver Nitrate			23			8	"	"
			149			8	"	"
Sodium Hypochlorite			23			8	"	"
			149			8	"	"
Sulfuric Acid		10	23			8	"	"
		10	23	7	0.25%	4		LNP; 20% gl.fib.
		10	23	7	0.25%	3		LNP; 20% carb.fib.
		10	23	7		6		"
		10	23	7		6		LNP; 20% gl.fib.
		10	82	3	0.25%	2		LNP; 20% carb.fib.
		10	82	3	0.25%	3		LNP; 20% gl.fib.
		10	82	3		4		"
		10	82	3		6		LNP; 20% carb.fib.
		10	121			8	recommended for use	Ausimont Halar
		10	149	1	0.25%	1		LNP; 20% carb.fib.
		10	149	1	0.25%	1		LNP; 20% gl.fib.
		10	149	1		1		"
		10	149	1		2		LNP; 20% carb.fib.

Ethylene Tetrafluoroethylene Copolymer

Reagent	Reagent Note	Conc. (%)	Temp. (°C)	Time (days)	Load	PDL Rating	Resistance Note	Material Note
Ammonia			150			8	exp. temp. is max. service temp.	DuPont Tefzel
		30	110			8	"	"
Benzaldehyde			100			8	"	"
Benzoic Acid			135			8	"	"
Benzyl Alcohol			120	7		9		"
			150			8	exp. temp. is max. service temp.	"
Bleach			100			8	"	"
			100			8	"	"
Boric Acid			150			8	"	"
Butyl Alcohol			150			8	"	"
Chlorine			100			8	"	"
			120			8	"	"
			120	7		4		"
Cresol			135			8	exp. temp. is max. service temp.	"
Ethyl Acetate			65			8	"	"
			77	7		6		"
Ethyl Alcohol			65			8	exp. temp. is max. service temp.	"
			150			8	"	"
Ethylene Chlorohydrin			65			8	"	"
Ethylene Glycol			23	7	0.25%	9		LNP; 20% carb.fib.
			23	7	0.25%	9		LNP; 20% gl.fib.
			23	7		9		LNP; 20% carb.fib.
			23	7		9		LNP; 20% gl.fib.
			82	3	0.25%	7		LNP; 20% carb.fib.
			82	3	0.25%	9		LNP; 20% gl.fib.
			82	3		9		"
			82	3		9		LNP; 20% carb.fib.
			149	1	0.25%	8		LNP; 20% gl.fib.
			149	1	0.25%	6		LNP; 20% carb.fib.
			149	1		9		LNP; 20% gl.fib.
			149	1		9		LNP; 20% carb.fib.
			150			8	exp. temp. is max. service temp.	DuPont Tefzel
Ethylene Oxide			110			8	"	"
Fluorine			40			8	"	"
Formaldehyde		37	110			8	"	"
Hydrogen Peroxide		30	23	7		9		"
		30	120			8	exp. temp. is max. service temp.	"
		90	65			8	"	"
Iodine			110			8	"	"
			110			8	"	"
Mercuric Chloride			135			8	"	"
Methyl Isobutyl Ketone			110			8	"	"
Morpholine			65			8	"	"

Ethylene Tetrafluoroethylene Copolymer (continued)

Reagent	Reagent Note	Conc. (%)	Temp. (°C)	Time (days)	Load	PDL Rating	Resistance Note	Material Note
Phenol			100			8	exp. temp. is max. service temp.	DuPont Tefzel
			100			8	"	"
		10	110			8	"	"
Potassium Permanganate			150			8	"	"
Propylene Oxide			65			8	"	"
Salicylaldehyde			100			8	"	"
Salicylic Acid			120			8	"	"
Silver Nitrate			150			8	"	"
Sodium Hypochlorite			150			8	"	"
Sulfuric Acid		10	23	7	0.25%	9		LNP; 20% carb.fib.
		10	23	7	0.25%	6		LNP; 20% gl.fib.
		10	23	7		9		LNP; 20% carb.fib.
		10	23	7		7		LNP; 20% gl.fib.
		10	82	3	0.25%	5		"
		10	82	3	0.25%	9		LNP; 20% carb.fib.
		10	82	3		5		LNP; 20% gl.fib.
		10	82	3		9		LNP; 20% carb.fib.
		10	149	1	0.25%	6		"
		10	149	1	0.25%	4		LNP; 20% gl.fib.
		10	149	1		4		"
		10	149	1		9		LNP; 20% carb.fib.

Fluorinated Ethylene Propylene Copolymer

Reagent	Reagent Note	Conc. (%)	Temp. (°C)	Time (days)	Load	PDL Rating	Resistance Note	Material Note
Ammonia						8	no visible attack up to b.p.	DuPont Teflon FEP
Benzaldehyde			179	7		9		DuPont Teflon FEP 160
Benzyl Alcohol						8	no visible attack up to b.p.	DuPont Teflon FEP
			204	7		9		DuPont Teflon FEP 160
Boric Acid						8	no visible attack up to b.p.	DuPont Teflon FEP
Chlorine						8	"	"
			120	7		8		DuPont Teflon FEP 160
Ethyl Acetate						8	no visible attack up to b.p.	DuPont Teflon FEP
			25	365		9		"
			50	365		8		"
			70	14		8		"
Ethyl Alcohol						8	no visible attack up to b.p.	"
		95	25	365		9		"
		95	50	365		9		"
		95	70	14		9		"
		95	200	0.33		9		"
Ethylene Glycol						8	no visible attack up to b.p.	"
			23	7	0.25%	9		LNP; 15% carb.fib.
			23	7	0.25%	9		LNP; 20% gl.fib.
			23	7		9		LNP; 15% carb.fib.
			23	7		9		LNP; 20% gl.fib.
			82	3	0.25%	7		"
			82	3	0.25%	5		LNP; 15% carb.fib.
			82	3		4		"
			82	3		8		LNP; 20% gl.fib.
			149	1	0.25%	4		LNP; 15% carb.fib.
			149	1	0.25%	7		LNP; 20% gl.fib.
			149	1		4		LNP; 15% carb.fib.
			149	1		8		LNP; 20% gl.fib.
Formaldehyde						8	no visible attack up to b.p.	DuPont Teflon FEP
Hydrogen Peroxide						8	"	"
Phenol						8	"	"
Potassium Permanganate						8	"	"
Sodium Hypochlorite						8	"	"
Sulfuric Acid		10	23	7	0.25%	6		LNP; 20% gl.fib.
		10	23	7	0.25%	3		LNP; 15% carb.fib.
		10	23	7		8		LNP; 20% gl.fib.
		10	23	7		9		LNP; 15% carb.fib.
		10	82	3	0.25%	3		"
		10	82	3	0.25%	5		LNP; 20% gl.fib.
		10	82	3		7		"
		10	82	3		9		LNP; 15% carb.fib.
		10	149	1	0.25%	3		"
		10	149	1	0.25%	5		LNP; 20% gl.fib.
		10	149	1		7		"
		10	149	1		9		LNP; 15% carb.fib.

Reagent	Reagent Note	Conc. (%)	Temp. (°C)	Time (days)	Load	PDL Rating	Resistance Note	Material Note
Perfluoroalkoxy Resin								
Ammonia						8	no visible attack up to b.p.	DuPont Teflon PFA
Benzaldehyde			179			9		"
			179	7		9		DuPont Teflon PFA 350
			179	7		9		DuPont Teflon PFA
Benzyl Alcohol						8	no visible attack up to b.p.	"
			204			9		"
			204	7		9		DuPont Teflon PFA 350
			205	7		9		DuPont Teflon PFA
Boric Acid						8	no visible attack up to b.p.	"
Chlorine						8	"	"
			120			8		"
			120	7		8		DuPont Teflon PFA 350
			120	7		9		DuPont Teflon PFA
Ethyl Acetate						8	no visible attack up to b.p.	"
			25	365		9		"
			50	365		8		"
			70	14		8		"
Ethyl Alcohol						8	no visible attack up to b.p.	"
		95	25	365		9		"
		95	50	365		9		"
		95	70	14		9		"
		95	100	0.33		9		"
		95	200	0.33		9		"
Ethylene Glycol						8	no visible attack up to b.p.	"
			23	7	0.25%	7		LNP; 20% mil.gl.
			23	7	0.25%	8		LNP; 20% carb.fib.
			23	7	0.25%	8		LNP; 20% gl.fib.
			23	7		7		LNP; 20% mil.gl.
			23	7		9		LNP; 20% carb.fib.
			23	7		9		LNP; 20% gl.fib.
			82	3	0.25%	6		LNP; 20% mil.gl.
			82	3	0.25%	7		LNP; 20% gl.fib.
			82	3	0.25%	6		LNP; 20% carb.fib.
			82	3		8		"
			82	3		8		LNP; 20% gl.fib.
			82	3		6		LNP; 20% mil.gl.
			149	1	0.25%	4		LNP; 20% carb.fib.
			149	1	0.25%	5		LNP; 20% gl.fib.
			149	1	0.25%	3		LNP; 20% mil.gl.
			149	1		3		"
			149	1		7		LNP; 20% carb.fib.
			149	1		6		LNP; 20% gl.fib.
Formaldehyde						8	no visible attack up to b.p.	DuPont Teflon PFA
Hydrogen Peroxide						8	"	"
		30	23	7		9		"
Phenol						8	no visible attack up to b.p.	"
Potassium Permanganate						8	"	"
Sodium Hypochlorite						8	"	"
Sulfuric Acid		10	23	7	0.25%	9		LNP; 20% gl.fib.
		10	23	7	0.25%	9		LNP; 20% mil.gl.
		10	23	7	0.25%	3		LNP; 20% carb.fib.
		10	23	7		7		"
		10	23	7		9		LNP; 20% gl.fib.
		10	23	7		9		LNP; 20% mil.gl.
		10	82	3	0.25%	8		LNP; 20% gl.fib.
		10	82	3	0.25%	3		LNP; 20% carb.fib.
		10	82	3	0.25%	4		LNP; 20% mil.gl.
		10	82	3		9		LNP; 20% carb.fib.
		10	82	3		5		LNP; 20% mil.gl.
		10	82	3		8		LNP; 20% gl.fib.
		10	149	1	0.25%	4		"
		10	149	1	0.25%	3		LNP; 20% carb.fib.
		10	149	1	0.25%	3		LNP; 20% mil.gl.
		10	149	1		5		LNP; 20% gl.fib.
		10	149	1		9		LNP; 20% carb.fib.
		10	149	1		3		LNP; 20% mil.gl.

Reagent	Reagent Note	Conc. (%)	Temp. (°C)	Time (days)	Load	PDL Rating	Resistance Note	Material Note
Polychlorotrifluoroethylene								
Ammonia			25	7		9	more crystalline forms have more chem. res.	3M Kel-F 81
Benzaldehyde			23	14		9	no visible chg.	Allied Sig. Aclar 22; film
			23	14		9	"	Allied Sig. Aclar 28; film
			23	14		9	"	Allied Sig. Aclar 33; film
			25	7		9	more crystalline forms have more chem. res.	3M Kel-F 81
Benzoic Acid			90	7		9	"	"
Benzyl Alcohol			25	7		9	"	"
Butyl Alcohol			23	14		9	no visible chg.	Allied Sig. Aclar 33; film
			25	7		9	more crystalline forms have more chem. res.	3M Kel-F 81
			70	7		9	"	"
			117	1		8	"	"
Chlorine						1	tends to plasticize the film	Allied Sig. Aclar 22; film
			-40	0.083		9	more crystalline forms have more chem. res.	3M Kel-F 81
			25	6		1	"	"
			25	60		9	"	"
			50	6		1	"	"
Cresol			25	7		9	"	"
			140	7		5	"	"
Ethyl Acetate			23	14		2	extremely flexible	Allied Sig. Aclar 28; film
			23	14		2	"	Allied Sig. Aclar 22; film
			23	14		3	very flexible	Allied Sig. Aclar 33; film
			25	7		6	more crystalline forms have more chem. res.	3M Kel-F 81
			25	30		2	"	"
			70	7		1	"	"
			77	1		2	"	"
Ethyl Alcohol			23	14		9	no visible chg.	Allied Sig. Aclar 28; film
			23	14		9	"	Allied Sig. Aclar 33; film
			23	14		9	"	Allied Sig. Aclar 22; film
			78	1		9	more crystalline forms have more chem. res.	3M Kel-F 81
			80	7		9	"	"
		50	25	7		2	"	"
		95	25	7		9	"	"
		95	135	7		9	"	"
Ethylene Glycol			175	7		9	"	"
Ethylene Oxide			23	14		3	clouded, extremely flexible	Allied Sig. Aclar 22; film
			23	14		3	"	Allied Sig. Aclar 28; film
			23	14		3	very flexible	Allied Sig. Aclar 33; film
			25			3	swelling	3M Kel-F 81
Fluorine			85	14		9	more crystalline forms have more chem. res.	"
Formaldehyde			135	7		8	"	"
Hydrogen Peroxide		3	25	7		9	"	"
		30	23	14		8	clouded	Allied Sig. Aclar 28; film
		30	23	14		8	"	Allied Sig. Aclar 22; film
		30	23	14		9	no visible chg.	Allied Sig. Aclar 33; film
		30	25	7		9	more crystalline forms have more chem. res.	3M Kel-F 81
		30	25	30		9	"	"
Mercuric Chloride			175	7		2	"	"
Phenol		5	70	7		9	"	"
Potassium Permanganate			25	30		9	"	"
Salicylic Acid			175	7		9	"	"
Sodium Hypochlorite			23	14		9	no visible chg.	Allied Sig. Aclar 22 & 28; film
			23	14		9	"	Allied Sig. Aclar 33; film
Sulfuric Acid		3	25	7		9	more crystalline forms have more chem. res.	3M Kel-F 81
Polytetrafluoroethylene								
Ammonia						8	no visible attack up to b.p.	DuPont Teflon PTFE
Benzyl Alcohol						8	"	"
Boric Acid						8	"	"
Chlorine						8	"	"
Ethyl Acetate						8	"	"
			25	365		9		"
			50	365		8		"
			70	14		8		"

Reagent	Reagent Note	Conc. (%)	Temp. (°C)	Time (days)	Load	PDL Rating	Resistance Note	Material Note
Polytetrafluoroethylene (continued)								
Ethyl Alcohol						8	no visible attack up to b.p.	DuPont Teflon PTFE
			50	28		9	resistant	ICI Fluon VG15; 15% gl.fib.
			50	28		9	"	ICI Fluon VX1; 30% gl.fib./misc.
			50	28		9	"	ICI Fluon VG25; 25% gl.fib.
		95	25	365		9		DuPont Teflon PTFE
		95	50	365		9		"
		95	70	14		9		"
		95	100	0.33		9		"
		95	200	0.33		9		"
Ethylene Glycol	up to b.p.					8	no visible attack up to b.p.	"
Formaldehyde	"					8	"	"
Hydrogen Peroxide	"					8	"	"
Phenol	"					8	"	"
			50	28		9	resistant	ICI Fluon VX1; 30% gl.fib./misc.
			50	28		9	"	ICI Fluon VG25; 25% gl.fib.
			50	28		9	"	ICI Fluon VG15; 15% gl.fib.
Potassium Permanganate	up to b.p.					8	no visible attack up to b.p.	DuPont Teflon PTFE
Sodium Hypochlorite	"					8	"	"
Polyvinylidene Fluoride								
Ammonia	gas		23			2	not recommended for use	Atochem Kynar
	liquid		23			2	"	"
	gas		150	7		7	appear. unchanged	Atochem Foraflon
Benzaldehyde			21			8	exp. temp. is max. service temp.	Atochem Kynar
			25	7		2	not recommended for use	Atochem Foraflon
Benzoic Acid			107			8	exp. temp. is max. service temp.	Atochem Kynar
	sat'd		125	7		7	appear. unchanged	Atochem Foraflon
Benzyl Alcohol			121			8	exp. temp. is max. service temp.	Atochem Kynar
Bleach	bleaching liquid; 100 degrees chloro		90	15	strained	2	not recommended for use	Atochem Foraflon
	"		90	15		8	resistant	"
	bleaching liquid; 48 degrees chloro		90	90	strained	2	not recommended for use	"
	48 degrees chloro		90	90		8	resistant	"
	bleaching liquid; 48 degrees chloro		130	90	strained	2	not recommended for use	"
	"		130	90		5	questionable, poss. use without pressure and stresses	"
	bleaching agents		135			8	exp. temp. is max. service temp.	Atochem Kynar
Boric Acid			135			8	"	"
			150	7		7	appear. unchanged	Atochem Foraflon
Butyl Alcohol			75	7		7	"	"
			107			8	exp. temp. is max. service temp.	Atochem Kynar
Chlorine	liquid		93			8	"	"
	gas		93			8	"	"
	dry		100	7		7	appear. unchanged	Atochem Foraflon
	moist		100	7		7	"	"
	chlorine water		107			8	exp. temp. is max. service temp.	Atochem Kynar
Cresol			66			8	"	"
			75	7		7	appear. unchanged	Atochem Foraflon
Ethyl Acetate			23			2	not recommended for use	Atochem Kynar
			25	7		7	appear. unchanged	Atochem Foraflon
Ethyl Alcohol				7		2	not recommended for use	"
	aq. sol'n or liquid		23			2	"	Atochem Kynar
			100	7		7	appear. unchanged	Atochem Foraflon
	aq. sol'n or liquid		141			8	exp. temp. is max. service temp.	Atochem Kynar
	in alcoholic spirits	40	93			8	"	"
Ethylene Chlorohydrin			24			8	"	"
Ethylene Glycol			23	7	0.25%	6		LNP; 15% carb.fib.
			23	7		7		"
			82	3	0.25%	3		"
			82	3		6		"
			130	365	strained	8	resistant	Atochem Foraflon
			130	365		8	"	"
	aq. sol'n or liquid		141			8	exp. temp. is max. service temp.	Atochem Kynar
			149	1	0.25%	1		LNP; 15% carb.fib.
			149	1		1		"
			150	7		7	appear. unchanged	Atochem Foraflon
Ethylene Oxide			50	7		7	"	"
			93			8	exp. temp. is max. service temp.	Atochem Kynar

Reagent	Reagent Note	Conc. (%)	Temp. (°C)	Time (days)	Load	PDL Rating	Resistance Note	Material Note
Polyvinylidene Fluoride (continued)								
Fluorine			24			8	exp. temp. is max. service temp.	Atochem Kynar
			25	7		7	appear. unchanged	Atochem Foraflon
Formaldehyde		30	50	7		7	"	"
		37	52			8	exp. temp. is max. service temp.	Atochem Kynar
Hydrogen Peroxide		<=30	93			8	"	"
		50	100	7		7	appear. unchanged	Atochem Foraflon
		90	21			8	exp. temp. is max. service temp.	Atochem Kynar
Iodine	gas		60			0	"	"
	dry		75	7		7	appear. unchanged	Atochem Foraflon
	moist		75	7		7	"	"
	in non-aq. solvent	10	66			8	exp. temp. is max. service temp.	Atochem Kynar
Isopropyl Alcohol	aq. sol'n or liquid; isopropanol		60			8	"	"
Mercuric Chloride			121			8	"	"
Methyl Isobutyl Ketone			23			2	not recommended for use	"
Morpholine				7		2	"	Atochem Foraflon
	aq. sol'n or liquid		24			8	exp. temp. is max. service temp.	Atochem Kynar
Phenol			50	7		7	appear. unchanged	Atochem Foraflon
			52			8	exp. temp. is max. service temp.	Atochem Kynar
	chlorinated		66			8	"	"
	"		66			8	"	"
	in water	5	79			8	"	"
		10	75	7		7	appear. unchanged	Atochem Foraflon
		10	90	365	strained	8	resistant	"
		10	90	365		8	"	"
Potassium Permanganate	aq. sol'n or solid		121			8	exp. temp. is max. service temp.	Atochem Kynar
Propylene Oxide				7		2	not recommended for use	Atochem Foraflon
			23			2	"	Atochem Kynar
Salicylaldehyde			52			8	exp. temp. is max. service temp.	"
Salicylic Acid			93			8	"	"
		50	50	7		7	appear. unchanged	Atochem Foraflon
Silver Nitrate	aq. sol'n or solid		141			8	exp. temp. is max. service temp.	Atochem Kynar
Sodium Hypochlorite		<=5	135			8	"	"
		6-15	93			8	"	"
Sulfuric Acid		10	23	7	0.25%	6		LNP; 15% carb.fib.
		10	23	7		8		"
		10	82	3	0.25%	6		"
		10	82	3		7		"
		10	149	1	0.25%	2		"
		10	149	1		7		"
	aq. sol'n	<=60	121			8	exp. temp. is max. service temp.	Atochem Kynar
Ionomer								
Ammonia	liquid		21			8	resistant	DuPont Surlyn; thickness >20 mils
	dry gas		21			8	"	"
	"		60			8	"	"
	liquid		60			8	"	"
Benzaldehyde			21			5	depends on use conditions, poss. stress cracks	"
			60			2	not recommended for use	"
Benzoic Acid			21			8	resistant	"
			60			8	"	"
Boric Acid			21			8	"	"
			60			8	"	"
Butyl Alcohol			21			4	depends on use conditions, stress cracks	"
			60			2	not recommended for use	"
Chlorine	liquid		21			2	"	"
	moist gas		21			2	"	"
	dry gas		21			2	"	"
	moist gas		60			2	"	"
	dry gas		60			2	"	"
	liquid		60			2	"	"
	chlorine water	2	21			6	resistant, poss. stress cracks	"
	"	2	60			5	depends on use conditions, poss. stress cracks	"
Cresol			21			2	not recommended for use	"
			60			2	"	"
Ethyl Acetate			21			6	resistant, poss. plasticizer	"
			60			2	not recommended for use	"

Reagent	Reagent Note	Conc. (%)	Temp. (°C)	Time (days)	Load	PDL Rating	Resistance Note	Material Note
Ionomer (continued)								
Ethyl Alcohol			21			4	depends on use conditions, stress cracks	DuPont Surlyn; thickness >20 mils
			60			2	not recommended for use	"
Ethylene Chlorohydrin			21			2	"	"
			60			2	"	"
Ethylene Glycol			21			4	depends on use conditions, stress cracks	"
			60			2	not recommended for use	"
Ethylene Oxide			21			8	resistant	"
			60			6	resistance depends on use conditions	"
Fluorine	wet gas		21			2	not recommended for use	"
	"		60			2	"	"
Formaldehyde			21			6	resistant, poss. stress cracks	"
			60			5	depends on use conditions, poss. stress cracks	"
Hydrogen Peroxide		3	21			8	resistant	"
		3	60			6	resistance depends on use conditions	"
Iodine			21			2	not recommended for use	"
			60			2	"	"
Isopropyl Alcohol			21			4	depends on use conditions, stress cracks	"
			60			2	not recommended for use	"
Mercuric Chloride			21			8	resistant	"
			60			8	"	"
Methyl Isobutyl Ketone			21			6	resistant, poss. plasticizer	"
			60			5	depends on use conditions, plasticizer	"
Phenol			21			2	not recommended for use	"
			60			2	"	"
Potassium Permanganate		20	21			8	resistant	"
		20	60			8	"	"
Salicylic Acid			21			8	"	"
			60			8	"	"
Silver Nitrate	sol'n		21			8	"	"
	"		60			8	"	"
Sodium Hypochlorite			21			6	resistant, poss. stress cracks	"
			60			5	depends on use conditions, poss. stress cracks	"
Modified Polyphenylene Ether								
Ammonia			26	7	41.4 MPa	8	no effect	GE Noryl GFN3; 30% gl.fib.
			26	7		8	"	"
Benzaldehyde			93	1		0		GE Noryl
Butyl Alcohol			23	2	19.6 MPa	8	no chg. in appear.	Mitsubishi Iupiace AH60
			23	2	19.6 MPa	8	"	Mitsubishi Iupiace AV60
			23	7	0.5%	9		GE GTX 901
			23	7	0.5%	9		GE GTX 910
			23	7	1%	9		GE GTX 901
			23	7	1%	9		GE GTX 910
			23	7		9		GE GTX 901
			23	7		9		GE GTX 910
			93	1		6		GE Noryl
Ethyl Acetate			23	2	19.6 MPa	3	creases/cracks appear	Mitsubishi Iupiace AV60
			23	2	19.6 MPa	3	"	Mitsubishi Iupiace AH60
			23	2		3	"	Mitsubishi Iupiace AV60
			23	2		3	"	Mitsubishi Iupiace AH60
			23	7	0.5%	3	incompatible, not recommended	GE Noryl 731
			23	7	1%	3	"	"
			23	7		8	compatible	"
			85	3	0.5%	3	incompatible, not recommended	"
			85	3	1%	3	"	"
			85	3		8	compatible	"
			93	1		0		GE Noryl
Ethyl Alcohol			23	2	19.6 MPa	8	no chg. in appear.	Mitsubishi Iupiace AH60
			23	2	19.6 MPa	8	"	Mitsubishi Iupiace AV60
			23	2		8	"	"
			23	2		8	"	Mitsubishi Iupiace AH60
			23	7	0.5%	8	compatible	GE Noryl 731
			23	7	0.5%	3	incompatible, not recommended	GE Noryl SE100

Reagent	Reagent Note	Conc. (%)	Temp. (°C)	Time (days)	Load	PDL Rating	Resistance Note	Material Note
Modified Polyphenylene Ether (continued)								
Ethyl Alcohol			23	7	1%	3	incompatible, not recommended	GE Noryl SE100
			23	7		8	compatible	"
			85	3	0.5%	3	incompatible, not recommended	GE Noryl 731
			85	3	1%	3	"	GE Noryl SE100
			85	3		8	compatible	GE Noryl 731
			85	3		8	"	GE Noryl SE100
		95	23	7	strained	8	appear. unchanged	GE Noryl N190
		95	23	7		9		"
Ethylene Glycol			23	2	19.6 MPa	8	no chg. in appear.	Mitsubishi Iupiace AV60
			23	2	19.6 MPa	8	"	Mitsubishi Iupiace AH60
			23	2		8	"	Mitsubishi Iupiace AV60
			23	2		8	"	Mitsubishi Iupiace AH60
			23	7	0.25%	9		LNP; 30% gl.fib.
			23	7		9		"
			26	7	24.8 MPa	8	no effect	GE Noryl 731
			26	7	24.8 MPa	8	"	GE Noryl SE1
			26	7	41.4 MPa	8	"	GE Noryl GFN3; 30% gl.fib.
			26	7		8	"	GE Noryl 731
			26	7		8	"	GE Noryl GFN3; 30% gl.fib.
			26	7		8	"	GE Noryl SE1
			81	3		9		LNP; 30% gl.fib.
			85	3	24.8 MPa	3	attacked	GE Noryl 731
			85	3	24.8 MPa	3	"	GE Noryl SE1
			85	3	41.4 MPa	3	"	GE Noryl GFN3; 30% gl.fib.
			85	3		8	no effect	GE Noryl SE1
			85	3		8	"	GE Noryl GFN3; 30% gl.fib.
			85	3		8	"	GE Noryl 731
			93	1		9		GE Noryl
Formaldehyde		37	93	1		9		"
Isopropyl Alcohol			23		1%	8	compatible	GE GTX
			23	2	19.6 MPa	8	no chg. in appear.	Mitsubishi Iupiace AH60
			23	2	19.6 MPa	8	"	Mitsubishi Iupiace AV60
			23	2		8	"	Mitsubishi Iupiace AH60
			23	2		8	"	Mitsubishi Iupiace AV60
			23	7	31 MPa	7	exp. stress is level where part stress cracks	GE Prevex
			23	7	0.5%	3	incompatible, not recommended	GE Noryl SE100
			23	7	0.5%	3	"	GE Noryl 731
			23	7	0.5%	9		GE GTX 910
			23	7	0.5%	9		GE GTX 901
			23	7	0.5%	9		GE GTX 830
			23	7	1%	3	incompatible, not recommended	GE Noryl SE100
			23	7	1%	3	"	GE Noryl 731
			23	7	1%	9		GE GTX 901
			23	7	1%	6		GE GTX 910
			23	7		8	compatible	GE Noryl SE100
			23	7		8	"	GE Noryl 731
			23	7		9		GE GTX 901
			23	7		9		GE GTX 830
			23	7		9		GE GTX 910
			26	7	24.8 MPa	8	no effect	GE Noryl SE1
			26	7	24.8 MPa	8	"	GE Noryl 731
			26	7	41.4 MPa	8	"	GE Noryl GFN3; 30% gl.fib.
			26	7		8	"	"
			26	7		8	"	GE Noryl SE1
			26	7		8	"	GE Noryl 731
			85	3	24.8 MPa	8	"	GE Noryl SE1
			85	3	24.8 MPa	8	"	GE Noryl 731
			85	3	41.4 MPa	8	"	GE Noryl GFN3; 30% gl.fib.
			85	3	0.5%	3	incompatible, not recommended	GE Noryl SE100
			85	3	0.5%	3	"	GE Noryl 731
			85	3	1%	3	"	GE Noryl SE100
			85	3	1%	3	"	GE Noryl 731
			85	3		8	compatible	GE Noryl SE100
			85	3		8	"	GE Noryl 731
			85	3		8	no effect	GE Noryl SE1
			85	3		8	"	GE Noryl GFN3; 30% gl.fib.
			85	3		8	"	GE Noryl 731
Morpholine			93	1		0		GE Noryl

Modified Polyphenylene Ether (continued)

Reagent	Reagent Note	Conc. (%)	Temp. (°C)	Time (days)	Load	PDL Rating	Resistance Note	Material Note
Phenol			93	1		0		GE Noryl
Potassium Permanganate		10	93	1		9		"
Sulfuric Acid		10	23	2	19.6 MPa	8	no chg. in appear.	Mitsubishi lupiace AH60
		10	23	2	19.6 MPa	8	"	Mitsubishi lupiace AV60
		10	23	2		8	"	"
		10	23	2		8	"	Mitsubishi lupiace AH60
		10	23	7	0.25%	9		LNP; 30% gl.fib.
		10	23	7	strained	8	appear. unchanged	GE Noryl N190
		10	23	7		8		"
		10	23	7		9		LNP; 30% gl.fib.
		10	26	7	24.8 MPa	8	no effect	GE Noryl SE1
		10	26	7	24.8 MPa	8	"	GE Noryl 731
		10	26	7	41.4 MPa	8	"	GE Noryl GFN3; 30% gl.fib.
		10	26	7		8	"	"
		10	26	7		8	"	GE Noryl 731
		10	26	7		8	"	GE Noryl SE1
		10	85	3	24.8 MPa	8	"	GE Noryl 731
		10	85	3	24.8 MPa	8	"	GE Noryl SE1
		10	85	3	41.4 MPa	8	"	GE Noryl GFN3; 30% gl.fib.
		10	85	3		8	"	"
		10	85	3		8	"	GE Noryl 731
		10	85	3		8	"	GE Noryl SE1

Nylon 11

Reagent	Reagent Note	Conc. (%)	Temp. (°C)	Time (days)	Load	PDL Rating	Resistance Note	Material Note
Ammonia	liquid or gas		20			8	good resistance	Atochem Rilsan B
	"		40			8	"	"
Benzaldehyde			20			8	"	"
			40			5	limited resistance, may swell or dissolve	"
			60			1	attacked	"
Benzyl Alcohol			20			5	limited resistance, may swell or dissolve	"
			40			1	attacked	"
			60			1	"	"
Butyl Alcohol			20			6	good resistance, some swelling	"
			40			5	limited resistance, may swell or dissolve	"
			60			1	attacked	"
Chlorine			20			1	"	"
			40			1	"	"
			60			1	"	"
Ethyl Acetate			20			8	good resistance	"
			40			8	"	"
			60			8	"	"
Ethyl Alcohol	pure		20			6	good resistance, some swelling	"
	"		40			5	limited resistance, may swell or dissolve	"
	"		60			1	attacked	"
Ethylene Chlorohydrin			20			1	"	"
			40			1	"	"
Ethylene Oxide			20			8	good resistance	"
			40			5	limited resistance, sl. browning or yellowing	"
			60			5	limited resistance, may swell or dissolve	"
Fluorine			20			1	attacked	"
			40			1	"	"
			60			1	"	"
Formaldehyde	technical		20			8	good resistance	"
	"		40			5	limited resistance, may swell or dissolve	"
	"		60			1	attacked	"
Hydrogen Peroxide	20 vol.		20			8	good resistance	"
	"		40			5	limited resistance, may swell or dissolve	"
Phenol			20			1	attacked	"
			40			1	"	"
			60			1	"	"
Potassium Permanganate		5	20			1	"	"
		5	40			1	"	"
Sulfuric Acid		1	20			8	good resistance	"
		1	40			5	limited resistance, may swell or dissolve	"

Reagent	Reagent Note	Conc. (%)	Temp. (°C)	Time (days)	Load	PDL Rating	Resistance Note	Material Note
Nylon 11 (continued)								
Sulfuric Acid		1	60			5	limited resistance, may swell or dissolve	Atochem Rilsan B
		10	20			8	good resistance	"
		10	40			5	limited resistance, may swell or dissolve	"
		10	60			1	attacked	"
Nylon 46								
Ammonia		10	23	365		5	little attack; some to consid. absorp.	DSM Stanyl
Benzaldehyde			23	365		6	little attack; some absorp.	"
Benzoic Acid			23	365		6	"	"
Boric Acid			23	365		6	"	"
Butyl Alcohol			23	365		8	no attack, little or no absorp.	"
Chlorine	gas		23	365		2	material decomposes	"
	chlorine water		23	365		2	"	"
Cresol			23	365		0	material dissolves	"
Ethyl Acetate			23	365		8	no attack, little or no absorp.	"
Ethyl Alcohol			23	365		7	little attack; little to some absorp.	"
Formaldehyde			23	365		8	no attack, little or no absorp.	"
Hydrogen Peroxide		3	23	365		2	material decomposes	"
Isopropyl Alcohol			23	365		8	no attack, little or no absorp.	"
Mercuric Chloride		10	23	365		3	material attacked; consid. absorp.	"
Phenol			23	365		2	material decomposes	"
Salicylic Acid			23	365		8	no attack, little or no absorp.	"
Silver Nitrate		10	23	365		8	"	"
Sulfuric Acid		10	23	365		2	material decomposes	"
Nylon 12								
Ammonia	concentrated sol'n		20			8	resistant	Huls Vestamid
	gas		20			8	"	"
	"		23	365		8	"	Emser Grilamid
	concentrated sol'n		60			8	"	Huls Vestamid
	gas		60			8	"	"
		10	20			8	"	"
		10	23	365		8	"	Emser Grilamid
		10	60			8	"	Huls Vestamid
Benzaldehyde			20			1	not resistant	"
	sat'd		20			8	resistant	"
			23	365		6	dimens./prop. chg. after prolonged contact	Emser Grilamid
			60			1	not resistant	Huls Vestamid
	sat'd		60			8	resistant	"
		0.3	23	365		8	"	Emser Grilamid
Benzoic Acid			20			8	"	Huls Vestamid
	sat'd		20			8	"	"
			23	365		6	dimens./prop. chg. after prolonged contact	Emser Grilamid
	sat'd		60			8	resistant	Huls Vestamid
			60			8	"	"
Benzyl Alcohol			23	365		3	not resistant except for brief contact	Emser Grilamid
Bleach	12.5% chlorine		20			5	limited resistance	Huls Vestamid
	"		60			1	not resistant	"
Boric Acid			20			8	resistant	"
	sat'd		20			8	"	"
			60			8	"	"
	sat'd		60			8	"	"
		10	23	365		8	"	Emser Grilamid
Butyl Alcohol			20			8	"	Huls Vestamid
			23	365		6	dimens./prop. chg. after prolonged contact	Emser Grilamid
			60			7	practically resistant	Huls Vestamid
Chlorine	chlorine water; sat'd sol'n		20			1	not resistant	"
	liquid		20			1	"	"
	dry gas		20			1	"	"
			23	365		1	soluble/ attacked after brief contact	Emser Grilamid
	chlorine water; sat'd sol'n		60			1	not resistant	Huls Vestamid
	dry gas		60			1	"	"
	liquid		60			1	"	"
	chlorine water	5	23	365		6	dimens./prop. chg. after prolonged contact	Emser Grilamid

Reagent	Reagent Note	Conc. (%)	Temp. (°C)	Time (days)	Load	PDL Rating	Resistance Note	Material Note
Nylon 12 (continued)								
Chlorine	gas	5	23	365		6	dimens./prop. chg. after prolonged contact	Emser Grilamid
	moist gas	10	20			1	not resistant	Huls Vestamid
	"	10	60			1	"	"
Cresol			20			1	"	"
	sol'n		20			8	resistant	"
			23	365		1	soluble/ attacked after brief contact	Emser Grilamid
			60			1	not resistant	Huls Vestamid
	sat'd	0.25	20			5	limited resistance	"
	"	0.25	60			1	not resistant	"
Ethyl Acetate			23	365		8	resistant	Emser Grilamid
Ethyl Alcohol			20			8	"	Huls Vestamid
			23	365		8	"	Emser Grilamid
			60			7	practically resistant	Huls Vestamid
		10	20			8	resistant	"
		10	60			8	"	"
		50	20			8	"	"
		50	60			8	"	"
		96	20			8	"	"
		96	60			7	practically resistant	"
Formaldehyde			20			7	"	"
			23	365		6	dimens./prop. chg. after prolonged contact	Emser Grilamid
		10	20			8	resistant	Huls Vestamid
		10	60			7	practically resistant	"
		30	20			7	"	"
		30	60			1	not resistant	"
		40	20			7	practically resistant	"
		40	23	365		6	dimens./prop. chg. after prolonged contact	Emser Grilamid
		40	60			1	not resistant	Huls Vestamid
Hydrogen Peroxide		2	23	365		6	dimens./prop. chg. after prolonged contact	Emser Grilamid
		3	20			8	resistant	Huls Vestamid
		3	60			8	"	"
		10	20			7	practically resistant	"
		10	60			3	little resistance	"
		19	23	365		6	dimens./prop. chg. after prolonged contact	Emser Grilamid
		30	20			7	practically resistant	Huls Vestamid
		30	23	365		1	soluble/ attacked after brief contact	Emser Grilamid
		30	60			3	little resistance	Huls Vestamid
Isopropyl Alcohol			20			8	resistant	"
			23	365		6	dimens./prop. chg. after prolonged contact	Emser Grilamid
			60			7	practically resistant	Huls Vestamid
Lime Chloride			20			5	limited resistance	"
			60			1	not resistant	"
Phenol			20			1	"	"
			23	365		3	not resistant except for brief contact	Emser Grilamid
			60			1	not resistant	Huls Vestamid
Potassium Permanganate	sat'd		20			3	little resistance	"
	"		60			1	not resistant	"
		1	23	365		1	soluble/ attacked after brief contact	Emser Grilamid
Salicylic Acid			23	365		8	resistant	"
Sodium Hypochlorite		5	20			7	practically resistant	Huls Vestamid
		5	23	365		6	dimens./prop. chg. after prolonged contact	Emser Grilamid
		5	60			3	little resistance	Huls Vestamid
Sulfuric Acid		2	23	365		6	dimens./prop. chg. after prolonged contact	Emser Grilamid
		10	20			7	practically resistant	Huls Vestamid
		10	23	365		6	dimens./prop. chg. after prolonged contact	Emser Grilamid
		10	60			1	not resistant	Huls Vestamid
Toothpaste			20			8	resistant	"
			23	365		8	"	Emser Grilamid
			60			8	"	Huls Vestamid

Nylon 6

Reagent	Reagent Note	Conc. (%)	Temp. (°C)	Time (days)	Load	PDL Rating	Resistance Note	Material Note
Ammonia	liquid		20	<360		8	no irreversible chg.	BASF Ultramid B
	gas		23			9	exc. resistance	Allied Sig. Capron
			23	365		8	resistant	Emser Grilon
			70	<360		6	noticeable chg. in mass, dimens., props.	BASF Ultramid B
			100			6	sl. changes after long expos.	Allied Sig. Capron
		10	23			6	"	"
		10	23	365		8	resistant	Emser Grilon
		20	23	<360		8	no irreversible chg.	BASF Ultramid B
Benzaldehyde			23			3	poor resistance, resistant for short period	Allied Sig. Capron
			23	<360		3	noticeable chg. in mass, dimens., props.	BASF Ultramid B
			23	365		3	not resistant except for brief contact	Emser Grilon
		0.3	23	365		8	resistant	"
Benzoic Acid	concentrated		23			3	poor resistance, resistant for short period	Allied Sig. Capron
	sat'd		23	<360		2	rapid attack, stress cracks, irreversible damage	BASF Ultramid B
			23	365		6	dimens./prop. chg. after prolonged contact	Emser Grilon
		20	23	<360		6	noticeable chg. in mass, dimens., props.	BASF Ultramid B
Benzyl Alcohol			23			3	poor resistance, resistant for short period	Allied Sig. Capron
			23	<360		3	noticeable chg. in mass, dimens., props.	BASF Ultramid B
			23	365		3	not resistant except for brief contact	Emser Grilon
			75	<360		0	polymer is soluble in the substance	BASF Ultramid B
Bleach	12.5% chlorine		23	<360		2	rapid attack, stress cracks, irreversible damage	"
	sat'd sol'n w\ calcium hypochlorite		23	<360		2	"	"
Boric Acid			23	<360		3	noticeable chg. in mass, dimens., props.	"
		10	23			6	sl. changes after long expos.	Allied Sig. Capron
		10	23	365		6	dimens./prop. chg. after prolonged contact	Emser Grilon
Butyl Alcohol			23			6	sl. changes after long expos.	Allied Sig. Capron
			23	<360		5	no irreversible chg.	BASF Ultramid B
			23	365		8	resistant	Emser Grilon
Chlorine	chlorine water; sat'd sol'n		23			3	poor resistance, resistant for short period	Allied Sig. Capron
	chlorine gas		23			3	"	"
			23	<360		2	rapid attack, stress cracks, irreversible damage	BASF Ultramid B
			23	365		1	soluble/ attacked after brief contact	Emser Grilon
	chlorine water	<5	23	365		3	not resistant except for brief contact	"
	chlorine gas	<5	23	365		3	"	"
Cresol			23			3	poor resistance, resistant for short period	Allied Sig. Capron
			23	<360		0	polymer is soluble in the substance	BASF Ultramid B
			23	365		1	soluble/ attacked after brief contact	Emser Grilon
Ethyl Acetate			23			9	exc. resistance	Allied Sig. Capron
			23	<360		7	no irreversible chg.	BASF Ultramid B
			23	365		8	resistant	Emser Grilon
Ethyl Alcohol			23	<360		3	noticeable chg. in mass, dimens., props.	BASF Ultramid B
			23	365		8	resistant	Emser Grilon
		40	23	<360		8	no irreversible chg.	BASF Ultramid B
		96	23			6	sl. changes after long expos.	Allied Sig. Capron
Ethylene Glycol			23	7	0.25%	9		LNP; 30% gl.fib.
			23	7		9		"
			23	<360		4	no irreversible chg.	BASF Ultramid B
			140	<360		0	polymer is soluble in the substance	"
		95	23			6	sl. changes after long expos.	Allied Sig. Capron
Ethylene Oxide			23	<360		6	noticeable chg. in mass, dimens., props.	BASF Ultramid B
			80	<360		2	rapid attack, stress cracks, irreversible damage	"
Fluorine			23	<360		2	"	"
Formaldehyde			23	365		6	dimens./prop. chg. after prolonged contact	Emser Grilon
		30	23			6	sl. changes after long expos.	Allied Sig. Capron
		40	23	365		3	not resistant except for brief contact	Emser Grilon

Reagent	Reagent Note	Conc. (%)	Temp. (°C)	Time (days)	Load	PDL Rating	Resistance Note	Material Note
Nylon 6 (continued)								
Hydrogen Peroxide		1	23			3	poor resistance, resistant for short period	Allied Sig. Capron
		1	23	<360		8	no irreversible chg.	BASF Ultramid B
		2	23	365		3	not resistant except for brief contact	Emser Grilon
		19	23	365		3	"	"
		30	23	<360		2	rapid attack, stress cracks, irreversible damage	BASF Ultramid B
		30	23	365		1	soluble/ attacked after brief contact	Emser Grilon
Iodine	tincture		23			3	poor resistance, resistant for short period	Allied Sig. Capron
	in alcohol		23	<360		2	rapid attack, stress cracks, irreversible damage	BASF Ultramid B
		3	23			3	poor resistance, resistant for short period	Allied Sig. Capron
Isopropyl Alcohol			23			6	sl. changes after long expos.	"
			23	<360		4	no irreversible chg.	BASF Ultramid B
			23	365		8	resistant	Emser Grilon
			60	<360		8	no irreversible chg.	BASF Ultramid B
Mercuric Chloride	sat'd sol'n		23	<360		2	rapid attack, stress cracks, irreversible damage	"
		6	23			1	severe attack, not recommened	Allied Sig. Capron
Phenol	molten					1	"	"
			23	365		3	not resistant except for brief contact	Emser Grilon
			43	<360		0	polymer is soluble in the substance	BASF Ultramid B
		6	23			3	poor resistance, resistant for short period	Allied Sig. Capron
	sol'n in ethanol	70	23	<360		0	polymer is soluble in the substance	BASF Ultramid B
		75	23			1	severe attack, not recommened	Allied Sig. Capron
		88	23	<360		0	polymer is soluble in the substance	BASF Ultramid B
Potassium Permanganate		1	23	<360		2	rapid attack, stress cracks, irreversible damage	"
		1	23	365		1	soluble/ attacked after brief contact	Emser Grilon
		10	23			3	poor resistance, resistant for short period	Allied Sig. Capron
Salicylic Acid			23			9	exc. resistance	"
			23	<360		8	no irreversible chg.	BASF Ultramid B
			23	365		8	resistant	Emser Grilon
Silver Nitrate			23			9	exc. resistance	Allied Sig. Capron
Sodium Hypochlorite			23			3	poor resistance, resistant for short period	"
		5	23	365		3	not resistant except for brief contact	Emser Grilon
		10	23	<360		4	not resist. to prolonged contact	BASF Ultramid B
Sulfuric Acid		2	23			3	poor resistance, resistant for short period	Allied Sig. Capron
		2	23	<360		2	rapid attack, stress cracks, irreversible damage	BASF Ultramid B
		2	23	365		1	soluble/ attacked after brief contact	Emser Grilon
		10	23			1	severe attack, not recommened	Allied Sig. Capron
		10	23	7	0.25%	1		LNP; 30% gl.fib.
		10	23	7		1		"
		10	23	365		1	soluble/ attacked after brief contact	Emser Grilon
Toothpaste			23	365		8	resistant	"
Nylon 610								
Ammonia			20	<360		8	no irreversible chg.	BASF Ultramid S
	sol'n		23	<360		8	"	"
			70	<360		6	noticeable chg. in mass, dimens., props.	"
Benzaldehyde			23	<360		4	not resist. to prolonged contact	"
Benzoic Acid			23	<360		2	rapid attack, stress cracks, irreversible damage	"
		20	23	<360		6	noticeable chg. in mass, dimens., props.	"
Benzyl Alcohol			23	<360		3	"	"
			75	<360		2	rapid attack, stress cracks, irreversible damage	"
Bleach	sat'd sol'n w\ calcium hypochlorite		23	<360		2	"	"
	bleaching liquid		23	<360		2	"	"
Boric Acid			23	<360		5	no irreversible chg.	"
Butyl Alcohol			23	<360		4	"	"
Chlorine			23	<360		2	rapid attack, stress cracks, irreversible damage	"
Cresol			23	<360		0	polymer is soluble in the substance	"

Reagent	Reagent Note	Conc. (%)	Temp. (°C)	Time (days)	Load	PDL Rating	Resistance Note	Material Note
Nylon 610 (continued)								
Ethyl Acetate			23	<360		7	no irreversible chg.	BASF Ultramid S
Ethyl Alcohol			23	<360		3	noticeable chg. in mass, dimens., props.	"
		40	23	<360		8	no irreversible chg.	"
Ethylene Glycol			23	7	0.25%	9		LNP; 30% gl.fib.
			23	7		9		"
			23	<360		6	no irreversible chg.	BASF Ultramid S
			81	3		4		LNP; 30% gl.fib.
			140	<360		2	rapid attack, stress cracks, irreversible damage	BASF Ultramid S
Ethylene Oxide			23	<360		6	noticeable chg. in mass, dimens., props.	"
			80	<360		2	rapid attack, stress cracks, irreversible damage	"
Fluorine			23	<360		2	"	"
Hydrogen Peroxide		1	23	<360		8	no irreversible chg.	"
		30	23	<360		2	rapid attack, stress cracks, irreversible damage	"
Iodine			23	<360		2	"	"
Isopropyl Alcohol			23	<360		4	no irreversible chg.	"
			60	<360		8	"	"
Mercuric Chloride	sat'd sol'n		23	<360		2	rapid attack, stress cracks, irreversible damage	"
Phenol			43	<360		0	polymer is soluble in the substance	"
	sol'n in ethanol	70	23	<360		0	"	"
		88	23	<360		0	"	"
Potassium Permanganate		1	23	<360		2	rapid attack, stress cracks, irreversible damage	"
Salicylic Acid			23	<360		8	no irreversible chg.	"
Sodium Hypochlorite		10	23	<360		6	noticeable chg. in mass, dimens., props.	"
Sulfuric Acid		2	23	<360		4	not resist. to prolonged contact	"
		10	23	7	0.25%	6		LNP; 30% gl.fib.
		10	23	7		6		"
Nylon 612								
Ammonia	concentrated sol'n		20			8	resistant	Huls Vestamid
	gas		20			8	"	"
	concentrated sol'n		60			8	"	"
	gas		60			8	"	"
		10	20			8	"	"
		10	60			8	"	"
Benzaldehyde			20			1	not resistant	"
	sat'd		20			8	resistant	"
			60			1	not resistant	"
	sat'd		60			8	resistant	"
Benzoic Acid			20			8	"	"
	sat'd		20			8	"	"
			60			8	"	"
	sat'd		60			8	"	"
Bleach	12.5% chlorine		20			5	limited resistance	"
	"		60			1	not resistant	"
Boric Acid	sat'd		20			8	resistant	"
			20			8	"	"
			60			8	"	"
	sat'd		60			8	"	"
Butyl Alcohol			20			8	"	"
			23	90		6	some chg. in physical props.	DuPont Zytel 158L
			50	45		8	little chg. in physical props.	DuPont Zytel 151L
			60			7	practically resistant	Huls Vestamid
Chlorine	liquid		20			1	not resistant	"
	chlorine water; sat'd sol'n		20			1	"	"
	dry gas		20			1	"	"
	"		60			1	"	"
	liquid		60			1	"	"
	chlorine water; sat'd sol'n		60			1	"	"
	moist gas	10	20			1	"	"
	"	10	60			1	"	"
Cresol			20			1	"	"
	sol'n		20			8	resistant	"
			60			1	not resistant	"

Reagent	Reagent Note	Conc. (%)	Temp. (°C)	Time (days)	Load	PDL Rating	Resistance Note	Material Note
Nylon 612 (continued)								
Cresol	sat'd	0.25	20			5	limited resistance	Huls Vestamid
	"	0.25	60			1	not resistant	"
Ethyl Alcohol			20			8	resistant	"
			60			7	practically resistant	"
		10	20			8	resistant	"
		10	60			8	"	"
		50	20			8	"	"
		50	60			8	"	"
		95	23	90		3	little chg. in physical props.	DuPont Zytel 158L
		96	20			8	resistant	Huls Vestamid
		96	60			7	practically resistant	"
Formaldehyde			20			7	"	"
		10	20			8	resistant	"
		10	60			7	practically resistant	"
		30	20			7	"	"
		30	60			1	not resistant	"
		40	20			7	practically resistant	"
		40	60			1	not resistant	"
Hydrogen Peroxide		3	20			8	resistant	"
		3	60			8	"	"
		10	20			7	practically resistant	"
		10	60			3	little resistance	"
		30	20			7	practically resistant	"
		30	60			3	little resistance	"
Isopropyl Alcohol			20			8	resistant	"
			60			7	practically resistant	"
Lime Chloride			20			5	limited resistance	"
			60			1	not resistant	"
Methyl Isobutyl Ketone			23	14		8	little chg. in physical props.	DuPont Zytel 151
Phenol			20			1	not resistant	Huls Vestamid
			60			1	"	"
Potassium Permanganate	sat'd		20			3	little resistance	"
	"		60			1	not resistant	"
Sodium Hypochlorite		5	20			7	practically resistant	"
		5	60			3	little resistance	"
Sulfuric Acid		10	20			7	practically resistant	"
		10	60			1	not resistant	"
Toothpaste			20			8	resistant	"
			60			8	"	"
Nylon 66								
Ammonia	liquid			7		8	little chg. in physical props.	DuPont Zytel 101
	"			14		8	"	"
			23			8	good resistance	BIP Beetle
	gas		23			4	may be suitable for limited exposure	Hoechst Cel. Nylon 1000
	"		23			4	"	Hoechst Cel. Nylon 1003
	liquid		24	200		6	some chg. in physical props.	DuPont Zytel 101
		10	23			8	little/ no absorp. or effect on mech. prop.	Hoechst Cel. Nylon 1000
		10	23			8	"	Hoechst Cel. Nylon 1003
	aq. sol'n	10	23			8	very good resistance	Rhone Pou. Technyl A 216
Benzaldehyde			23			6	some absorp. and sl. reduction in mech. prop.	Hoechst Cel. Nylon 1003
			23			6	"	Hoechst Cel. Nylon 1000
			93	1		9		DuPont Zytel 101
Benzoic Acid	sat'd		23			4	may be suitable for limited exposure	Hoechst Cel. Nylon 1003
	"		23			4	"	Hoechst Cel. Nylon 1000
		10	24			2	signif. chg. in physical props.	DuPont Zytel 101
Benzyl Alcohol			23			5	limited resistance	Rhone Pou. Technyl A 216
			23			4	poor resistance	BIP Beetle
			23			4	may be suitable for limited exposure	Hoechst Cel. Nylon 1003
			23			4	"	Hoechst Cel. Nylon 1000
Bleach	sat'd sol'n w\ calcium hypochlorite		23	<360		2	rapid attack, stress cracks, irreversible damage	BASF Ultramid A
Boric Acid		7	35	316		2	signif. chg. in physical props.	DuPont Zytel 101
		10	23			8	little/ no absorp. or effect on mech. prop.	Hoechst Cel. Nylon 1000
		10	23			8	"	Hoechst Cel. Nylon 1003

Nylon 66 (continued)

Reagent	Reagent Note	Conc. (%)	Temp. (°C)	Time (days)	Load	PDL Rating	Resistance Note	Material Note
Butyl Alcohol			23			7	good resistance	Rhone Pou. Technyl A 216
			23			8	"	BIP Beetle
			23			8	little/ no absorp. or effect on mech. prop.	Hoechst Cel. Nylon 1003
			23			8	"	Hoechst Cel. Nylon 1000
			40			6	limited resistance	BIP Beetle
			93	1		7		DuPont Zytel 101
Chlorine	chlorine water		23			1	decomposes in short time	Hoechst Cel. Nylon 1003
	"		23			1	"	Hoechst Cel. Nylon 1000
			23			2	poor resistance	Rhone Pou. Technyl A 216
			23			4	"	BIP Beetle
	chlorine water; dilute		23			6	some chg. in physical props.	DuPont Zytel 101
	chlorine water; concentrated		23			2	signif. chg. in physical props.	"
		10	23			1	decomposes in short time	Hoechst Cel. Nylon 1000
		10	23			1	"	Hoechst Cel. Nylon 1003
Cresol			23			1	"	Hoechst Cel. Nylon 1000
			23			1	"	Hoechst Cel. Nylon 1003
			23			4	poor resistance	BIP Beetle
Ethyl Acetate			23		41.3 MPa	8	no stress cracks observed	DuPont Minlon 10B; min.
			23			8	little/ no absorp. or effect on mech. prop.	Hoechst Cel. Nylon 1003
			23			8	"	Hoechst Cel. Nylon 1000
			23			8	very good resistance	Rhone Pou. Technyl A 216
			23	0.0035	93 MPa	8	no stress cracks observed	DuPont Zytel 70G33L; 33% gl.fib.
			23	21		9		DuPont Minlon 10B40; 40% min.
			23	63		5		DuPont Zytel 70G33L; 33% gl.fib.
			93	1		7		DuPont Zytel 101
		95	50	365		9	little chg. in physical props.	"
Ethyl Alcohol			23			8	good resistance	BIP Beetle
			23	21		9		DuPont Minlon 10B40; 40% min.
		95	23	365		3	little chg. in physical props.	DuPont Zytel 101
		95	50	365		3	"	"
		96	23			5	limited resistance	Rhone Pou. Technyl A 216
		96	23			6	some absorp. and sl. reduction in mech. prop.	Hoechst Cel. Nylon 1003
		96	23			6	"	Hoechst Cel. Nylon 1000
Ethylene Chlorohydrin			23			1	decomposes in short time	Hoechst Cel. Nylon 1003
			23			1	"	Hoechst Cel. Nylon 1000
Ethylene Glycol			23		41.3 MPa	8	no stress cracks observed	DuPont Minlon 10B; min.
			23			6	some absorp. and sl. reduction in mech. prop.	Hoechst Cel. Nylon 1000
			23			6	"	Hoechst Cel. Nylon 1003
			23	0.0035	93 MPa	8	no stress cracks observed	DuPont Zytel 70G33L; 33% gl.fib.
			23	7	0.25%	7		LNP; 30% gl.fil.
			23	7	0.25%	3		LNP
			23	7		6		LNP; 30% gl.fil.
			23	7		3		LNP
			23	21		9		DuPont Minlon 10B40; 40% min.
			23	56		7	little chg. in physical props.	DuPont Zytel 101
			81	3		2		LNP; 30% gl.fil.
			81	3		1		LNP
			93	1		9		DuPont Zytel 101
		50	23	60	0.25%	3		LNP Maranyl A127HS; heat stab.
		50	23	60	0.25%	3		LNP Maranyl A322S; 33% gl.fib.; heat stab.
		50	23	60	0.25%	4		LNP Verton RF-7006 HS; 30% lng.gl.fib.; heat stab.
		50	23	60	0.25%	5		LNP Maranyl PDX-R-86474; heat stab., toughened
		50	23	60	0.25%	3		LNP Verton RF-700-10 HS; 50% lng.gl.fib.; heat stab.
		50	23	60		3		LNP Maranyl A127HS
		50	23	60		4		LNP Verton RF-7006 HS; 30% lng.gl.fib.
		50	23	60		4		LNP Maranyl PDX-R-86474
		50	23	60		5		LNP Verton RF-700-10 HS; 50% lng.gl.fib.
		50	23	60		3		LNP Maranyl A322S; 33% gl.fib.
		50	82	30	0.25%	2		LNP Maranyl A322S; 33% gl.fib.; heat stab.
		50	82	30	0.25%	4		LNP Verton RF-7006 HS; 30% lng.gl.fib.; heat stab.

Reagent	Reagent Note	Conc. (%)	Temp. (°C)	Time (days)	Load	PDL Rating	Resistance Note	Material Note
Nylon 66 (continued)								
Ethylene Glycol		50	82	30	0.25%	4		LNP Maranyl PDX-R-86474; heat stab., toughened
		50	82	30	0.25%	2		LNP Verton RF-700-10 HS; 50% lng.gl.fib.; heat stab.
		50	82	30	0.25%	3		LNP Maranyl A127HS; heat stab.
		50	82	30		2		LNP Verton RF-700-10 HS; 50% lng.gl.fib.
		50	82	30		2		LNP Maranyl A322S; 33% gl.fib.
		50	82	30		3		LNP Maranyl PDX-R-86474
		50	82	30		3		LNP Verton RF-7006 HS; 30% lng.gl.fib.
		50	82	30		2		LNP Maranyl A127HS
		50	127	7		1		chem. resist., heat stab., water resist. grade
		50	127	28		1		"
		50	150	7	0.25%	1	specimen cracked	LNP Verton RF-700-10 HS; 50% lng.gl.fib.; heat stab.
		50	150	7	0.25%	1	"	LNP Maranyl PDX-R-86474; heat stab., toughened
		50	150	7	0.25%	1	"	LNP Verton RF-7006 HS; 30% lng.gl.fib.; heat stab.
		50	150	7	0.25%	1	"	LNP Maranyl A322S; 33% gl.fib.; heat stab.
		50	150	7	0.25%	1	"	LNP Maranyl A127HS; heat stab.
		50	150	7		1	"	LNP Maranyl PDX-R-86474; heat stab., toughened
		50	150	7		1	"	LNP Verton RF-7006 HS; 30% lng.gl.fib.
		50	150	7		1	"	LNP Maranyl A127HS; heat stab.
		50	150	7		1	"	LNP Maranyl A322S; 33% gl.fib.; heat stab.
		50	150	7		2	"	LNP Verton RF-700-10 HS; 50% lng.gl.fib.
Ethylene Oxide			23			8	good resistance	BIP Beetle
Fluorine			23			1	decomposes in short time	Hoechst Cel. Nylon 1003
			23			1	"	Hoechst Cel. Nylon 1000
	gas		23			4	poor resistance	BIP Beetle
Formaldehyde			23			6	limited resistance	"
	aq.	30	23			5	"	Rhone Pou. Technyl A 216
		37	93	1		5		DuPont Zytel 101
		38	23	14		8	little chg. in physical props.	"
		40	23			8	little/ no absorp. or effect on mech. prop.	Hoechst Cel. Nylon 1000
		40	23			8	"	Hoechst Cel. Nylon 1003
Hydrogen Peroxide		0.5	23			4	may be suitable for limited exposure	"
		0.5	23			4	"	Hoechst Cel. Nylon 1000
		3	23			1	decomposes in short time	Hoechst Cel. Nylon 1003
		3	23			1	"	Hoechst Cel. Nylon 1000
	aq. sol'n	3	23			2	poor resistance	Rhone Pou. Technyl A 216
		5	43	30		2	signif. chg. in physical props.	DuPont Zytel 101
		20	23			8	good resistance	BIP Beetle
		20	40			6	limited resistance	"
Iodine	in alcohol		23			1	decomposes in short time	Hoechst Cel. Nylon 1003
	"		23			1	"	Hoechst Cel. Nylon 1000
	tincture		23			4	poor resistance	BIP Beetle
	in KI sol'n	3	23			1	decomposes in short time	Hoechst Cel. Nylon 1003
	"	3	23			1	"	Hoechst Cel. Nylon 1000
Isopropyl Alcohol			22	30		8		33% gl.fib.
			23			5	limited resistance	Rhone Pou. Technyl A 216
Mercuric Chloride	sat'd sol'n		23	<360		2	rapid attack, stress cracks, irreversible damage	BASF Ultramid A
		6	23			4	may be suitable for limited exposure	Hoechst Cel. Nylon 1000
		6	23			4	"	Hoechst Cel. Nylon 1003
		10	23			5	limited resistance	Rhone Pou. Technyl A 216
Morpholine			93	1		8		DuPont Zytel 101
Phenol			23			1	decomposes in short time	Hoechst Cel. Nylon 1003
			23			4	poor resistance	BIP Beetle
	aq. sol'n		23			0	soluble in reagent	Rhone Pou. Technyl A 216
	molten		23			0	"	"
	sat'd aq. sol'n		23	63		1	sample underwent serious attack	DuPont Zytel 70G33L; 33% gl.fib.
			93	1		0		DuPont Zytel 101
	sol'n in ethanol	70	23	<360		0	polymer is soluble in the substance	BASF Ultramid A
		90	23			0	solvent for Nylon 66	DuPont Zytel 101

Reagent	Reagent Note	Conc. (%)	Temp. (°C)	Time (days)	Load	PDL Rating	Resistance Note	Material Note

Nylon 66 (continued)

Reagent	Reagent Note	Conc. (%)	Temp. (°C)	Time (days)	Load	PDL Rating	Resistance Note	Material Note
Potassium Permanganate		1	23			1	decomposes in short time	Hoechst Cel. Nylon 1003
		1	23			1	"	Hoechst Cel. Nylon 1000
	aq. sol'n	1	23			2	poor resistance	Rhone Pou. Technyl A 216
		5	23			4	"	BIP Beetle
		5	23	10		2	signif. chg. in physical props.	DuPont Zytel 101
		10	93	1		1		"
Salicylic Acid			23			8	good resistance	BIP Beetle
			23			8	little/ no absorp. or effect on mech. prop.	Hoechst Cel. Nylon 1003
			23			8	"	Hoechst Cel. Nylon 1000
			23			8	very good resistance	Rhone Pou. Technyl A 216
Silver Nitrate			23			8	little/ no absorp. or effect on mech. prop.	Hoechst Cel. Nylon 1003
			23			8	"	Hoechst Cel. Nylon 1000
Sodium Hypochlorite	15% chlorine		23			1	decomposes in short time	Hoechst Cel. Nylon 1003
	"		23			1	"	Hoechst Cel. Nylon 1000
Sulfuric Acid		2	23			4	may be suitable for limited exposure	"
		2	23			4	"	Hoechst Cel. Nylon 1003
	aq. sol'n	3	23	365		1	poor resistance	Rhone Pou. Technyl A 216
		5	23			1	decomposes in short time	Hoechst Cel. Nylon 1003
		5	23			1	"	Hoechst Cel. Nylon 1000
		10	23			8	good resistance	BIP Beetle
		10	23	7	0.25%	2		LNP; 30% gl.fil.
		10	23	7	0.25%	2		LNP
		10	23	7		2		LNP; 30% gl.fil.
		10	23	7		3		LNP
		10	40			6	limited resistance	BIP Beetle

Nylon MXD6

Reagent	Reagent Note	Conc. (%)	Temp. (°C)	Time (days)	Load	PDL Rating	Resistance Note	Material Note
Ammonia	aq. sol'n	10	20	7		9		Mitsubishi Reny 1002; 30% gl.fib.
Butyl Alcohol			20	7		9		"
Ethyl Acetate			20	7		9		"
Formaldehyde		37	20	7		9		"
Phenol		5	20	7		2		"

Amorphous Nylon

Reagent	Reagent Note	Conc. (%)	Temp. (°C)	Time (days)	Load	PDL Rating	Resistance Note	Material Note
Ammonia	sol'n	25	23	180		6	conditionally stable	Huls Trogamid T
Benzaldehyde			23	30		6	not stable	"
			23	180		5	not stable, crazing	"
Benzoic Acid	sat'd		23	30		5	conditionally stable	"
	"		23	180		4	conditionally stable, swelling, color chg.	"
Benzyl Alcohol			23	180		2	not stable	"
Boric Acid	sat'd		23	30		6	stable	"
	"		23	180		4	stable, swelling, color/ transpar. chg.	"
Butyl Alcohol			23	180		1	not stable, dissolving	"
Chlorine	chlorine water; 2 mg chlorine per liter		23	180		3	not stable, swelling, crazing	"
Ethyl Acetate			23	180		9	stable	"
Ethyl Alcohol	denatured	96	23	180		2	not stable, swelling, crazing	"
Ethylene Chlorohydrin			23	180		1	not stable, dissolving, crazing	"
Ethylene Glycol			23	30		6	not stable	"
			23	180		6	not stable, crazing	"
Formaldehyde			23	30		6	conditionally stable	"
			23	180		5	conditionally stable, swelling	"
Isopropyl Alcohol			23	30		1	not stable, dissolving	"
			23	180		1	"	"
Phenol			23	180		2	not stable	"
Silver Nitrate		10	23	180		8	stable	"
Sodium Hypochlorite		1	23	180		8	"	"
Sulfuric Acid		1	23	180		3	not stable, swelling	"
		10	23	180		7	stable, swelling	"

Polyarylamide

Reagent	Reagent Note	Conc. (%)	Temp. (°C)	Time (days)	Load	PDL Rating	Resistance Note	Material Note
Benzyl Alcohol			23	30		7	variation in tens. str. <10%	Solvay Ixef 1002; 30% gl.fib.
Ethyl Alcohol			23	30		7	"	"
Formaldehyde			23	30		5		"

Reagent	Reagent Note	Conc. (%)	Temp. (°C)	Time (days)	Load	PDL Rating	Resistance Note	Material Note

Polybenzimidazole

Reagent	Reagent Note	Conc. (%)	Temp. (°C)	Time (days)	Load	PDL Rating	Resistance Note	Material Note
Ammonia	300 psi		93	1		4	large effect	Hoechst Cel. Celazole U60
	"		93	7		4	"	"
	"		93	30		4	"	"
Chlorine	dry		24	1		9	no effect	"
	"		24	7		9	"	"
	"		66	1		9	"	"
	"		66	7		9	"	"
Ethylene Glycol	ambient pressure		93	7		9	"	"
	"		93	30		9	"	"
	reflux temp., ambient pressure		196	7		2	severe attack	"
	"		196	30		2	"	"
Phenol			93	1		9	no effect	"
			93	7		9	"	"
			93	30		9	"	"
Sodium Hypochlorite		5	93	1		9	"	"
		5	93	7		6	small effect	"
		5	93	30		6	"	"

Polybutadiene

Reagent	Reagent Note	Conc. (%)	Temp. (°C)	Time (days)	Load	PDL Rating	Resistance Note	Material Note
Butyl Alcohol			25	1		7	insoluble	Jap. Synth. JSR RB820
			35	1		7	"	"
			60	1		7	"	"

Polycarbonate

Reagent	Reagent Note	Conc. (%)	Temp. (°C)	Time (days)	Load	PDL Rating	Resistance Note	Material Note
Ammonia	gas		23			2	attacked	GE Lexan
Benzaldehyde			93	1		0		GE Lexan 141
Benzoic Acid			23			2	not resistant	Miles Makrolon
Benzyl Alcohol			23			2	"	"
			23	0.125	0.7%	1	severe cracking or breaking	gen. purp. grade
Betadine			23	0.125	0.7%	8	no chg. in appear.	"
	povidone iodine sol'n		23	168	0.5%	6		Calibre MegaRad 2081-15
	"		23	168	0.5%	8		Calibre 2061-15
	"		23	168	1.5%	5		Calibre MegaRad 2081-15
	"		23	168	1.5%	4		Calibre 2061-15
	"		23	168		6		Calibre MegaRad 2081-15
	"		23	168		9		Calibre 2061-15
Bleach	Clorox	10	23	168	0.5%	7		Calibre MegaRad 2081-15
	"	10	23	168	0.5%	6		Calibre 2061-15
	"	10	23	168	1.5%	0	specimens fractured	"
	"	10	23	168	1.5%	0	"	Calibre MegaRad 2081-15
	"	10	23	168		7		"
	"	10	23	168		8		Calibre 2061-15
Butyl Alcohol			23			8	resistant	Miles Makrolon
			93	1		9		GE Lexan 141
Carbolic Acid			23			2	not resistant	Miles Makrolon
Chlorine	gas		23	4		8	no effect	GE Lexan
Cidex 7			93.3	0.0097	20.7 MPa	0	exp. time is time to failure	Lexan 141; irradiated @ 4 Mrads gamma rays
			93.3	0.024	20.7 MPa	0	"	Lexan 141
			93.3	0.0125	20.7 MPa	0	"	Lexan 141; irradiated @ 6 Mrads gamma rays
			93.3	0.015	20.7 MPa	0	"	Lexan 141; irradiated @ 2.5 Mrads gamma rays
Cresol			23			2	attacked	GE Lexan
			23			0	dissolved	Miles Makrolon
Ethyl Acetate			23			2	attacked	GE Lexan
			93	1		0		GE Lexan 141
Ethyl Alcohol			23	30		6	no effect	GE Lexan
			85	30	0.5%	6	crazing after 1 month	"
		96	23			8	resistant	Miles Makrolon
Ethylene Chlorohydrin			23			2	not resistant	"
Ethylene Glycol			23	5	0.5%	7	not attacked	GE Lexan 101
			23	5	1%	7	"	"
			23	5		7	"	"
			23	7	0.25%	9		
			23	7	0.25%	5		LNP; 30% gl.fib.
			23	7		7		
			23	7		9		LNP; 30% gl.fib.
			23	365		8	dulling	GE Lexan

Reagent	Reagent Note	Conc. (%)	Temp. (°C)	Time (days)	Load	PDL Rating	Resistance Note	Material Note
Polycarbonate (continued)								
Ethylene Glycol			50	5	0.5%	7	not attacked	GE Lexan 101
			50	5	1%	7	"	"
			50	5		7	"	"
			70	5	0.5%	7	"	"
			70	5	1%	3	attacked	"
			70	5		7	not attacked	"
			81	3		7		
			81	3		9		LNP; 30% gl.fib.
			93	1		9		GE Lexan 141
			93	1		9		"
Ethylene Oxide		100	54	.17-.25		9		gen. purp. grade
Formaldehyde			23			8	resistant	Miles Makrolon
			23	365	0.5%	8	no effect	GE Lexan
			23	365		8	"	"
		37	93	1		9		GE Lexan 141
Glutaraldehyde		50	23	0.125	0.7%	8	no chg. in appear.	gen. purp. grade
Hydrogen Peroxide			23	3	1%	8	no effect	GE Lexan
			23	3		8	"	"
			85	3	1%	4	crazing	"
			85	3		8	no effect	"
		30	23			8	resistant	Miles Makrolon
		30	23	425		9	no effect	GE Lexan
Iodine	tincture	5	23			8	resistant	Miles Makrolon
Isopropyl Alcohol			23	0.125	0.7%	8	no chg. in appear.	gen. purp. grade
			23	3	1%	8	no effect	GE Lexan
			23	6		8	"	"
			85	3	1%	4	crazing	"
			85	3		8	no effect	"
		70	23	168	0.5%	9		Calibre MegaRad 2081-15
		70	23	168	0.5%	7		Calibre 2061-15
		70	23	168	1.5%	3		Calibre MegaRad 2081-15
		70	23	168	1.5%	3		Calibre 2061-15
		70	23	168		7		Calibre MegaRad 2081-15
		70	23	168		7		Calibre 2061-15
Lyposin	IV sol'n lyposin II		23	0.125	0.7%	8	no chg. in appear.	gen. purp. grade
Mercuric Chloride			23			8	resistant	Miles Makrolon
Methyl Isobutyl Ketone			23	0.125	0.7%	1	severe cracking or breaking	gen. purp. grade
Morpholine			93	1		0		GE Lexan 141
Omnicide	2% glutaraldehyde		23	168	0.5%	9		Calibre 2061-15
	"		23	168	0.5%	7		Calibre MegaRad 2081-15
	"		23	168	1.5%	2	some samples fractured	Calibre 2061-15
	"		23	168	1.5%	3	"	Calibre MegaRad 2081-15
	"		23	168		7		Calibre 2061-15
	"		23	168		8		Calibre MegaRad 2081-15
Phenol			23			2	not resistant	Miles Makrolon
			93	1		0		GE Lexan 141
		5	23	3	0.5%	2	attacked	GE Lexan
		5	23	3		8	no effect	"
		5	85	3		2	attacked	"
Potassium Permanganate			23			8	resistant	Miles Makrolon
		10	93	1		9		GE Lexan 141
Silver Nitrate			23			8	resistant	Miles Makrolon
Sodium Hypochlorite	bleaches	5.25	23	0.125	0.7%	8	no chg. in appear.	gen. purp. grade
Sulfuric Acid		10	23			8	resistant	Miles Makrolon
		10	23	0.125	0.7%	8	no chg. in appear.	gen. purp. grade
		10	23	7	0.25%	6		
		10	23	7	0.25%	8		LNP; 30% gl.fib.
		10	23	7		9		
		10	23	7		7		LNP; 30% gl.fib.
Toothpaste	Arm & Hammer					1		
Polycarbonate Copolymer								
Betadine	povidone iodine sol'n		23	168	0.5%	4		Apec 9350
	"		23	168	1.5%	3		"
	"		23	168		8		"
Bleach	Clorox	10	23	168	0.5%	7		"
	"	10	23	168	1.5%	0	specimens fractured	"
	"	10	23	168		7		"

Reagent	Reagent Note	Conc. (%)	Temp. (°C)	Time (days)	Load	PDL Rating	Resistance Note	Material Note
Polycarbonate Copolymer (continued)								
Isopropyl Alcohol		70	23	168	0.5%	8		"
		70	23	168	1.5%	2		"
		70	23	168		8		"
Omnicide	2% glutaraldehyde		23	168	0.5%	3		"
	"		23	168	1.5%	0	specimens fractured	"
	"		23	168		6		"
Polyethylene Terephthalate								
Benzyl Alcohol			23	28		5	no visible chg.	Eastman Ektar PET 7352
			23	28		8	sl. discolor.	Eastman Ektar FB EG001; 30% gl.fib.
			23	31		9	retains 100% of tear strength	DuPont Mylar 92A; film
			75	1		9	"	"
Bleach	powder and liquid		23	0.67		9	no staining tendency	"
	Clorox	5	23	28		7	no visible chg.	Eastman Ektar FB EG001; 30% gl.fib.
	"	5	23	28		5	"	Eastman Ektar PET 7352
Ethyl Acetate			23	0.67		7	sl. mark	DuPont Mylar 92A; film
			23	21		6		DuPont Rynite 545; 45% gl.fib.
			23	21		6		DuPont Rynite 530; 30% gl.fib.
			23	28		5	no visible chg.	Eastman Ektar FB EG001; 30% gl.fib.
			23	28		4	sl. swelling	Eastman Ektar PET 7352
			23	31		8	retains 100% of tear strength	DuPont Mylar 92A; film
			75	1		9	"	"
Ethyl Alcohol			23	21		9		DuPont Rynite 530; 30% gl.fib.
			23	21		9		DuPont Rynite 545; 45% gl.fib.
			23	28		9	no visible chg.	Eastman Ektar PET 7352
			23	28		9	"	Eastman Ektar FB EG001; 30% gl.fib.
			23	31		9	retains 100% of tear strength	DuPont Mylar 92A; film
			23	365		9		DuPont Rynite 545; 45% gl.fib.
			75	1		8	retains 100% of tear strength	DuPont Mylar 92A; film
		50	23	28		7	no visible chg.	Eastman Ektar PET 7352
		50	23	28		8	sl. discolor.	Eastman Ektar FB EG001; 30% gl.fib.
Ethylene Glycol			23	21		9		DuPont Rynite 530; 30% gl.fib.
			23	21		9		DuPont Rynite 545; 45% gl.fib.
Hydrogen Peroxide		28	23	31		9	retains 100% of tear strength	DuPont Mylar 92A; film
Iodine		10	23	0.67		9	no staining tendency	"
Isopropyl Alcohol			22	30		9		30% gl.fib.
Methyl Isobutyl Ketone			23	28		9	no visible chg.	Eastman Ektar FB EG001; 30% gl.fib.
			23	28		5	"	Eastman Ektar PET 7352
Phenol		5	23	31		9	retains 100% of tear strength	DuPont Mylar 92A; film
		5	75	1		9	"	"
Potassium Permanganate		1	23	0.67		9	no staining tendency	"
Sulfuric Acid		3	23	31		9	retains 100% of tear strength	"
		3	75	1		9	"	"
		10	23	21		8		DuPont Rynite 530; 30% gl.fib.
		10	23	21		8		DuPont Rynite 545; 45% gl.fib.
Polybutylene Terephthalate								
Ammonia	liquid		20	<360		6	changes in mass; dimen.; props.	BASF Ultradur
	concentrated sol'n		23			5	limited resistance	Bayer Pocan
	sol'n		23	<360		8	no irreversible chg.	BASF Ultradur
	concentrated sol'n		60			2	not resistant	Bayer Pocan
	liquid		70	<360		2	rapid attack, stress cracks, irreversible damage	BASF Ultradur
	sol'n	10	23			8	resistant	Bayer Pocan
	"	10	60			2	not resistant	"
Benzoic Acid	sat'd		23	<360		8	no irreversible chg.	BASF Ultradur
		20	23	<360		8	"	"
Bleach	sat'd sol'n w\ calcium hypochlorite		23	<360		6	noticeable chg. in mass, dimen., props.	"
	12.5% chlorine		23	<360		6	changes in mass; dimen.; props.	"
Boric Acid			23	<360		8	no irreversible chg.	"
Butyl Alcohol			23			8	resistant	Bayer Pocan
			23	<360		8	no irreversible chg.	BASF Ultradur
			60			8	resistant	Bayer Pocan

Polybutylene Terephthalate (continued)

Reagent	Reagent Note	Conc. (%)	Temp. (°C)	Time (days)	Load	PDL Rating	Resistance Note	Material Note
Chlorine			23	<360		2	rapid attack, stress cracks, irreversible damage	BASF Ultradur
Cresol			23			2	not resistant	Bayer Pocan
			23	<360		0	dissolved	BASF Ultradur
			60			2	not resistant	Bayer Pocan
Ethyl Acetate			23			8	unaffected	Hoechst Cel. Celanex; gl.fib.
			23			5	limited resistance	Bayer Pocan
			23	30		5		GE Valox 310-SEO
			23	30		9		GE Valox 420; 30% gl.fib.
			23	30		1		GE Valox 735; 40% gl.fib./min.
			23	30		5		GE Valox 325
			23	30		5		GE Valox 760; 25% min.
			23	90		1		GE Valox 735; 40% gl.fib./min.
			23	90		3		GE Valox 325
			23	90		3		GE Valox 310-SEO
			23	90		7		GE Valox 420; 30% gl.fib.
			23	90		2		GE Valox 760; 25% min.
			23	180		3		GE Valox 325
			23	180		2		GE Valox 310-SEO
			23	180		5		GE Valox 420; 30% gl.fib.
			23	<360		4	not resistant to prolonged contact	BASF Ultradur
			60			2	not resistant	Bayer Pocan
Ethyl Alcohol			23			8	resistant	"
			23	30		9		GE Valox 325
			23	30		9		GE Valox 760; 25% min.
			23	30		9		GE Valox 735; 40% gl.fib./min.
			23	30		9		GE Valox 420; 30% gl.fib.
			23	90		9		GE Valox 760; 25% min.
			23	90		9		GE Valox 735; 40% gl.fib./min.
			23	90		9		GE Valox 420; 30% gl.fib.
			23	90		9		GE Valox 325
			23	180		9		"
			23	180		9		GE Valox 420; 30% gl.fib.
			23	<360		8	no irreversible chg.	BASF Ultradur
			60			8	resistant	Bayer Pocan
		40	23	<360		8	no irreversible chg.	BASF Ultradur
		95	23	90		9		Hoechst Cel. Celanex; gl.fib.
		95	23	180		8		"
		95	23	360		9		"
		95	82	24		4		"
Ethylene Glycol			23			8	resistant	Bayer Pocan
			23	7	0.25%	9		LNP; 40% gl.fib.
			23	7	0.25%	7		LNP; 30% gl.fib.
			23	7		8		"
			23	7		9		LNP; 40% gl.fib.
			23	<360		8	no irreversible chg.	BASF Ultradur
			60			5	limited resistance	Bayer Pocan
			82	3	0.25%	2	crazed or cracked	LNP; 40% gl.fib.
			82	3		2	"	"
			93	0.21	82.7 MPa	8	no stress cracks observed	Hoechst Cel. Celanex 3300; 30% gl.fib.
			120	16		1		Hoechst Cel. Celanex; gl.fib.
			140	<360		2	rapid attack, stress cracks, irreversible damage	BASF Ultradur
			149	1	0.25%	2	crazed or cracked	LNP; 40% gl.fib.
			149	1		2	"	"
		50	23	60	0.25%	8		LNP Thermocomp WF-1006; 30% gl.fib.
		50	23	60		9		"
		50	23	90		9		Hoechst Cel. Celanex; gl.fib.
		50	23	180		9		"
		50	23	360		9		"
		50	82	30	0.25%	1		LNP Thermocomp WF-1006; 30% gl.fib.
		50	82	30		5		"
		50	150	7	0.25%	1	specimen cracked	"
		50	150	7		5	"	"
Ethylene Oxide			23	<360		8	no irreversible chg.	BASF Ultradur
Fluorine			23	<360		2	rapid attack, stress cracks, irreversible damage	"

Reagent	Reagent Note	Conc. (%)	Temp. (°C)	Time (days)	Load	PDL Rating	Resistance Note	Material Note

Polybutylene Terephthalate (continued)

Reagent	Reagent Note	Conc. (%)	Temp. (°C)	Time (days)	Load	PDL Rating	Resistance Note	Material Note
Hydrogen Peroxide		1	23	<360		8	no irreversible chg.	BASF Ultradur
		20	23			8	resistant	Bayer Pocan
		20	60			5	limited resistance	"
		30	23	<360		8	no irreversible chg.	BASF Ultradur
Isopropyl Alcohol			23			8	resistant	Bayer Pocan
			23	30		9		GE Valox 735; 40% gl.fib./min.
			23	30		9		GE Valox 325
			23	30		9		GE Valox 420; 30% gl.fib.
			23	30		9		GE Valox 760; 25% min.
			23	30		9		GE Valox 310-SEO
			23	90		9		GE Valox 420; 30% gl.fib.
			23	90		9		GE Valox 760; 25% min.
			23	90		9		GE Valox 310-SEO
			23	90		9		GE Valox 325
			23	90		9		GE Valox 735; 40% gl.fib./min.
			23	180		9		GE Valox 325
			23	180		9		GE Valox 420; 30% gl.fib.
			23	180		9		GE Valox 310-SEO
			23	<360		8	no irreversible chg.	BASF Ultradur
			60			5	limited resistance	Bayer Pocan
			60	<360		6	changes in mass; dimen.; props.	BASF Ultradur
		50	23	30		9		GE Valox 420; 30% gl.fib.
		50	23	30		9		GE Valox 310-SEO
		50	23	30		6		GE Valox 735; 40% gl.fib./min.
		50	23	30		9		GE Valox 325
		50	23	30		9		GE Valox 760; 25% min.
		50	23	90		9		GE Valox 325
		50	23	90		9		GE Valox 310-SEO
		50	23	90		9		GE Valox 760; 25% min.
		50	23	90		9		GE Valox 420; 30% gl.fib.
		50	23	90		4		GE Valox 735; 40% gl.fib./min.
		50	23	180		9		GE Valox 310-SEO
		50	23	180		9		GE Valox 420; 30% gl.fib.
		50	23	180		9		GE Valox 325
Phenol			43	<360		2	rapid attack, stress cracks, irreversible damage	BASF Ultradur
		10	23			2	not resistant	Bayer Pocan
		10	60			2	"	"
Potassium Permanganate		1	23	<360		8	no irreversible chg.	BASF Ultradur
		10	23			8	resistant	Bayer Pocan
		10	60			5	limited resistance	"
Salicylic Acid			23	<360		6	changes in mass; dimen.; props.	BASF Ultradur
Sodium Hypochlorite		10	23	<360		6	"	"
Sulfuric Acid		2	23	<360		8	no irreversible chg.	"
		3	23	90		9		Hoechst Cel. Celanex; gl.fib.
		3	23	180		9		"
		3	23	360		9		"
		3	82	24		8		"
		3	82	64		7		"
		10	23			8	resistant	Bayer Pocan
		10	23	7	0.25%	3		LNP; 40% gl.fib.
		10	23	7	0.25%	5		LNP; 30% gl.fib.
		10	23	7		5		"
		10	23	7		5		LNP; 40% gl.fib.
		10	23	30		9		GE Valox 760; 25% min.
		10	23	30		9		GE Valox 325
		10	23	30		9		GE Valox 420; 30% gl.fib.
		10	23	30		5		GE Valox 735; 40% gl.fib./min.
		10	23	90		9		GE Valox 420; 30% gl.fib.
		10	23	90		4		GE Valox 735; 40% gl.fib./min.
		10	23	90		9		GE Valox 760; 25% min.
		10	23	90		9		GE Valox 310-SEO
		10	23	180		8		GE Valox 420; 30% gl.fib.
		10	23	180		8		GE Valox 325
		10	23	180		9		GE Valox 310-SEO
		10	60			5	limited resistance	Bayer Pocan
		10	82	3	0.25%	2		LNP; 40% gl.fib.
		10	82	3		2		"
		10	149	1	0.25%	2	crazed or cracked	"
		10	149	1		2	"	"

Polycyclohexylenedimethylene Terephthalate

Reagent	Reagent Note	Conc. (%)	Temp. (°C)	Time (days)	Load	PDL Rating	Resistance Note	Material Note
Benzyl Alcohol			23	28		9	no visible chg.	Eastman Ektar FB CG002; 30% gl.fib.
			23	28		3	sl. swelling	Eastman Ektar FB CG001; 30% gl.fib.
Bleach	Clorox	5	23	28		9	no visible chg.	Eastman Ektar FB CG002; 30% gl.fib.
	"	5	23	28		7	"	Eastman Ektar FB CG001; 30% gl.fib.
Ethyl Acetate			23	28		7	"	Eastman Ektar FB CG002; 30% gl.fib.
			23	28		7	"	Eastman Ektar FB CG001; 30% gl.fib.
Ethyl Alcohol			23	28		9	"	Eastman Ektar FB CG002; 30% gl.fib.
			23	28		9	"	Eastman Ektar FB CG001; 30% gl.fib.
		50	23	28		9	"	"
		50	23	28		9	"	Eastman Ektar FB CG002; 30% gl.fib.
Methyl Isobutyl Ketone			23	28		7	"	Eastman Ektar FB CG001; 30% gl.fib.
			23	28		9	"	Eastman Ektar FB CG002; 30% gl.fib.

Glycol Modified Polycyclohexylenedimethylene Terephthal

Reagent	Reagent Note	Conc. (%)	Temp. (°C)	Time (days)	Load	PDL Rating	Resistance Note	Material Note
Benzyl Alcohol			23	0.125	0.7%	6	discoloration but no cracking	Eastman Ektar DN001
			23	28		5	no visible chg.	Eastman Ektar FB DG002; 20% gl.fib.
			23	28		5	"	Eastman Ektar 5445
			23	28		5	"	Eastman Ektar DN001
Betadine			23	0.125	0.7%	8	no chg. in appear.	"
Bleach	Clorox	5	23	28		9	no visible chg.	"
	"	5	23	28		9	"	Eastman Ektar 5445
	"	5	23	28		9	"	Eastman Ektar FB DG002; 20% gl.fib.
Ethyl Acetate			23	1		2	hazy	Eastman Kodar 5445; 0.25 mm film
			23	7		2	"	"
			23	28		3	no visible chg.	Eastman Ektar FB DG002; 20% gl.fib.
			23	28		4	sl. discolor.	Eastman Ektar 5445
			23	28		5	"	Eastman Ektar DN001
Ethyl Alcohol			23	1		5	clear appear. retained	Eastman Kodar 5445; 0.25 mm film
			23	7		5	"	"
			23	28		9	no visible chg.	Eastman Ektar 5445
			23	28		9	"	Eastman Ektar FB DG002; 20% gl.fib.
			23	28		9	"	Eastman Ektar DN001
		50	23	28		9	"	"
		50	23	28		8	"	Eastman Ektar FB DG002; 20% gl.fib.
		50	23	28		9	"	Eastman Ektar 5445
Glutaraldehyde		50	23	0.125	0.7%	8	no chg. in appear.	Eastman Ektar DN001
Hydrogen Peroxide		28	23	1		5	clear appear. retained	Eastman Kodar 5445; 0.25 mm film
		28	23	7		4	sl. yellowing	"
Isopropyl Alcohol			23	0.125	0.7%	6	crazing or sl. cracking	Eastman Ektar DN001
Lyposin	IV sol'n lyposin II		23	0.125	0.7%	8	no chg. in appear.	"
Methyl Isobutyl Ketone			23	0.125	0.7%	3	severe cracking or breaking	"
			23	28		5	sl. swelling	"
			23	28		8	no visible chg.	Eastman Ektar FB DG002; 20% gl.fib.
			23	28		5	sl. swelling	Eastman Ektar 5445
Sodium Hypochlorite		3.5	23	1		5	clear appear. retained	Eastman Kodar 5445; 0.25 mm film
		3.5	23	7		6	"	"
	bleaches	5.25	23	0.125	0.7%	8	no chg. in appear.	Eastman Ektar DN001
Sulfuric Acid		10	23	0.125	0.7%	8	"	"

Polycyclohexylenedimethylene Ethylene Terephthalate

Reagent	Reagent Note	Conc. (%)	Temp. (°C)	Time (days)	Load	PDL Rating	Resistance Note	Material Note
Benzyl Alcohol			23	28		0	strongly swollen	Eastman Ektar GN001
			23	28		0	"	Eastman Ektar 6763
Bleach	Clorox	5	23	28		8	no visible chg.	Eastman Ektar GN001
	"	5	23	28		8	"	Eastman Ektar 6763
Ethyl Acetate			23	28		0	strongly swollen	Eastman Ektar GN001
			23	28		0	"	Eastman Ektar 6763
			23	365		1	discolored, swollen, rubber-like	Eastman Kodar 6763

Reagent	Reagent Note	Conc. (%)	Temp. (°C)	Time (days)	Load	PDL Rating	Resistance Note	Material Note
Polycyclohexylenedimethylene Ethylene Terephthalate (continued)								
Ethyl Alcohol			23	28		8	no visible chg.	Eastman Ektar GN001
			23	28		8	"	Eastman Ektar 6763
			23	365		6	sl. yellowing	Eastman Kodar 6763
		50	23	28		9	no visible chg.	Eastman Ektar GN001
		50	23	28		9	"	Eastman Ektar 6763
		50	23	365		6	sl. yellowing	Eastman Kodar 6763
Hydrogen Peroxide		3	23	365		6	"	"
		28	23	365		6	"	"
Methyl Isobutyl Ketone			23	28		0	strongly swollen	Eastman Ektar 6763
			23	28		0	"	Eastman Ektar GN001
Phenol		5	23	365		2	blackened	Eastman Kodar 6763
Sodium Hypochlorite		3.5	23	365		6	sl. yellowing	"
Sulfuric Acid		3	23	365		6	"	"
Toothpaste	Arm & Hammer					1		
Liquid Crystal Polymer								
Chlorine	dry gas		23	30		9	unaffected	Hoechst Cel. Vectra A950
	"		23	30		9	"	Hoechst Cel. Vectra A625; 25% min.
	chlorine water; sat'd		23	30		9	"	Hoechst Cel. Vectra A950
	"		23	30		9	"	Hoechst Cel. Vectra A625; 25% min.
	dry gas		23	30		9	"	Hoechst Cel. Vectra A130; 30% gl.fib.
	"		23	60		9	"	Hoechst Cel. Vectra A950
	"		23	60		9	"	Hoechst Cel. Vectra A625; 25% min.
	"		23	60		9	"	Hoechst Cel. Vectra A130; 30% gl.fib.
	chlorine water; sat'd		23	60		9	"	Hoechst Cel. Vectra A950
	"		23	60		9	"	Hoechst Cel. Vectra A625; 25% min.
	"		23	60		9	"	Hoechst Cel. Vectra A130; 30% gl.fib.
Ethyl Acetate			77	30		6	resistant	Hoechst Cel. Vectra
			77	180		5		Hoechst Cel. Vectra A130; 30% gl.fib.
			77	180		5		Hoechst Cel. Vectra A625; 25% min.
			77	180		5		Hoechst Cel. Vectra A950
Ethyl Alcohol			52	30		6	resistant	Hoechst Cel. Vectra
			52	30		5		Hoechst Cel. Vectra A950
Ethylene Glycol			200	1		8		Amoco Xydar MG350; 50% gl./min.
Isopropyl Alcohol			50	5		9		Amoco Xydar
Morpholine	200 ppm/steam		132	10		5		Hoechst Cel. Vectra A130; 30% gl.fib.
Phenol			100	180		5		Hoechst Cel. Vectra A950
			100	180		4		Hoechst Cel. Vectra A130; 30% gl.fib.
			100	180		5		Hoechst Cel. Vectra A625; 25% min.
Sodium Hypochlorite			50	5		9		Amoco Xydar
		20	88	30		6	resistant	Hoechst Cel. Vectra
Polyimide								
Ammonia						3	should be avoided	DuPont Vespel
Benzoic Acid			23			7	acceptable for use	
			65			7	"	
Boric Acid			23			7	"	
			65			7	"	
Butyl Alcohol			23			8	exc. resistance	
			65			8	"	
Chlorine	wet		23			7	acceptable for use	
	dry		23			7	"	
	wet		65			7	"	
	dry		65			7	"	
Ethyl Acetate			23			8	exc. resistance	
			65			8	"	
Ethyl Alcohol			23			8	"	
			65			8	"	
			99	79		9		DuPont Vespel SP1

Polyimide (continued)

Reagent	Reagent Note	Conc. (%)	Temp. (°C)	Time (days)	Load	PDL Rating	Resistance Note	Material Note
Ethylene Glycol						8	exc. resistance	Lenzing Lenzing P84; coating
			23	7	0.25%	9		LNP; 30% gl.fib.
			23	7		9		"
			82	3	0.25%	3		"
			82	3		4		"
			149	1	0.25%	2		"
			149	1		3		"
Formaldehyde		37	23			7	acceptable for use	
		37	65			7	"	
Isopropyl Alcohol			23	0.0069		9		DuPont Kapton HN; 0.025 mm film
Phenol			23			7	acceptable for use	
			65			7	"	
Potassium Permanganate			23			1	unacceptable, attacked	
			65			1	"	
Sodium Hypochlorite			23			1	"	
			65			1	"	
Sulfuric Acid		5	95	4.2		3		Lenzing Lenzing P84; fiber
		10	20	4.2		8		"
		10	23			7	acceptable for use	
		10	23	7	0.25%	4		LNP; 30% gl.fib.
		10	23	7		5		"
		10	65			7	acceptable for use	
		10	82	3	0.25%	3		LNP; 30% gl.fib.
		10	82	3		3		"
		10	149	1	0.25%	2		"
		10	149	1		2		"

Polyamideimide

Reagent	Reagent Note	Conc. (%)	Temp. (°C)	Time (days)	Load	PDL Rating	Resistance Note	Material Note
Benzaldehyde			93	1		9		Amoco Torlon 4203L; 3.5% TiO2/fluoro.
Butyl Alcohol			93	1		9		"
Ethyl Acetate			93	1		9		"
Ethylene Glycol			23	7		9		Amoco Torlon 4203
			82	3		9		"
			93	1		9		Amoco Torlon 4203L; 3.5% TiO2/fluoro.
			149	1		9		Amoco Torlon 4203
Formaldehyde		37	93	1		9		Amoco Torlon 4203L; 3.5% TiO2/fluoro.
Morpholine			93	1		9		"
Potassium Permanganate		10	93	1		9		"
Sodium Hypochlorite		10	93	1		9		"
Sulfuric Acid		10	23	7		9		Amoco Torlon 4203
		10	82	3		7		"
		10	149	1		2		"

Polyetherimide

Reagent	Reagent Note	Conc. (%)	Temp. (°C)	Time (days)	Load	PDL Rating	Resistance Note	Material Note
Betadine	povidone iodine sol'n		23	168	0.5%	9		Ultem 1000
	"		23	168	1.5%	8		"
	"		23	168		7		"
	"	100	21	14	34.5 MPa	9	no stress cracks or crazing	GE Ultem
Bleach	Clorox	10	23	168	0.5%	8		Ultem 1000
	"	10	23	168	1.5%	3		"
	"	10	23	168		9		"
Butyl Alcohol			22	14	12.4 MPa	8	no cracking for duration of test	GE Ultem 1000
			22	14	17.2 MPa	8	"	"
			22	14	4.1 MPa	8	"	"
			22	14	0.25%	8	"	"
			22	14	0.5%	8	"	"
			22	14		8	"	"
Cidex 7	glutaraldehyde type; disinfectant	100	21	24	34.5 MPa	9	no stress cracks or crazing	GE Ultem
Ethyl Acetate			22	<1	12.4 MPa	3	cracks at time shown	GE Ultem 1000
			22	<1	17.2 MPa	3	"	"
			22	<1	0.5%	3	"	"
			22	14	4.1 MPa	8	no cracking for duration of test	"
			22	14	8.3 MPa	8	"	"
			22	14	0.25%	8	"	"
			22	14		8	"	"

Polyetherimide (continued)

Reagent	Reagent Note	Conc. (%)	Temp. (°C)	Time (days)	Load	PDL Rating	Resistance Note	Material Note
Ethyl Alcohol			22	5	12.4 MPa	8	no cracking for duration of test	GE Ultem 1000
			22	5	17.2 MPa	8	"	"
			22	5	4.1 MPa	8	"	"
			22	5	8.3 MPa	8	"	"
			22	5	0.25%	8	"	"
			22	5	0.5%	8	"	"
			22	5		8	"	"
Ethylene Glycol			23	7	0.25%	9		LNP; 30% carb.fib.
			23	7		9		"
			82	3	0.25%	6		"
			82	3		9		"
			149	1	0.25%	2		"
			149	1		9		"
		50	23	60	0.25%	7		LNP Thermocomp EF-1006; 30% gl.fib.
		50	23	60		9		"
		50	82	30	0.25%	7		"
		50	82	30		8		"
		50	150	7	0.25%	3		"
		50	150	7		5		"
Formaldehyde		37	21	14	34.5 MPa	9	no stress cracks or crazing	GE Ultem
Isopropyl Alcohol			21	14	34.5 MPa	9	"	"
			22	14	12.4 MPa	8	no cracking for duration of test	GE Ultem 1000
			22	14	17.2 MPa	8	"	"
			22	14	4.1 MPa	8	"	"
			22	14	8.3 MPa	8	"	"
			22	14	0.5%	8	"	"
			22	14		8	"	"
		70	23	168	0.5%	9		Ultem 1000
		70	23	168	1.5%	8		"
		70	23	168		8		"
Omnicide	2% glutaraldehyde		23	168	0.5%	7		"
	"		23	168	1.5%	6	some samples fractured	"
	"		23	168		9		"
	"	100	21	18	24.8 MPa	9	no stress cracks or crazing	GE Ultem
	"	100	70	7	12.4 MPa	6	cracking or crazing in 168 hrs.	"
	"	100	70	7	23.4 MPa	6	"	"
Phenol	sat'd aq. sol'n		22	<0.667	12.4 MPa	3	cracks at time shown	GE Ultem 1000
	"		22	<0.667	17.2 MPa	3	"	"
	"		22	<0.667	0.5%	3	"	"
	"		22	0.667	8.3 MPa	3	"	"
	"		22	0.667	0.25%	3	"	"
	"		22	14	4.1 MPa	8	no cracking for duration of test	"
	"		22	14		8	"	"
Sodium Hypochlorite	5% chlorine		21	31	34.5 MPa	9	no stress cracks or crazing	GE Ultem
Sporicidin	buffered glutaraldehyde/ phenolic type	6.25	21	24	34.5 MPa	9	"	"
	"	100	21	<1	34.5 MPa	3	cracking or crazing in <24 hrs.	"
	"	100	21	<2	27.6 MPa	4	cracking or crazing in <48 hrs.	"
	2% glutaraldehyde, phenol, sodium phenate	100	21	18	24.8 MPa	9	no stress cracks or crazing	"
	buffered glutaraldehyde/ phenolic type	100	21	24	22.3 MPa	9	"	"
	2% glutaraldehyde, phenol, sodium phenate	100	70	<1	12.4 MPa	3	cracking or crazing in <24 hrs.	"
Sulfuric Acid		10	23	7	0.25%	0		LNP; 30% carb.fib.
		10	23	7		0		"
		10	82	3	0.25%	0		"
Tergiquat	quaternary ammonia/ detergent type	1.56	21	6	34.5 MPa	5	cracking or crazing in 144 hrs.	GE Ultem
	"	1.56	21	8	27.6 MPa	6	cracking or crazing in 192 hrs.	"
	"	1.56	21	24	22.3 MPa	9	no stress cracks or crazing	"
	"	100	21	5	34.5 MPa	6	cracking or crazing in 120 hrs.	"
	"	100	21	8	27.6 MPa	6	cracking or crazing in 192 hrs.	"
	"	100	21	24	22.3 MPa	9	no stress cracks or crazing	"
Wavecide-01	2% glutaraldehyde	100	21	18	24.8 MPa	9	"	"
	"	100	70	7	12.4 MPa	6	cracking or crazing in 168 hrs.	"
	"	100	70	7	23.4 MPa	6	"	"

Reagent	Reagent Note	Conc. (%)	Temp. (°C)	Time (days)	Load	PDL Rating	Resistance Note	Material Note
Polyketone								
Butyl Alcohol			130	7		9		Amoco Kadel E-1140; 40% gl.fib.
			130	7		9		Amoco Kadel E-3140; 40% gl.fib.
Ethyl Acetate			100	7		9		Amoco Kadel E-1140; 40% gl.fib.
			100	7		9		Amoco Kadel E-3140; 40% gl.fib.
Ethylene Glycol			130	7		8		"
			130	7		9		Amoco Kadel E-1140; 40% gl.fib.
Isopropyl Alcohol			100	7		7		"
			100	7		6		Amoco Kadel E-1000
			100	7		7		Amoco Kadel E-3140; 40% gl.fib.
			100	90		6		Amoco Kadel E-1000
			100	90		8		Amoco Kadel E-3140; 40% gl.fib.
			100	90		8		Amoco Kadel E-1140; 40% gl.fib.
Phenol			130	7		1		"
			130	7		6		Amoco Kadel E-3140; 40% gl.fib.
Polyetherketone								
Sodium Hypochlorite			23	7		9		LNP Victrex 220GL30; 30% gl.fib.
			23	7		9		LNP Victrex 220G
			23	30		9		"
			23	30		9		LNP Victrex 220GL30; 30% gl.fib.
Polyetheretherketone								
Benzaldehyde			23	7		9		LNP Victrex 450G
Chlorine	chlorine water; sat'd sol'n		23	30		2	degraded	"
Ethyl Acetate			23	0.042	3%	8	no effect	"
			23	7		9		"
Ethyl Alcohol			23	7		9		"
Ethylene Glycol			23	7	0.25%	5		LNP; 20% carb.fib.
			23	7	0.25%	9		LNP; 30% gl.fib.
			23	7		8		LNP Victrex 450G
			23	7		9		LNP; 30% gl.fib.
			23	7		6		LNP; 20% carb.fib.
			82	3	0.25%	6		"
			82	3	0.25%	7		LNP; 30% gl.fib.
			82	3		7		LNP; 20% carb.fib.
			82	3		8		LNP; 30% gl.fib.
			149	1	0.25%	2		"
			149	1	0.25%	4		LNP; 20% carb.fib.
			149	1		6		"
			149	1		6		LNP; 30% gl.fib.
Formaldehyde			23	7		7		LNP Victrex 450G
Hydrogen Peroxide		28	23	7		9		"
		30	23	30		9		"
Isopropyl Alcohol			23	0.015	3%	8	no effect	"
Sulfuric Acid		3	23	7		8		"
		10	23	7	0.25%	7		LNP; 30% gl.fib.
		10	23	7	0.25%	7		LNP; 20% carb.fib.
		10	23	7		8		"
		10	23	7		8		LNP; 30% gl.fib.
		10	82	3	0.25%	6		LNP; 20% carb.fib.
		10	82	3	0.25%	4		LNP; 30% gl.fib.
		10	82	3		7		LNP; 20% carb.fib.
		10	82	3		5		LNP; 30% gl.fib.
		10	149	1	0.25%	6		LNP; 20% carb.fib.
		10	149	1	0.25%	2		LNP; 30% gl.fib.
		10	149	1		6		LNP; 20% carb.fib.
		10	149	1		5		LNP; 30% gl.fib.
Low Density Polyethylene								
Ammonia	dry gas	100	21			8	resistant	DuPont Can. Sclair
	"	100	60			8	"	"
Benzoic Acid			21			8	"	Quantum Petrothene
	all concentrations		21			8	"	DuPont Can. Sclair
			60			8	"	Quantum Petrothene
	all concentrations		60			8	"	DuPont Can. Sclair
Boric Acid	concentrated sol'n		21			8	"	Quantum Petrothene
	all concentrations		21			8	"	DuPont Can. Sclair
	dilute		21			8	"	Quantum Petrothene
	all concentrations		60			8	"	DuPont Can. Sclair

Reagent	Reagent Note	Conc. (%)	Temp. (°C)	Time (days)	Load	PDL Rating	Resistance Note	Material Note

Low Density Polyethylene (continued)

Reagent	Reagent Note	Conc. (%)	Temp. (°C)	Time (days)	Load	PDL Rating	Resistance Note	Material Note
Boric Acid	concentrated sol'n		60			8	resistant	Quantum Petrothene
	dilute		60			8	“	“
Butyl Alcohol			21			8	“	“
			60			8	“	“
		100	21			6	stress crack agent; otherwise acceptable	DuPont Can. Sclair
		100	60			6	“	“
Chlorine	liquid		21			2	not recommended for use	“
	moist gas		21			5	some attack	Quantum Petrothene
	liquid		21			2	unsatisf. resistance	“
	“		60			2	not recommended for use	DuPont Can. Sclair
	“		60			2	unsatisf. resistance	Quantum Petrothene
	moist gas		60			2	“	“
	chlorine water; sat'd sol'n	2	21			8	resistant	DuPont Can. Sclair
	chlorine water	2	21			2	unsatisf. resistance	Quantum Petrothene
	chlorine water; sat'd sol'n	2	60			8	resistant	DuPont Can. Sclair
	chlorine water	2	60			2	unsatisf. resistance	Quantum Petrothene
	dry gas	100	21			5	resistance is conditional; oxidizer	DuPont Can. Sclair
	“	100	60			2	not recommended for use	“
Cresol			21			8	resistant	Quantum Petrothene
			60			8	“	“
Ethyl Acetate			21			5	some attack	“
			23	365		6	no chg. in appear., poss. stress crack agent	Eastman Tenite
			60			2	unsatisf. resistance	Quantum Petrothene
		100	21			4	suspected stress crack agent, plasticizer	DuPont Can. Sclair
		100	60			2	stress crack agent, plasticizer	“
Ethyl Alcohol			21			6	resistant, poss. stress cracks	Quantum Petrothene
			60			6	“	“
		35	21			6	stress crack agent; otherwise acceptable	DuPont Can. Sclair
		35	21			6	resistant, poss. stress cracks	Quantum Petrothene
		35	60			6	stress crack agent; otherwise acceptable	DuPont Can. Sclair
		35	60			6	resistant, poss. stress cracks	Quantum Petrothene
		50	23	365		8	no chg. in appear., poss. stress crack agent	Eastman Tenite
		95	23	365		7	no chg. in appear., stress crack agent	“
		100	21			6	stress crack agent; otherwise acceptable	DuPont Can. Sclair
		100	60			6	“	“
Ethylene Chlorohydrin			21			2	unsatisf. resistance	Quantum Petrothene
			60			2	“	“
Ethylene Glycol			21			6	stress crack agent; otherwise acceptable	DuPont Can. Sclair
			21			6	resistant, poss. stress cracks	Quantum Petrothene
			23	365		7	no chg. in appear., stress crack agent	Eastman Tenite
			60			6	stress crack agent; otherwise acceptable	DuPont Can. Sclair
			60			6	resistant, poss. stress cracks	Quantum Petrothene
Fluorine			21			8	resistant	“
			60			2	unsatisf. resistance	“
Formaldehyde		35	23	365		8	no chg. in appear., poss. stress crack agent	Eastman Tenite
		40	21			6	resistant, poss. stress cracks	Quantum Petrothene
		40	60			6	“	“
Hydrogen Peroxide		30	21			8	resistant	“
		30	23	365		9	no chg. in appear.	Eastman Tenite
		30	60			8	resistant	Quantum Petrothene
		90	21			8	“	“
		90	60			2	unsatisf. resistance	“
Iodine	in KI sol'n		21			5	resistance is conditional; oxidizer	DuPont Can. Sclair
	“		21			5	some attack	Quantum Petrothene
	“		60			2	not recommended for use	DuPont Can. Sclair
			60			2	unsatisf. resistance	Quantum Petrothene
Mercuric Chloride	sat'd		21			8	resistant	“
	“		60			8	“	“

Reagent	Reagent Note	Conc. (%)	Temp. (°C)	Time (days)	Load	PDL Rating	Resistance Note	Material Note
Low Density Polyethylene (continued)								
Phenol		5	23	365		8	no chg. in appear., poss. stress crack agent	Eastman Tenite
		90	21			2	unsatisf. resistance	Quantum Petrothene
		90	60			2	"	"
Potassium Permanganate		20	21			8	resistant	DuPont Can. Sclair
		20	21			8	"	Quantum Petrothene
		20	60			8	"	DuPont Can. Sclair
		20	60			8	"	Quantum Petrothene
Salicylic Acid	sat'd		21			8	"	DuPont Can. Sclair
	"		60			8	"	"
Silver Nitrate	sol'n		21			8	"	"
	"		21			8	"	Quantum Petrothene
	"		60			8	"	DuPont Can. Sclair
			60			8	"	Quantum Petrothene
Sodium Hypochlorite			21			8	"	"
			21			6	variable resistance, excercise caution	DuPont Can. Sclair
			60			8	resistant	Quantum Petrothene
			60			6	variable resistance, excercise caution	DuPont Can. Sclair
		5	23	365		9	no chg. in appear.	Eastman Tenite
Sulfuric Acid		3	23	365		9	"	"
		0-50	21			8	resistant	DuPont Can. Sclair
		0-50	60			8	"	"
Medium Density Polyethylene								
Ammonia	dry gas	100	21			8	"	"
	"	100	60			8	"	"
Benzoic Acid	all concentrations		21			8	"	"
	"		60			8	"	"
Boric Acid	"		21			8	"	"
	"		60			8	"	"
Butyl Alcohol		100	21			6	stress crack agent; otherwise acceptable	"
		100	60			6	"	"
Chlorine	liquid		21			2	not recommended for use	"
	"		60			2	"	"
	chlorine water; sat'd sol'n	2	21			8	resistant	"
	"	2	60			8	"	"
	dry gas	100	21			5	resistance is conditional; oxidizer	"
	"	100	60			2	not recommended for use	"
Ethyl Acetate		100	21			4	suspected stress crack agent, plasticizer	"
		100	60			2	stress crack agent, plasticizer	"
Ethyl Alcohol		35	21			6	stress crack agent; otherwise acceptable	"
		35	60			6	"	"
		100	21			6	"	"
		100	60			6	"	"
Ethylene Glycol			21			6	"	"
			60			6	"	"
Iodine	in KI sol'n		21			5	resistance is conditional; oxidizer	"
	"		60			2	not recommended for use	"
Potassium Permanganate		20	21			8	resistant	"
		20	60			8	"	"
Salicylic Acid	sat'd		21			8	"	"
	"		60			8	"	"
Silver Nitrate	sol'n		21			8	"	"
	"		60			8	"	"
Sodium Hypochlorite			21			6	variable resistance, excercise caution	"
			60			6	"	"
Sulfuric Acid		0-50	21			8	resistant	"
		0-50	60			8	"	"
High Density Polyethylene								
Ammonia	ammonia water		20	60		8	"	Hoechst Cel. Hostalen HMW
	liquid		20	60		8	"	"
	sol'n		20	60		8	"	"
	gas		20	60		8	"	"

High Density Polyethylene (continued)

Reagent	Reagent Note	Conc. (%)	Temp. (°C)	Time (days)	Load	PDL Rating	Resistance Note	Material Note
Ammonia	dry gas		21			8	resistant	Phillips Marlex
	"		21			8	"	Solvay Fortiflex
	"		60			8	"	Phillips Marlex
	"		60			8	"	Solvay Fortiflex
	ammonia water		60	60		8	"	Hoechst Cel. Hostalen HMW
	gas		60	60		8	"	"
	sol'n		60	60		8	"	"
	dry gas	100	21			8	"	DuPont Can. Sclair
	"	100	60			8	"	"
Benzaldehyde			20	60		8	"	Hoechst Cel. Hostalen HMW
			21			8	"	Solvay Fortiflex
			27	90		6		Phillips Marlex
			49	90		4		"
			60			5	some attack	Solvay Fortiflex
			60	60		7	limited resistance to resistant	Hoechst Cel. Hostalen HMW
			27-65	<=90		8	exc. resistance, <3% perm. loss/year	Phillips Marlex
			66	90		3		"
	in isopropyl alcohol	1	20	60		8	resistant	Hoechst Cel. Hostalen HMW
	"	1	60	60		8	"	"
Benzoic Acid			20	60		8	"	"
	all concentrations		21			8	"	DuPont Can. Sclair
	crystals		21			8	"	Solvay Fortiflex
	sat'd		21			8	"	"
	all concentrations		60			8	"	DuPont Can. Sclair
	crystals		60			8	"	Solvay Fortiflex
	sat'd		60			8	"	"
			60	60		8	"	Hoechst Cel. Hostalen HMW
Benzyl Alcohol			20	60		8	"	"
			60	60		8	"	"
Bleach	12.5% chlorine		20	60		5	limited resistance	"
	"		20	60		7	limited resistance to resistant	"
	bleaching powder; chloride of lime		20	60		8	resistant	"
	12.5% chlorine		60	60		3	not resistant	"
	"		60	60		3	"	"
	bleaching powder; chloride of lime		60	60		8	resistant	"
			27-65	<=90		8	exc. resistance, <3% perm. loss/year	Phillips Marlex
Boric Acid	all concentrations		20	60		8	resistant	Hoechst Cel. Hostalen HMW
	dilute		21			8	"	Phillips Marlex
	all concentrations		21			8	"	DuPont Can. Sclair
	concentrated sol'n		21			8	"	Solvay Fortiflex
	dilute		21			8	"	"
	all concentrations		60			8	"	DuPont Can. Sclair
	dilute		60			8	"	Phillips Marlex
	"		60			8	"	Solvay Fortiflex
	concentrated sol'n		60			8	"	"
	all concentrations		60	60		8	"	Hoechst Cel. Hostalen HMW
	concentrated sol'n		27-65	<=90		8	exc. resistance, <3% perm. loss/year	Phillips Marlex
Butyl Alcohol			20	60		8	resistant	Hoechst Cel. Hostalen HMW
			21			8	"	Solvay Fortiflex
			50	28		9		bottles
			60			8	resistant	Solvay Fortiflex
			60	60		8	"	Hoechst Cel. Hostalen HMW
			27-65	<=90		8	exc. resistance, <3% perm. loss/year	Phillips Marlex
		100	21			6	stress crack agent; otherwise acceptable	DuPont Can. Sclair
		100	60			6	"	"
Carbolic Acid			20	60		8	resistant	Hoechst Cel. Hostalen HMW
			60	60		7	resistant, discolors	"
Chlorine	moist gas		20	60		5	limited resistance	"
	dry gas		20	60		5	"	"
	liquid		20	60		3	not resistant	"
	chlorine water		20	60		8	resistant	"
	liquid		21			2	not recommended for use	DuPont Can. Sclair
	"		21			5	some attack	Phillips Marlex
	"		21			5	"	Solvay Fortiflex
	"		60			2	unsatisf. resistance	"

High Density Polyethylene (continued)

Reagent	Reagent Note	Conc. (%)	Temp. (°C)	Time (days)	Load	PDL Rating	Resistance Note	Material Note
Chlorine	liquid		60			2	unsatisf. resistance	Phillips Marlex
	"		60			2	not recommended for use	DuPont Can. Sclair
	chlorine water		60	60		5	limited resistance	Hoechst Cel. Hostalen HMW
	dry gas		60	60		3	not resistant	"
	moist gas		60	60		3	"	"
	chlorine water; sat'd sol'n	2	21			8	resistant	DuPont Can. Sclair
	"	2	60			8	"	"
Cresol	diluted		20	60		8	"	Hoechst Cel. Hostalen HMW
			21			8	"	Solvay Fortiflex
			60			5	some attack	"
	diluted		60	60		7	resistant, discolors	Hoechst Cel. Hostalen HMW
		100	20	60		8	resistant	"
		100	60	60		4	limited resistance, discolors	"
Ethyl Acetate			20	60		8	resistant	"
			21			5	some attack	Solvay Fortiflex
			27	90		4		Phillips Marlex
			50	28		3		bottles
			60			5	some attack	Solvay Fortiflex
			60	60		5	limited resistance	Hoechst Cel. Hostalen HMW
			27-65	<=90		4	softening and deform., 9% perm. loss/year	Phillips Marlex
		100	21			4	suspected stress crack agent, plasticizer	DuPont Can. Sclair
Ethyl Alcohol			21			8	resistant	Solvay Fortiflex
			27	90		8		Phillips Marlex
			60			8	resistant	Solvay Fortiflex
			27-65	<=90		7	exc. resistance, <3% perm. loss/year, poss. stress cracks	Phillips Marlex
		35	21			6	stress crack agent; otherwise acceptable	DuPont Can. Sclair
		35	21			8	resistant	Solvay Fortiflex
		35	60			8	"	"
		35	60			6	stress crack agent; otherwise acceptable	DuPont Can. Sclair
		96	20	60		8	resistant	Hoechst Cel. Hostalen HMW
		96	60	60		8	"	"
		100	21			6	stress crack agent; otherwise acceptable	DuPont Can. Sclair
		100	60			6	"	"
Ethylene Glycol			20	60		8	resistant	Hoechst Cel. Hostalen HMW
			21			6	stress crack agent; otherwise acceptable	DuPont Can. Sclair
			21			8	resistant	Solvay Fortiflex
			23	7	0.25%	9		LNP; 30% gl.fib.
			23	7		9		"
			27	90		9		Phillips Marlex
			49	90		8		"
			60			8	resistant	Solvay Fortiflex
			60			6	stress crack agent; otherwise acceptable	DuPont Can. Sclair
			60	60		8	resistant	Hoechst Cel. Hostalen HMW
			27-65	<=90		8	exc. resistance, <3% perm. loss/year	Phillips Marlex
			66	90		7		"
			81	3		8		LNP; 30% gl.fib.
Ethylene Oxide	gas		20	60		8	resistant	Hoechst Cel. Hostalen HMW
	"		60	60		8	"	"
Fluorine	"		20	60		3	not resistant	"
			21			8	resistant	Phillips Marlex
			21			8	"	Solvay Fortiflex
			60			2	unsatisf. resistance	"
			60			2	"	Phillips Marlex
Formaldehyde		30	21			8	resistant	Solvay Fortiflex
		30	60			8	"	"
		40	20	60		8	"	Hoechst Cel. Hostalen HMW
		40	21			8	"	Solvay Fortiflex
		40	60			5	some attack	"
		40	60	60		8	resistant	Hoechst Cel. Hostalen HMW
		40	27-65	<=90		8	exc. resistance, <3% perm. loss/year	Phillips Marlex
Hydrogen Peroxide		3	27-65	<=90		8	"	"
		10	20	60		8	resistant	Hoechst Cel. Hostalen HMW

High Density Polyethylene (continued)

Reagent	Reagent Note	Conc. (%)	Temp. (°C)	Time (days)	Load	PDL Rating	Resistance Note	Material Note
Hydrogen Peroxide		10	60	60		8	resistant	Hoechst Cel. Hostalen HMW
		30	20	60		8	"	"
		30	21			8	"	Solvay Fortiflex
		30	27	90		9		Phillips Marlex
		30	49	90		8		"
		30	60			8	resistant	Solvay Fortiflex
		30	60	60		8	"	Hoechst Cel. Hostalen HMW
		30	66	90		7		Phillips Marlex
		90	20	60		8	resistant	Hoechst Cel. Hostalen HMW
		90	21			8	"	Solvay Fortiflex
		90	60			5	some attack	"
		90	60	60		3	not resistant	Hoechst Cel. Hostalen HMW
Iodine	tincture		20	60		8	resistant	"
	crystals		21			5	some attack	Solvay Fortiflex
	in KI sol'n		21			5	resistance is conditional; oxidizer	DuPont Can. Sclair
	crystals		60			5	some attack	Solvay Fortiflex
	tincture		60	60		4	limited resistance, discolors	Hoechst Cel. Hostalen HMW
	"		27-65	<=90		6	stain/embrittlement poss., <3% perm. loss/year	Phillips Marlex
Isopropyl Alcohol			20	60		8	resistant	Hoechst Cel. Hostalen HMW
			20	60		8	"	"
			21			8	"	Solvay Fortiflex
			60			8	"	"
			60	60		8	"	Hoechst Cel. Hostalen HMW
			60	60		8	"	"
Lime Chloride	bleaching powder		20	60		8	"	"
	"		60	60		8	"	"
Mercuric Chloride			20	60		8	"	"
			21			8	"	Phillips Marlex
			21			8	"	Solvay Fortiflex
			60			8	"	Phillips Marlex
			60			8	"	Solvay Fortiflex
			60	60		8	"	Hoechst Cel. Hostalen HMW
Methyl Isobutyl Ketone			20	60		8	"	"
			50	28		5		bottles
			60	60		4	limited to not resistant	Hoechst Cel. Hostalen HMW
Morpholine			20	60		8	resistant	"
			60	60		8	"	"
Phenol			20	60		8	"	"
			21			8	"	Solvay Fortiflex
			27	90		7		Phillips Marlex
			49	90		8		"
			60			8	resistant	Solvay Fortiflex
			60	60		7	resistant, discolors	Hoechst Cel. Hostalen HMW
			66	90		7		Phillips Marlex
		90	27-65	<=90		8	exc. resistance, <3% perm. loss/year	"
Potassium Permanganate			20	60		8	resistant	Hoechst Cel. Hostalen HMW
			60	60		8	"	"
		<=6	20	60		8	"	"
		<=6	60	60		7	resistant, discolors	"
		20	21			8	resistant	Phillips Marlex
		20	21			8	"	DuPont Can. Sclair
		20	21			8	"	Solvay Fortiflex
		20	60			8	"	Phillips Marlex
		20	60			8	"	DuPont Can. Sclair
		20	60			8	"	Solvay Fortiflex
Propylene Oxide			20	60		8	"	Hoechst Cel. Hostalen HMW
			60	60		8	"	"
Salicylic Acid			20	60		8	"	"
	sat'd		21			8	"	DuPont Can. Sclair
			21			8	"	Solvay Fortiflex
	sat'd		60			8	"	DuPont Can. Sclair
			60			8	"	Solvay Fortiflex
			60	60		8	"	Hoechst Cel. Hostalen HMW
Silver Nitrate	all concentrations		20	60		8	"	"
	sol'n		21			8	"	DuPont Can. Sclair
	"		21			8	"	Solvay Fortiflex
	"		60			8	"	DuPont Can. Sclair
	"		60			8	"	Solvay Fortiflex

Reagent	Reagent Note	Conc. (%)	Temp. (°C)	Time (days)	Load	PDL Rating	Resistance Note	Material Note
High Density Polyethylene (continued)								
Silver Nitrate	all concentrations		60	60		8	resistant	Hoechst Cel. Hostalen HMW
	sol'n		27-65	<=90		8	exc. resistance, <3% perm. loss/year	Phillips Marlex
Sodium Hypochlorite	dry		20	60		8	resistant	Hoechst Cel. Hostalen HMW
	12% chlorine		20	60		8	"	"
			21			8	"	Solvay Fortiflex
			21			6	variable resistance, excercise caution	DuPont Can. Sclair
			60			8	resistant	Solvay Fortiflex
			60			6	variable resistance, excercise caution	DuPont Can. Sclair
	12% chlorine		60	60		3	not resistant	Hoechst Cel. Hostalen HMW
Sulfuric Acid		10	23	7	0.25%	6		LNP; 30% gl.fib.
		10	23	7		7		"
		<50	21			8	resistant	Phillips Marlex
		<50	60			8	"	"
		0-50	21			6	oxidizer; no other indication of impaired service	DuPont Can. Sclair
		0-50	60			8	resistant	"
Polyethylene Copolymer								
Ethyl Acetate			20	1		8	no changes observed	Arco Arcel
			20	30		8	"	"
Ethyl Alcohol			20	1		8	"	"
			20	30		8	"	"
Hydrogen Peroxide			20	1		8	"	"
			20	30		8	"	"
Polypropylene								
Ammonia	dry gas		20	30		9	should be suitable for all applications	Himont Pro-Fax
	anhydrous		21			8	"	Solvay Fortilene
			23			8	no signif. effect	Phillips Marlex
	anhydrous		60			8	"	Solvay Fortilene
	dry gas		60	30		9	should be suitable for all applications	Himont Pro-Fax
		10	20	180		9		Network Pol. Polyfine
	aq. sol'n	30	20	30		9	should be suitable for all applications	Himont Pro-Fax
		30	20	180		9		Network Pol. Polyfine
	aq. sol'n	30	21			8	resistant	Solvay Fortilene
	"	30	60			8	"	"
Benzaldehyde			21			8	"	"
			23			8	no signif. effect	Phillips Marlex
			60			8	resistant	Solvay Fortilene
Benzoic Acid			20	30		9	should be suitable for all applications	Himont Pro-Fax
			23			8	no signif. effect	Phillips Marlex
			60	30		9	should be suitable for all applications	Himont Pro-Fax
Benzyl Alcohol			20	30		9	"	"
			21			8	"	Solvay Fortilene
			23			8	no signif. effect	Phillips Marlex
			23	28		9	no visible chg.	Eastman Ektar FB PG001; 10% gl.fib.
			60			8	"	Solvay Fortilene
			80	30		9	should be suitable for all applications	Himont Pro-Fax
Bleach	Clorox	5	23	28		9	no visible chg.	Eastman Ektar FB PG001; 10% gl.fib.
Boric Acid			20	30		9	should be suitable for all applications	Himont Pro-Fax
			21			8	resistant	Solvay Fortilene
			23			8	no signif. effect	Phillips Marlex
			60			8	resistant	Solvay Fortilene
			60	30		9	should be suitable for all applications	Himont Pro-Fax
Butyl Alcohol			21			8	"	Solvay Fortilene
			23			8	no signif. effect	Phillips Marlex
			60			8	"	Solvay Fortilene
			100	30		9	should be suitable for all applications	Himont Pro-Fax
Chlorine	gas		20	30		1	specimen dissolves or disintegrates	"
	wet		21			5	some attack	Solvay Fortilene

Polypropylene (continued)

Reagent	Reagent Note	Conc. (%)	Temp. (°C)	Time (days)	Load	PDL Rating	Resistance Note	Material Note
Chlorine	dry		21			2	unsatisf. resistance	Solvay Fortilene
	gas		23			4	may attack polymer causing eventual embrittlement	Phillips Marlex
	dry		60			2	unsatisf. resistance	Solvay Fortilene
	wet		60			2	"	"
	gas		60	30		1	specimen dissolves or disintegrates	Himont Pro-Fax
	moist gas		70	30		1	"	"
Cresol			21			8	resistant	Solvay Fortilene
			23			7	effect from elevated temp. is poss.	Phillips Marlex
			60			8	resistant	Solvay Fortilene
Ethyl Acetate			20	30		7	limited absorp. or attack	Himont Pro-Fax
			20	180		4		Network Pol. Polyfine
			21			8	resistant	Solvay Fortilene
			23	28		4	no visible chg.	Eastman Ektar FB PG001; 10% gl.fib.
			23	30		4		Phillips Marlex
			23	365		4	slightly swollen	Eastman Tenite 4231
			60			8	"	Solvay Fortilene
			60	30		7	limited absorp. or attack	Himont Pro-Fax
			60	180		4		Network Pol. Polyfine
Ethyl Alcohol			20	30		9	should be suitable for all applications	Himont Pro-Fax
			21			8	"	Solvay Fortilene
			21			8	"	"
			23			8	no signif. effect	Phillips Marlex
			23	28		9	no visible chg.	Eastman Ektar FB PG001; 10% gl.fib.
			23	30		9		Phillips Marlex
			60			8	"	Solvay Fortilene
			60			8	"	"
			60	30		9	should be suitable for all applications	Himont Pro-Fax
			100			8	resistant	Solvay Fortilene
		50	23	28		9	no visible chg.	Eastman Ektar FB PG001; 10% gl.fib.
		50	23	365		9	no chg. in appear.	Eastman Tenite 4231
		95	23	365		9	"	"
		96	20	30		9	should be suitable for all applications	Himont Pro-Fax
		96	20	180		9		Network Pol. Polyfine
		96	60	30		9	should be suitable for all applications	Himont Pro-Fax
		96	60	180		9		Network Pol. Polyfine
		96	80	30		9	should be suitable for all applications	Himont Pro-Fax
		96	100	180		9		Network Pol. Polyfine
Ethylene Chlorohydrin			21			8	resistant	Solvay Fortilene
			60			8	"	"
Ethylene Glycol			20	30		9	should be suitable for all applications	Himont Pro-Fax
			21			8	"	Solvay Fortilene
			23			8	no signif. effect	Phillips Marlex
			23	7	0.25%	9		LNP; 30% gl.fib.
			23	7		9		"
			23	365		9	no chg. in appear.	Eastman Tenite 4231
			60			8	"	Solvay Fortilene
			60	30		9	should be suitable for all applications	Himont Pro-Fax
			81	3		8		LNP; 30% gl.fib.
Ethylene Oxide			10	30		7	limited absorp. or attack	Himont Pro-Fax
			21			8	resistant	Solvay Fortilene
			23			5	softening and swelling possible	Phillips Marlex
Fluorine			21			2	unsatisf. resistance	Solvay Fortilene
			23			4	may attack polymer causing eventual embrittlement	Phillips Marlex
			60			2	unsatisf. resistance	Solvay Fortilene
Formaldehyde			20	180		9		Network Pol. Polyfine
			21			8	resistant	Solvay Fortilene
			60			8	"	"
			100			5	some attack	"
		35	23			8	no signif. effect	Phillips Marlex
		35	23	365		8	slightly yellowed	Eastman Tenite 4231
		40	20	30		9	should be suitable for all applications	Himont Pro-Fax

Polypropylene (continued)

Reagent	Reagent Note	Conc. (%)	Temp. (°C)	Time (days)	Load	PDL Rating	Resistance Note	Material Note
Formaldehyde		40	23			8	no signif. effect	Phillips Marlex
		40	60	30		9	should be suitable for all applications	Himont Pro-Fax
Hydrogen Peroxide		3	20	30		9	"	"
		3	20	180		9		Network Pol. Polyfine
		10	20	30		9	should be suitable for all applications	Himont Pro-Fax
		10	20	180		9		Network Pol. Polyfine
		10	60	30		7	limited absorp. or attack	Himont Pro-Fax
		10	60	180		4		Network Pol. Polyfine
		30	20	30		9	should be suitable for all applications	Himont Pro-Fax
		30	20	180		9		Network Pol. Polyfine
		30	21			8	resistant	Solvay Fortilene
		30	23	365		8	yellowed	Eastman Tenite 4231
		30	60			5	some attack	Solvay Fortilene
		30	100	30		1	specimen dissolves or disintegrates	Himont Pro-Fax
		30	100	180		1	not resistant	Network Pol. Polyfine
		90	21			5	some attack	Solvay Fortilene
		90	60			5	"	"
		90	100			2	unsatisf. resistance	"
Iodine	tincture		20	30		9	should be suitable for all applications	Himont Pro-Fax
	dry		21			8	resistant	Solvay Fortilene
	moist		21			2	unsatisf. resistance	"
	dry		23			8	no signif. effect	Phillips Marlex
	moist		23			4	may attack polymer causing eventual embrittlement	"
	dry		60			8	resistant	Solvay Fortilene
Isopropyl Alcohol			20	30		9	should be suitable for all applications	Himont Pro-Fax
			20	180		9		Network Pol. Polyfine
			21			8	resistant	Solvay Fortilene
			23			8	no signif. effect	Phillips Marlex
			60			8	resistant	Solvay Fortilene
			60	30		9	should be suitable for all applications	Himont Pro-Fax
			60	180		9		Network Pol. Polyfine
Mercuric Chloride			21			8	resistant	Solvay Fortilene
			23			8	no signif. effect	Phillips Marlex
			60			8	resistant	Solvay Fortilene
		40	20	30		9	should be suitable for all applications	Himont Pro-Fax
		40	60	30		9	"	"
Methyl Isobutyl Ketone			21			8	resistant	Solvay Fortilene
			23			5	softening and swelling possible	Phillips Marlex
			23	28		4	no visible chg.	Eastman Ektar FB PG001; 10% gl.fib.
			60			8	resistant	Solvay Fortilene
Phenol			20	30		9	should be suitable for all applications	Himont Pro-Fax
			23			8	no signif. effect	Phillips Marlex
			60	30		9	should be suitable for all applications	Himont Pro-Fax
		5	23	365		9	no chg. in appear.	Eastman Tenite 4231
Potassium Permanganate			21			8	resistant	Solvay Fortilene
			23			8	no signif. effect	Phillips Marlex
			60			5	some attack	Solvay Fortilene
		20	20	30		9	should be suitable for all applications	Himont Pro-Fax
		20	60	30		9	"	"
Silver Nitrate			21			8	resistant	Solvay Fortilene
			23			8	no signif. effect	Phillips Marlex
			60			8	resistant	Solvay Fortilene
			100			8	"	"
			23	365		9	no chg. in appear.	Eastman Tenite 4231
		20	20	30		9	should be suitable for all applications	Himont Pro-Fax
		20	20	180		9		Network Pol. Polyfine
		20	60	30		7	limited absorp. or attack	Himont Pro-Fax
		20	60	180		4		Network Pol. Polyfine
		20	100	30		7	limited absorp. or attack	Himont Pro-Fax

Reagent	Reagent Note	Conc. (%)	Temp. (°C)	Time (days)	Load	PDL Rating	Resistance Note	Material Note

Polypropylene (continued)

Reagent	Reagent Note	Conc. (%)	Temp. (°C)	Time (days)	Load	PDL Rating	Resistance Note	Material Note
Sulfuric Acid		2	20	180		9		Network Pol. Polyfine
		2	60	180		9		"
		2	100	180		9		"
		3	23			8	no signif. effect	Phillips Marlex
		3	23	365		9	no chg. in appear.	Eastman Tenite 4231
		10	20	30		9	should be suitable for all applications	Himont Pro-Fax
		10	20	180		9		Network Pol. Polyfine
		10	21			8	resistant	Solvay Fortilene
		10	23			8	no signif. effect	Phillips Marlex
		10	23	7	0.25%	9		LNP; 30% gl.fib.
		10	23	7		9		"
		10	60			8	resistant	Solvay Fortilene
		10	60	30		9	should be suitable for all applications	Himont Pro-Fax
		10	60	180		9		Network Pol. Polyfine
		10	100			8	resistant	Solvay Fortilene
		10	100	30		9	should be suitable for all applications	Himont Pro-Fax
		10	100	180		9		Network Pol. Polyfine

Polyallomer

Reagent	Reagent Note	Conc. (%)	Temp. (°C)	Time (days)	Load	PDL Rating	Resistance Note	Material Note
Ethyl Acetate			23	365		5	no chg. in appear.	Eastman Tenite 5020
Ethyl Alcohol			23	365		9	"	"
		50	23	365		9	"	"
Hydrogen Peroxide		3	23	365		9	"	"
		28	23	365		9	"	"
Iodine	tincture		23	365		7	stained	Eastman Tenite 5020; 0.25 mm thick
	"		60	365		3	"	"
Phenol		5	23	365		9	no chg. in appear.	Eastman Tenite 5020
Sodium Hypochlorite		3.5	23	365		8	discolored	"
Sulfuric Acid		3	23	365		9	no chg. in appear.	"

Fluorinated Polyethylene

Reagent	Reagent Note	Conc. (%)	Temp. (°C)	Time (days)	Load	PDL Rating	Resistance Note	Material Note
Ethyl Acetate			50	28		4		bottles; surface treated
Methyl Isobutyl Ketone			50	28		8		▪

Polymethylpentene

Reagent	Reagent Note	Conc. (%)	Temp. (°C)	Time (days)	Load	PDL Rating	Resistance Note	Material Note
Ammonia			20	90		8	should be suitable for all applications	Mitsui TPX
			60	90		8	▪	▪
Benzaldehyde			20	90		8	▪	▪
			60	90		6	limited absorp. or attack	▪
Benzoic Acid			20	90		8	should be suitable for all applications	▪
			60	90		8	▪	▪
Benzyl Alcohol			20	90		8	▪	▪
			60	90		8	▪	▪
Chlorine	chlorine water		20	90		1	extensive attack	▪
	▪		60	90		1	▪	▪
Ethyl Acetate			20	90		6	limited absorp. or attack	▪
			60	90		4	extensive absorp.	▪
Ethyl Alcohol			20	90		8	should be suitable for all applications	▪
			60	90		6	limited absorp. or attack	▪
Ethylene Glycol			20	90		8	should be suitable for all applications	▪
			60	90		8	▪	▪
		50	127	7		7		Phillips Crystalor HBG30; 30% gl.fib.
		50	127	28		6		▪
Formaldehyde			20	90		8	should be suitable for all applications	Mitsui TPX
			60	90		8	▪	▪
Hydrogen Peroxide		10	20	90		8	▪	▪
		10	60	90		8	▪	▪
		35	20	90		8	▪	▪
		35	60	90		6	limited absorp. or attack	▪
		70	20	90		8	should be suitable for all applications	▪
		70	60	90		6	limited absorp. or attack	▪

Reagent	Reagent Note	Conc. (%)	Temp. (°C)	Time (days)	Load	PDL Rating	Resistance Note	Material Note
Polymethylpentene (continued)								
Iodine	in ethanol sol'n		20	90		8	should be suitable for all applications	Mitsui TPX
	▪		60	90		8	▪	▪
Isopropyl Alcohol			20	90		8	▪	▪
			60	90		6	limited absorp. or attack	▪
Mercuric Chloride			20	90		8	should be suitable for all applications	▪
			60	90		8	▪	▪
Phenol			20	90		8	▪	▪
			60	90		8	▪	▪
Potassium Permanganate			20	90		8	▪	▪
			60	90		8	▪	▪
Sodium Hypochlorite			20	90		8	▪	▪
			60	90		8	▪	▪
Sulfuric Acid		10	20	90		8	▪	▪
		10	60	90		8	▪	▪
Polyphenylene Sulfide								
Ammonia	aq. sol'n		21			8	minimal prop. chg.	Hoechst Cel. Fortron; 40% gl.fib.
	anhydrous		93			8	acceptable for use	Phillips Ryton
Benzaldehyde			60			8	minimal prop. chg.	Hoechst Cel. Fortron; 40% gl.fib.
			93			3	questionable to not recommended	Phillips Ryton
			93	1		9		Phillips Ryton R-4; 40% gl.fib.
			93	1		9		GE Supec 401; 40% gl.fib.
			93	1		9		GE Supec 402; 40% gl.fib.
			93	1		6		Phillips Ryton
			93	90		1		Phillips Ryton R-4; 40% gl.fib.
			93	90		1		GE Supec 401; 40% gl.fib.
			93	90		1		GE Supec 402; 40% gl.fib.
			93	365		1		Phillips Ryton R-4; 40% gl.fib.
Benzoic Acid			93			8	minimal prop. chg.	Hoechst Cel. Fortron; 40% gl.fib.
Butyl Alcohol			93			8	▪	▪
			93			8	acceptable for use	Phillips Ryton
			93	1		9		Phillips Ryton R-4; 40% gl.fib.
			93	1		9		GE Supec 402; 40% gl.fib.
			93	1		9		GE Supec 401; 40% gl.fib.
			93	1		9		Phillips Ryton
			93	90		9		GE Supec 401; 40% gl.fib.
			93	90		9		Phillips Ryton R-4; 40% gl.fib.
			93	90		9		GE Supec 402; 40% gl.fib.
			93	365		6		Phillips Ryton R-4; 40% gl.fib.
Chlorine	dry		93			5	questionable resistance	Phillips Ryton
Ethyl Acetate			93			8	minimal prop. chg.	Hoechst Cel. Fortron; 40% gl.fib.
			93			8	acceptable for use	Phillips Ryton
			93	1		9		Phillips Ryton R-4; 40% gl.fib.
			93	1		9		GE Supec 401; 40% gl.fib.
			93	1		9		GE Supec 402; 40% gl.fib.
			93	1		9		Phillips Ryton
			93	90		7		Phillips Ryton R-4; 40% gl.fib.
			93	90		8		GE Supec 402; 40% gl.fib.
			93	90		8		GE Supec 401; 40% gl.fib.
			93	365		2		Phillips Ryton R-4; 40% gl.fib.
Ethyl Alcohol			93			8	minimal prop. chg.	Hoechst Cel. Fortron; 40% gl.fib.
			93			8	▪	▪
			93			8	acceptable for use	Phillips Ryton
Ethylene Glycol			23	7	0.25%	8		LNP; 40% gl.fib.
			23	7		9		▪
			82	3	0.25%	7		▪
			82	3		8		▪
			93			8	minimal prop. chg.	Hoechst Cel. Fortron; 40% gl.fib.
			93			8	acceptable for use	Phillips Ryton
			149	1	0.25%	4		LNP; 40% gl.fib.
			149	1		5		▪
		50	23	60		9		LNP Lubricomp OCL-4036; 45% carb./TFE
		50	23	60		9		LNP Lubricomp OFL-4036; 45% gl.fib./TFE
		50	82	30		9		LNP Lubricomp OCL-4036; 45% carb./TFE
		50	82	30		9		LNP Lubricomp OFL-4036; 45% gl.fib./TFE

Reagent	Reagent Note	Conc. (%)	Temp. (°C)	Time (days)	Load	PDL Rating	Resistance Note	Material Note

Polyphenylene Sulfide (continued)

Reagent	Reagent Note	Conc. (%)	Temp. (°C)	Time (days)	Load	PDL Rating	Resistance Note	Material Note
Ethylene Glycol		50	150	7		4		LNP Lubricomp OFL-4036; 45% gl.fib./TFE
		50	150	7		4		LNP Lubricomp OCL-4036; 45% carb./TFE
Formaldehyde		37	93			8	acceptable for use	Phillips Ryton
		37	93			8	minimal prop. chg.	Hoechst Cel. Fortron; 40% gl.fib.
		37	93	1		9		Phillips Ryton
Hydrogen Peroxide		30	93			5	questionable resistance	•
		50	23			2	not recommended for use	Hoechst Cel. Fortron; 40% gl.fib.
Methyl Isobutyl Ketone			21			8	minimal prop. chg.	•
			93			8	acceptable for use	Phillips Ryton
Morpholine			93			6	acceptable to questionable	•
			93	1		6		•
Phenol			93			8	acceptable for use	•
			93	1		9		GE Supec 401; 40% gl.fib.
			93	1		9		GE Supec 402; 40% gl.fib.
			93	1		9		Phillips Ryton R-4; 40% gl.fib.
			93	1		9		Phillips Ryton
			93	90		8		GE Supec 401; 40% gl.fib.
			93	90		8		Phillips Ryton R-4; 40% gl.fib.
			93	90		8		GE Supec 402; 40% gl.fib.
			93	365		3		Phillips Ryton R-4; 40% gl.fib.
Potassium Permanganate			93			8	minimal prop. chg.	Hoechst Cel. Fortron; 40% gl.fib.
			93			8	acceptable for use	Phillips Ryton
		10	93	1		9		•
Silver Nitrate			93			8	minimal prop. chg.	Hoechst Cel. Fortron; 40% gl.fib.
Sodium Hypochlorite	sol'n		93			8	•	•
	•		93			5	questionable resistance	Phillips Ryton
		5	93	1		9		GE Supec 402; 40% gl.fib.
		5	93	1		9		GE Supec 401; 40% gl.fib.
		5	93	90		9		•
Sulfuric Acid		10	23	7	0.25%	6		LNP; 40% gl.fib.
		10	23	7		7		•
		10	82	3	0.25%	2		•
		10	82	3		4		•
		10	93			8	minimal prop. chg.	Hoechst Cel. Fortron; 40% gl.fib.
		10	149	1	0.25%	1		LNP; 40% gl.fib.
		10	149	1		1		•

Polyphenylene Sulfide Sulfone

Reagent	Reagent Note	Conc. (%)	Temp. (°C)	Time (days)	Load	PDL Rating	Resistance Note	Material Note
Butyl Alcohol				1		4		Phillips Ryton S
Ethyl Acetate				1		6		•
				14		9		•
				56		3		•
				168		2		•

Polyphthalamide

Reagent	Reagent Note	Conc. (%)	Temp. (°C)	Time (days)	Load	PDL Rating	Resistance Note	Material Note
Isopropyl Alcohol			22	30		9		Amoco Amodel A-1133HS; 33% gl.fib.

General Purpose Polystyrene

Reagent	Reagent Note	Conc. (%)	Temp. (°C)	Time (days)	Load	PDL Rating	Resistance Note	Material Note
Ammonia	aq. sol'n		22	7		8	resistant	Monsanto Lustrex
	•		50	7		6	heat reduces room temp. resist.	•
	•	25	20	<=365		9	resistant for yrs.	BASF Polystyrol 168N
	•	25	50	<=365		8	resistant for mths.	•
Benzaldehyde			20	<=365		3	resistant for hrs.	•
	undiluted		22	7		2	plastic severely crazed, softened or dissolved	Monsanto Lustrex
			50	<=365		3	resistant for hrs.	BASF Polystyrol 168N
Benzoic Acid			20	<=365		9	resistant for yrs.	•
			22	7		8	resistant	Monsanto Lustrex
			50	7		6	heat reduces room temp. resist.	•
			50	<=365		8	resistant for mths.	BASF Polystyrol 168N
Benzoic Acid Ointment	ointment; w\ salicylic acid		22	7		8	resistant	Monsanto Lustrex
	•		50	7		6	heat reduces room temp. resist.	•
Benzyl Alcohol			20	<=365		0	dissolved	BASF Polystyrol 168N
			50	<=365		0	•	•
Bleach	sat'd sol'n	12	20	<=365		8	resistant for mths.	•
	•	12	50	<=365		6	resistant for wks.	•

General Purpose Polystyrene (continued)

Reagent	Reagent Note	Conc. (%)	Temp. (°C)	Time (days)	Load	PDL Rating	Resistance Note	Material Note
Boric Acid	sat'd		20	<=365		9	resistant for yrs.	*
			22	7		8	resistant	Monsanto Lustrex
			50	7		6	heat reduces room temp. resist.	*
	sat'd		50	<=365		8	resistant for mths.	BASF Polystyrol 168N
Butyl Alcohol			20	<=365		9	resistant for yrs.	*
			22	7		8	resistant	Monsanto Lustrex
			50	7		6	heat reduces room temp. resist.	*
			50	<=365		8	resistant for mths.	BASF Polystyrol 168N
Carbolic Acid			22	7		2	plastic severely crazed, softened or dissolved	Monsanto Lustrex
		50	22	7		5	some discolor./ crazing/ checking occurs	*
		50	50	7		4	heat reduces moder. room temp. resist.	*
Chlorine	chlorine water		20	<=365		5	resistant for days	BASF Polystyrol 168N
	dry gas		20	<=365		3	resistant for hrs.	*
	liquid		20	<=365		3	*	*
	sat'd, water		22	7		2	plastic severely crazed, softened or dissolved	Monsanto Lustrex
	liquid		50	<=365		3	resistant for hrs.	BASF Polystyrol 168N
	dry gas		50	<=365		3	*	*
Cresol	cresolic compounds		22	7		5	some discolor./ crazing/ checking occurs	Monsanto Lustrex
Ethyl Acetate			20	<=365		0	dissolved	BASF Polystyrol 168N
			50	<=365		0	*	*
		90	22	7		2	plastic severely crazed, softened or dissolved	Monsanto Lustrex
Ethyl Alcohol	formula 30		22	7		5	some discolor./ crazing/ checking occurs	*
	2B-95		22	7		2	plastic severely crazed, softened or dissolved	*
	denaturated		22	7		8	resistant	*
	2B absolute		22	7		8	*	*
	formula 30		50	7		4	heat reduces moder. room temp. resist.	*
	denaturated		50	7		6	heat reduces room temp. resist.	*
	2B absolute		50	7		6	*	*
		40	20	<=365		7	resistant for yrs., poss. stress cracks	BASF Polystyrol 168N
		40	50	<=365		6	resistant for mths.; poss. stress cracks	*
		95	20	<=365		7	resistant for yrs., poss. stress cracks	*
		95	22	7		5	some discolor./ crazing/ checking occurs	Monsanto Lustrex
		95	50	7		4	heat reduces moder. room temp. resist.	*
		95	50	<=365		6	resistant for mths.; poss. stress cracks	BASF Polystyrol 168N
Ethylene Glycol			20	<=365		9	resistant for yrs.	*
			22	7		8	resistant	Monsanto Lustrex
			23	7	0.25%	5		LNP; 30% gl.fib.
			23	7		8		*
			50	7		8	resistant	Monsanto Lustrex
			50	<=365		9	resistant for yrs.	BASF Polystyrol 168N
			81	3		7		LNP; 30% gl.fib.
Ethylene Oxide			22	7		2	plastic severely crazed, softened or dissolved	Monsanto Lustrex
Formaldehyde		30	20	<=365		8	resistant for mths.	BASF Polystyrol 168N
		30	22	7		2	plastic severely crazed, softened or dissolved	Monsanto Lustrex
		30	50	<=365		8	resistant for mths.	BASF Polystyrol 168N
Hydrogen Peroxide		3	20	<=365		9	resistant for yrs.	*
		3	22	7		8	resistant	Monsanto Lustrex
		3	50	7		6	heat reduces room temp. resist.	*
		3	50	<=365		8	resistant for mths.	BASF Polystyrol 168N
		30	22	7		8	resistant	Monsanto Lustrex
		30	50	7		6	heat reduces room temp. resist.	*
Iodine	tincture		20	<=365		6	resistant for wks.	BASF Polystyrol 168N
		3	22	7		5	some discolor./ crazing/ checking occurs	Monsanto Lustrex
		3	50	7		5	*	*
	in alcohol	7	22	7		5	*	*
Isopropyl Alcohol			22	7		8	resistant	*
			50	7		6	heat reduces room temp. resist.	*

General Purpose Polystyrene (continued)

Reagent	Reagent Note	Conc. (%)	Temp. (°C)	Time (days)	Load	PDL Rating	Resistance Note	Material Note
Mercuric Chloride	sat'd sol'n		20	<=365		8	resistant for mths.	BASF Polystyrol 168N
	sat'd		22	7		5	some discolor./ crazing/ checking occurs	Monsanto Lustrex
	"		50	7		4	heat reduces moder. room temp. resist.	"
	sat'd sol'n		50	<=365		8	resistant for mths.	BASF Polystyrol 168N
	powder	5	22	7		8	resistant	Monsanto Lustrex
	"	5	50	7		6	heat reduces room temp. resist.	"
Methyl Isobutyl Ketone			20	<=365		0	dissolved	BASF Polystyrol 168N
			22	7		2	plastic severely crazed, softened or dissolved	Monsanto Lustrex
			50	<=365		0	dissolved	BASF Polystyrol 168N
Phenol			20	<=365		5	resistant for days	"
			50	<=365		3	resistant for hrs.	"
		5	22	7		5	some discolor./ crazing/ checking occurs	Monsanto Lustrex
		5	50	7		4	heat reduces moder. room temp. resist.	"
		10	20	<=365		8	resistant for mths.	BASF Polystyrol 168N
		10	50	<=365		6	resistant for wks.	"
Potassium Permanganate			20	<=365		8	resistant for mths.	"
	sat'd		22	7		5	some discolor./ crazing/ checking occurs	Monsanto Lustrex
	"		50	7		5	"	"
			50	<=365		8	resistant for mths.	BASF Polystyrol 168N
Propylene Oxide			20	<=365		3	resistant for hrs.	"
			50	<=365		3	"	"
Salicylic Acid			20	<=365		9	resistant for yrs.	"
	sat'd		22	7		8	resistant	Monsanto Lustrex
	ointment; w\ benzoic acid		22	7		8	"	"
	sat'd		50	7		6	heat reduces room temp. resist.	"
	ointment; w\ benzoic acid		50	7		6	"	"
			50	<=365		8	resistant for mths.	BASF Polystyrol 168N
Silver Nitrate			20	<=365		8	"	"
	sat'd		22	7		8	resistant	Monsanto Lustrex
	"		50	7		6	heat reduces room temp. resist.	"
			50	<=365		8	resistant for mths.	BASF Polystyrol 168N
Sodium Hypochlorite	12% chlorine		20	<=365		8	"	"
			22	7		8	resistant	Monsanto Lustrex
			50	7		8	"	"
	12% chlorine		50	<=365		6	resistant for wks.	BASF Polystyrol 168N
Sulfuric Acid		10	20	<=365		9	resistant for yrs.	"
		10	23	7	0.25%	3		LNP; 30% gl.fib.
		10	23	7		4		"
		10	50	<=365		9	resistant for yrs.	BASF Polystyrol 168N
Thymol			20	<=365		5	resistant for days	"
			22	7		2	plastic severely crazed, softened or dissolved	Monsanto Lustrex
			50	<=365		3	resistant for hrs.	BASF Polystyrol 168N
Toothpaste	Ipana		22	7		8	resistant	Monsanto Lustrex
	"		50	7		8	"	"

Impact Polystyrene

Reagent	Reagent Note	Conc. (%)	Temp. (°C)	Time (days)	Load	PDL Rating	Resistance Note	Material Note
Ammonia	aq. sol'n		22	7		8	"	"
	"	25	20	<=365		9	resistant for yrs.	BASF Polystyrol 456M
	"	25	50	<=365		8	resistant for mths.	"
Benzaldehyde			20	<=365		3	resistant for hrs.	"
	undiluted		22	7		2	plastic severely crazed, softened or dissolved	Monsanto Lustrex
			50	<=365		3	resistant for hrs.	BASF Polystyrol 456M
Benzoic Acid			20	<=365		8	resistant for mths.	"
			22	7		6	resistance less than gen. purp. Lustrex	Monsanto Lustrex
			50	<=365		8	resistant for mths.	BASF Polystyrol 456M
Benzoic Acid Ointment	ointment; w\ salicylic acid		22	7		8	resistant	Monsanto Lustrex
Benzyl Alcohol			20	<=365		0	dissolved	BASF Polystyrol 456M
			50	<=365		0	"	"
Bleach	sat'd sol'n	12	20	<=365		8	resistant for mths.	"
	"	12	50	<=365		6	resistant for wks.	"

Reagent	Reagent Note	Conc. (%)	Temp. (°C)	Time (days)	Load	PDL Rating	Resistance Note	Material Note

Impact Polystyrene (continued)

Reagent	Reagent Note	Conc. (%)	Temp. (°C)	Time (days)	Load	PDL Rating	Resistance Note	Material Note
Boric Acid	sat'd		20	<=365		9	resistant for yrs.	BASF Polystyrol 456M
			22	7		8	resistant	Monsanto Lustrex
	sat'd		50	<=365		8	resistant for mths.	BASF Polystyrol 456M
Butyl Alcohol						4	may cause cracks on stressed parts	Dow Styron
			20	<=365		8	resistant for mths.	BASF Polystyrol 456M
			22	7		6	resistance less than gen. purp. Lustrex	Monsanto Lustrex
			50	<=365		6	resistant for wks.	BASF Polystyrol 456M
Carbolic Acid			22	7		2	plastic severely crazed, softened or dissolved	Monsanto Lustrex
		50	22	7		4	visual changes, less resist. than gen. purp. Lustrex	"
Chlorine	chlorine water		20	<=365		5	resistant for days	BASF Polystyrol 456M
	dry gas		20	<=365		3	resistant for hrs.	"
	liquid		20	<=365		3	"	"
	sat'd, water		22	7		2	plastic severely crazed, softened or dissolved	Monsanto Lustrex
	dry gas		50	<=365		3	resistant for hrs.	BASF Polystyrol 456M
	liquid		50	<=365		3	"	"
Cresol	cresolic compounds		22	7		4	visual changes, less resist. than gen. purp. Lustrex	Monsanto Lustrex
Ethyl Acetate			20	<=365		0	dissolved	BASF Polystyrol 456M
			50	<=365		0	"	"
		90	22	7		2	plastic severely crazed, softened or dissolved	Monsanto Lustrex
Ethyl Alcohol	2B absolute					7	working stress reduced by 1/8	Dow Styron
	formula 30		22	7		5	some discolor./ crazing/ checking occurs	Monsanto Lustrex
	2B-95		22	7		2	plastic severely crazed, softened or dissolved	"
	2B absolute		22	7		8	resistant	"
		40	20	<=365		7	resistant for yrs., poss. stress cracks	BASF Polystyrol 456M
		40	50	<=365		6	resistant for mths.; poss. stress cracks	"
		50	22	1	0.55%	4	exp. strain is level where part stress cracks	Monsanto Lustrex; 6.4 mm samples
		95	20	<=365		7	resistant for yrs., poss. stress cracks	BASF Polystyrol 456M
		95	50	<=365		6	resistant for mths.; poss. stress cracks	"
		100	22	1	0.45%	4	exp. strain is level where part stress cracks	Monsanto Lustrex; 6.4 mm samples
		100	22	17.1		8	no chg. in surface or elongation	Monsanto Lustrex
Ethylene Glycol			20	<=365		7	resistant for yrs., poss. stress cracks	BASF Polystyrol 456M
			22	7		8	resistant	Monsanto Lustrex
			50	<=365		7	resistant for yrs., poss. stress cracks	BASF Polystyrol 456M
Ethylene Oxide			22	7		2	plastic severely crazed, softened or dissolved	Monsanto Lustrex
Formaldehyde		30	20	<=365		8	resistant for mths.	BASF Polystyrol 456M
		30	22	7		2	plastic severely crazed, softened or dissolved	Monsanto Lustrex
		30	50	<=365		8	resistant for mths.	BASF Polystyrol 456M
Hydrogen Peroxide		3	20	<=365		9	resistant for yrs.	"
		3	22	7		6	resistance less than gen. purp. Lustrex	Monsanto Lustrex
		3	22	17.1		9	no chg. in surface or tensile props.	"
		3	50	<=365		8	resistant for mths.	BASF Polystyrol 456M
		30	22	7		6	resistance less than gen. purp. Lustrex	Monsanto Lustrex
Iodine	tincture		20	<=365		6	resistant for wks.	BASF Polystyrol 456M
		3	22	7		5	some discolor./ crazing/ checking occurs	Monsanto Lustrex
	in alcohol	7	22	7		5	"	"
Isopropyl Alcohol						7	working stress reduced by 1/8	Dow Styron
						5	poss. stress crack agent	BASF Polystyrol 456M
			22	7		6	resistance less than gen. purp. Lustrex	Monsanto Lustrex
Mercuric Chloride	sat'd sol'n		20	<=365		8	resistant for mths.	BASF Polystyrol 456M
	sat'd		22	7		4	visual changes, less resist. than gen. purp. Lustrex	Monsanto Lustrex
	sat'd sol'n		50	<=365		6	resistant for wks.	BASF Polystyrol 456M
	powder	5	22	7		6	resistance less than gen. purp. Lustrex	Monsanto Lustrex

Impact Polystyrene (continued)

Reagent	Reagent Note	Conc. (%)	Temp. (°C)	Time (days)	Load	PDL Rating	Resistance Note	Material Note
Methyl Isobutyl Ketone			20	<=365		0	dissolved	BASF Polystyrol 456M
			22	7		2	plastic severely crazed, softened or dissolved	Monsanto Lustrex
			50	<=365		0	dissolved	BASF Polystyrol 456M
Phenol			20	<=365		5	resistant for days	"
			50	<=365		3	resistant for hrs.	"
		10	20	<=365		8	resistant for mths.	"
		10	50	<=365		6	resistant for wks.	"
		50	22	17.1		3	softened	Monsanto Lustrex
Potassium Permanganate			20	<=365		6	resistant for wks.	BASF Polystyrol 456M
	sat'd		22	7		4	visual changes, less resist. than gen. purp. Lustrex	Monsanto Lustrex
	"		22	17.1		8	surface darkened, no chg. in tensile props.	"
			50	<=365		5	resistant for days	BASF Polystyrol 456M
Propylene Oxide			20	<=365		3	resistant for hrs.	"
			50	<=365		3	"	"
Salicylic Acid			20	<=365		9	resistant for yrs.	"
	sat'd		22	7		6	resistance less than gen. purp. Lustrex	Monsanto Lustrex
	ointment; w\ benzoic acid		22	7		8	resistant	"
			50	<=365		8	resistant for mths.	BASF Polystyrol 456M
Silver Nitrate			20	<=365		8	"	"
	sat'd		22	7		8	resistant	Monsanto Lustrex
			50	<=365		6	resistant for wks.	BASF Polystyrol 456M
Sodium Hypochlorite	12% chlorine		20	<=365		8	resistant for mths.	"
			22	7		6	resistance less than gen. purp. Lustrex	Monsanto Lustrex
			50	<=365		6	resistant for wks.	BASF Polystyrol 456M
Sulfuric Acid		10	20	<=365		9	resistant for yrs.	"
		10	22	17.1		6	surface darkened, no chg. in tensile strength	Monsanto Lustrex
		10	50	<=365		9	resistant for yrs.	BASF Polystyrol 456M
Thymol			20	<=365		3	resistant for hrs.	"
			22	7		2	plastic severely crazed, softened or dissolved	Monsanto Lustrex
			50	<=365		3	resistant for hrs.	BASF Polystyrol 456M
Toothpaste	Ipana		22	7		8	resistant	Monsanto Lustrex

Polysulfone

Reagent	Reagent Note	Conc. (%)	Temp. (°C)	Time (days)	Load	PDL Rating	Resistance Note	Material Note
Ammonia		29	22	1	13.79 MPa	8	OK at 24 hrs.	Amoco Udel
		29	22	1	27.6 MPa	8	"	"
Benzaldehyde			93	1		0		"
Butyl Alcohol			22	<1	20.68 MPa	4	crazing in <24 hrs.	"
			22	<1	27.6 MPa	1	ruptured in less than 24 hrs.	"
			22	1	13.79 MPa	8	OK at 24 hrs.	"
			22	7		8	no visible chg.	"
			93	1		9		"
Chlorine	wet		22	102		2	severe surface attack	"
			85			0	ruptured under no stree	"
Cidex 7			93.3	0.033	20.7 MPa	0	exp. time is time to failure	Udel P1700; irradiated @ 2.5 Mrads gamma rays
			93.3	0.021	20.7 MPa	0	"	Udel P1700; irradiated @ 6 Mrads gamma rays
			93.3	0.021	20.7 MPa	0	"	Udel P1700; irradiated @ 4 Mrads gamma rays
			93.3	0.04	20.7 MPa	0	"	Udel P1700
Ethyl Acetate			22	<.001	13.79 MPa	1	ruptured in less than 0.01 hrs.	Amoco Udel
			22	7		2	softened, swollen	"
			93	1		0		"
Ethyl Alcohol			22	<0.083	20.68 MPa	1	ruptured in 2 hrs.	"
			22	7		9	no visible chg.	"
			22	20.8	13.79 MPa	8	OK at 500 hrs.	"
			22	110		9	no visible chg.	"
Ethylene Glycol			22	12	27.6 MPa	8	OK at 288 hrs.	"
			22	110		9	no visible chg.	"
			23	7	0.25%	9		LNP; 30% gl.fib.
			23	7	0.25%	9		LNP; 40% gl.fib.
			23	7		9		"
			23	7		9		LNP; 30% gl.fib.

Reagent	Reagent Note	Conc. (%)	Temp. (°C)	Time (days)	Load	PDL Rating	Resistance Note	Material Note
Polysulfone (continued)								
Ethylene Glycol			81	3		9		LNP; 30% gl.fib.
			82	3	0.25%	7		LNP; 40% gl.fib.
			82	3		9		"
			93	1		9		Amoco Udel
			149	1	0.25%	2		LNP; 40% gl.fib.
			149	1		3		"
	w/ 2% anticorrosion oil	50	129	1		1		BASF Ultrason S 2010
	"	50	129	1		9		BASF Ultrason S 2010G6; 30% gl.fib.
	"	50	129	6.6		5		"
	"	50	129	66		2		"
Formaldehyde			150	92		3	severely blistered	Amoco Udel
		37	93	1		9		"
Hydrogen Peroxide			22	4.2	34.47 MPa	8	OK at 100 hrs.	"
			22	7		8	no visible chg.	"
Isopropyl Alcohol			22	0.208	20.68 MPa	8	OK at 5 hrs.	"
			22	<1	13.79 MPa	4	crazing in <24 hrs.	"
Morpholine			22	110		1	immediate attack	"
			93	1		0		"
		0.2	93	<0.333	6.89 MPa	1	ruptured in less than 8 hrs.	"
Omnicide	2% glutaraldehyde	100	21	18	24.8 MPa	9	no stress cracks or crazing	
	"	100	70	<1	12.4 MPa	3	cracking or crazing in <24 hrs.	
Phenol			93	1		0		Amoco Udel
Potassium Permanganate		10	93	1		9		"
Sodium Hypochlorite	Clorox	0.17	149			8	acceptable for use	"
Sporicidin	2% glutaraldehyde, phenol, sodium phenate	100	21	<1	24.8 MPa	3	cracking or crazing in <24 hrs.	
	"	100	21	18	18.3 MPa	9	no stress cracks or crazing	
	"	100	70	<1	14.5 MPa	3	cracking or crazing in <24 hrs.	
	"	100	70	2	12.4 MPa	4	cracking or crazing in 48 hrs.	
Sulfuric Acid		5	85	102		8	clouded	Amoco Udel
		5	149	92		8	"	"
		10	23	7	0.25%	5		LNP; 30% gl.fib.
		10	23	7	0.25%	5		LNP; 40% gl.fib.
		10	23	7		8		"
		10	23	7		7		LNP; 30% gl.fib.
		10	82	3	0.25%	5		LNP; 40% gl.fib.
		10	82	3		7		"
		10	149	1	0.25%	1		"
		10	149	1		1		"
Tergitol 15S12			22	<0.042	27.6 MPa	4	crazing in <1 hr.	Amoco Udel
			22	2	6.89 MPa	4	crazing at 48 hrs.	"
Wavecide-01	2% glutaraldehyde	100	21	18	24.8 MPa	9	no stress cracks or crazing	
	"	100	70	<1	23.4 MPa	3	cracking or crazing in <24 hrs.	
	"	100	70	2	12.4 MPa	4	cracking or crazing in 48 hrs.	
Polyethersulfone								
Ammonia	ammonia-880		20			8	no attack, little or no absorp.	Victrex
Benzaldehyde			20			1	severe attack, should not be used	"
Benzoic Acid			20			8	no attack, little or no absorp.	"
Bleach			20			8	"	"
Boric Acid			20			7	no attack predicted	"
Butyl Alcohol			20			8	no attack, little or no absorp.	"
			23	0.0139	28.4 MPa	4	crazed after 20 minutes	Victrex 4100G
			23	0.0139	37.9 MPa	4	"	Victrex 4100G
			23	0.0139	37.9 MPa	4	"	Victrex 4800G
			23	0.014	19 MPa	8	unaffected after 20 minutes	"
			23	0.014	19 MPa	8	"	Victrex 4101GL30; 30% gl.fib.
			23	0.014	19 MPa	8	"	Victrex 4100G
			23	0.014	28.4 MPa	8	"	Victrex 4101GL30; 30% gl.fib.
			23	0.014	37.9 MPa	8	"	"
			23	0.014	9.5 MPa	8	"	Victrex 4800G
			23	0.014	9.5 MPa	8	"	Victrex 4101GL30; 30% gl.fib.
			23	0.014	9.5 MPa	8	"	Victrex 4100G
Chlorine	wet		20			1	severe attack, should not be used	Victrex
	dry		20			1	"	"
			23	28		6	cracks observed	Victrex 4100G
			23	30		2	cracked	Victrex 4800G

Reagent	Reagent Note	Conc. (%)	Temp. (°C)	Time (days)	Load	PDL Rating	Resistance Note	Material Note
Polyethersulfone (continued)								
Cresol			20			1	severe attack, should not be used	Victrex
Ethyl Acetate			20			1	"	"
			23	<0.0001	28.4 MPa	1	ruptured after 6 seconds	Victrex 4100G
			23	<0.0001	37.9 MPa	1	ruptured after 2 seconds	"
			23	<0.0002	19 MPa	1	ruptured after 17 seconds	"
			23	0.00036	9.5 MPa	1	ruptured after 31 seconds	"
			23	0.0014	37.9 MPa	1	ruptured after 2 minutes	Victrex 4800G
			23	0.0028	28.4 MPa	1	ruptured after 4 minutes	"
			23	0.0049	19 MPa	1	ruptured after 7 minutes	"
			23	0.0139	9.5 MPa	4	crazed after 20 minutes	"
			23	0.014	19 MPa	8	unaffected after 20 minutes	Victrex 4101GL30; 30% gl.fib.
			23	0.014	28.4 MPa	8	"	"
			23	0.014	37.9 MPa	8	"	"
			23	0.014	9.5 MPa	8	"	"
			23	60		2	softened	Victrex 4800G
Ethyl Alcohol			20			8	no attack, little or no absorp.	Victrex
			23	0.014	19 MPa	8	unaffected after 20 minutes	Victrex 4101GL30; 30% gl.fib.
			23	0.014	19 MPa	8	"	Victrex 4800G
			23	0.014	19 MPa	8	"	Victrex 4100G
			23	0.014	28.4 MPa	8	"	Victrex 4800G
			23	0.014	28.4 MPa	8	"	Victrex 4101GL30; 30% gl.fib.
			23	0.014	28.4 MPa	8	"	Victrex 4100G
			23	0.014	37.9 MPa	8	"	Victrex 4800G
			23	0.014	37.9 MPa	8	"	Victrex 4100G
			23	0.014	37.9 MPa	8	"	Victrex 4101GL30; 30% gl.fib.
			23	0.014	9.5 MPa	8	"	Victrex 4100G
			23	0.014	9.5 MPa	8	"	Victrex 4800G
			23	0.014	9.5 MPa	8	"	Victrex 4101GL30; 30% gl.fib.
			23	7		9		Victrex 4100G
			23	30		9		"
			23	90		7		"
			23	180		6		Victrex 4800G
			23	180		6		Victrex 4100G
Ethylene Glycol			20			8	no attack, little or no absorp.	Victrex
			23	0.014	19 MPa	8	unaffected after 20 minutes	Victrex 4800G
			23	0.014	19 MPa	8	"	Victrex 4100G
			23	0.014	19 MPa	8	"	Victrex 4101GL30; 30% gl.fib.
			23	0.014	28.4 MPa	8	"	Victrex 4100G
			23	0.014	28.4 MPa	8	"	Victrex 4800G
			23	0.014	28.4 MPa	8	"	Victrex 4101GL30; 30% gl.fib.
			23	0.014	37.9 MPa	8	"	Victrex 4800G
			23	0.014	37.9 MPa	8	"	Victrex 4100G
			23	0.014	37.9 MPa	8	"	Victrex 4101GL30; 30% gl.fib.
			23	0.014	9.5 MPa	8	"	"
			23	0.014	9.5 MPa	8	"	Victrex 4800G
			23	0.014	9.5 MPa	8	"	Victrex 4100G
			23	7	0.25%	9		Victrex; 40% gl.fib.
			23	7		9		"
			23	120		9		Victrex 4800G
			82	7	0.25%	8		Victrex; 40% gl.fib.
			82	7		9		"
			149	7	0.25%	3		"
			149	7		5		"
		50	23	60	0.25%	9		Victrex PES 4100G
		50	23	60	0.25%	9		Victrex PES 4101 GL30; 30% gl.fib.
		50	23	60		8		Victrex PES 4100G
		50	23	60		8		Victrex PES 4101 GL30; 30% gl.fib.
		50	82	30	0.25%	8		"
		50	82	30	0.25%	9		Victrex PES 4100G
		50	82	30		7		"
		50	82	30		7		Victrex PES 4101 GL30; 30% gl.fib.
	w/ 2% anticorrosion oil	50	129	1		9		BASF Ultrason E 2010G4; 20% gl.fib.
	"	50	129	1		9		BASF Ultrason E 3010
	"	50	129	6.6		7		BASF Ultrason E 2010G4; 20% gl.fib.
	"	50	129	6.6		8		BASF Ultrason E 3010
	"	50	129	66		4		BASF Ultrason E 2010G4; 20% gl.fib.
	"	50	129	66		6		BASF Ultrason E 3010
		50	150	7	0.25%	9		Victrex PES 4100G

Polyethersulfone (continued)

Reagent	Reagent Note	Conc. (%)	Temp. (°C)	Time (days)	Load	PDL Rating	Resistance Note	Material Note
Ethylene Glycol		50	150	7	0.25%	3		Victrex PES 4101 GL30; 30% gl.fib.
		50	150	7		5		▪
		50	150	7		7		Victrex PES 4100G
Ethylene Oxide			20			8	no attack, little or no absorp.	Victrex
			23	7	1%	8	no effect	Victrex 4100G
			23	30		7		▪
			23	180		3		▪
			23	190		2	some stress cracks at 2000 psi	Victrex 4800G
Fluorine			20			1	severe attack, should not be used	Victrex
Formaldehyde			20			8	no attack, little or no absorp.	▪
Hydrogen Peroxide	concentrated		20			8	▪	▪
	100 vol.		23	120		4		Victrex 4100G
Iodine	in potassium iodide		20			6	sl. attack, some absorp.	Victrex
Isopropyl Alcohol			20			8	no attack, little or no absorp.	▪
			23	0.0139	19 MPa	5	sl. crazing after 20 minutes	Victrex 4100G
			23	0.014	19 MPa	8	unaffected after 20 minutes	Victrex 4101GL30; 30% gl.fib.
			23	0.014	19 MPa	8	▪	Victrex 4800G
			23	0.014	28.4 MPa	5	slightly crazed after 20 minutes	▪
			23	0.014	28.4 MPa	8	unaffected after 20 minutes	Victrex 4100G
			23	0.014	28.4 MPa	8	▪	Victrex 4101GL30; 30% gl.fib.
			23	0.014	37.9 MPa	8	▪	Victrex 4100G
			23	0.014	37.9 MPa	8	▪	Victrex 4800G
			23	0.014	37.9 MPa	8	▪	Victrex 4101GL30; 30% gl.fib.
			23	0.014	9.5 MPa	8	▪	Victrex 4100G
			23	0.014	9.5 MPa	8	▪	Victrex 4101GL30; 30% gl.fib.
			23	0.014	9.5 MPa	8	▪	Victrex 4800G
Mercuric Chloride			20			7	no attack predicted	Victrex
Phenol			20			1	severe attack, should not be used	▪
		5	23	7		4		Victrex 4100G
		5	23	30		3		▪
		5	23	90		2		▪
		5	23	90		1		Victrex 4800G
Silver Nitrate			20			8	no attack, little or no absorp.	Victrex
Sodium Hypochlorite	sat'd		23	180		6		Victrex 4100G
	▪		25	30		8		Victrex 4800G
	▪		25	90		6		▪
	▪		25	180		6		▪
		25	90	90		8		▪
		25	90	180		8		▪
		50	20			8	no attack, little or no absorp.	Victrex
Sulfuric Acid		10	20			8	▪	▪
		10	23	7	0.25%	6		Victrex; 40% gl.fib.
		10	23	7		7		▪
		10	23	30		7		Victrex 4800G
		10	23	90		6		▪
		10	23	180		3		▪
		10	23	180		5		Victrex 4100G
		10	82	7	0.25%	6		Victrex; 40% gl.fib.
		10	82	7		7		▪
		10	149	7	0.25%	1		▪
		10	149	7		2		▪

Polyurethane

Reagent	Reagent Note	Conc. (%)	Temp. (°C)	Time (days)	Load	PDL Rating	Resistance Note	Material Note
Ammonia						2	unsatisf. for use	
						2	▪	
	liquid					2	▪	
Benzaldehyde						2	▪	
Benzoic Acid						2	▪	
Benzyl Alcohol						2	▪	
Bleach	sol'ns					2	▪	
Boric Acid						8	recommended for use	
Butyl Alcohol						2	unsatisf. for use	
						2	▪	
Chlorine	dry					2	▪	
	wet					2	▪	
Ethyl Acetate	organic ester					2	▪	
Ethyl Alcohol						2	▪	

Reagent	Reagent Note	Conc. (%)	Temp. (°C)	Time (days)	Load	PDL Rating	Resistance Note	Material Note
Polyurethane (continued)								
Ethylene Chlorohydrin						2	unsatisf. for use	
Ethylene Glycol						6	minor to moder. effect	
Ethylene Oxide						2	unsatisf. for use	
Formaldehyde						2	"	
Isopropyl Alcohol						2	"	
						2	"	
Methyl Isobutyl Ketone						2	"	
Phenol						2	"	
		70				2	"	
		85				2	"	
Propylene Oxide						2	"	
Silver Nitrate						8	recommended for use	
Sodium Hypochlorite						2	unsatisf. for use	
Rigid Thermoplastic Urethane								
Ammonia	concentrated sol'n		23	28		8	exc. resistance	Dow Isoplast 301
	"		23	28		8	"	Dow Isoplast 101
	"		23	28		8	"	Dow Isoplast 201
Betadine	povidone iodine sol'n		23	168	0.5%	7		Isoplast 2530
	"		23	168	0.5%	6		Isoplast 2520
	"		23	168	0.5%	9		Isoplast 2540
	"		23	168	1.5%	7		Isoplast 2520
	"		23	168	1.5%	7		Isoplast 2530
	"		23	168	1.5%	0	specimens fractured	Isoplast 2540
	"		23	168		5		Isoplast 2520
	"		23	168		9		Isoplast 2540
	"		23	168		6		Isoplast 2530
Bleach	Clorox		23	28		9	exc. resistance	Dow Isoplast 101
	"		23	28		9	"	Dow Isoplast 201
	"		23	28		9	"	Dow Isoplast 301
	"	10	23	168	0.5%	5		Isoplast 2530
	"	10	23	168	0.5%	6		Isoplast 2520
	"	10	23	168	0.5%	8		Isoplast 2540
	"	10	23	168	1.5%	6		Isoplast 2520
	"	10	23	168	1.5%	6		Isoplast 2530
	"	10	23	168	1.5%	0	specimens fractured	Isoplast 2540
	"	10	23	168		6		Isoplast 2530
	"	10	23	168		9		Isoplast 2540
	"	10	23	168		6		Isoplast 2520
Boric Acid	sol'n		23	28		8	exc. resistance	Dow Isoplast 101
Ethyl Acetate			23	28		1	poor resistance, swelling/ softening observed	Dow Isoplast 101
			23	28		1	poor resistance, color chg., swelling/ softening	Dow Isoplast 301
			23	28		1	poor resistance, swelling/ softening observed	Dow Isoplast 201
Rigid Thermoplastic Urethane								
Ethyl Alcohol		50	23	28		8	exc. resistance	Dow Isoplast 101
		50	23	28		8	"	Dow Isoplast 201
		50	23	28		9	"	Dow Isoplast 301
		95	23	28		4	fair resistance	"
		95	23	28		5	"	Dow Isoplast 201
		95	23	28		6	good resistance	Dow Isoplast 101
Ethylene Glycol			23	28		9	exc. resistance	Dow Isoplast 101
			23	28		9	"	Dow Isoplast 201
			23	28		9	"	Dow Isoplast 301
		50	23	28		9	"	"
		50	23	28		8	"	Dow Isoplast 201
		50	23	28		9	"	Dow Isoplast 101
Hydrogen Peroxide		3	23	28		8	exc. resistance	Dow Isoplast 101
		3	23	28		8	"	Dow Isoplast 301
		3	23	28		8	"	Dow Isoplast 201
		30	23	28		8	exc. resistance	Dow Isoplast 301
		30	23	28		9	"	Dow Isoplast 201
		30	23	28		8	"	Dow Isoplast 101
Isopropyl Alcohol		70	23	168	0.5%	7		Isoplast 2530
		70	23	168	0.5%	5		Isoplast 2520
		70	23	168	0.5%	9		Isoplast 2540
		70	23	168	1.5%	5		Isoplast 2530

Reagent	Reagent Note	Conc. (%)	Temp. (°C)	Time (days)	Load	PDL Rating	Resistance Note	Material Note

Rigid Thermoplastic Urethane (continued)

Reagent	Reagent Note	Conc. (%)	Temp. (°C)	Time (days)	Load	PDL Rating	Resistance Note	Material Note
Isopropyl Alcohol		70	23	168	1.5%	3		Isoplast 2520
		70	23	168	1.5%	0	specimens fractured	Isoplast 2540
		70	23	168		8		Isoplast 2520
		70	23	168		8		Isoplast 2530
		70	23	168		9		Isoplast 2540
Omnicide	2% glutaraldehyde		23	168	0.5%	7		Isoplast 2520
	"		23	168	0.5%	8		Isoplast 2530
	"		23	168	0.5%	9		Isoplast 2540
	"		23	168	1.5%	8		Isoplast 2530
	"		23	168	1.5%	5		Isoplast 2520
	"		23	168	1.5%	0	specimens fractured	Isoplast 2540
	"		23	168		6		Isoplast 2520
	"		23	168		8		Isoplast 2530
	"		23	168		6		Isoplast 2540
Phenol		5	23	28		1	poor resistance, changes observed	Dow Isoplast 201
		5	23	28		1	"	Dow Isoplast 301
		5	23	28		3	"	Dow Isoplast 101

Styrene Acrylonitrile Copolymer

Reagent	Reagent Note	Conc. (%)	Temp. (°C)	Time (days)	Load	PDL Rating	Resistance Note	Material Note
Ammonia	aq. sol'n		22	7		8	resistant	Monsanto Lustran
	"	25	20	<=365		9	resistant for yrs.	BASF Luran 368 R
	"	25	50	<=365		7	resistant for mths.	"
		100	23	28		7	sl. clouding/discoloration, expected life: mths. to yrs.	Dow Tyril
		100	52	28		7	"	"
Benzaldehyde			20	<=365		0	dissolved	BASF Luran 368 R
	undiluted		22	7		2	plastic severely crazed, softened or dissolved	Monsanto Lustran
			50	<=365		0	dissolved	BASF Luran 368 R
	in isopropyl alcohol	1	23	28		5	moder. effect; expected life: wks. to mths.	Dow Tyril
	"	1	52	28		2	severe attack, softened in few hrs.	"
	"	5	23	28		3	consid. chg.; expected life: days	"
	"	5	52	28		2	severe attack, softened in few hrs.	"
	"	10	23	28		2	"	"
	"	10	52	28		2	"	"
		100	23	28		1	solvent	"
		100	52	28		1	"	"
Benzalkonium Chloride		100	23	28		9	exc. resistance	"
		100	52	28		5	moder. effect; expected life: wks. to mths.	"
Benzoic Acid			20	<=365		9	resistant for yrs.	BASF Luran 368 R
			22	7		8	resistant	Monsanto Lustran
	crystals		23	28		9	exc. resistance	Dow Tyril
			50	7		8	resistant	Monsanto Lustran
			50	<=365		9	resistant for yrs.	BASF Luran 368 R
	crystals		52	28		9	exc. resistance	Dow Tyril
Benzyl Alcohol			20	<=365		0	dissolved	BASF Luran 368 R
			50	<=365		0	"	"
		1.5	23	28		5	moder. effect; expected life: wks. to mths.	Dow Tyril
		1.5	52	28		5	"	"
Betadine	liquid		23		strained	8	resistant	Lustran SAN 31
Bleach	sol'ns	12	20	<=365		9	resistant for yrs.	BASF Luran 368 R
	"	12	50	<=365		9	"	"
Boric Acid	sol'n		20	<=365		9	"	"
			22	7		8	resistant	Monsanto Lustran
	sat'd		23	28		9	exc. resistance	Dow Tyril
			50	7		8	resistant	Monsanto Lustran
	sol'n		50	<=365		9	resistant for yrs.	BASF Luran 368 R
	sat'd		52	28		9	exc. resistance	Dow Tyril
		10	23	28		9	"	"
		10	52	28		9	"	"
Butyl Alcohol			20	<=365		7	resistant for mths.	BASF Luran 368 R
			22	7		8	resistant	Monsanto Lustran
			50	7		6	heat reduces room temp. resist.	"
			50	<=365		5	resistant for wks.	BASF Luran 368 R
		100	23	28		7	sl. clouding/discoloration, expected life: mths. to yrs.	Dow Tyril
		100	52	28		2	severe attack, softened in few hrs.	"

Reagent	Reagent Note	Conc. (%)	Temp. (°C)	Time (days)	Load	PDL Rating	Resistance Note	Material Note
Styrene Acrylonitrile Copolymer (continued)								
Carbolic Acid			22	7		2	plastic severely crazed, softened or dissolved	Monsanto Lustran
		50	22	7		2	"	"
Chlorine	dry gas		20	<=365		3	resistant for hrs.	BASF Luran 368 R
	chlorine water		20	<=365		5	resistant for wks.	"
	liquid		20	<=365		3	resistant for hrs.	"
	dry gas		50	<=365		3	"	"
	liquid		50	<=365		3	"	"
	in air	10	23	28		5	moder. effect; expected life: wks. to mths.	Dow Tyril
	"	10	52	28		5	"	"
		100	23	28		2	severe attack, softened in few hrs.	"
		100	52	28		2	"	"
Cresol		100	23	28		2	"	"
		100	52	28		2	"	"
Ethyl Acetate			20	<=365		0	dissolved	BASF Luran 368 R
			50	<=365		0	"	"
		90	22	7		2	plastic severely crazed, softened or dissolved	Monsanto Lustran
		100	23	28		2	severe attack, softened in few hrs.	Dow Tyril
		100	52	28		2	"	"
Ethyl Alcohol	2B absolute		22	7		2	plastic severely crazed, softened or dissolved	Monsanto Lustran
	formula 30		22	7		8	resistant	"
	2B-95		22	7		8	"	"
	"		50	7		6	heat reduces room temp. resist.	"
	formula 30		50	7		6	"	"
		40	20	<=365		9	resistant for yrs.	BASF Luran 368 R
		40	23	28		7	sl. clouding/discoloration, expected life: mths. to yrs.	Dow Tyril
		40	50	<=365		7	resistant for mths.	BASF Luran 368 R
		40	52	28		2	severe attack, softened in few hrs.	Dow Tyril
		95	20	<=365		7	resistant for mths.	BASF Luran 368 R
		95	23	28		2	severe attack, softened in few hrs.	Dow Tyril
		95	50	<=365		4	resistant for days	BASF Luran 368 R
		95	52	28		2	severe attack, softened in few hrs.	Dow Tyril
Ethylene Chlorohydrin		100	23	28		5	moder. effect; expected life: wks. to mths.	"
		100	52	28		2	severe attack, softened in few hrs.	"
Ethylene Glycol			20	<=365		9	resistant for yrs.	BASF Luran 368 R
			22	7		8	resistant	Monsanto Lustran
			23	7	0.25%	7		LNP; 30% gl.fil.
			23	7		9		"
			50	<=365		9	resistant for yrs.	BASF Luran 368 R
		100	23	28		9	exc. resistance	Dow Tyril
		100	52	28		9	"	"
Ethylene Oxide		100	23	28		2	severe attack, softened in few hrs.	"
		100	52	28		2	"	"
Fluorine		100	23	28		2	"	"
		100	52	28		2	"	"
Formaldehyde		30	20	<=365		7	resistant for mths.	BASF Luran 368 R
		30	22	7		8	resistant	Monsanto Lustran
		30	23	28		9	exc. resistance	Dow Tyril
		30	50	7		6	heat reduces room temp. resist.	Monsanto Lustran
		30	50	<=365		5	resistant for wks.	BASF Luran 368 R
		30	52	28		5	moder. effect; expected life: wks. to mths.	Dow Tyril
		37		28		3	consid. chg.; expected life: days	"
		37		28		2	severe attack, softened in few hrs.	"
Hydrogen Peroxide		3	20	<=365		9	resistant for yrs.	BASF Luran 368 R
		3	22	7		8	resistant	Monsanto Lustran
		3	23	28		9	exc. resistance	Dow Tyril
		3	50	7		8	resistant	Monsanto Lustran
		3	50	<=365		9	resistant for yrs.	BASF Luran 368 R
		3	52	28		9	exc. resistance	Dow Tyril
		30	23	28		9	"	"
		30	52	28		9	"	"
Iodine	tincture		20	<=365		5	resistant for wks.	BASF Luran 368 R
	crystals		23	28		2	severe attack, softened in few hrs.	Dow Tyril
	"		52	28		2	"	"
		3	22	7		8	resistant	Monsanto Lustran

Styrene Acrylonitrile Copolymer (continued)

Reagent	Reagent Note	Conc. (%)	Temp. (°C)	Time (days)	Load	PDL Rating	Resistance Note	Material Note
Iodine		3	50	7		8	resistant	Monsanto Lustran
		83	22	7		2	plastic severely crazed, softened or dissolved	"
Isopropyl Alcohol			22	7		8	resistant	"
			50	7		6	heat reduces room temp. resist.	"
		100	23	28		9	exc. resistance	Dow Tyril
		100	52	28		3	consid. chg.; expected life: days	"
Mercuric Chloride	sol'n		20	<=365		9	resistant for yrs.	BASF Luran 368 R
	sat'd		22	7		8	resistant	Monsanto Lustran
	"		50	7		8	"	"
	sol'n		50	<=365		9	resistant for yrs.	BASF Luran 368 R
	powder	5	22	7		8	resistant	Monsanto Lustran
		5	23	28		9	exc. resistance	Dow Tyril
	powder	5	50	7		8	resistant	Monsanto Lustran
		5	52	28		7	sl. clouding/discoloration, expected life: mths. to yrs.	Dow Tyril
		100	23	28		9	exc. resistance	"
		100	52	28		9	"	"
Methyl Isobutyl Ketone			20	<=365		0	dissolved	BASF Luran 368 R
			22	7		2	plastic severely crazed, softened or dissolved	Monsanto Lustran
			50	<=365		0	dissolved	BASF Luran 368 R
		100	23	28		1	solvent	Dow Tyril
		100	52	28		1	"	"
Morpholine			22	7		2	plastic severely crazed, softened or dissolved	Monsanto Lustran
		100	23	28		2	severe attack, softened in few hrs.	Dow Tyril
		100	52	28		2	"	"
Phenol			20	<=365		3	resistant for hrs.	BASF Luran 368 R
	crystals		23	28		2	severe attack, softened in few hrs.	Dow Tyril
			50	<=365		3	resistant for hrs.	BASF Luran 368 R
	crystals		52	28		2	severe attack, softened in few hrs.	Dow Tyril
		5	22	7		2	plastic severely crazed, softened or dissolved	Monsanto Lustran
		10	20	<=365		4	resistant for days	BASF Luran 368 R
		10	50	<=365		4	"	"
Potassium Permanganate			20	<=365		7	resistant for mths.	"
	sat'd		22	7		8	resistant	Monsanto Lustran
	"		50	7		6	heat reduces room temp. resist.	"
			50	<=365		4	resistant for days	BASF Luran 368 R
		100	23	28		7	sl. clouding/discoloration, expected life: mths. to yrs.	Dow Tyril
		100	52	28		5	moder. effect; expected life: wks. to mths.	"
Propylene Oxide			20	<=365		3	resistant for hrs.	BASF Luran 368 R
			50	<=365		3	"	"
		100	23	28		2	severe attack, softened in few hrs.	Dow Tyril
		100	52	28		2	"	"
Salicylaldehyde		100	23	28		2	"	"
		100	52	28		2	"	"
Salicylic Acid			20	<=365		9	resistant for yrs.	BASF Luran 368 R
	sat'd		22	7		8	resistant	Monsanto Lustran
	"		23	28		9	exc. resistance	Dow Tyril
	"		50	7		8	resistant	Monsanto Lustran
			50	<=365		7	resistant for mths.	BASF Luran 368 R
	sat'd		52	28		7	sl. clouding/discoloration, expected life: mths. to yrs.	Dow Tyril
Silver Nitrate			20	<=365		7	resistant for mths.	BASF Luran 368 R
	sat'd		22	7		8	resistant	Monsanto Lustran
	"		23	28		9	exc. resistance	Dow Tyril
	"		50	7		6	heat reduces room temp. resist.	Monsanto Lustran
			50	<=365		7	resistant for mths.	BASF Luran 368 R
	sat'd		52	28		5	moder. effect; expected life: wks. to mths.	Dow Tyril
Sodium Hypochlorite	12% chlorine		20	<=365		9	resistant for yrs.	BASF Luran 368 R
			22	7		8	resistant	Monsanto Lustran
	15% chlorine		23	28		9	exc. resistance	Dow Tyril
			50	7		8	resistant	Monsanto Lustran
	12% chlorine		50	<=365		7	resistant for mths.	BASF Luran 368 R
	15% chlorine		52	28		9	exc. resistance	Dow Tyril
Sulfuric Acid		1-6	23	28		9	"	"
		1-6	52	28		9	"	"

Styrene Acrylonitrile Copolymer (continued)

Reagent	Reagent Note	Conc. (%)	Temp. (°C)	Time (days)	Load	PDL Rating	Resistance Note	Material Note
Sulfuric Acid		10	20	<=365		9	resistant for yrs.	BASF Luran 368 R
		10	23	7	0.25%	6		LNP; 30% gl.fil.
		10	23	7		7		"
		10	23	28		9	exc. resistance	Dow Tyril
		10	50	<=365		9	resistant for yrs.	BASF Luran 368 R
		10	52	28		7	sl. clouding/discoloration, expected life: mths. to yrs.	Dow Tyril
Thymol			20	<=365		3	resistant for hrs.	BASF Luran 368 R
			50	<=365		3	"	"

Styrene Butadiene Copolymer

Reagent	Reagent Note	Conc. (%)	Temp. (°C)	Time (days)	Load	PDL Rating	Resistance Note	Material Note
Ethyl Acetate			23	7		0	dissolved in 1 hr.	Phillips K Resin KR03
Ethyl Alcohol	concentrated		23	7		5	no chg. in appear.	"
		50	23	7		7	"	"
		50	50	7		7	clouded	"
Hydrogen Peroxide		3	23	7		9	no chg. in appear.	"
		3	50	7		9	"	"
		28	23	7		9	"	"
Sodium Hypochlorite			23	7		8	"	"
			50	7		9	"	"
Sulfuric Acid		3	23	7		8	"	"
		3	50	7		9	"	"

Styrene Maleic Anhydride Copolymer

Reagent	Reagent Note	Conc. (%)	Temp. (°C)	Time (days)	Load	PDL Rating	Resistance Note	Material Note
Ammonia	aq. sol'n		22	5	13.8 MPa	8	no effect	Arco Dylark 132; high heat, transp. grade
	"		22	5	13.8 MPa	8	"	Arco Dylark 700; med. heat, high impact grade
	"		22	5	27.6 MPa	8	"	Arco Dylark 250P20; 20% gl.fib.; med. heat grade
	"		22	5		8	"	"
	"		22	5		8	"	Arco Dylark 700; med. heat, high impact grade
	"		22	5		8	"	Arco Dylark 132; high heat, transp. grade
Bleach	Clorox		22	5	13.8 MPa	8	"	"
	"		22	5	13.8 MPa	8	"	Arco Dylark 700; med. heat, high impact grade
	"		22	5	27.6 MPa	8	"	Arco Dylark 250P20; 20% gl.fib.; med. heat grade
	"		22	5		8	"	"
	"		22	5		8	"	Arco Dylark 700; med. heat, high impact grade
	"		22	5		8	"	Arco Dylark 132; high heat, transp. grade
Ethyl Acetate			22	5	13.8 MPa	3	attacked	Arco Dylark 700; med. heat, high impact grade
			22	5	13.8 MPa	3	"	Arco Dylark 132; high heat, transp. grade
			22	5	27.6 MPa	3	"	Arco Dylark 250P20; 20% gl.fib.; med. heat grade
			22	5		3	"	Arco Dylark 700; med. heat, high impact grade
			22	5		3	"	Arco Dylark 132; high heat, transp. grade
Ethyl Alcohol			22	5	13.8 MPa	8	no effect	"
			22	5	13.8 MPa	8	"	Arco Dylark 700; med. heat, high impact grade
			22	5	27.6 MPa	8	"	Arco Dylark 250P20; 20% gl.fib.; med. heat grade
			22	5		8	"	Arco Dylark 132; high heat, transp. grade
			22	5		8	"	Arco Dylark 250P20; 20% gl.fib.; med. heat grade
			22	5		8	"	Arco Dylark 700; med. heat, high impact grade
Ethylene Glycol			22	5	13.8 MPa	8	"	Arco Dylark 132; high heat, transp. grade
			22	5	13.8 MPa	8	"	Arco Dylark 700; med. heat, high impact grade
			22	5	27.6 MPa	8	"	Arco Dylark 250P20; 20% gl.fib.; med. heat grade
			22	5		8	"	Arco Dylark 132; high heat, transp. grade
			22	5		8	"	Arco Dylark 250P20; 20% gl.fib.; med. heat grade
			22	5		8	"	Arco Dylark 700; med. heat, high impact grade

Reagent	Reagent Note	Conc. (%)	Temp. (°C)	Time (days)	Load	PDL Rating	Resistance Note	Material Note
Polyvinyl Chloride								
Ammonia	liquid		22			3	not recommended for use	Geon Co. Geon
	dry gas		22			8	recommended for use	"
	liquid		60			3	not recommended for use	"
	dry gas		60			8	recommended for use	"
Benzaldehyde			22			3	not recommended for use	"
			60			3	"	"
		10	23			3	"	Geon Co. Geon; Type II - impact modified
		10	23			8	recommended for use	Geon Co. Geon; Type I - normal impact
		10	60			3	not recommended for use	Geon Co. Geon; Type II - impact modified
		10	60			3	"	Geon Co. Geon; Type I - normal impact
Benzalkonium Chloride			22			8	recommended for use	Geon Co. Geon
Benzoic Acid			22			8	"	"
			60			8	"	"
Bleach	12% chlorine		22			8	"	"
	Clorox		22			8	"	"
	12% chlorine		60			8	"	"
Boric Acid			22			8	"	"
			60			8	"	"
Butyl Alcohol			22			8	"	"
	primary		23			3	not recommended for use	Geon Co. Geon; Type II - impact modified
	"		23			8	recommended for use	Geon Co. Geon; Type I - normal impact
			23			8	"	"
			23			8	"	Geon Co. Geon; Type II - impact modified
			60			3	not recommended for use	Geon Co. Geon
			60			3	"	Geon Co. Geon; Type II - impact modified
	primary		60			3	"	"
	"		60			8	recommended for use	Geon Co. Geon; Type I - normal impact
			60			8	"	"
Chlorine	dry gas		22			3	not recommended for use	Geon Co. Geon
	liquid, dry		22			3	"	"
	liquid, pressurized		22			3	"	"
	moist gas		22			3	"	"
	chlorine water		22			8	recommended for use	"
	dry gas		23	30		1		Geon Co. Geon 8750
	moist gas		23	30		1		"
	"		23	30		1		Geon Co. Geon 8700A
	dry gas		23	30		1		"
	"		60			3	not recommended for use	Geon Co. Geon
	liquid, dry		60			3	"	"
	moist gas		60			3	"	"
	chlorine water		60			8	recommended for use	"
	moist gas		60	30		1		Geon Co. Geon 8700A
	dry gas		60	30		1		"
Cresol			23			3	not recommended for use	Geon Co. Geon; Type I - normal impact
			23			3	"	Geon Co. Geon; Type II - impact modified
			60			3	"	Geon Co. Geon; Type I - normal impact
			60			3	"	Geon Co. Geon; Type II - impact modified
Ethyl Acetate			22			3	"	Geon Co. Geon
			60			3	"	"
Ethyl Alcohol			23			8	recommended for use	Geon Co. Geon; Type II - impact modified
			23			8	"	Geon Co. Geon; Type I - normal impact
			60			3	not recommended for use	Geon Co. Geon; Type II - impact modified
			60			8	recommended for use	Geon Co. Geon; Type I - normal impact
		95	23	7	strained	8	appear. unchanged	Georgia Gulf HF-2230
		95	23	7		8		"
Ethylene Chlorohydrin			22			3	not recommended for use	Geon Co. Geon
			60			3	"	"

Polyvinyl Chloride (continued)

Reagent	Reagent Note	Conc. (%)	Temp. (°C)	Time (days)	Load	PDL Rating	Resistance Note	Material Note
Ethylene Glycol			22			8	recommended for use	"
			23	7	0.25%	9		LNP; 15% gl.fib.
			23	7		9		"
			60			8	recommended for use	Geon Co. Geon
Ethylene Oxide			22			3	not recommended for use	"
			60			3	"	"
Fluorine	wet gas		22			8	recommended for use	"
	gas		23			3	not recommended for use	Geon Co. Geon; Type II - impact modified
	"		23			8	recommended for use	Geon Co. Geon; Type I - normal impact
	"		60			3	not recommended for use	Geon Co. Geon; Type II - impact modified
	"		60			3	"	Geon Co. Geon; Type I - normal impact
	wet gas		60			3	"	Geon Co. Geon
Formaldehyde			23			3	"	Geon Co. Geon; Type II - impact modified
			23			8	recommended for use	Geon Co. Geon; Type I - normal impact
			60			3	not recommended for use	Geon Co. Geon; Type II - impact modified
			60			8	recommended for use	Geon Co. Geon; Type I - normal impact
Hydrogen Peroxide		30	22			8	"	Geon Co. Geon
		30	60			8	"	"
		30	60	30		3		Geon Co. Geon 8700A
		50	22			8	recommended for use	Geon Co. Geon
		50	23	30		7		Geon Co. Geon 8700A
		50	23	30		9		Geon Co. Geon 8750
		50	60			8	recommended for use	Geon Co. Geon
		50	60	30		9		Geon Co. Geon 8750
		50	60	30		3		Geon Co. Geon 8700A
		90	22			8	recommended for use	Geon Co. Geon
		90	60			8	"	"
Iodine			22			3	not recommended for use	"
			60			3	"	"
	sol'n	10	22			3	"	"
	"	10	60			3	"	"
Mercuric Chloride			22			8	recommended for use	"
			60			8	"	"
Methyl Isobutyl Ketone			22			3	not recommended for use	"
			60			3	"	"
Phenol			22			3	"	"
			60			3	"	"
Potassium Permanganate		10	22			8	recommended for use	"
		10	60			8	"	"
		25	23			8	"	Geon Co. Geon; Type I - normal impact
		25	52			8	"	Geon Co. Geon; Type II - impact modified
		25	60			3	not recommended for use	Geon Co. Geon; Type I - normal impact
Propylene Oxide			22			3	"	Geon Co. Geon
			60			3	"	"
Salicylic Acid			22			8	recommended for use	"
			23	30		9		Geon Co. Geon 8700A
			23	30		9		Geon Co. Geon 8750
			60			8	recommended for use	Geon Co. Geon
			60	30		7		Geon Co. Geon 8700A
			60	30		7		Geon Co. Geon 8750
Silver Nitrate			22			8	recommended for use	Geon Co. Geon
			60			8	"	"
Sodium Hypochlorite			22			8	"	"
Sulfuric Acid		3	22			8	"	"
		3	60			8	"	"
		10	22			8	"	"
		10	23	7	0.25%	9		LNP; 15% gl.fib.
		10	23	7	strained	8	appear. unchanged	Georgia Gulf HF-2230
		10	23	7		9		LNP; 15% gl.fib.
		10	23	7		7		Georgia Gulf HF-2230
		10	60			8	recommended for use	Geon Co. Geon

Reagent	Reagent Note	Conc. (%)	Temp. (°C)	Time (days)	Load	PDL Rating	Resistance Note	Material Note

Chlorinated Polyvinyl Chloride

Reagent	Reagent Note	Conc. (%)	Temp. (°C)	Time (days)	Load	PDL Rating	Resistance Note	Material Note
Benzoic Acid			23			8	recommended for use	
			66			8	"	
Bleach	household		23			8	"	BF Good. TempRite
	"		23	90		9		BF Good. TempRite; pipe
	"		82			8	recommended for use	BF Good. TempRite
Boric Acid	sat'd		23			8	"	"
			23			8	"	"
	sat'd		23	90		9		BF Good. TempRite; pipe
			82			8	recommended for use	BF Good. TempRite
Butyl Alcohol			23			7	good resistance	
			66			2	not recommended for use	
Cresol			23			2	"	BF Good. TempRite
			82			2	"	"
Ethyl Acetate			23			2	"	"
			82			2	not recommended for use	BF Good. TempRite
Ethyl Alcohol			23			4	caution, suspect with certain stress levels	"
			23			7	good resistance	
			23	90		8	poss. stress crack agent	BF Good. TempRite; pipe
			66			7	good resistance	
Ethylene Glycol			23			4	caution, suspect with certain stress levels	BF Good. TempRite
			23	90		8	poss. stress crack agent	BF Good. TempRite; pipe
			82			4	caution, suspect with certain stress levels	BF Good. TempRite
			82	90		7	poss. stress crack agent	BF Good. TempRite; pipe
Formaldehyde		37	23			4	fair resistance	
		37	66			2	not recommended for use	
Hydrogen Peroxide		30	23			8	recommended for use	BF Good. TempRite
		30	23	90		9		BF Good. TempRite; pipe
Isopropyl Alcohol			23			4	caution, suspect with certain stress levels	BF Good. TempRite
			23	90		8	poss. stress crack agent	BF Good. TempRite; pipe
			82			4	caution, suspect with certain stress levels	BF Good. TempRite
Mercuric Chloride			23			8	recommended for use	"
Methyl Isobutyl Ketone			23			2	not recommended for use	"
			82			2	"	"
Phenol			23			4	fair resistance	
			66			2	not recommended for use	
Potassium Permanganate		10	23			8	recommended for use	BF Good. TempRite
		10	23	90		9		BF Good. TempRite; pipe
		10	82			4	caution, suspect with certain stress levels	BF Good. TempRite
		10	82	90		6		BF Good. TempRite; pipe
Propylene Oxide			23			2	not recommended for use	BF Good. TempRite
			82			2	"	"
Silver Nitrate			23			8	recommended for use	"
Sodium Hypochlorite	bleaches	4-6	23	90		9		BF Good. TempRite; pipe
		15	23			8	recommended for use	BF Good. TempRite
		15	23	90		9		BF Good. TempRite; pipe
		15	82			8	recommended for use	BF Good. TempRite
		15	82	90		9		BF Good. TempRite; pipe

Polyvinylidene Chloride

Reagent	Reagent Note	Conc. (%)	Temp. (°C)	Time (days)	Load	PDL Rating	Resistance Note	Material Note
Ammonia			23			2	severe attack, softening in few hrs. expected	Dow Saran
			52			2	"	"
Benzaldehyde			23			5	moder. effect; expected life: wks. to mths.	"
			52			5	"	"
	in isopropyl alcohol	1	23			7	sl. dicoloring, expected life: mths. to yrs.	"
	"	1	52			7	"	"
	"	5	23			7	"	"
	"	5	52			7	"	"
	"	10	23			7	"	"
	"	10	52			7	"	"
Benzalkonium Chloride			23			8	plastic expected to be unaffected	"
			52			8	"	"

Polyvinylidene Chloride (continued)

Reagent	Reagent Note	Conc. (%)	Temp. (°C)	Time (days)	Load	PDL Rating	Resistance Note	Material Note
Benzoic Acid	sat'd		23			8	plastic expected to be unaffected	Dow Saran
	crystals		23	28		9	plastic was unaffected for duration of test	"
	sat'd		52			8	plastic expected to be unaffected	"
	crystals		52	28		9	plastic was unaffected for duration of test	"
Benzyl Alcohol			23			8	plastic expected to be unaffected	"
			52			5	moder. effect; expected life: wks. to mths.	"
		1.5	23			8	plastic expected to be unaffected	"
		1.5	52			7	sl. dicoloring, expected life: mths. to yrs.	"
Boric Acid			23			8	plastic expected to be unaffected	"
	sat'd		23	28		9	plastic was unaffected for duration of test	"
			52			8	plastic expected to be unaffected	"
	sat'd		52	28		9	plastic was unaffected for duration of test	"
		10	23	28		9	"	"
		10	52	28		9	"	"
Butyl Alcohol			23	28		9	"	"
			52	28		9	"	"
Chlorine			23			2	severe attack, softening in few hrs. expected	"
			52			2	"	"
	moist	10	23			5	moder. effect; expected life: wks. to mths.	"
	in air	10	23	28		9	plastic was unaffected for duration of test	"
	"	10	52	28		9	"	"
	moist	10	52	28		3	consid. chg.; expected life: days	"
Cresol			23			5	moder. effect; expected life: wks. to mths.	"
			52			5	"	"
Ethyl Acetate			23	28		5	"	"
			52	28		5	"	"
		85-88	23			7	sl. dicoloring, expected life: mths. to yrs.	"
		85-88	52			5	moder. effect; expected life: wks. to mths.	"
Ethyl Alcohol	formula 30		23	28		9	plastic was unaffected for duration of test	"
	absolute		23	28		5	moder. effect; expected life: wks. to mths.	"
	"		52	28		5	"	"
	formula 30		52	28		7	sl. clouding/discoloration, expected life: mths. to yrs.	"
		40	23			8	plastic expected to be unaffected	"
		40	52			8	"	"
	2B-95	95	23			7	sl. dicoloring, expected life: mths. to yrs.	"
	"	95	52			7	"	"
Ethylene Chlorohydrin			23			5	moder. effect; expected life: wks. to mths.	"
			52	28		3	consid. chg.; expected life: days	"
Ethylene Glycol			23			7	sl. dicoloring, expected life: mths. to yrs.	"
			52			7	"	"
Ethylene Oxide			23	28		9	plastic was unaffected for duration of test	"
			52	28		5	moder. effect; expected life: wks. to mths.	"
Fluorine			23			2	severe attack, softening in few hrs. expected	"
			52			2	"	"
Formaldehyde		10	23			8	plastic expected to be unaffected	"
		10	52			8	"	"
		30	23			8	"	"
		30	52			8	"	"
		37	23	28		9	plastic was unaffected for duration of test	"
		37	52	28		9	"	"
		40	23			8	plastic expected to be unaffected	"
		40	52			8	"	"

Polyvinylidene Chloride (continued)

Reagent	Reagent Note	Conc. (%)	Temp. (°C)	Time (days)	Load	PDL Rating	Resistance Note	Material Note
Hydrogen Peroxide		1	23			8	plastic expected to be unaffected	Dow Saran
		1	52			7	sl. dicoloring, expected life: mths. to yrs.	"
		8	23	28		9	plastic was unaffected for duration of test	"
		8	52			7	sl. dicoloring, expected life: mths. to yrs.	"
		30	23			8	plastic expected to be unaffected	"
		30	52			7	sl. dicoloring, expected life: mths. to yrs.	"
		90	23			8	plastic expected to be unaffected	"
		90	52			7	sl. dicoloring, expected life: mths. to yrs.	"
Iodine	crystals		23			5	moder. effect; expected life: wks. to mths.	"
	"		52			5	"	"
Isopropyl Alcohol			23			7	sl. dicoloring, expected life: mths. to yrs.	"
			52			7	"	"
Mercuric Chloride			23			8	plastic expected to be unaffected	"
	sat'd		23			8	"	"
			52			8	"	"
	sat'd		52			8	"	"
		5	23			8	"	"
		5	52			8	"	"
Methyl Isobutyl Ketone			23	28		5	moder. effect; expected life: wks. to mths.	"
			52	28		5	"	"
Morpholine			23	28		3	consid. chg.; expected life: days	"
			52			2	severe attack, softening in few hrs. expected	"
Phenol	crystals		23	28		7	sl. clouding/discoloration, expected life: mths. to yrs.	"
	"		52			5	moder. effect; expected life: wks. to mths.	"
		5	23	28		7	sl. clouding/discoloration, expected life: mths. to yrs.	"
		5	52			5	moder. effect; expected life: wks. to mths.	"
Potassium Permanganate			23			7	sl. dicoloring, expected life: mths. to yrs.	"
	sat'd		23	28		7	sl. clouding/discoloration, expected life: mths. to yrs.	"
	"		52			7	sl. dicoloring, expected life: mths. to yrs.	"
			52			7	"	"
Propylene Oxide			23	28		5	moder. effect; expected life: wks. to mths.	"
			52	28		3	consid. chg.; expected life: days	"
Salicylaldehyde			23			7	sl. dicoloring, expected life: mths. to yrs.	"
			52			7	"	"
Salicylic Acid	sat'd		23			8	plastic expected to be unaffected	"
	powder		23			8	"	"
	ointment		23	28		9	plastic was unaffected for duration of test	"
	sat'd		52			8	plastic expected to be unaffected	"
	ointment		52			8	"	"
	powder		52			8	"	"
Silver Nitrate	sat'd		23			8	"	"
			23	28		9	plastic was unaffected for duration of test	"
	sat'd		52			8	plastic expected to be unaffected	"
			52	28		9	plastic was unaffected for duration of test	"
Sodium Hypochlorite	5% chlorine		23			7	sl. dicoloring, expected life: mths. to yrs.	"
	"		52			7	"	"
		15	23	28		9	plastic was unaffected for duration of test	"
		15	52	28		7	sl. clouding/discoloration, expected life: mths. to yrs.	"
Sulfuric Acid		10	23	28		9	plastic was unaffected for duration of test	"
		10	52	28		7	sl. clouding/discoloration, expected life: mths. to yrs.	"

Reagent	Reagent Note	Conc. (%)	Temp. (°C)	Time (days)	Load	PDL Rating	Resistance Note	Material Note

Acrylic PVC Alloy

Reagent	Reagent Note	Conc. (%)	Temp. (°C)	Time (days)	Load	PDL Rating	Resistance Note	Material Note
Ammonia	household		23	30		8	resistant, no staining tendency	Kleerdex Kydex 100; sheet
	"		60	30		5	sl. attack, sl. staining tendency	"
Ethyl Acetate			23	7		2	attacked	Kleerdex Kydex 100; sheet
Ethyl Alcohol		50	23	7		8	no chg. observed	Kleerdex Kydex 100; sheet
		95	23	7		8	"	"
Hydrogen Peroxide		3	23	7		8	no chg. observed	Kleerdex Kydex 100; sheet
Isopropyl Alcohol			23	30		8	resistant, no staining tendency	Kleerdex Kydex 100; sheet
Phenol		5	23	7		7	very slightly whitened	Kleerdex Kydex 100; sheet
Sulfuric Acid		3	23	7		8	no chg. observed	Kleerdex Kydex 100; sheet

ASA PVC Alloy

Reagent	Reagent Note	Conc. (%)	Temp. (°C)	Time (days)	Load	PDL Rating	Resistance Note	Material Note
Butyl Alcohol			25	7	0.5%	8		GE Geloy XP2001
			25	7	1%	8		"
			25	7		8		"
Isopropyl Alcohol			25	7	0.5%	8		GE Geloy XP2001
			25	7	1%	7		"
			25	7		8		"

Polycarbonate ABS Alloy

Reagent	Reagent Note	Conc. (%)	Temp. (°C)	Time (days)	Load	PDL Rating	Resistance Note	Material Note
Ethylene Glycol			23	0.33		9	exc. resistance	Dow Pulse 1350
Sulfuric Acid		5	23	0.33		8	"	"

Polycarbonate Acrylic Resin Alloy

Reagent	Reagent Note	Conc. (%)	Temp. (°C)	Time (days)	Load	PDL Rating	Resistance Note	Material Note
Ethyl Alcohol						8	"	Cyrex 200-8005
Isopropyl Alcohol						8	"	"

Polycarbonate Polyester Alloy

Reagent	Reagent Note	Conc. (%)	Temp. (°C)	Time (days)	Load	PDL Rating	Resistance Note	Material Note
Ammonia			23	1	28.96 MPa	9	exp. strain is level where part stress cracks	Miles Makroblend UT 1018
		1	23	1		9	exc. resistance	Dow Sabre 1664
Benzyl Alcohol			23	0.125	0.7%	5	crazing or sl. cracking	Eastman Ektar MB DA003
			23	28		0	soluble in reagent	"
Betadine			23	0.125	0.7%	8	no chg. in appear.	"
Bleach	Clorox	5	23	28		8	no visible chg.	"
	"	5	23	28		8	"	"
Butyl Alcohol			23	3	0.5%	6	marginally compatible	GE Xenoy 1102
			23	3	0.5%	6	"	GE Xenoy 2720
			23	3	1%	5	incompatible, not recommended	"
			23	3	1%	4	"	GE Xenoy 1102
			23	3		7	marginally compatible	GE Xenoy 2720
			23	3		5	incompatible, not recommended	GE Xenoy 1102
Ethyl Acetate			23	28		0	strongly swollen	Eastman Ektar MB DA003
			25	30		3	incompatible, not recommended	GE Xenoy 6620
			25	30		1	"	GE Xenoy 6370; 30% gl.fib.
			25	30		3	"	GE Xenoy 6123
			25	30		5	"	GE Xenoy 6120
			25	90		2	"	GE Xenoy 6123
			25	90		3	"	GE Xenoy 6620
			25	90		1	"	GE Xenoy 6370; 30% gl.fib.
			25	90		3	"	GE Xenoy 6120
Ethyl Alcohol			23	28		9	no visible chg.	Eastman Ektar MB DA003
			25	30		9	compatible	GE Xenoy 6370; 30% gl.fib.
			25	30		9	"	GE Xenoy 6123
			25	30		9	"	GE Xenoy 6120
			25	30		9	"	GE Xenoy 6620
			25	90		8	"	GE Xenoy 6370; 30% gl.fib.
			25	90		8	"	GE Xenoy 6123
			25	90		8	"	GE Xenoy 6120
			25	90		7	marginally compatible	GE Xenoy 6620
		50	23	28		8	no visible chg.	Eastman Ektar MB DA003
Ethylene Glycol			23	1	28.96 MPa	9	exp. strain is level where part stress cracks	Miles Makroblend UT 1018
			23	1		9	exc. resistance	Dow Sabre 1664
			25	30		9	compatible	GE Xenoy 6123, 6620
			25	30		9	"	GE Xenoy 6370; 30% gl.fib.
			25	30		9	"	GE Xenoy 6120, 6620
			25	90		9	"	GE Xenoy 6370; 30% gl.fib.
			25	90		9	"	GE Xenoy 6123
			25	90		9	"	GE Xenoy 6120

Reagent	Reagent Note	Conc. (%)	Temp. (°C)	Time (days)	Load	PDL Rating	Resistance Note	Material Note
Polycarbonate Polyester Alloy								
Glutaraldehyde		50	23	0.125	0.7%	8	no chg. in appear.	Eastman Ektar MB DA003
Isopropyl Alcohol			23	0.125	0.7%	6	crazing or sl. cracking	"
			23	3	0.5%	9	compatible	GE Xenoy 2720
			23	3	0.5%	6	marginally compatible	GE Xenoy 1102
			23	3	1%	9	compatible	GE Xenoy 2720
			23	3	1%	5	incompatible, not recommended	GE Xenoy 1102
			23	3		9	compatible	GE Xenoy 2720
			23	3		6	marginally compatible	GE Xenoy 1102
			25	30		9	compatible	GE Xenoy 6120
			25	30		9	"	GE Xenoy 6620
			25	30		9	"	GE Xenoy 6370; 30% gl.fib.
			25	30		9	"	GE Xenoy 6123
			25	90		9	"	GE Xenoy 6370; 30% gl.fib.
			25	90		9	"	GE Xenoy 6120
			25	90		9	"	GE Xenoy 6123
			25	90		9	"	GE Xenoy 6620
		50	25	30		9	"	"
		50	25	30		9	"	GE Xenoy 6123
		50	25	30		8	"	GE Xenoy 6370; 30% gl.fib.
		50	25	30		9	"	GE Xenoy 6120
		50	25	90		8	"	GE Xenoy 6123
		50	25	90		9	"	GE Xenoy 6120
		50	25	90		8	"	GE Xenoy 6370; 30% gl.fib.
		50	25	90		9	"	GE Xenoy 6620
Lyposin	IV sol'n lyposin II		23	0.125	0.7%	8	no chg. in appear.	Eastman Ektar MB DA003
Methyl Isobutyl Ketone			23	0.125	0.7%	1	severe cracking or breaking	"
			23	28		0	strongly swollen	"
Sodium Hypochlorite	bleaches	5.25	23	0.125	0.7%	8	no chg. in appear.	"
Sulfuric Acid		5	23	1		9	exc. resistance	Dow Sabre 1664
		10	23	0.125	0.7%	8	no chg. in appear.	Eastman Ektar MB DA003
		10	25	30		9	compatible	GE Xenoy 6120
		10	25	30		9	"	GE Xenoy 6123
		10	25	30		9	"	GE Xenoy 6620
		10	25	30		7	marginally compatible	GE Xenoy 6370; 30% gl.fib.
		10	25	90		9	compatible	GE Xenoy 6123
		10	25	90		9	"	GE Xenoy 6620
		10	25	90		9	"	GE Xenoy 6120
		10	25	90		7	marginally compatible	GE Xenoy 6370; 30% gl.fib.
Polycarbonate Polyester Copolymer Alloy								
Toothpaste	Arm & Hammer					1		
Polycarbonate Polyester PCTG Alloy								
Benzyl Alcohol			23	28		1	swelled	Eastman Ektar MB DA001
			23	28		0	strongly swollen	Eastman Ektar MB EA001
			23	28		0	"	Eastman Ektar MB DA002
			23	28		0	soluble in reagent	Eastman Ektar MB DA003
Bleach	Clorox	5	23	28		9	no visible chg.	Eastman Ektar MB EA001
	"	5	23	28		9	"	Eastman Ektar MB DA001
	"	5	23	28		8	"	Eastman Ektar MB DA003
	"	5	23	28		9	"	Eastman Ektar MB DA002
Ethyl Acetate			23	28		0	strongly swollen	Eastman Ektar MB DA003
			23	28		2	sl. swelling	Eastman Ektar MB EA001
			23	28		2	"	Eastman Ektar MB DA002
			23	28		1	swelled	Eastman Ektar MB DA001
Ethyl Alcohol			23	28		9	no visible chg.	Eastman Ektar MB EA001
			23	28		9	"	Eastman Ektar MB DA001
			23	28		9	"	Eastman Ektar MB DA003
			23	28		9	"	Eastman Ektar MB DA002
		50	23	28		9	"	"
		50	23	28		9	"	Eastman Ektar MB DA001
		50	23	28		9	"	Eastman Ektar MB EA001
		50	23	28		9	"	Eastman Ektar MB DA003
Methyl Isobutyl Ketone			23	28		2	sl. swelling	Eastman Ektar MB EA001
			23	28		0	strongly swollen	Eastman Ektar MB DA002
			23	28		0	"	Eastman Ektar MB DA003
			23	28		1	swelled	Eastman Ektar MB DA001

Reagent	Reagent Note	Conc. (%)	Temp. (°C)	Time (days)	Load	PDL Rating	Resistance Note	Material Note

SMA PC Alloy

Reagent	Reagent Note	Conc. (%)	Temp. (°C)	Time (days)	Load	PDL Rating	Resistance Note	Material Note
Bleach			23	5	41.4 MPa	8	no visible chg.	Arco Arloy 1000
			23	5		8	"	"
Ethyl Acetate			23	5	41.4 MPa	3	attacked	"
			23	5		3	"	"
Ethyl Alcohol			23	5	41.4 MPa	8	no visible chg.	"
			23	5		8	"	"
Ethylene Glycol			23	5	41.4 MPa	8	"	"
			23	5		8	"	"
			66	5	41.4 MPa	8	"	"
			66	5		8	"	"

Olefinic Thermoplastic Elastomer

Reagent	Reagent Note	Conc. (%)	Temp. (°C)	Time (days)	Load	PDL Rating	Resistance Note	Material Note
Ammonia	ammonia liquor					8	good resistance	Adv. Elast. Santoprene
	ammonia water					8	"	"
Benzaldehyde			24			2	likely to have severe effect	DuPont Alcryn
Benzoic Acid						8	good resistance	Adv. Elast. Santoprene
Boric Acid			24			6	likely to have minor effect	DuPont Alcryn
Butyl Alcohol			23	7		7		Santoprene 281-73
Ethyl Acetate	acetic ether					6	fair resistance	Adv. Elast. Santoprene
			23	7		7		Santoprene 281-73
			24			2	likely to have severe effect	DuPont Alcryn
Ethyl Alcohol						8	good resistance	Adv. Elast. Santoprene
						8	"	"
			24			8	fluid has little to no effect	DuPont Alcryn
		95	23	7		7		Adv. Elast. Santoprene 201-55
		95	23	7		8		Adv. Elast. Santoprene 203-40
		95	23	7		8		Adv. Elast. Santoprene 101-64
		95	23	7		7		Adv. Elast. Santoprene 101-55
		95	23	7		8		Adv. Elast. Santoprene 101-80
		95	23	7		8		Adv. Elast. Santoprene 101-87
		95	23	7		8		Adv. Elast. Santoprene 203-50
		95	23	7		8		Adv. Elast. Santoprene 201-87
		95	23	7		8		Adv. Elast. Santoprene 201-73
		95	23	7		8		Adv. Elast. Santoprene 201-64
		95	23	7		8		Adv. Elast. Santoprene 103-40
		95	23	7		8		Adv. Elast. Santoprene 201-80
		95	23	7		8		Adv. Elast. Santoprene 101-73
		95	23	7		8		Adv. Elast. Santoprene 103-50
		95	23	7		8		Santoprene 281-73
		95	24	7		7		DuPont Alcryn 1201 B-70
Ethylene Chlorohydrin						6	fair resistance	Adv. Elast. Santoprene
Ethylene Glycol						8	good resistance	"
			23	7		9		Adv. Elast. Santoprene 281-64
			23	7		9		Adv. Elast. Santoprene 181-55
			23	7		9		Adv. Elast. Santoprene 281-87
			23	7		9		Adv. Elast. Santoprene 181-64
			23	7		9		Adv. Elast. Santoprene 281-55
			24			8	fluid has little to no effect	DuPont Alcryn
			100	2.9		7		Adv. Elast. Santoprene 181-64
			100	2.9		9		Adv. Elast. Santoprene 281-87
			100	2.9		8		Adv. Elast. Santoprene 181-55
			100	2.9		8		Adv. Elast. Santoprene 281-55
			100	2.9		7		Adv. Elast. Santoprene 281-64
Formaldehyde						8	good resistance	Adv. Elast. Santoprene
	aq.	37	23	7		9		Santoprene 281-73
		40	24			2	likely to have severe effect	DuPont Alcryn
Hydrogen Peroxide						4	resistance is conditional	Adv. Elast. Santoprene
	5% ammonium persulfate	3	60	7		8		Santoprene 281-73
Isopropyl Alcohol			23	7		6		"
			24			8	fluid has little to no effect	DuPont Alcryn
			24	7		8		DuPont Alcryn 1201 B-70
			24	7		8		DuPont Alcryn 1201 B-60
			24	7		7		DuPont Alcryn ALR-6387
			24	7		8		DuPont Alcryn 1201 B-80
Methyl Isobutyl Ketone			24			2	fluid has severe effect	DuPont Alcryn
			24	7		1		DuPont Alcryn 1201 B-70
Phenol						6	fair resistance	Adv. Elast. Santoprene

Reagent	Reagent Note	Conc. (%)	Temp. (°C)	Time (days)	Load	PDL Rating	Resistance Note	Material Note
Olefinic Thermoplastic Elastomer (continued)								
Sodium Hypochlorite						8	good resistance	Adv. Elast. Santoprene
Sulfuric Acid		10	66			8	"	"
Polyamide Thermoplastic Elastomer								
Ammonia		25	23	50		8		Huls Vestamid E62M-S3
		25	23	50		8		Huls Vestamid E47M-S3
		25	23	50		7		Huls Vestamid E40-S3
Butyl Alcohol			23	7		5	moder. effect	Atochem Pebax 6333
			23	7		5	"	Atochem Pebax 6312
			23	7		5	"	Atochem Pebax 5512
			23	7		5	"	Atochem Pebax 5533
			23	7		2	severe attack	Atochem Pebax 2533
			23	7		2	"	Atochem Pebax 4033
			23	7		2	"	Atochem Pebax 3533
			23	7		8		DuPont Zytel FN 714
			23	7		8		DuPont Zytel FN 718
			23	7		9		DuPont Zytel FN 726
			23	7		8		DuPont Zytel FN 716
Chlorine	chlorine water	16	23	50		9		Huls Vestamid E62M-S3
	"	16	23	50		9		Huls Vestamid E47M-S3
	"	16	23	50		8		Huls Vestamid E40-S3
Ethyl Acetate			23	7		9		DuPont Zytel FN 716
			23	7		9		DuPont Zytel FN 714
			23	7		9		DuPont Zytel FN 718
Ethyl Alcohol			23	7		8	little or no effect	Atochem Pebax 4033
			23	7		8	"	Atochem Pebax 6333
			23	7		8	"	Atochem Pebax 5533
			23	7		5	moder. effect	Atochem Pebax 5512
			23	7		5	"	Atochem Pebax 6312
			23	7		5	"	Atochem Pebax 3533
			23	7		2	severe attack	Atochem Pebax 2533
			23	7		9		DuPont Zytel FN 714
			23	7		9		DuPont Zytel FN 716
			23	7		9		DuPont Zytel FN 718
Ethylene Glycol			23	7		8	little or no effect	Atochem Pebax 5533
			23	7		8	"	Atochem Pebax 6312
			23	7		8	"	Atochem Pebax 4033
			23	7		8	"	Atochem Pebax 6333
			23	7		8	"	Atochem Pebax 3533
			23	7		8	"	Atochem Pebax 5512
			23	7		8	"	Atochem Pebax 2533
		50	100	7		9		DuPont Zytel FN 726
		50	100	7		9		DuPont Zytel FN 718
		50	100	7		9		DuPont Zytel FN 716
		50	100	7		9		DuPont Zytel FN 714
Sulfuric Acid		10	23	7		8	little or no effect	Atochem Pebax 2533
		10	23	7		8	"	Atochem Pebax 5533
		10	23	7		8	"	Atochem Pebax 3533
		10	23	7		8	"	Atochem Pebax 4033
		10	23	7		8	"	Atochem Pebax 6333
		10	23	7		2	severe attack	Atochem Pebax 6312
		10	23	7		2	"	Atochem Pebax 5512
Polyester Thermoplastic Elastomer								
Benzyl Alcohol			23	28		0	dissolved	Eastman Ecdel
Boric Acid			22			8	fluid has little to no effect	DuPont Hytrel
Butyl Alcohol			22	7		7		DuPont Hytrel 55D
			22	7		7		DuPont Hytrel 40D
			22	7		8		DuPont Hytrel 72D
Chlorine	moist gas		22			3	likely to have severe effect	DuPont Hytrel
	dry gas		22			3	"	"
Ethyl Acetate			22			6	fluid has minor to moderate effect	DuPont Hytrel 40D
			22			8	fluid has little to no effect	DuPont Hytrel 72D
			22			6	fluid has minor to moderate effect	DuPont Hytrel 55D
			23	28		3	strongly swollen	Eastman Ecdel
Ethyl Alcohol			22			8	fluid has little to no effect	DuPont Hytrel
			23	28		6	sl. swelling	Eastman Ecdel
		50	23	28		6	"	"

Reagent	Reagent Note	Conc. (%)	Temp. (°C)	Time (days)	Load	PDL Rating	Resistance Note	Material Note
Polyester Thermoplastic Elastomer (continued)								
Ethylene Glycol			22			8	fluid has little to no effect	DuPont Hytrel
			22	7		9		DuPont Hytrel 72D, 55D
			22	7		7		DuPont Hytrel 40D
			23	28		8	no visible chg.	Eastman Ecdel
Ethylene Oxide			22			8	fluid has little to no effect	DuPont Hytrel
Formaldehyde		40	22			6	fluid has minor to moderate effect	"
Iodine						5	plastic is stained by reagent	Eastman Ecdel
Isopropyl Alcohol			22			8	fluid has little to no effect	DuPont Hytrel
Mercuric Chloride			22			7	likely to have minor effect	"
Methyl Isobutyl Ketone			22	7		7		DuPont Hytrel 72D, 55D
			22	7		4		DuPont Hytrel 40D
			23	28		3	strongly swollen	Eastman Ecdel
Phenol			22			2	fluid has severe effect	DuPont Hytrel
Sodium Hypochlorite		5	22			8	fluid has little to no effect	"
	bleaches	5	23	28		8	no visible chg.	Eastman Ecdel
Sulfuric Acid		10	22	7		9		DuPont Hytrel 72D
		<50	22			8	fluid has little to no effect	DuPont Hytrel
		>50	22			2	fluid has severe effect	"
Styrenic Thermoplastic Elastomer								
Ammonia	liquid or gas					8	acceptable for use	Conc. Pol. C-Flex
Benzaldehyde						3	not acceptable	"
Benzoic Acid						3	"	"
Benzyl Alcohol						3	"	"
Bleach	bleaching liquors					8	acceptable for use	"
Boric Acid						8	"	"
		3.1	23	14		9		Shell Kraton D1101; 0.25-0.3 mm films
Butyl Alcohol						6	testing recommended before using	Conc. Pol. C-Flex
Chlorine	wet					8	acceptable for use	"
	dry					8	"	"
Cresol						3	not acceptable	"
Ethyl Acetate						8	acceptable for use	"
Ethyl Alcohol						6	testing recommended before using	"
						6	"	"
Ethylene Chlorohydrin						8	acceptable for use	"
Ethylene Glycol						6	testing recommended before using	"
Ethylene Oxide						8	acceptable for use	"
Formaldehyde						8	"	"
Hydrogen Peroxide	diluted					8	"	"
	concentrated					8	"	"
Iodine						6	testing recommended before using	"
Phenol						3	not acceptable	"
Silver Nitrate						8	acceptable for use	"
Sodium Hypochlorite		<5				8	"	"
		>5				8	"	"
Sulfuric Acid		5	23	14		8		Shell Kraton D1101; 0.25-0.3 mm films
		10	23	14		8		"
Urethane Thermoplastic Elastomer								
Benzalkonium Chloride	sol'n					3	permanent discolor.; not recommended as sterilant	Pellethane 2363
Betadine	povidone iodine sol'n		23	168	0.5%	5		Pellethane 2363-75D
	"		23	168	1.5%	4		"
	"		23	168		6		"
Bleach	Clorox	10	23	168	0.5%	4		"
	"	10	23	168	1.5%	6		"
	"	10	23	168		4		"
Ethylene Glycol		97	100	2.9		6	100% modulus decreases 15%	Parker Han. Parker P4700A90
Ethylene Oxide			23	0.25		7		Texin 5286
			23	0.5		6		"
Isopropyl Alcohol		70	23	168	0.5%	5		Pellethane 2363-75D
		70	23	168	1.5%	3		"
		70	23	168		6		"
Omnicide	2% glutaraldehyde		23	168	0.5%	7		"
	"		23	168	1.5%	9		"
	"		23	168		7		"

Thermoplastic Polyester Urethane Elastomer

Reagent	Reagent Note	Conc. (%)	Temp. (°C)	Time (days)	Load	PDL Rating	Resistance Note	Material Note
Ammonia	3n		23	7		9	exc. resistance	BASF Elastollan C85A
	"		23	7		9	"	BASF Elastollan C64D
		10	22	270		4	snow white, retains 62.3% of tear strength	Miles Texin 591AR
		10	22	365		3	blistered, whitened, retains 10.7% tear str.	Miles Texin 355DR
		10	22	365		4	snow white, retains 20.5% of tear strength	Miles Texin 480AR
Benzyl Alcohol			23	7		0	dissolved	BASF Elastollan C85A
			23	7		0	"	BASF Elastollan C64D
Bleach	undiluted					8	little or no effect	Dow Pellethane 2102-65A
	"					6	minor effect	Dow Pellethane 2102-55D
	"					8	little or no effect	Dow Pellethane 2102-90A
	"					6	minor effect	Dow Pellethane 2102-80A
	sodium hypochlorite	40	23	7		9	exc. resistance	BASF Elastollan C85A
	"	40	23	7		9	"	BASF Elastollan C64D
Boric Acid		4				8	little or no effect	Dow Pellethane 2102-80A
		4				8	"	Dow Pellethane 2102-90A
		4				6	minor effect	Dow Pellethane 2102-55D
		4				6	"	Dow Pellethane 2102-65A
Butyl Alcohol			23	7		5	fair resistance	BASF Elastollan C64D
			23	7		5	"	BASF Elastollan C85A
Ethyl Acetate			20			6	parts functional only under certain condit.	BASF Elastollan C64D
			22	365		2	retains 11.4% of tear strength	Miles Texin 480AR
			22	365		4	white, retains 22.7% of tear strength	Miles Texin 355DR
			23	7		4	poor resistance	BASF Elastollan C85A
			23	7		4	"	BASF Elastollan C64D
Ethyl Alcohol		96	20			7	extended expos. has some effect on weight/dimen.	"
		96	23	7		5	fair resistance	BASF Elastollan C85A
		96	23	7		5	"	BASF Elastollan C64D
Ethylene Glycol		50	23			6	minor effect	Dow Pellethane 2102-90A
		50	23			8	little or no effect	Dow Pellethane 2102-80A
		50	23			8	"	Dow Pellethane 2102-55D
		50	70			6	minor effect	Dow Pellethane 2102-80A
		50	70			8	little or no effect	Dow Pellethane 2102-55D
		50	70			4	moder. effect	Dow Pellethane 2102-90A
		100	23			6	minor effect	"
		100	23			6	"	Dow Pellethane 2102-65A
		100	23			8	little or no effect	Dow Pellethane 2102-80A
		100	70			8	"	Dow Pellethane 2102-55D
		100	70			4	moder. effect	Dow Pellethane 2102-80A
		100	70			4	"	Dow Pellethane 2102-90A
Hydrogen Peroxide		3	22	270		6	retains 100% of tear strength	Miles Texin 591AR
		3	22	365		5	retains 76.1% of tear strength	Miles Texin 480AR
		3	22	365		6	retains 65.3% of tear strength	Miles Texin 355DR
		3	23	7		9	exc. resistance	BASF Elastollan C85A
		3	23	7		9	"	BASF Elastollan C64D
Isopropyl Alcohol			20			7	extended expos. has some effect on weight/dimen.	BASF Elastollan C85A
			20			7	"	BASF Elastollan C64D
			23	7		5	fair resistance	BASF Elastollan C85A
			23	7		7	good resistance	BASF Elastollan C64D
		50	22	270		4	white, retains 67% of tear strength	Miles Texin 591AR
		50	22	365		4	tan, retains 40.7% of tear strength	Miles Texin 355DR
		50	22	365		4	white coating, retains 69.3% of tear strength	Miles Texin 480AR
Potassium Permanganate		5	23	7		5	fair resistance	BASF Elastollan C64D
		5	23	7		5	"	BASF Elastollan C85A
Sodium Hypochlorite			23	7		9	exc. resistance	"
			23	7		9	"	BASF Elastollan C64D
		5	22	270		6	retains 92.5% of tear strength	Miles Texin 591AR
		5	22	365		4	retains 60% of tear strength	Miles Texin 355DR
		5	22	365		6	retains 67% of tear strength	Miles Texin 480AR
Sulfuric Acid		1	20			9	resistant	BASF Elastollan C85A
		10	20			9	"	"

Reagent	Reagent Note	Conc. (%)	Temp. (°C)	Time (days)	Load	PDL Rating	Resistance Note	Material Note

Thermoplastic Polyether Urethane Elastomer

Reagent	Reagent Note	Conc. (%)	Temp. (°C)	Time (days)	Load	PDL Rating	Resistance Note	Material Note
Ammonia	3n		23	7		9	exc. resistance	BASF Elastollan 1164D
	"		23	7		9	"	BASF Elastollan 1185A
Benzyl Alcohol			23	7		0	dissolved	"
			23	7		0	"	BASF Elastollan 1164D
Bleach	undiluted					6	minor effect	Dow Pellethane 2103-55D
	"					2	severe attack, not recommened	Dow Pellethane 2103-90A
	"					2	"	Dow Pellethane 2103-90AE
	"					2	"	Dow Pellethane 2103-80AE
	sodium hypochlorite	40	23	7		9	exc. resistance	BASF Elastollan 1185A
	"	40	23	7		9	"	BASF Elastollan 1164D
Boric Acid		4				8	little or no effect	Dow Pellethane 2103-55D
		4				8	"	Dow Pellethane 2103-80AE
		4				6	minor effect	Dow Pellethane 2103-90AE
		4				8	little or no effect	Dow Pellethane 2103-90A
Butyl Alcohol			23	7		2	poor resistance	BASF Elastollan 1164D
			23	7		2	"	BASF Elastollan 1185A
Ethyl Acetate			23	7		2	"	"
			23	7		2	"	BASF Elastollan 1164D
Ethyl Alcohol			23	7		5	fair resistance	"
		96	23	7		5	"	BASF Elastollan 1185A
Ethylene Glycol		50	23			6	minor effect	Dow Pellethane 2103-80AE
		50	23			8	little or no effect	Dow Pellethane 2103-90A
		50	23			8	"	Dow Pellethane 2103-90AE
		50	70			6	minor effect	Dow Pellethane 2103-90A
		50	70			6	"	Dow Pellethane 2103-80AE
		50	70			8	little or no effect	Dow Pellethane 2103-55D
		50	70			8	"	Dow Pellethane 2103-90AE
		100	23			6	minor effect	Dow Pellethane 2103-80AE
		100	23			6	"	Dow Pellethane 2103-90A
		100	23			6	"	Dow Pellethane 2103-90AE
		100	23			6	"	Dow Pellethane 2103-55D
		100	70			6	"	Dow Pellethane 2103-90A
		100	70			6	"	Dow Pellethane 2103-90AE
		100	70			8	little or no effect	Dow Pellethane 2103-55D
		100	70			4	moder. effect	Dow Pellethane 2103-80AE
Hydrogen Peroxide		3	23	7		9	exc. resistance	BASF Elastollan 1164D
		3	23	7		9	"	BASF Elastollan 1185A
Isopropyl Alcohol			23	7		7	good resistance	BASF Elastollan 1164D
			23	7		5	fair resistance	BASF Elastollan 1185A
Potassium Permanganate		5	23	7		2	poor resistance	BASF Elastollan 1164D
		5	23	7		2	"	BASF Elastollan 1185A
Sodium Hypochlorite			23	7		9	exc. resistance	"
			23	7		9	"	BASF Elastollan 1164D

Polyvinyl Chloride Polyol

Reagent	Reagent Note	Conc. (%)	Temp. (°C)	Time (days)	Load	PDL Rating	Resistance Note	Material Note
Ammonia	liquid		23			5	fair resistance	Colorite; PVC tubing
	gas		23			5	"	"
Benzaldehyde			23			2	poor resistance	"
			23	0.0208		1	not recommended	Ans. Edm. Ever-Flex; lined glove film
Benzoic Acid			23			8	good resistance	Colorite; PVC tubing
Benzyl Alcohol			23			5	fair resistance	"
Boric Acid			23			8	good resistance	"
Butyl Alcohol			23			5	fair resistance	"
			23			7	very good perm. rate; 1-5 drops/hr.	Ans. Edm. Monkey Grip; lined glove film
			23	0.0208		8	material is well suited for use with chemical	Ans. Edm. Ever-Flex; lined glove film
Carbolic Acid						6		Pioneer Pylox V-20; glove film
Chlorine	wet		23			5	fair resistance	Colorite; PVC tubing
	dry		23			8	good resistance	"
Cresol			23			2	poor resistance	"
Ethyl Acetate			23	0.0208		1	not recommended	Ans. Edm. Ever-Flex; lined glove film
Ethyl Alcohol			23			5	fair resistance	Colorite; PVC tubing
			23			7	very good perm. rate; 1-5 drops/hr.	Ans. Edm. Monkey Grip; lined glove film
			23	0.0208		8	material is well suited for use with chemical	Ans. Edm. Ever-Flex; lined glove film

Reagent	Reagent Note	Conc. (%)	Temp. (°C)	Time (days)	Load	PDL Rating	Resistance Note	Material Note
Polyvinyl Chloride Polyol								
Ethylene Chlorohydrin			23			2	poor resistance	Colorite; PVC tubing
Ethylene Glycol						9	no perm. detected	Pioneer Pylox V-20; glove film
			23			5	fair resistance	Colorite; PVC tubing
			23	0.0035		8	no degradation	Pioneer Pylox V-20; glove film
			23	0.0208		9	exc. resistance	Ans. Edm. Ever-Flex; lined glove film
			23	0.0208		8	no degradation	Pioneer Pylox V-20; glove film
			23	0.0417		8	"	"
			23	0.167		8	"	"
			23	0.25		9	low permeation rate; 0-1/2 drops/hr.	Ans. Edm. Monkey Grip; lined glove film
Formaldehyde			23			5	fair resistance	Colorite; PVC tubing
			23			7	very good perm. rate; 1-5 drops/hr.	Ans. Edm. Monkey Grip; lined glove film
			23	0.0208		9	exc. resistance	Ans. Edm. Ever-Flex; lined glove film
		37				9	no perm. detected	Pioneer Pylox V-20; glove film
		37	23	0.0035		6	good resistance	"
		37	23	0.0208		6	"	"
		37	23	0.0417		6	"	"
		37	23	0.167		6	"	"
Hydrogen Peroxide	concentrated		23			5	fair resistance	Colorite; PVC tubing
	diluted		23			8	good resistance	"
		30	23	0.0208		9	exc. resistance	Ans. Edm. Ever-Flex; lined glove film
		30	23	0.25		9	no perm. detected during 6 hr. test	Ans. Edm. Monkey Grip; lined glove film
Iodine			23			8	good resistance	Colorite; PVC tubing
Isopropyl Alcohol						8	no perm. detected	Pioneer Pylox V-20; glove film
			23			8	low permeation rate; 0-1/2 drops/hr.	Ans. Edm. Monkey Grip; lined glove film
			23	0.0035		8	no degradation	Pioneer Pylox V-20; glove film
			23	0.0208		8	"	"
			23	0.0208		8	material is well suited for use with chemical	Ans. Edm. Ever-Flex; lined glove film
			23	0.0417		8	no degradation	Pioneer Pylox V-20; glove film
			23	0.167		8	"	"
Methyl Isobutyl Ketone			23	0.0208		1	not recommended	Ans. Edm. Ever-Flex; lined glove film
Morpholine			23	0.0208		1	"	"
Phenol	sat'd					6		Pioneer Pylox V-20; glove film
			23			5	fair resistance	Colorite; PVC tubing
			23			7	very good perm. rate; 1-5 drops/hr.	Ans. Edm. Monkey Grip; lined glove film
			23	0.0208		8	material is well suited for use with chemical	Ans. Edm. Ever-Flex; lined glove film
Propylene Oxide			23	0.0208		1	not recommended	"
Silver Nitrate			23			8	good resistance	Colorite; PVC tubing
Sodium Hypochlorite			23			5	fair resistance	"
Butyl Rubber								
Ammonia	hot					6	minor to moder. effect	
						6	"	
	liquid					8	recommended for use	
						8	"	
	cold					8	"	
Benzaldehyde						8	"	
Benzoic Acid						2	unsatisf. for use	
Benzyl Alcohol						6	minor to moder. effect	
Bleach	sol'ns					8	recommended for use	
Boric Acid						8	"	
Butyl Alcohol						6	minor to moder. effect	
						6	"	
Carbolic Acid						2	unsatisf. for use	
Chlorine	wet					4	moder. to severe effect	
	dry					2	unsatisf. for use	
Ethyl Acetate	organic ester					6	minor to moder. effect	
Ethyl Alcohol						8	recommended for use	
Ethylene Chlorohydrin						6	minor to moder. effect	
Ethylene Glycol						8	recommended for use	
Ethylene Oxide						4	moder. to severe effect	

Reagent	Reagent Note	Conc. (%)	Temp. (°C)	Time (days)	Load	PDL Rating	Resistance Note	Material Note
Butyl Rubber								
Fluorine	liquid					4	moder. to severe effect	
Formaldehyde						8	recommended for use	
Hydrogen Peroxide						8	"	
		90				4	moder. to severe effect	
Iodine						6	minor to moder. effect	
Isopropyl Alcohol						8	recommended for use	
						8	"	
Mercuric Chloride						8	"	
Methyl Isobutyl Ketone						4	moder. to severe effect	
Phenol						2	unsatisf. for use	
		70				2	"	
		85				2	"	
Propylene Oxide						6	minor to moder. effect	
Salicylic Acid						8	recommended for use	
Silver Nitrate						8	"	
Sodium Hypochlorite						6	minor to moder. effect	
Chlorosulfonated Polyethylene Rubber								
Ammonia	hot					6	minor to moder. effect	DuPont Hypalon
	cold					8	recommended for use	"
						2	unsatisf. for use	"
	liquid					2	"	"
						2	"	"
	anhydrous		23			5	fluid has minor to moderate effect	"
Benzaldehyde						2	unsatisf. for use	"
			23			2	fluid has severe effect	"
Benzoic Acid						2	unsatisf. for use	"
Benzyl Alcohol						6	minor to moder. effect	"
Bleach	sol'ns					8	recommended for use	"
Boric Acid						8	"	"
	sol'n		93			8	fluid has little to no effect	"
Butyl Alcohol						8	recommended for use	"
						8	"	"
Carbolic Acid						4	moder. to severe effect	"
Chlorine	wet					4	"	"
	dry					6	minor to moder. effect	"
	dry gas		23			5	fluid has minor to moderate effect	"
	moist gas		23			5	"	"
Ethyl Acetate	organic ester					2	unsatisf. for use	"
			23			2	fluid has severe effect	"
Ethyl Alcohol						8	recommended for use	"
			93			8	fluid has little to no effect	"
Ethylene Chlorohydrin						6	minor to moder. effect	"
Ethylene Glycol						8	recommended for use	"
			93			8	fluid has little to no effect	"
Ethylene Oxide						2	unsatisf. for use	"
			23			3	fluid is not likely to be compatible	"
Formaldehyde						4	moder. to severe effect	"
		40	23			8	fluid has little to no effect	"
		40	70			2	fluid has severe effect	"
Hydrogen Peroxide						6	minor to moder. effect	"
		90				4	moder. to severe effect	"
		90	23			8	fluid has little to no effect	"
Iodine						6	minor to moder. effect	"
Isopropyl Alcohol						8	recommended for use	"
						8	"	"
			93			8	fluid has little to no effect	"
Mercuric Chloride						8	recommended for use	"
	sol'n		23			8	fluid has little to no effect	"
Methyl Isobutyl Ketone						2	unsatisf. for use	"
Phenol						2	"	"
			23			2	fluid has severe effect	"
		70				2	unsatisf. for use	"
		85				2	"	"
Propylene Oxide						2	"	"
Silver Nitrate						8	recommended for use	"

Reagent	Reagent Note	Conc. (%)	Temp. (°C)	Time (days)	Load	PDL Rating	Resistance Note	Material Note
Chlorosulfonated Polyethylene Rubber								
Sodium Hypochlorite						6	minor to moder. effect	DuPont Hypalon
		5	23			8	fluid has little to no effect	"
		20	70			8	"	"
Sulfuric Acid		>5	23			8	"	"
Epichlorohydrin Rubber								
Benzaldehyde						2	unsatisf. for use	
Benzyl Alcohol						2	"	
Boric Acid						8	recommended for use	
Chlorine	wet					6	minor to moder. effect	
	dry					6	"	
Ethyl Acetate	organic ester					2	unsatisf. for use	
Ethyl Alcohol						8	recommended for use	
Ethylene Glycol						8	"	
Ethylene Oxide						2	unsatisf. for use	
Formaldehyde						6	minor to moder. effect	
Isopropyl Alcohol						8	recommended for use	
Mercuric Chloride						8	"	
Methyl Isobutyl Ketone						2	unsatisf. for use	
Phenol						2	"	
		70				2	"	
		85				2	"	
Silver Nitrate						8	recommended for use	
Sodium Hypochlorite						8	"	
Ethylene Acrylate Rubber								
Benzaldehyde			23			2	fluid has severe effect	DuPont Vamac
Boric Acid	sol'n		23			7	fluid is likely to be compatible	"
Ethyl Acetate			23			2	fluid has severe effect	"
			70			3	fluid is not likely to be compatible	"
Ethyl Alcohol			23			5	fluid has minor to moderate effect	"
Ethylene Glycol			100			8	fluid has little to no effect	"
Ethylene Oxide			23			3	fluid is not likely to be compatible	"
Isopropyl Alcohol			23			7	fluid is likely to be compatible	"
Mercuric Chloride	sol'n		23			7	"	"
Phenol			23			3	fluid is not likely to be compatible	"
Sodium Hypochlorite		5	23			8	fluid has little to no effect	"
		20	23			8	"	"
Sulfuric Acid		>5	23			8	"	"
Ethylene Propylene Diene Methylene Terpolymer								
Ammonia	hot					6	minor to moder. effect	
						8	recommended for use	
	cold					8	"	
	liquid					8	"	
	gas, hot		23			6	may cause sl. visible swell/loss of prop.	
	gas, cold		23			8	little/no effect	
	anhydrous		23			8	"	
	"		23			7	fluid is likely to be compatible	DuPont Nordel
Benzaldehyde						8	recommended for use	
			23			5	fluid has minor to moderate effect	DuPont Nordel
			23			8	little/no effect	
Benzoic Acid						2	unsatisf. for use	
			23			4	moder./ severe swell and/or loss of prop.	
Benzyl Alcohol						6	minor to moder. effect	
			23			8	little/no effect	
Bleach	sol'ns					8	recommended for use	
	lime bleach		23			8	little/no effect	
	sol'ns		23			8	"	
Boric Acid						8	recommended for use	
	sol'n		23			8	fluid has little to no effect	DuPont Nordel
			23			8	little/no effect	
Butyl Alcohol						6	minor to moder. effect	
						6	"	
			23			6	may cause sl. visible swell/loss of prop.	

Ethylene Propylene Diene Methylene Terpolymer (continued)

Reagent	Reagent Note	Conc. (%)	Temp. (°C)	Time (days)	Load	PDL Rating	Resistance Note	Material Note
Carbolic Acid						6	minor to moder. effect	
	phenol		23			6	may cause sl. visible swell/loss of prop.	
Chlorine	wet					4	moder. to severe effect	
	dry					8	recommended for use	
	moist gas		23			3	fluid is not likely to be compatible	DuPont Nordel
	dry gas		23			3	"	"
	wet		23			4	moder./ severe swell and/or loss of prop.	
	dry		23			1	not suitable for service	
Cresol	methyl phenol		23			1	"	
Ethyl Acetate	organic ester					6	minor to moder. effect	
			23			8	fluid has little to no effect	DuPont Nordel
			23			6	may cause sl. visible swell/loss of prop.	
			70			5	fluid has minor to moderate effect	DuPont Nordel
Ethyl Alcohol						8	recommended for use	
			23			8	fluid has little to no effect	DuPont Nordel
			23			6	may cause sl. visible swell/loss of prop.	
			23			8	little/no effect	
			23			8	"	
Ethylene Chlorohydrin						6	minor to moder. effect	
			23			6	may cause sl. visible swell/loss of prop.	
Ethylene Glycol						8	recommended for use	
			23			8	fluid has little to no effect	DuPont Nordel
			23			8	little/no effect	
Ethylene Oxide						4	moder. to severe effect	
			23			3	fluid is not likely to be compatible	DuPont Nordel
			23			4	moder./ severe swell and/or loss of prop.	
Fluorine	liquid					4	moder. to severe effect	
	"		23			1	not suitable for service	
Formaldehyde						8	recommended for use	
			23			8	little/no effect	
		40	23			8	fluid has little to no effect	DuPont Nordel
Hydrogen Peroxide						8	recommended for use	
		90				4	moder. to severe effect	
		90	23			6	may cause sl. visible swell/loss of prop.	
		90	23			7	fluid is likely to be compatible	DuPont Nordel
Iodine						6	minor to moder. effect	
Isopropyl Alcohol						8	recommended for use	
						8	"	
			23			8	little/no effect	
			23			7	fluid is likely to be compatible	DuPont Nordel
Mercuric Chloride						8	recommended for use	
	sol'n		23			8	fluid has little to no effect	DuPont Nordel
			23			8	little/no effect	
Methyl Isobutyl Ketone						4	moder. to severe effect	
Phenol						6	minor to moder. effect	
			23			5	fluid has minor to moderate effect	DuPont Nordel
			23			6	may cause sl. visible swell/loss of prop.	
		70				2	unsatisf. for use	
		85				2	"	
Propylene Oxide						6	minor to moder. effect	
			23			6	may cause sl. visible swell/loss of prop.	
Salicylic Acid						8	recommended for use	
			23			8	little/no effect	
Silver Nitrate						8	recommended for use	
			23			8	little/no effect	
Sodium Hypochlorite						6	minor to moder. effect	
			23			6	may cause sl. visible swell/loss of prop.	
		5	23			8	fluid has little to no effect	DuPont Nordel
		20	23			8	"	"
Sulfuric Acid		>5	23			8	"	"

Reagent	Reagent Note	Conc. (%)	Temp. (°C)	Time (days)	Load	PDL Rating	Resistance Note	Material Note
Natural Rubber								
Ammonia	cold					8	recommended for use	
						2	unsatisf. for use	
	liquid					2	"	
	hot					2	"	
						2	"	
Benzaldehyde						2	"	
			23			6	very good perm. rate; 1-5 drops/hr.	Ans. Edm. Canners and Handlers 392; 0.48 mm glove film
			23	0.0208		5	fluid has minor degrading effect	"
Benzoic Acid						2	unsatisf. for use	
Benzyl Alcohol						2	"	
Bleach	sol'ns					2	"	
Boric Acid						8	recommended for use	
Butyl Alcohol						8	"	
						8	"	
			23			7	very good perm. rate; 1-5 drops/hr.	Ans. Edm. Canners and Handlers 392; 0.48 mm glove film
			23	0.0208		6	fluid has very little degrading effect	"
Carbolic Acid						9	no perm. detected	Pioneer L-118; glove film
						2	unsatisf. for use	
			23	0.0035		8	no degradation	Pioneer L-118; glove film
			23	0.0208		8	"	"
			23	0.0417		8	"	"
			23	0.167		6	good resistance	"
Chlorine	dry					2	unsatisf. for use	
	wet					2	"	
Ethyl Acetate	organic ester					2	"	
			23			3	fair permeation rate	Ans. Edm. Canners and Handlers 392; 0.48 mm glove film
			23	0.0208		5	fluid has minor degrading effect	"
Ethyl Alcohol						8	recommended for use	
			23			6	very good perm. rate; 1-5 drops/hr.	Ans. Edm. Canners and Handlers 392; 0.48 mm glove film
			23	0.0208		6	fluid has very little degrading effect	"
Ethylene Chlorohydrin						6	minor to moder. effect	
Ethylene Glycol						9	no perm. detected	Pioneer L-118; glove film
						8	recommended for use	
			23	0.0035		8	no degradation	Pioneer L-118; glove film
			23	0.0208		8	"	"
			23	0.0208		9	exc. resistance	Ans. Edm. Canners and Handlers 392; 0.48 mm glove film
			23	0.0417		8	no degradation	Pioneer L-118; glove film
			23	0.167		8	"	"
			23	0.25		9	low permeation rate; 0-1/2 drops/hr.	Ans. Edm. Canners and Handlers 392; 0.48 mm glove film
Ethylene Oxide						2	unsatisf. for use	

Natural Rubber (continued)

Reagent	Reagent Note	Conc. (%)	Temp. (°C)	Time (days)	Load	PDL Rating	Resistance Note	Material Note
Formaldehyde						6	minor to moder. effect	
			23			5	good perm. rate; 6 to 50 drops/hr.	Ans. Edm. Canners and Handlers 392; 0.48 mm glove film
			23	0.0208		6	fluid has very little degrading effect	"
		37				9	no perm. detected	Pioneer L-118; glove film
		37	23	0.0035		8	no degradation	"
		37	23	0.0208		8	"	"
		37	23	0.0417		8	"	"
		37	23	0.167		8	"	"
Hydrogen Peroxide						6	minor to moder. effect	
		30	23	0.0208		9	exc. resistance	Ans. Edm. Canners and Handlers 392; 0.48 mm glove film
		30	23	0.25		9	no perm. detected during 6 hr. test	"
		90				2	unsatisf. for use	
Isopropyl Alcohol						8	no perm. detected	Pioneer L-118; glove film
						8	recommended for use	
			23			7	very good perm. rate; 1-5 drops/hr.	Ans. Edm. Canners and Handlers 392; 0.48 mm glove film
			23	0.0035		8	no degradation	Pioneer L-118; glove film
			23	0.0208		8	"	"
			23	0.0208		6	fluid has very little degrading effect	Ans. Edm. Canners and Handlers 392; 0.48 mm glove film
			23	0.0417		8	no degradation	Pioneer L-118; glove film
			23	0.167		8	"	"
Mercuric Chloride						8	recommended for use	
Methyl Isobutyl Ketone						2	unsatisf. for use	
			23	0.0208		2	fluid has pronounced degrading effect	Ans. Edm. Canners and Handlers 392; 0.48 mm glove film
Morpholine			23			6	good perm. rate; 6 to 50 drops/hr.	"
			23	0.0208		5	fluid has minor degrading effect	"
Phenol	sat'd					9	no perm. detected	Pioneer L-118; glove film
						2	unsatisf. for use	
			23			6		Ans. Edm. Canners and Handlers 392; 0.48 mm glove film
	sat'd		23	0.0035		8	no degradation	Pioneer L-118; glove film
			23	0.0208		9	exc. resistance	Ans. Edm. Canners and Handlers 392; 0.48 mm glove film
	sat'd		23	0.0208		8	no degradation	Pioneer L-118; glove film
	"		23	0.0417		8	"	"
	"		23	0.167		6	good resistance	"
		70				2	unsatisf. for use	
		85				2	"	
Propylene Oxide						2	"	
			23	0.0208		2	fluid has pronounced degrading effect	Ans. Edm. Canners and Handlers 392; 0.48 mm glove film
Salicylic Acid						8	recommended for use	
Silver Nitrate						8	"	
Sodium Hypochlorite						4	moder. to severe effect	

Reagent	Reagent Note	Conc. (%)	Temp. (°C)	Time (days)	Load	PDL Rating	Resistance Note	Material Note
Fluoroelastomer								
Ammonia						2	unsatisf. for use	
	cold					2	"	
	hot					2	"	
						2	"	
	liquid					2	"	
Benzaldehyde						2	"	
Benzoic Acid						8	recommended for use	
Benzyl Alcohol						8	"	
Bleach	sol'ns					8	"	
Boric Acid						8	"	
Butyl Alcohol						8	"	
						8	"	
Carbolic Acid						8	"	
Chlorine	wet					8	"	
	dry					8	"	
Ethyl Acetate	organic ester					2	unsatisf. for use	
Ethyl Alcohol						8	recommended for use	
Ethylene Chlorohydrin						8	"	
Ethylene Glycol						8	"	
Ethylene Oxide						2	unsatisf. for use	
Fluorine	liquid					6	minor to moder. effect	
Formaldehyde						2	unsatisf. for use	
Hydrogen Peroxide						8	recommended for use	
		90				6	minor to moder. effect	
Iodine						8	recommended for use	
Isopropyl Alcohol						8	"	
						8	"	
Mercuric Chloride						8	"	
Methyl Isobutyl Ketone						2	unsatisf. for use	
Phenol						8	recommended for use	
		70				8	"	
		85				8	"	
Propylene Oxide						2	unsatisf. for use	
Salicylic Acid						8	recommended for use	
Silver Nitrate						8	"	
Sodium Hypochlorite						8	"	
Tetrafluoroethylene Perfluoromethyl Vinyl Ether Copolymer								
Ammonia	gas, cold		100			8	severe condit. may cause sl. swell/prop. loss	DuPont Kalrez
	gas, hot		100			8	"	"
	anhydrous		100			8	"	"
Benzaldehyde			100			8	"	"
Benzoic Acid			100			8	"	"
Benzyl Alcohol			100			8	"	"
Bleach	lime bleach		100			8	"	"
	sol'ns		100			8	"	"
Boric Acid			100			8	"	"
Butyl Alcohol			100			8	"	"
Carbolic Acid	phenol		100			8	"	"
Chlorine	dry		100			8	"	"
	wet		100			6	may cause sl. visible swell/loss of prop.	"
Cresol	methyl phenol		100			8	severe condit. may cause sl. swell/prop. loss	"
Ethyl Acetate			23	7		9		"
			100			8	severe condit. may cause sl. swell/prop. loss	"
Ethyl Alcohol			23	7		9		"
			100			8	severe condit. may cause sl. swell/prop. loss	"
Ethylene Chlorohydrin			100			8	"	"
Ethylene Glycol			100			8	"	"
Ethylene Oxide			100			8	"	"
Fluorine	liquid		100			6	may cause sl. visible swell/loss of prop.	"
Formaldehyde			100			8	severe condit. may cause sl. swell/prop. loss	"

Reagent	Reagent Note	Conc. (%)	Temp. (°C)	Time (days)	Load	PDL Rating	Resistance Note	Material Note

Tetrafluoroethylene Perfluoromethyl Vinyl Ether Copolymer (continued)

Reagent	Reagent Note	Conc. (%)	Temp. (°C)	Time (days)	Load	PDL Rating	Resistance Note	Material Note
Hydrogen Peroxide		90	100			8	severe condit. may cause sl. swell/prop. loss	DuPont Kalrez
Iodine			100			8	"	"
Isopropyl Alcohol			100			8	"	"
Mercuric Chloride			100			8	"	"
Morpholine			100			8	"	"
Phenol			100			8	"	"
Potassium Permanganate			100			8	"	"
Propylene Oxide			100			8	"	"
Salicylic Acid			100			8	"	"
Silver Nitrate			100			8	"	"
Sodium Hypochlorite			100			8	"	"

Vinylidene Fluoride Hexafluoropropylene Copolymer

Reagent	Reagent Note	Conc. (%)	Temp. (°C)	Time (days)	Load	PDL Rating	Resistance Note	Material Note
Ammonia	gas, cold		23			1	not suitable for service	
	gas, hot		23			1	"	
	anhydrous		23			1	"	
	"		25	1		6	not recommended for use	3M Fluorel
Benzaldehyde			23			1	not suitable for service	
			25	3		2	not recommended for use	3M Fluorel
Benzoic Acid			23			8	little/no effect	
Benzyl Alcohol			23			8	"	
Bleach	sol'ns		23			8	"	
	lime bleach		23			8	"	
Boric Acid			23			8	"	
Butyl Alcohol			23			8	"	
Carbolic Acid	phenol		23			8	"	
Chlorine	wet		23			6	may cause sl. visible swell/loss of prop.	
	dry		23			8	little/no effect	
	"		100	5		8	exc. resistance	3M Fluorel
Cresol	methyl phenol		23			8	little/no effect	
Ethyl Acetate			23			1	not suitable for service	
			23	7		1		
			25	7		6	good to exc. resistance, moder. effect	3M Fluorel FLS 2330
			25	7		1	not recommended for use	3M Fluorel
Ethyl Alcohol			23			4	moder./ severe swell and/or loss of prop.	
			23			1	not suitable for service	
			23	7		8		
			25	7		9	exc. resistance	3M Fluorel
Ethylene Chlorohydrin			23			8	little/no effect	
Ethylene Glycol			23			8	"	
			100	14		9	exc. resistance	3M Fluorel
		50	100	3		9	"	3M Fluorel FLS 2330
		50	100	7		7	good to exc. resistance, moder. effect	3M Fluorel
Ethylene Oxide			23			1	not suitable for service	
			70	5		1	not recommended for use	3M Fluorel
Fluorine	liquid		23			6	may cause sl. visible swell/loss of prop.	
Formaldehyde			23			1	not suitable for service	
Hydrogen Peroxide	90% active		24	7		9	exc. resistance	3M Fluorel
		90	23			6	may cause sl. visible swell/loss of prop.	
Isopropyl Alcohol			23			8	little/no effect	
Mercuric Chloride			23			8	"	
Phenol			23			8	"	
			25	3		9	exc. resistance	3M Fluorel
			100	28		8	"	"
			149	28		6	good to exc. resistance, moder. effect	"
Propylene Oxide			23			1	not suitable for service	
Salicylic Acid			23			8	little/no effect	
Silver Nitrate			23			8	"	
Sodium Hypochlorite			23			8	"	

Neoprene Rubber

Reagent	Reagent Note	Conc. (%)	Temp. (°C)	Time (days)	Load	PDL Rating	Resistance Note	Material Note
Ammonia	hot					6	minor to moder. effect	DuPont Neoprene
	liquid					8	recommended for use	"
						8	"	"
	cold					8	"	"
	anhydrous		23			8	fluid has little to no effect	"
Benzaldehyde						2	unsatisf. for use	"
			23			2	fluid has severe effect	"
			23	0.0208		1	not recommended	Ans. Edm. Neox; lined glove film
			23	0.0208		1	"	Ans. Edm. Neoprene 29-840; 0.38 mm glove film
Benzoic Acid						2	unsatisf. for use	DuPont Neoprene
Benzyl Alcohol						6	minor to moder. effect	"
Bleach	sol'ns					4	moder. to severe effect	"
Boric Acid						8	recommended for use	"
	sol'n		70			8	fluid has little to no effect	"
Butyl Alcohol						8	recommended for use	"
						8	"	"
			23			9	low permeation rate; 0-1/2 drops/hr.	Ans. Edm. Neox; lined glove film
			23			7	very good perm. rate; 1-5 drops/hr.	Ans. Edm. Neoprene 29-840; 0.38 mm glove film
			23	0.0208		9	exc. resistance	"
			23	0.0208		9	"	Ans. Edm. Neox; lined glove film
Carbolic Acid						4	moder. to severe effect	DuPont Neoprene
						9	no perm. detected	Pioneer Stanzoil N-44; glove film
			23	0.0035		8	no degradation	"
			23	0.0208		8	"	"
			23	0.0417		8	"	"
			23	0.167		8	"	"
Chlorine	dry					4	moder. to severe effect	DuPont Neoprene
	wet					2	unsatisf. for use	"
	moist gas		23			2	fluid has severe effect	"
	dry gas		23			5	fluid has minor to moderate effect	"
Ethyl Acetate	organic ester					2	unsatisf. for use	"
						4		Pioneer Stanzoil N-44; glove film
			23			6	good perm. rate; 6 to 50 drops/hr.	Ans. Edm. Neox; lined glove film
			23			5	"	Ans. Edm. Neoprene 29-840; 0.38 mm glove film
			23			2	fluid has severe effect	DuPont Neoprene
			23	0.0035		8	no degradation	Pioneer Stanzoil N-44; glove film
			23	0.0208		5	fluid has moder. degrading effect	Ans. Edm. Neoprene 29-840; 0.38 mm glove film
			23	0.0208		8	no degradation	Pioneer Stanzoil N-44; glove film
			23	0.0208		5	fluid has moder. degrading effect	Ans. Edm. Neox; lined glove film
			23	0.0417		8	no degradation	Pioneer Stanzoil N-44; glove film
			23	0.167		8	"	"
Ethyl Alcohol						8	recommended for use	DuPont Neoprene
			23			7	very good perm. rate; 1-5 drops/hr.	Ans. Edm. Neox; lined glove film
			23			7	"	Ans. Edm. Neoprene 29-840; 0.38 mm glove film
			23	0.0208		9	exc. resistance	"
			23	0.0208		9	"	Ans. Edm. Neox; lined glove film
			70			8	fluid has little to no effect	DuPont Neoprene
Ethylene Chlorohydrin						6	minor to moder. effect	"
Ethylene Glycol						9	no perm. detected	Pioneer Stanzoil N-44; glove film
						8	recommended for use	DuPont Neoprene
			23	0.0208		9	exc. resistance	Ans. Edm. Neoprene 29-840; 0.38 mm glove film
			23	0.0208		9	"	Ans. Edm. Neox; lined glove film
			23	0.25		9	low permeation rate; 0-1/2 drops/hr.	Ans. Edm. Neoprene 29-840; 0.38 mm glove film
			23	0.25		9	"	Ans. Edm. Neox; lined glove film
			70			8	fluid has little to no effect	DuPont Neoprene
Ethylene Oxide						2	unsatisf. for use	"
						6		Pioneer Stanzoil N-44; glove film
			23			3	fluid is not likely to be compatible	DuPont Neoprene
			23	0.0035		8	no degradation	Pioneer Stanzoil N-44; glove film
			23	0.0208		8	"	"
			23	0.0417		8	"	"
			23	0.167		8	"	"

Reagent	Reagent Note	Conc. (%)	Temp. (°C)	Time (days)	Load	PDL Rating	Resistance Note	Material Note
Neoprene Rubber								
Formaldehyde						4	moder. to severe effect	DuPont Neoprene
			23			8	low permeation rate; 0-1/2 drops/hr.	Ans. Edm. Neoprene 29-840; 0.38 mm glove film
			23			7	very good perm. rate; 1-5 drops/hr.	Ans. Edm. Neox; lined glove film
			23	0.0208		9	exc. resistance	Ans. Edm. Neoprene 29-840; 0.38 mm glove film
			23	0.0208		9	"	Ans. Edm. Neox; lined glove film
		37				9	no perm. detected	Pioneer Stanzoil N-44; glove film
		37	23	0.0035		8	no degradation	"
		37	23	0.0208		8	"	"
		37	23	0.0417		8	"	"
		37	23	0.167		8	"	"
		40	23			8	fluid has little to no effect	DuPont Neoprene
		40	70			2	fluid has severe effect	"
Hydrogen Peroxide						8	recommended for use	"
		30	23			3		Ans. Edm. Neoprene 29-840; 0.38 mm glove film
		30	23			4		Ans. Edm. Neox; lined glove film
		30	23	0.0208		6	fluid has very little degrading effect	Ans. Edm. Neoprene 29-840; 0.38 mm glove film
		30	23	0.0208		6	"	Ans. Edm. Neox; lined glove film
		90				2	unsatisf. for use	DuPont Neoprene
		90	23			5	fluid has minor to moderate effect	"
Iodine						2	unsatisf. for use	"
Isopropyl Alcohol						8	no perm. detected	Pioneer Stanzoil N-44; glove film
						6	minor to moder. effect	DuPont Neoprene
						6	"	"
			23			8	fluid has little to no effect	"
			23	0.0035		8	no degradation	Pioneer Stanzoil N-44; glove film
			23	0.0208		9	exc. resistance	Ans. Edm. Neox; lined glove film
			23	0.0208		9	"	Ans. Edm. Neoprene 29-840; 0.38 mm glove film
			23	0.0208		8	no degradation	Pioneer Stanzoil N-44; glove film
			23	0.0417		8	"	"
			23	0.167		8	"	"
			23	0.25		9	low permeation rate; 0-1/2 drops/hr.	Ans. Edm. Neoprene 29-840; 0.38 mm glove film
			23	0.25		9	"	Ans. Edm. Neox; lined glove film
Mercuric Chloride						8	recommended for use	DuPont Neoprene
	sol'n		23			8	fluid has little to no effect	"
Methyl Isobutyl Ketone						2	unsatisf. for use	"
			23	0.0208		1	not recommended	Ans. Edm. Neox; lined glove film
			23	0.0208		1	"	Ans. Edm. Neoprene 29-840; 0.38 mm glove film
Morpholine			23	0.0208		2	fluid has pronounced degrading effect	"
			23	0.0208		2	"	Ans. Edm. Neox; lined glove film
Phenol	sat'd					9	no perm. detected	Pioneer Stanzoil N-44; glove film
						2	unsatisf. for use	DuPont Neoprene
			23			6	good perm. rate; 6 to 50 drops/hr.	Ans. Edm. Neoprene 29-840; 0.38 mm glove film
			23			9	low permeation rate; 0-1/2 drops/hr.	Ans. Edm. Neox; lined glove film
			23			2	fluid has severe effect	DuPont Neoprene
	sat'd		23	0.0035		8	no degradation	Pioneer Stanzoil N-44; glove film
			23	0.0208		9	exc. resistance	Ans. Edm. Neoprene 29-840; 0.38 mm glove film
	sat'd		23	0.0208		8	no degradation	Pioneer Stanzoil N-44; glove film
			23	0.0208		9	exc. resistance	Ans. Edm. Neox; lined glove film
	sat'd		23	0.0417		8	no degradation	Pioneer Stanzoil N-44; glove film
	"		23	0.167		8	"	"
		70				2	unsatisf. for use	DuPont Neoprene
		85				2	"	"
Propylene Oxide						2	"	"
			23	0.0208		1	not recommended	Ans. Edm. Neox; lined glove film
			23	0.0208		1	"	Ans. Edm. Neoprene 29-840; 0.38 mm glove film
Silver Nitrate						8	recommended for use	DuPont Neoprene
Sodium Hypochlorite						6	minor to moder. effect	"
		5	23			8	fluid has little to no effect	"
		20	23			5	fluid has minor to moderate effect	"
Sulfuric Acid		>5	23			8	fluid has little to no effect	"

Reagent	Reagent Note	Conc. (%)	Temp. (°C)	Time (days)	Load	PDL Rating	Resistance Note	Material Note
Nitrile Rubber								
Ammonia						6	minor to moder. effect	
	liquid					6	▪	
						6	▪	
	cold					8	recommended for use	
	hot					2	unsatisf. for use	
	anhydrous		23			6	may cause sl. visible swell/loss of prop.	
	gas, cold		23			8	little/no effect	
	gas, hot		23			1	not suitable for service	
Benzaldehyde						2	unsatisf. for use	
			23			1	not suitable for service	
			23	0.0208		1	not recommended	Ans. Edm. Sol-Vex 37-145; 0.28 mm glove film
Benzoic Acid						2	unsatisf. for use	
			23			4	moder./ severe swell and/or loss of prop.	
Benzyl Alcohol						2	unsatisf. for use	
			23			1	not suitable for service	
Bleach	sol'ns					2	unsatisf. for use	
	lime bleach		23			8	little/no effect	
	sol'ns		23			1	not suitable for service	
Boric Acid						8	recommended for use	
			23			8	little/no effect	
Butyl Alcohol						8	recommended for use	
						8	▪	
			23			8	little/no effect	
			23	0.0208		9	exc. resistance	Ans. Edm. Sol-Vex 37-145; 0.28 mm glove film
			23	0.25		9	low permeation rate; 0-1/2 drops/hr.	Ans. Edm. Sol-Vex 37-165; 0.54 mm glove film
Carbolic Acid						9	no perm. detected	Pioneer Stansolv A-14; glove film
						2	unsatisf. for use	
	phenol		23			1	not suitable for service	
Chlorine	wet					2	unsatisf. for use	
	dry					2	▪	
	▪		23			1	not suitable for service	
	wet		23			1	▪	
Cresol	methyl phenol		23			1	▪	
Ethyl Acetate	organic ester					2	unsatisf. for use	
			23			1	not suitable for service	
			23	0.0208		1	not recommended	Ans. Edm. Sol-Vex 37-145; 0.28 mm glove film
Ethyl Alcohol						4	moder. to severe effect	
			23			8	little/no effect	
			23			6	may cause sl. visible swell/loss of prop.	
			23			1	not suitable for service	
			23			7	very good perm. rate; 1-5 drops/hr.	Ans. Edm. Sol-Vex 37-165; 0.54 mm glove film
			23	0.0208		9	exc. resistance	Ans. Edm. Sol-Vex 37-145; 0.28 mm glove film
Ethylene Chlorohydrin						2	unsatisf. for use	
			23			1	not suitable for service	
Ethylene Glycol						9	no perm. detected	Pioneer Stansolv A-14; glove film
						8	recommended for use	
			23			8	little/no effect	
			23	0.0035		8	no degradation	Pioneer Stansolv A-14; glove film
			23	0.0208		9	exc. resistance	Ans. Edm. Sol-Vex 37-145; 0.28 mm glove film
			23	0.0208		8	no degradation	Pioneer Stansolv A-14; glove film
			23	0.0417		8	▪	▪
			23	0.167		8	▪	▪
			23	0.25		9	low permeation rate; 0-1/2 drops/hr.	Ans. Edm. Sol-Vex 37-165; 0.54 mm glove film
Ethylene Oxide						2	unsatisf. for use	
						5		Pioneer Stansolv A-14; glove film
			23			1	not suitable for service	
Fluorine	liquid		23			1	▪	
Formaldehyde						4	moder. to severe effect	
			23			4	moder./ severe swell and/or loss of prop.	
			23	0.0208		9	exc. resistance	Ans. Edm. Sol-Vex 37-145; 0.28 mm glove film

Reagent	Reagent Note	Conc. (%)	Temp. (°C)	Time (days)	Load	PDL Rating	Resistance Note	Material Note

Nitrile Rubber

Reagent	Reagent Note	Conc. (%)	Temp. (°C)	Time (days)	Load	PDL Rating	Resistance Note	Material Note
Formaldehyde			23	0.25		9	low permeation rate; 0-1/2 drops/hr.	Ans. Edm. Sol-Vex 37-165; 0.54 mm glove film
		37				9	no perm. detected	Pioneer Stansolv A-14; glove film
		37	23	0.0035		8	no degradation	"
		37	23	0.0208		8	"	"
		37	23	0.0417		8	"	"
		37	23	0.167		8	"	"
Hydrogen Peroxide						6	minor to moder. effect	
		30	23	0.0208		9	exc. resistance	Ans. Edm. Sol-Vex 37-145; 0.28 mm glove film
		30	23	0.25		9	no perm. detected during 6 hr. test	Ans. Edm. Sol-Vex 37-165; 0.54 mm glove film
		90				2	unsatisf. for use	
		90	23			1	not suitable for service	
Iodine						6	minor to moder. effect	
Isopropyl Alcohol						9	no perm. detected	Pioneer Stansolv A-14; glove film
						6	minor to moder. effect	
						6	"	
			23			6	may cause sl. visible swell/loss of prop.	
			23	0.0035		8	no degradation	Pioneer Stansolv A-14; glove film
			23	0.0208		8	"	"
			23	0.0208		9	exc. resistance	Ans. Edm. Sol-Vex 37-145; 0.28 mm glove film
			23	0.0417		8	no degradation	Pioneer Stansolv A-14; glove film
			23	0.167		8	"	"
			23	0.25		9	low permeation rate; 0-1/2 drops/hr.	Ans. Edm. Sol-Vex 37-165; 0.54 mm glove film
Mercuric Chloride						8	recommended for use	
			23			8	little/no effect	
Methyl Isobutyl Ketone						2	unsatisf. for use	
			23	0.0208		2	fluid has pronounced degrading effect	Ans. Edm. Sol-Vex 37-145; 0.28 mm glove film
Morpholine		70	23	0.0208		1	not recommended	"
Phenol	sat'd					9	no perm. detected	Pioneer Stansolv A-14; glove film
						2	unsatisf. for use	
			23			1	not suitable for service	
			23	0.0208		1	not recommended	Ans. Edm. Sol-Vex 37-145; 0.28 mm glove film
		70				2	unsatisf. for use	
		85				2	"	
Propylene Oxide						2	"	
			23			1	not suitable for service	
			23	0.0208		1	not recommended	Ans. Edm. Sol-Vex 37-145; 0.28 mm glove film
Salicylic Acid						6	minor to moder. effect	
			23			6	may cause sl. visible swell/loss of prop.	
Silver Nitrate						6	minor to moder. effect	
			23			6	may cause sl. visible swell/loss of prop.	
Sodium Hypochlorite						6	minor to moder. effect	
			23			6	may cause sl. visible swell/loss of prop.	

Polysulfide Rubber

Reagent	Reagent Note	Conc. (%)	Temp. (°C)	Time (days)	Load	PDL Rating	Resistance Note	Material Note
Ammonia	cold					8	recommended for use	
						2	unsatisf. for use	
						2	"	
	liquid					2	"	
	hot					2	"	
Benzaldehyde						2	"	
Benzyl Alcohol						2	"	
Bleach	sol'ns					2	"	
Boric Acid						2	"	
Butyl Alcohol						6	minor to moder. effect	
						6	"	
Carbolic Acid						2	unsatisf. for use	
Chlorine	wet					2	"	
	dry					2	"	
Ethyl Acetate	organic ester					6	minor to moder. effect	

Polysulfide Rubber

Reagent	Reagent Note	Conc. (%)	Temp. (°C)	Time (days)	Load	PDL Rating	Resistance Note	Material Note
Ethyl Alcohol						8	recommended for use	
Ethylene Chlorohydrin						6	minor to moder. effect	
Ethylene Glycol						4	moder. to severe effect	
Fluorine	liquid					2	unsatisf. for use	
Formaldehyde						6	minor to moder. effect	
Hydrogen Peroxide						4	moder. to severe effect	
		90				2	unsatisf. for use	
Isopropyl Alcohol						8	recommended for use	
						8	"	
Methyl Isobutyl Ketone						6	minor to moder. effect	
Phenol						2	unsatisf. for use	
		70				2	"	
		85				2	"	
Silver Nitrate						6	minor to moder. effect	
Sodium Hypochlorite						2	unsatisf. for use	

Tetrafluoroethylene Propylene Copolymer

Reagent	Reagent Note	Conc. (%)	Temp. (°C)	Time (days)	Load	PDL Rating	Resistance Note	Material Note
Ammonia		28				8	exc. prop. retention	3M Aflas
Benzaldehyde						6	good prop. retention, moder. vol. swell	"
Benzyl Alcohol						8	exc. prop. retention	3M Aflas
			23			6	may cause sl. visible swell/loss of prop.	
Bleach	sol'ns					6	minor to moder. effect	
	"		23			6	may cause sl. visible swell/loss of prop.	
	lime bleach		23			6	"	
	bleaching powder	10				8	exc. prop. retention	3M Aflas
Butyl Alcohol						8	exc. prop. retention	3M Aflas
Cresol						8	exc. prop. retention	3M Aflas
Ethyl Acetate						2	poor prop. retention, volume swell >40%	3M Aflas
			23	7		2		
Ethyl Alcohol						8	exc. prop. retention	3M Aflas
			23	7		9		
Ethylene Chlorohydrin						8	exc. prop. retention	3M Aflas
Ethylene Oxide	w/ rust inhibitor					8	exc. prop. retention	3M Aflas
Formaldehyde	technical					8	exc. prop. retention	3M Aflas
Hydrogen Peroxide		30				8	exc. prop. retention	3M Aflas
Phenol						8	exc. prop. retention	3M Aflas
Sodium Hypochlorite		10				8	exc. prop. retention	3M Aflas

Silicone

Reagent	Reagent Note	Conc. (%)	Temp. (°C)	Time (days)	Load	PDL Rating	Resistance Note	Material Note
Ammonia	liquid					4	moder. to severe effect	
						6	minor to moder. effect	
	cold					8	recommended for use	
	hot					8	"	
						2	unsatisf. for use	
	gas, hot		23			8	little/no effect	
	gas, cold		23			8	"	
	anhydrous		23			4	moder./ severe swell and/or loss of prop.	
Benzaldehyde						2	unsatisf. for use	
			23			6	may cause sl. visible swell/loss of prop.	
Benzoic Acid						6	minor to moder. effect	
			23			4	moder./ severe swell and/or loss of prop.	
Boric Acid						8	recommended for use	
			23			8	little/no effect	
Butyl Alcohol						6	minor to moder. effect	
						6	"	
			23			6	may cause sl. visible swell/loss of prop.	
Carbolic Acid						2	unsatisf. for use	
	phenol		23			1	not suitable for service	
Chlorine	dry					2	unsatisf. for use	
	"		23			1	not suitable for service	
	wet		23			1	"	
Cresol	methyl phenol		23			1	not suitable for service	

Silicone

Reagent	Reagent Note	Conc. (%)	Temp. (°C)	Time (days)	Load	PDL Rating	Resistance Note	Material Note
Ethyl Acetate	organic ester					6	minor to moder. effect	
			23			6	may cause sl. visible swell/loss of prop.	
Ethyl Alcohol						6	minor to moder. effect	
			23			8	little/no effect	
			23			6	may cause sl. visible swell/loss of prop.	
Ethylene Chlorohydrin						4	moder. to severe effect	
			23			4	moder./ severe swell and/or loss of prop.	
Ethylene Glycol						8	recommended for use	
			23			8	little/no effect	
Ethylene Oxide						2	unsatisf. for use	
			23			1	not suitable for service	
Fluorine	liquid					2	unsatisf. for use	
	"		23			1	not suitable for service	
Formaldehyde						6	minor to moder. effect	
			23			6	may cause sl. visible swell/loss of prop.	
Hydrogen Peroxide						8	recommended for use	
		90				6	minor to moder. effect	
		90	23			6	may cause sl. visible swell/loss of prop.	
Isopropyl Alcohol						8	recommended for use	
						8	"	
			23			8	little/no effect	
Methyl Isobutyl Ketone						2	unsatisf. for use	
Phenol						2	"	
			23			1	not suitable for service	
		70				2	unsatisf. for use	
		85				2	"	
Propylene Oxide						2	"	
			23			1	not suitable for service	
Silver Nitrate						8	recommended for use	
			23			8	little/no effect	
Sodium Hypochlorite						6	minor to moder. effect	
			23			6	may cause sl. visible swell/loss of prop.	

Methylsilicone

Reagent	Reagent Note	Conc. (%)	Temp. (°C)	Time (days)	Load	PDL Rating	Resistance Note	Material Note
Ammonia			24	7		9		Dow Corn. Silastic
Butyl Alcohol			24	7		6		"
Ethyl Alcohol			24	7		9		"
Ethylene Glycol		60	135	7		4		"
Ethylene Oxide			24	3		6		"
			71	14		4		"
Phenol		70	100	7		5		"
		85	24	7		7		"
Propylene Oxide			24	7		2		"

Methylphenylsilicone

Reagent	Reagent Note	Conc. (%)	Temp. (°C)	Time (days)	Load	PDL Rating	Resistance Note	Material Note
Butyl Alcohol			24	7		5		"
Hydrogen Peroxide		3	24	7		9	no effect	"
		30	24	7		9	"	"
Isopropyl Alcohol			24	7		6		"

Methylphenylvinylsilicone

Reagent	Reagent Note	Conc. (%)	Temp. (°C)	Time (days)	Load	PDL Rating	Resistance Note	Material Note
Ammonia			24	7		8		"
Butyl Alcohol			24	7		5		"
Ethyl Alcohol			24	7		6		"
Ethylene Glycol		50	121	7		8		"
Ethylene Oxide			110	1.3		7		"

Methylvinylfluorosilicone

Reagent	Reagent Note	Conc. (%)	Temp. (°C)	Time (days)	Load	PDL Rating	Resistance Note	Material Note
Ammonia	300 lb. pressure		110	1		2	deteriorated	"
Butyl Alcohol			24	7		9		"
Ethyl Alcohol			24	7		7		"
			121	2.9		2	deteriorated	"

Reagent	Reagent Note	Conc. (%)	Temp. (°C)	Time (days)	Load	PDL Rating	Resistance Note	Material Note
Methylvinylfluorosilicone								
Ethylene Glycol			24	7		9		•
		50	83	7		9		•
Ethylene Oxide			24	7		2		•
Hydrogen Peroxide		90	65	7		8		•
Methylvinylsilicone								
Ammonia			24	7		8		•
	300 lb. pressure		110	1		0		•
Butyl Alcohol			24	7		6		•
Ethyl Alcohol			24	2.9		8		•
			24	7		8		•
			38	2.9		7		•
			121	2.9		5		•
Ethylene Glycol	1/3 ethyl alcohol, 1/3 water	33	100	7		9		•
		50	83	7		9		•
		50	100	2.9		9		•
		50	100	7		9		•
		50	100	14		8		•
Ethylene Oxide			71	7		1		•
Fluorosilicone								
Ammonia	cold					8	recommended for use	
						2	unsatisf. for use	
	hot					2	•	
						2	•	
	liquid					2	•	
	gas, hot		23			1	not suitable for service	
	anhydrous		23			1	•	
	gas, cold		23			1	•	
Benzaldehyde						2	unsatisf. for use	
			23			4	moder./ severe swell and/or loss of prop.	
Benzoic Acid						6	minor to moder. effect	
			23			6	may cause sl. visible swell/loss of prop.	
Benzyl Alcohol						6	minor to moder. effect	
			23			6	may cause sl. visible swell/loss of prop.	
Bleach	sol'ns					6	minor to moder. effect	
	lime bleach		23			8	little/no effect	
	sol'ns		23			6	may cause sl. visible swell/loss of prop.	
Boric Acid						8	recommended for use	
			23			8	little/no effect	
Butyl Alcohol						8	recommended for use	
						8	•	
			23			6	may cause sl. visible swell/loss of prop.	
Carbolic Acid						8	recommended for use	
	phenol		23			8	little/no effect	
Chlorine	wet					6	minor to moder. effect	
	dry					8	recommended for use	
	•		23			8	little/no effect	
	wet		23			6	may cause sl. visible swell/loss of prop.	
Cresol	methyl phenol		23			6	•	
Ethyl Acetate	organic ester					2	unsatisf. for use	
			23			1	not suitable for service	
Ethyl Alcohol						8	recommended for use	
			23			8	little/no effect	
			23			1	not suitable for service	
Ethylene Chlorohydrin						6	minor to moder. effect	
			23			6	may cause sl. visible swell/loss of prop.	
Ethylene Glycol						8	recommended for use	
			23			8	little/no effect	
Ethylene Oxide						2	unsatisf. for use	
			23			1	not suitable for service	

Reagent	Reagent Note	Conc. (%)	Temp. (°C)	Time (days)	Load	PDL Rating	Resistance Note	Material Note

Fluorosilicone

Reagent	Reagent Note	Conc. (%)	Temp. (°C)	Time (days)	Load	PDL Rating	Resistance Note	Material Note
Formaldehyde						2	unsatisf. for use	
			23			1	not suitable for service	
Hydrogen Peroxide						8	recommended for use	
		90				6	minor to moder. effect	
		90	23			6	may cause sl. visible swell/loss of prop.	
Iodine						8	recommended for use	
Isopropyl Alcohol						6	minor to moder. effect	
						6	"	
			23			6	may cause sl. visible swell/loss of prop.	
Methyl Isobutyl Ketone						2	unsatisf. for use	
Phenol						6	minor to moder. effect	
			23			8	little/no effect	
		70				6	minor to moder. effect	
		85				6	"	
Propylene Oxide						2	unsatisf. for use	
			23			1	not suitable for service	
Salicylic Acid						8	recommended for use	
			23			8	little/no effect	
Silver Nitrate						8	recommended for use	
			23			8	little/no effect	
Sodium Hypochlorite						6	minor to moder. effect	
			23			6	may cause sl. visible swell/loss of prop.	

Styrene Butadiene Rubber

Reagent	Reagent Note	Conc. (%)	Temp. (°C)	Time (days)	Load	PDL Rating	Resistance Note	Material Note
Ammonia	cold					8	recommended for use	
	hot					2	unsatisf. for use	
	liquid					2	"	
						2	"	
						2	"	
Benzaldehyde						2	"	
Benzoic Acid						2	"	
Benzyl Alcohol						2	"	
Bleach	sol'ns					2	"	
Boric Acid						8	recommended for use	
Butyl Alcohol						8	"	
						8	"	
Carbolic Acid						2	unsatisf. for use	
Chlorine	wet					2	"	
	dry					2	"	
Ethyl Acetate	organic ester					2	"	
Ethyl Alcohol						8	recommended for use	
Ethylene Chlorohydrin						6	minor to moder. effect	
Ethylene Glycol						8	recommended for use	
Ethylene Oxide						2	unsatisf. for use	
Formaldehyde						4	moder. to severe effect	
Hydrogen Peroxide						6	minor to moder. effect	
		90				2	unsatisf. for use	
Iodine						6	minor to moder. effect	
Isopropyl Alcohol						6	"	
						6	"	
Mercuric Chloride						8	recommended for use	
Methyl Isobutyl Ketone						2	unsatisf. for use	
Phenol						2	"	
		70				2	"	
		85				2	"	
Propylene Oxide						2	"	
Salicylic Acid						6	minor to moder. effect	
Silver Nitrate						8	recommended for use	
Sodium Hypochlorite						4	moder. to severe effect	

Polyvinyl Alcohol

Reagent	Reagent Note	Conc. (%)	Temp. (°C)	Time (days)	Load	PDL Rating	Resistance Note	Material Note
Benzaldehyde			23	0.0208		8	material is well suited for use with chemical	Ans. Edm. PVA; lined glove film
			23	0.25		8	low permeation rate; 0-1/2 drops/hr.; PVA is water soluble	"

Reagent	Reagent Note	Conc. (%)	Temp. (°C)	Time (days)	Load	PDL Rating	Resistance Note	Material Note
Polyvinyl Alcohol								
Butyl Alcohol			23			6	good perm. rate; 6 to 50 drops/hr.	Ans. Edm. PVA; lined glove film
			23	0.0208		5	fluid has moder. degrading effect	"
Ethyl Acetate			23	0.0208		5	"	"
			23	0.25		7	low permeation rate; 0-1/2 drops/hr.; PVA is water soluble	"
Ethyl Alcohol			23	0.0208		1	not recommended; PVA is water soluble	Ans. Edm. PVA; lined glove film
Ethylene Glycol			23			7	very good perm. rate; 1-5 drops/hr.; PVA is water soluble	Ans. Edm. PVA; lined glove film
			23	0.0208		5	fluid has moder. degrading effect	"
Isopropyl Alcohol			23	0.0208		1	not recommended; PVA is water soluble	Ans. Edm. PVA; lined glove film
Methyl Isobutyl Ketone			23	0.0208		5	fluid has moder. degrading effect	Ans. Edm. PVA; lined glove film
			23	0.25		7	low permeation rate; 0-1/2 drops/hr.; PVA is water soluble	"
Morpholine			23			6	good perm. rate; 6 to 50 drops/hr.; PVA is water soluble	"
			23	0.0208		8	material is well suited for use with chemical	"
Phenol			23	0.0208		5	fluid has moder. degrading effect	Ans. Edm. PVA; lined glove film
			23	0.25		7	low permeation rate; 0-1/2 drops/hr.; PVA is water soluble	"
Propylene Oxide			23			6	good perm. rate; 6 to 50 drops/hr.; PVA is water soluble	Ans. Edm. PVA; lined glove film
			23	0.0208		5	fluid has minor degrading effect	"

ABS PC alloy See *acrylonitrile butadiene styrene polymer polycarbonate alloy.*

ABS See *acrylonitrile butadiene styrene polymer.*

ABS nylon alloy See *acrylonitrile butadiene styrene polymer nylon alloy.*

ABS resin See *acrylonitrile butadiene styrene polymer.*

acetal resins Thermoplastics prepared by polymerization of formaldehyde or its trioxane trimer. Acetals have high impact strength and stiffness, low friction coefficient and permeability, good dimensional stability and dielectric properties, and high fatigue strength and thermal stability. Acetals have poor acid and UV resistance and are flammable. Processed by injection and blow molding and extrusion. Used in mechanical parts such as gears and bearings, automotive components, appliances, and plumbing and electronic applications. Also called acetals.

acetals See *acetal resins.*

acrylate styrene acrylonitrile polymer Acrylic rubber-modified thermoplastic with high weatherability. ASA has good heat and chemical resistance, toughness, rigidity, and antistatic properties. Processed by extrusion, thermoforming, and molding. Used in construction, leisure, and automotive applications such as siding, exterior auto trim, and outdoor furniture. Also called ASA.

acrylic resins Thermoplastic polymers of alkyl acrylates such as methyl methacrylates. Acrylic resins have good optical clarity, weatherability, surface hardness, chemical resistance, rigidity, impact strength, and dimensional stability. They have poor solvent resistance, resistance to stress cracking, flexibility, and thermal stability. Processed by casting, extrusion, injection molding, and thermoforming. Used in transparent parts, auto trim, household items, light fixtures, and medical devices. Also called polyacrylates.

acrylonitrile butadiene styrene polymer ABS resins are thermoplastics comprised of a mixture of styrene-acrylonitrile copolymer (SAN) and SAN-grafted butadiene rubber. They have high impact resistance, toughness, rigidity and processability, but low dielectric strength, continuous service temperature, and elongation. Outdoor use requires protective coatings in some cases. Plating grades provide excellent adhesion to metals. Processed by extrusion, blow molding, thermoforming, calendaring and injection molding. Used in household appliances, tools, nonfood packaging, business machinery, interior automotive parts, extruded sheet, pipe and pipe fittings. Also called ABS, ABS resin, acrylonitrile-butadiene-styrene polymer.

acrylonitrile butadiene styrene polymer nylon alloy A thermoplastic processed by injection molding, with properties similar to ABS but higher elongation at yield. Also called ABS nylon alloy.

acrylonitrile butadiene styrene polymer polycarbonate alloy A thermoplastic processed by injection molding and extrusion, with properties similar to ABS. Used in automotive applications. Also called ABS PC alloy.

acrylonitrile copolymer A thermoplastic prepared by copolymerization of acrylonitrile with small amounts of other unsaturated monomers. Has good gas barrier properties and chemical resistance. Processed by extrusion, injection molding, and thermoforming. Used in food packaging.

acrylonitrile-butadiene-styrene polymer See *acrylonitrile butadiene styrene polymer.*

alcohols A class of hydroxy compounds in which a hydroxy group(s) is attached to a carbon chain or ring. Alcohols are produced synthetically from petroleum stock, e.g., by hydration of ethylene, or derived from natural products, e.g., by fermentation of grain. The alcohols are divided in the following groups: monohydric, dihydric, trihydric and polyhydric. Used in organic synthesis, as solvents, plasticizers, fuels, beverages, detergents, etc.

amorphous nylon Transparent aromatic polyamide thermoplastics.

aromatic polyester estercarbonate A thermoplastic block copolymer of an aromatic polyester with polycarbonate. Has higher heat distortion temperature than regular polycarbonate.

aromatic polyesters Engineering thermoplastics prepared by polymerization of aromatic polyol with aromatic dicarboxylic anhydride. They are tough with somewhat low chemical resistance. Processed by injection and blow molding, extrusion, and thermoforming. Drying is required. Used in automotive housings and trim, electrical wire jacketing, printed circuit boards, and appliance enclosures.

ASA See *acrylate styrene acrylonitrile polymer.*

ASTM D256 The method for determination of the resistance to breakage by flexural shock of plastics and electrical insulating materials, as indicated by the energy extracted from standard pendulum-type hammers in breaking standard specimens with one pendulum swing. The hammers are mounted on standard machines of either Izod or Charpy type. **Note:** Impact properties determined include Izod or Charpy impact energy normalized per width of the specimen. Also called ASTM method D256-84. See also *impact energy.*

ASTM D3763 The method for determination of the resistance of plastics, including films, to high-speed puncture over a broad range of test velocities using load and displacement sensors. **Note:** Puncture properties determined include maximum load, deflection to maximum load point, energy to maximum load point and total energy. Also called ASTM method D3763-86. See also *impact energy.*

ASTM D638 The method for determination of tensile properties of unreinforced and reinforced plastics in the form of standard dumbbell-shaped specimens under defined conditions of pretreatment, temperature, humidity and testing machine speed. **Note:** Tensile properties determined include tensile stress (strength) at yield and at break, percentage elongation at yield or at break and modulus of elasticity. Also called ASTM method D638-84. See also *tensile strength.*

ASTM method D256-84 See *ASTM D256.*

ASTM method D3763-86 See *ASTM D3763.*

ASTM method D638-84 See *ASTM D638.*

autoclave sterilization Sterilization by steam under pressure in an autoclave. Also called steam sterilization.

B

bending properties See *flexural properties.*

bending strength See *flexural strength.*

bending stress See *flexural stress.*

bisphenol A polyester A thermoset unsaturated polyester based on bisphenol A and fumaric acid.

bleaching Complete loss of color of the material as a result of degradation or removal of colored substances present on its surface. Bleaching can be caused by chemical reactions, radiation, etc.

boiler additive See *boiler additives.*

boiler additives A chemical used to treat boiler water to prevent corrosion, scaling and fouling of heat-exchanging surfaces; foaming and contamination of steam. Also called boiler compounds, boiler water additives, boiler additive, boiler water treatment. See also *morpholine.*

boiler compounds See *boiler additives.*

boiler water additives See *boiler additives.*

boiler water treatment See *boiler additives.*

breaking elongation See *elongation.*

C

CA See *cellulose acetate.*

CAB See *cellulose acetate butyrate.*

cellulose acetate Thermoplastic esters of cellulose with acetic acid. Have good toughness, gloss, clarity, processability, stiffness, hardness, and dielectric properties, but poor chemical, fire and water resistance and compressive strength. Processed by injection and blow molding and extrusion. Used for appliance cases, steering wheels, pens, handles, containers, eyeglass frames, brushes, and sheeting. Also called CA.

cellulose acetate butyrate Thermoplastic mixed esters of cellulose with acetic and butyric acids. Have good toughness, gloss, clarity, processability, dimensional stability, weatherability, and dielectric properties, but poor chemical, fire and water resistance and compressive strength. Processed by injection and blow molding and extrusion. Used for appliance cases, steering wheels, pens, handles, containers, eyeglass frames, brushes, and sheeting. Also called CAB.

cellulose propionate Thermoplastic esters of cellulose with propionic acid. Have good toughness, gloss, clarity, processability, dimensional stability, weatherability, and dielectric properties, but poor chemical, fire and water resistance and compressive strength. Processed by injection and blow molding and extrusion. Used for appliance cases, steering wheels, pens, handles, containers, eyeglass frames, brushes, and sheeting. Also called CP.

cellulosic plastics Thermoplastic cellulose esters and ethers. Have good toughness, gloss, clarity, processability, and dielectric properties, but poor chemical, fire and water resistance and compressive strength. Processed by injection and blow molding and extrusion. Used for appliance cases, steering wheels, pens, handles, containers, eyeglass frames, brushes, and sheeting.

chemical sterilization agent hydrolysis products Chemicals produced as a result of hydrolysis of chemisterilants during sterilization at elevated temperatures in the presence of water. Also called chemisterilant hydrolysis products. See also *ethylene glycol, ethylene chlorohydrin.*

chemical sterilization agents A chemical used to kill all microorganisms, including spores, on or in an object. Also called chemisterilants.

chemisterilant hydrolysis products See *chemical sterilization agent hydrolysis products.*

chemisterilants See *chemical sterilization agents.*

chlorendic polyester A chlorendic anhydride-based unsaturated polyester.

chlorinated polyvinyl chloride Thermoplastic produced by chlorination of polyvinyl chloride. Has increased glass transition temperature, chemical and fire resistance, rigidity, tensile strength, and weatherability as compared to PVC. Processed by extrusion, injection molding, casting, and calendering. Used for pipes, auto parts, waste disposal devices, and outdoor applications. Also called CPVC.

chloroethyl alcohol(2-) See *ethylene chlorohydrin.*

chlorohydrins Halohydrins with chlorine as a halogen atom. One of the most reactive of halohydrins. Dichlorohydrins are used in the preparation of epichlorohydrins, important monomers in the manufacture of epoxy resins. Most chlorohydrins are reactive colorless liquids, soluble in polar solvents such as alcohols. **Note:** Chlorohydrins are a class of organic compounds, not to be mixed with a specific member of this class, 1-chloropropane-2,3-diol sometimes called chlorohydrin.

chlorosulfonated polyethylene rubber Thermosetting elastomers containing 20- 40% chlorine. Have good weatherability and heat and chemical resistance. Used for hoses, tubes, sheets, footwear soles, and inflatable boats.

Co-60 See *cobalt-60.*

cobalt-60 (Co) One of the unstable isotopes of Co used widely as a source of gamma radiation. Also called Co-60.

cold sterilants See *cold sterilization agents.*

cold sterilization agents A chemical or physical agent used to kill all microorganisms, including spores, on or in an object under ambient conditions. Also called cold sterilants.

color The wavelength composition of light, specifically of the light reflected or emitted by the material and its visual appearance. Also called hue, tint, coloration.

color change See *discoloration.*

color difference The square root of the sum of the squares of the chromaticity difference and the lightness difference. Also called delta E, delta E color change.

coloration See *color.*

concentration units The units for measuring the content of a distinct material or substance in a medium other than this material or substance, such as solvent. **Note:** The concentration units are usually expressed in the units of mass or volume of substance per one unit of mass or volume of medium. When the units of substance and medium are the same, the percentage is often used.

conditioning Process of bringing the material or apparatus to a certain condition, e.g., moisture content or temperature, prior to further processing, treatment, etc. Also called conditioning cycle.

conditioning cycle See *conditioning.*

CP See *cellulose propionate.*

CPVC See *chlorinated polyvinyl chloride.*

cracking Appearance of external and/or internal cracks in the material as a result of stress that exceeds the strength of the material. The stress can be external and/or internal and can be caused by a variety of adverse conditions: structural defects, impact, aging, corrosion, etc. or a combination of thereof. Also called cracks. See also *processing defects.*

cracks See *cracking.*

crazes See *crazing.*

crazing Appearance of thin cracks on the surface of the material or, sometimes, minute frost-like internal cracks, as a result of stress that exceeds the strength of the material. Also called crazes.

crosslinked polyethylene Polyethylene thermoplastics partially photochemically or chemically crosslinked. Have improved tensile strength, dielectric properties, and impact strength at low and elevated temperatures.

crosslinking when chemical links set up between molecular chains, the plastic is said to crosslinked. In thermosets, crosslinking makes one infusable molecule of all the chains, contributing to strength, rigidity and high temperature resistance. Thermoplastics can also be crosslinked (e.g., by irradiation or chenically through formulation) to produce three dimensional structures that are thermoset in nature and offer improved tensile strength and stress crack resistance.

crystal polystyrene See *general purpose polystyrene.*

CTFE See *polychlorotrifluoroethylene.*

cycle time See *processing time.*

cyclic compounds A broad class of organic compounds consisting of carbon rings that are saturated, partially unsaturated or aromatic, in which some carbon atoms may be replaced by other atoms such as oxygen, sulfur and nitrogen.

D

DAP See *diallyl phthalate resins.*

decoloration Complete or partial loss of color of the material as a result of degradation or removal of colored substances present in it. Also called decoloring.

defects See *processing defects.*

deflection temperature under load See *heat deflection temperature.*

degassing rate The rate of the removal of undesirable gases adsorbed or dissolved in a liquid or solid material. Degassing can be either natural, e.g., due to desorption or dissipation under normal conditions, or forced as in purging or washing with air (aeration), evacuation (creating vacuum) or heating. Also called outgassing rate.

degradation Loss or undesirable change in the properties, such as color, of a material as a result of aging, chemical reaction, wear, exposure, etc. See also *stability.*

delta E See *color difference.*

delta E color change See *color difference.*

diallyl phthalate resins Thermosets supplied as diallyl phthalate prepolymer or monomer. Have high chemical, heat and water resistance, dimensional stability, and strength. Shrink during peroxide curing. Processed by injection, compression and transfer molding. Used in glass-reinforced tubing, auto parts, and electrical components. Also called DAP.

dihydric alcohols See *glycols.*

dihydroxy alcohols See *glycols.*

discoloration A change in color due to chemical or physical changes in the material. Also called color change.

displacement Process of removing one object, e.g., a medium in an apparatus, or its part, and replacing it with another. Also called displacement cycle.

displacement cycle See *displacement.*

drop dart impact See *falling weight impact energy.*

drop dart impact energy See *falling weight impact energy.*

drop dart impact strength See *falling weight impact energy.*

drop weight impact See *falling weight impact energy.*

drop weight impact energy See *falling weight impact energy.*

drop weight impact strength See *falling weight impact energy.*

dwell time The time during which a material, a substance or a particle remains in the atmosphere or other media, or any defined space such as an apparatus before leaving it. Also called residence time. See also *sterilization time.*

E

ECTFE See *ethylene chlorotrifluoroethylene copolymer.*

electromagnetic properties The response of materials to electromagnetic fields and their ability to produce these fields.

electromagnetic radiation Waves of electric charges propagated through space by oscillating electromagnetic fields and associated energy. See also *electromagnetic radiation units.*

electromagnetic radiation units The units for measuring quantities related to emission, propagation and absorption of electromagnetic radiation, such as radioactivity and transmittance. See also *electromagnetic radiation.*

electron beam See *electron beam radiation.*

electron beam radiation Ionizing radiation propagated by electrons that move forward in a narrow stream with approximately equal velocity. Also called electron beams, electron beam. See also *physical sterilization agents.*

elongation The increase in gauge length of a specimen in tension, measured at or after the fracture, depending on the viscoelastic properties of the material. **Note:** Elongation is usually expressed as a percentage of the original gauge length. Also called tensile elongation, elongation at break, ultimate elongation, breaking elongation, elongation at rupture. See also *tensile strain.*

elongation at break See *elongation.*

elongation at rupture See *elongation.*

EMAC See *ethylene methyl acrylate copolymer.*

EPDM See *EPDM rubber.*

EPDM rubber Sulfur-vulcanizable thermosetting elastomers produced from ethylene, propylene, and a small amount of nonconjugated diene such as hexadiene. Have good weatherability and chemical and heat resistance. Used as impact modifiers and for weather stripping, auto parts, cable insulation, conveyor belts, hoses, and tubing. Also called EPDM.

epoxides Organic compounds containing three-membered cyclic group(s) in which two carbon atoms are linked with an oxygen atom as in an ether. This group is called an epoxy group and is quite reactive, allowing the use of epoxides as intermediates in preparation of certain fluorocarbons and cellulose derivatives and as monomers in preparation of epoxy resins. Also called epoxy compounds.

epoxies See *epoxy resins.*

epoxy compounds See *epoxides.*

epoxy resins Thermosetting polyethers containing crosslinkable glycidyl groups. Usually prepared by polymerization of bisphenol A and epichlorohydrin or reacting phenolic novolaks with epichlorohydrin. Can be made unsaturated by acrylation. Unmodified varieties are cured at room or elevated temperatures with polyamines or anhydrides. Bisphenol A epoxy resins have excellent adhesion and very low shrinkage during curing. Cured novolak epoxies have good UV stability and dielectric properties. Cured acrylated epoxies have high strength and chemical resistance. Processed by molding, casting, coating, and lamination. Used as protective coatings, adhesives, potting compounds, and binders in laminates and composites. Also called epoxies.

epoxyethane See *ethylene oxide.*

EPR See *ethylene propene rubber.*

ETFE See *ethylene tetrafluoroethylene copolymer.*

ethanediol(1,2-) See *ethylene glycol.*

ethers A class of organic compounds in which an oxygen atom is interposed between two carbon atoms in a chain or a ring. Ethers are derived mainly by catalytic hydration of olefins. The lower molecular weight ethers are dangerous fire and explosion hazards. **Note:** Major types of ethers include aliphatic, cyclic and polymeric ethers.

ethylene acrylic rubber Copolymers of ethylene and acrylic esters. Have good toughness, low temperature properties, and resistance to heat, oil, and water. Used in auto and heavy equipment parts.

ethylene alcohol See *ethylene glycol.*

ethylene chlorohydrin (C2H5ClO) Ethylene chlorohydrin, ClCH2CH2OH, is a colorless liquid easily soluble in most organic liquids and water. It has an autoignition temperature of 425 °C (797 °F) and is a moderate fire hazard. Derived by reaction of hydrochlorous acid with ethylene. It is a strong irritant, deadly via inhalation, skin absorption, etc. with TLV of 1 ppm in air. Penetrates through rubber gloves. Used as a solvent for cellulose derivatives, intermediate in organic synthesis (e.g., for ethylene oxide) and sprouting activator. **Note:** Hydrolysis of ethylene oxide during sterilization can result in the formation of ethylene chlorohydrin and its residual presence in sterilized goods. Also called 2-chloroethyl alcohol, glycol chlorohydrin. See also *chemical sterilization agent hydrolysis products.*

ethylene chlorotrifluoroethylene copolymer Thermoplastic alternating copolymer of ethylene and chlorotrifluoroethylene. Has superior strength and creep and wear resistance compared to other fluoropolymers, and good dielectric properties and chemical and fire resistance. Processed by molding, extrusion, and powder coating. Used in electric insulation, pump parts, filter housings, tubing, linings, and release films. Also called ECTFE.

ethylene copolymers See *ethylene polymers.*

ethylene glycol (C2H6O2) Ethylene glycol, CH2OHCH2OH, is a colorless heavy liquid soluble in polar organic solvents such as alcohols and water. It has an autoignition temperature of 413 °C (775 °F). Its TLV is 50 ppm and it is toxic by ingestion and inhalation. Derived by air oxidation of ethylene followed by hydration of resultant ethylene oxide, from synthesis gas, by oxirane process, etc. Used as coolant and antifreeze, as a monomer for the production of polyesters, solvent additive, foam stabilizer, brake fluids and many other products. **Note:** Hydrolysis of ethylene oxide during sterilization can result in the formation of ethylene glycol and its residual presence in sterilized goods. Also called ethylene alcohol, glycol, ethanediol(1,2-). See also *chemical sterilization agent hydrolysis products.*

ethylene methyl acrylate copolymer Thermoplastic copolymers of ethylene with <40% methyl acrylate. Have good dielectric properties, toughness, thermal stability, stress crack resistance, and compatibility with other polyolefins. Transparency decreases with increasing content of acrylate. Processed by blow film extrusion and blow and injection molding. Used in heat-sealable films, disposable gloves, and packaging. Some grades are FDA-approved for food packaging. Also called EMAC.

ethylene oxide (C2H4O) The simplest, unsubstituted, saturated epoxide with molecular formula CH2OCH2. Ethylene oxide is a colorless gas at room temperature with autoignition temperature 429 °C (805 °F). It is an eye and skin irritant and a suspected human carcinogen. Its TLV is 1ppm in air. It is dangerous fire and explosion hazard. Ethylene oxide is derived by catalytic oxidation of ethylene or by alkaline hydrolysis of ethylene chlorohydrin. It is used in manufacture of ethylene glycol and other important chemicals such as polyethylene oxide surfactants, and as sterilant, fumigant and rocket propellant. Also called ETO, epoxyethane, methylene oxide, ethylene oxide gas, oxirane. See also *gaseous chemical sterilization agents, rate of residual ethylene oxide dissipation.*

ethylene oxide gas See *ethylene oxide.*

ethylene polymers Ethylene polymers include ethylene homopolymers and copolymers with other unsaturated monomers, most importantly olefins such as propylene and polar substances such as vinyl acetate. The properties and uses of ethylene polymers depend on the molecular structure and weight. Also called ethylene copolymers.

ethylene propene rubber Stereospecific copolymers of ethylene with propylene. Used as impact modifiers for plastics. Also called EPR.

ethylene tetrafluoroethylene copolymer Thermoplastic alternating copolymer of ethylene and tetrafluoroethylene. Has good impact strength, abrasion and chemical resistance, weatherability, and dielectric properties. Processed by molding, extrusion, and powder coating. Used in tubing, cables, pump parts, and tower packing in a wide temperature range. Also called ETFE.

ethylene vinyl alcohol copolymer Thermoplastics prepared by hydrolysis of ethylene-vinyl acetate polymers. Have good barrier properties, mechanical strength, gloss, elasticity, weatherability, clarity, and abrasion resistance. Barrier properties and processibility improve with increasing content of ethylene due to lower absorption of moisture. Processed by extrusion, coating, blow and blow film molding, and thermoforming. Used as packaging films and container liners. Also called EVOH.

ETO See *ethylene oxide.*

EVOH See *ethylene vinyl alcohol copolymer.*

F

falling dart impact See *falling weight impact energy.*

falling dart impact energy See *falling weight impact energy.*

falling dart impact strength See *falling weight impact energy.*

falling weight impact See *falling weight impact energy.*

falling weight impact energy The mean energy of a free-falling dart or weight (tup) that will cause 50% failures after 50 tests to a directly or indirectly stricken specimen. The energy is calculated by multiplying dart mass, gravitational acceleration and drop height. Also called falling weight impact strength, falling weight impact, falling dart impact energy, falling dart impact strength, falling dart impact, drop dart impact energy, drop dart impact strength, drop dart impact, drop weight impact energy, drop weight impact strength, drop weight impact.

falling weight impact strength See *falling weight impact energy.*

FEP See *fluorinated ethylene propylene copolymer.*

five-membered heterocyclic compounds A class of heterocyclic compounds containing rings that consist of five atoms.

five-membered heterocyclic nitrogen compounds A class of heterocyclic compounds containing rings that consist of five atoms, some of which is a nitrogen.

five-membered heterocyclic oxygen compounds A class of heterocyclic compounds containing rings that consist of five atoms, some of which is an oxygen.

flaw See *processing defects.*

flexural properties Properties describing the reaction of physical systems to flexural stress and strain. Also called bending properties.

flexural strength The maximum stress in the extreme fiber of a specimen loaded to failure in bending. **Note:** Flexural strength is calculated as a function of load, support span and specimen geometry. Also called modulus of rupture in bending, modulus of rupture, bending strength.

flexural stress The maximum stress in the extreme fiber of a specimen in bending. **Note:** Flexural stress is calculated as a function of load at a given strain or at failure, support span and specimen geometry. Also called bending stress.

fluorinated ethylene propylene copolymer Thermoplastic copolymer of tetrafluoroethylene and hexafluoropropylene. Has decreased tensile strength and wear and creep resistance, but good weatherability, dielectric properties, fire and chemical resistance, and friction. Decomposes above 204 °C (400 °F), releasing toxic products. Processed by molding, extrusion, and powder coating. Used in chemical apparatus liners, pipes, containers, bearings, films, coatings, and cables. Also called FEP.

fluoro rubber See *fluoroelastomers.*

fluoroelastomers Fluorine-containing synthetic rubber with good chemical and heat resistance. Used in underhood applications such as fuel lines, oil and coolant seals, and fuel pumps, and as a flow additive for polyolefins. Also called fluoro rubber.

fluoroplastics See *fluoropolymers.*

fluoropolymers Polymers prepared from unsaturated fluorine-containing hydrocarbons. Have good chemical resistance, weatherability, thermal stability, antiadhesive properties and low friction and flammability, but low creep resistance and strength and poor processibility. The properties vary with the fluorine content. Processed by extrusion and molding. Used as liners in chemical apparatus, in bearings, films, coatings, and containers. Also called fluoroplastics.

fluorosilicones Polymers with chains of alternating silicon and oxygen atoms and trifluoropropyl pendant groups. Most are rubbers.

FMQ See *methylfluorosilicones.*

G

gamma radiation Ionizing radiation propagated by high-energy protons, e.g., emitted by a nucleus in transition between two energy levels. **Note:** A dosage of 2.5 megarad is considered the industry standard for disposable medical parts. Also called gamma rays. See also *physical sterilization agents.*

gamma rays See *gamma radiation.*

gaseous chemical sterilization agents A chemical, gaseous under ambient conditions, used to kill all microorganisms, including spores, on or in an object. Also called gaseous chemisterilants. See also *ethylene oxide.*

gaseous chemisterilants See *gaseous chemical sterilization agents.*

general purpose polystyrene General purpose polystyrene is an amorphous thermoplastic prepared by homopolymerization of styrene. It has good tensile and flexural strengths, high light transmission and adequate resistance to water, detergents and inorganic chemicals. It is attached by hydrocarbons and has a relatively low impact resistance. Processed by injection molding and foam extrusion. Used to manufacture containers, health care items such as pipettes, kitchen and bathroom housewares, stereo and camera parts and foam sheets for food packaging. Also called crystal polystyrene.

glycol See *ethylene glycol.*

glycol chlorohydrin See *ethylene chlorohydrin.*

glycol modified polycyclohexylenedimethylene terephthalate Thermoplastic polyester prepared from glycol, cyclohexylenedimethanol, and terephthalic acid. Has good impact strength and other mechanical properties, chemical resistance, and clarity. Processed by injection molding and extrusion. Can be blended with polycarbonate. Also called PCTG.

glycols Aliphatic alcohols with two hydroxy groups attached to a carbon chain. Can be produced by oxidation of alkenes followed by hydration. Also called dihydric alcohols, dihydroxy alcohols.

gravity displacement Process of displacement of one object in an enclosed space, e.g., a medium in an apparatus, with another by the force of gravity, i.e., using higher density of the displacing medium as compared to that of the displaced medium. **Note:** Usually gravity displacement is applied to gases or vapors. Also called gravity displacement cycle. See also *sterilization*.

gravity displacement cycle See *gravity displacement.*

H

halogen compounds A class of organic compounds containing halogen atoms such as chlorine. A simple example is halocarbons but many other subclasses with various functional groups and of different molecular structure exist as well.

halohydrins Halogen compounds that contain a halogen atom(s) and a hydroxy (OH) group(s) attached to a carbon chain or ring. Can be prepared by reaction of halogens with alkenes in the presence of water or by reaction of halogens with triols. Halohydrins can be easily dehydrochlorinated in the presence of a base to give an epoxy compound.

HDPE See *high density polyethylene.*

HDT See *heat deflection temperature.*

heat deflection point See *heat deflection temperature.*

heat deflection temperature The temperature at which a material specimen (standard bar) is deflected by a certain degree under specified load. Also called heat distortion temperature, heat distortion point, heat deflection point, deflection temperature under load, tensile heat distortion temperature, HDT.

heat distortion point See *heat deflection temperature.*

heat distortion temperature See *heat deflection temperature.*

heterocyclic compounds A class of cyclic compounds containing rings with some carbon atoms replaced by other atoms such as oxygen, sulfur and nitrogen.

high density polyethylene A linear polyethylene with density 0.94-0.97 g/cm3. Has good toughness at low temperatures, chemical resistance, and dielectric properties and high softening temperature, but poor weatherability. Processed by extrusion, blow and injection molding, and powder coating. Used in houseware, containers, food packaging, liners, cable insulation, pipes, bottles, and toys. Also called HDPE.

high impact polystyrene See *impact polystyrene.*

high molecular weight low density polyethylene Thermoplastic with improved abrasion and stress crack resistance and impact strength, but poor processibility and reduced tensile strength. Also called HMWLDPE.

HIPS See *impact polystyrene.*

HMWLDPE See *high molecular weight low density polyethylene.*

hue See *color.*

hydroxy compounds A broad class of organic compounds that contain a hydroxy (OH) group(s) that is not part of another functional group such as carboxylic group. Also called hydroxyl-containing compounds.

hydroxyl-containing compounds See *hydroxy compounds.*

I

impact energy The energy required to break a specimen, equal to the difference between the energy in the striking member of the impact apparatus at the instant of impact and the energy remaining after complete fracture of the specimen. Also called impact strength. See also *ASTM D256, ASTM D3763.*

impact polystyrene Impact polystyrene is a thermoplastic produced by polymerizing styrene dissolved in butadiene rubber. Impact polystyrene has good dimensional stability, high rigidity and good low temperature impact strength, but poor barrier properties, grease resistance and heat resistance. Processed by extrusion, injection molding, thermoforming and structural foam molding. Used in food packaging, kitchen housewares, toys, small appliances, personal care items and audio products. Also called IPS, high impact polystyrene, HIPS, impact PS.

impact property tests Names and designations of the methods for impact testing of materials. Also called impact tests. See also *impact toughness.*

impact PS See *impact polystyrene.*

impact strength See *impact energy.*

impact tests See *impact property tests.*

impact toughness Property of a material indicating its ability to absorb energy of a high-speed impact by plastic deformation rather than crack or fracture. See also *impact property tests.*

ionization radiation See *ionizing radiation.*

ionizing radiation Electromagnetic radiation with sufficient energy to ionize the material exposed to it. Also called ionization radiation. See also *radioisotopes, ionizing radiation units.*

ionizing radiation dosage See *radiation dose.*

ionizing radiation dose See *radiation dose.*

ionizing radiation units The units for measuring quantities related to emission, propagation and absorption of ionizing radiation, such as radioactivity, radiation dose and dose rate. See also *ionizing radiation.*

ionomers Thermoplastics containing a relatively small amount of pendant ionized acid groups. Have good flexibility and impact strength in a wide temperature range, puncture and chemical resistance, adhesion, and dielectric properties, but poor weatherability, fire resistance, and thermal stability. Processed by injection, blow and rotational molding, blown film extrusion, and extrusion coating. Used in food packaging, auto bumpers, sporting goods, and foam sheets.

IPS See *impact polystyrene.*

isophthalate polyester An unsaturated polyester based on isophthalic acid.

isotopes One of two or more atoms having the same atomic number but different mass numbers.

Izod See *Izod impact energy.*

Izod impact See *Izod impact energy.*

Izod impact energy The energy required to break a specimen equal to the difference between the energy in the striking member of the Izod-type impact apparatus at the instant of impact and the energy remaining after complete fracture of the specimen. Also called Izod impact, Izod impact strength, Izod.

Izod impact strength See *Izod impact energy.*

L

LCP See *liquid crystal polymers.*

LDPE See *low density polyethylene.*

light transmission See *transmittance.*

linear low density polyethylene Linear polyethylenes with density 0.91-0.94 g/cm3. Has better tensile, tear, and impact strength and crack resistance properties, but poorer haze and gloss than branched low-density polyethylene. Processed by extrusion at increased pressure and higher melt temperatures compared to branched low-density polyethylene, and by molding. Used to manufacture film, sheet, pipe, electrical insulation, liners, bags and food wraps. Also called LLDPE, LLDPE resin.

linear polyethylenes Linear polyethylenes are polyolefins with linear carbon chains. They are prepared by copolymerization of ethylene with small amounts of higher alfa-olefins such as 1-butene. Linear polyethylenes are stiff, tough and have good resistance to environmental cracking and low temperatures. Processed by extrusion and molding. Used to manufacture film, bags, containers, liners, profiles and pipe.

liquid crystal polymers Thermoplastic aromatic copolyesters with highly ordered structure. Have good tensile and flexural properties at high temperatures, chemical, radiation and fire resistance, and weatherability. Processed by sintering and injection molding. Used to substitute ceramics and metals in electrical components, electronics, chemical apparatus, and aerospace and auto parts. Also called LCP.

LLDPE See *linear low density polyethylene.*

LLDPE resin See *linear low density polyethylene.*

low density polyethylene A branched-chain thermoplastic with density 0.91-0.94 g/cm3. Has good impact strength, flexibility, transparency, chemical resistance, dielectric properties, and low water permeability and brittleness temperature, but poor heat, stress cracking and fire resistance and weatherability. Processed by extrusion coating, injection and blow molding, and film extrusion. Can be crosslinked. Used in packaging and shrink films, toys, bottle caps, cable insulation, and coatings. Also called LDPE.

luminous transmittance See *transmittance.*

M

macroscopic properties See *thermodynamic properties.*

mechanical properties Properties describing the reaction of physical systems to stress and strain.

megarad One rad is equivalent to an energy absorption per unit mass of 0.01 joule per kilogram of irradiated material; one megarad is 1E+06 rads. Also called Mrad.

melamine resins Thermosetting resins prepared by condensation of formaldehyde with melamine. Have good hardness, scratch and fire resistance, clarity, colorability, rigidity, dielectric properties, and tensile strength, but poor impact strength. Molding grades are filled. Processed by compression, transfer, and injection molding, impregnation, and coating. Used in cosmetic containers, appliances, tableware, electrical insulators, furniture laminates, adhesives, and coatings.

methylene oxide See *ethylene oxide.*

methylfluorosilicones Silicone rubbers containing pendant fluorine and methyl groups. Have good chemical and heat resistance. Used in gasoline lines, gaskets, and seals. Also called FMQ.

methylphenylsilicones Silicone rubbers containing pendant phenyl and methyl groups. Have good resistance to heat, oxidation, and radiation, and compatibility with plastics.

methylsilicone Silicone rubbers containing pendant methyl groups. Have good heat and oxidation resistance. Used in electrical insulation and coatings. Also called MQ.

methylvinylfluorosilicone Silicone rubbers containing pendant vinyl, methyl, and fluorine groups. Can be additionally crosslinked via vinyl groups. Have good resistance to petroleum products at elevated temperatures.

methylvinylsilicone Silicone rubbers containing pendant methyl and vinyl groups. Can be additionally crosslinked via vinyl groups. vulcanized to high degrees of crosslinking. Used in sealants, adhesives, coatings, cables, gaskets, tubing, and electrical tape.

modified polyphenylene ether Thermoplastic polyphenylene ether alloys with impact polystyrene. Have good impact strength, resistance to heat and fire, but poor resistance to solvents. Processed by injection and structural foam molding and extrusion. Used in auto parts, appliances, and telecommunication devices. Also called MPE, MPO, modified polyphenylene oxide.

modified polyphenylene oxide See *modified polyphenylene ether.*

modulus of rupture See *flexural strength.*

modulus of rupture in bending See *flexural strength.*

molding defects Structural and other defects in material caused inadvertently during molding by using wrong tooling, process parameters, ingredients, part design, etc. Usually preventable. Also called molding flaw.

molding flaw See *molding defects.*

molecular weight The sum of the atomic weights of all atoms in a molecule. Also called MW.

molecular weight distribution The relative amounts of polymeric molecules of different weights in a specimen. **Note:** The molecular weight distribution can be expressed in terms of the ratio between weight- and number-average molecular weights. Also called polydispersity, MWD, molecular weight ratio.

molecular weight ratio See *molecular weight distribution.*

morpholine (C4H9ON) Morpholine is a saturated oxazoline. It is Colorless hygroscopic liquid, has autoignition temperature of 310 °C (590 °F) and TLV 20 ppm in air. It is moderate fire hazard, toxic by inhalation and ingestion and irritant to skin. Derived by dehydration of diethanolamine. Used as vulcanization accelerator, boiler water additive, corrosion inhibitor, paper preservant, optical brightener and organic synthesis intermediate. Also called tetrahydro-1,4-oxazine. See also *boiler additives.*

MPE See *modified polyphenylene ether.*

MPO See *modified polyphenylene ether.*

MQ See *methylsilicone.*

Mrad See *megarad.*

MW See *molecular weight.*

MWD See *molecular weight distribution.*

N

neoprene rubber Polychloroprene rubbers with good resistance to petroleum products, heat, and ozone, weatherability, and toughness.

nitrile rubber Rubbers prepared by free-radical polymerization of acrylonitrile with butadiene. Have good resistance to petroleum products, heat, and abrasion. Used in fuel hoses, shoe soles, gaskets, oil seals, and adhesives.

nonelastomeric thermoplastic polyurethanes See *rigid thermoplastic polyurethanes.*

nonelastomeric thermosetting polyurethane Curable mixtures of isocyanate prepolymers or monomers. Have good abrasion resistance and low-temperature stability, but poor heat, fire, and solvent resistance and weatherability. Processed by reaction injection and structural foam molding, casting, potting, encapsulation, and coating. Used in heat insulation, auto panels and trim, and housings for electronic devices.

notched Izod See *notched Izod impact energy.*

notched Izod impact See *notched Izod impact energy.*

notched Izod impact energy The energy required to break a notched specimen equal to the difference between the energy in the striking member of the Izod-type impact apparatus at the instant of impact and the energy remaining after complete fracture of the specimen. **Note:** Energy depends on geometry (e.g., width, depth, shape) of the notch, on the cross-sectional area of the specimen and on the place of impact (on the side of the notch or on the opposite side). In some tests notch is made on both sides of the specimen. Also called notched Izod impact strength, notched Izod impact, notched Izod.

notched Izod impact strength See *notched Izod impact energy.*

nylon Thermoplastic polyamides often prepared by ring-opening polymerization of lactam. Have good resistance to most chemicals, abrasion, and creep, good impact and tensile strengths, barrier properties, and low friction, but poor resistance to moisture and light. Have high mold shrinkage. Processed by injection, blow, and rotational molding, extrusion, and powder coating. Used in fibers, auto parts, electrical devices, gears, pumps, appliance housings, cable jacketing, pipes, and films.

nylon 11 Thermoplastic polymer of 11-aminoundecanoic acid having good impact strength, hardness, abrasion resistance, processability, and dimensional stability. Processed by powder coating, rotational molding, extrusion, and injection molding. Used in electric insulation, tubing, profiles, bearings, and coatings.

nylon 12 Thermoplastic polymer of lauric lactam having good impact strength, hardness, abrasion resistance, and dimensional stability. Processed by powder coating, rotational molding, extrusion, and injection molding. Used in sporting goods and auto parts.

nylon 46 Thermoplastic copolymer of 2-pyrrolidone and caprolactam.

nylon 6 Thermoplastic polymer of caprolactam. Has good weldability and mechanical properties but rapidly picks up moisture which results in strength losses. Processed by injection, blow, and rotational molding and extrusion. Used in fibers, tire cord, and machine parts.

nylon 610 Thermoplastic polymer of hexamethylenediamine and sebacic acid having decreased melting point and water absorption and good retention of mechanical properties. Processed by injection molding and extrusion. Used in fibers and machine parts.

nylon 612 Thermoplastic polymer of 1,12-dodecanedioic acid and hexamethylenediamine having good dimensional stability, low moisture absorption, and good retention of mechanical properties. Processed by injection molding and extrusion. Used in wire jackets, cable sheath, packaging film, fibers, bushings, and housings.

nylon 66 Thermoplastic polymer of adipic acid and hexamethylenediamine having good tensile strength, elasticity, toughness, heat resistance, abrasion resistance, and solvent resistance but low weatherability and color resistance. Processed by injection molding and extrusion. Used in fibers, bearings, gears, rollers, and wire jackets.

nylon 666 Thermoplastic polymer of adipic acid, caprolactam, and hexamethylenediamine having good strength, toughness, abrasion and fatigue resistance, and low friction but high moisture absorption and low dimensional stability. Processed by injection molding and extrusion. Used in electrical devices and auto and mechanical parts.

nylon MXD6 Thermoplastic polymer of m-xylyleneadipamide having good flexural strength and chemical resistance but decreased tensile strength.

O

olefin resins See *polyolefins.*

olefinic resins See *polyolefins.*

olefinic thermoplastic elastomers Blends of EPDM or EP rubbers with polypropylene or polyethylene, optionally crosslinked. Have low density, good dielectric and mechanical properties, and processibility but low oil resistance and high flammability. Processed by extrusion, injection and blow molding, thermoforming, and calendering. Used in auto parts, construction, wire jackets, and sporting goods. Also called TPO.

optical characteristics See *optical properties.*

optical properties The effects of a material or medium on light or other electromagnetic radiation passing through it, such as absorption, reflection, etc. Also called optical characteristics, optical property.

optical property See *optical properties.*

optical transmittance See *transmittance.*

organic compounds See *halogen compounds.*

organic compounds Chemical compounds based on carbon chains and rings and also containing hydrogen that can be entirely or partially substituted with oxygen, nitrogen and other elements. Also called organic substances.

organic substances See *organic compounds.*

outgassing rate See *degassing rate.*

oxazolines Heterocyclic compounds containing five-membered rings in which one carbon is replaced with an oxygen atom and another with a nitrogen atom. Oxazolines are colorless liquids soluble in organic solvents and water. Used as intermediates, e.g., in synthesis of surfactants.

oxirane See *ethylene oxide.*

P

PA See *polyamides.*

PABM See *polyaminobismaleimide resins.*

PBI See *polybenzimidazoles.*

PBT See *polybutylene terephthalate.*

PC See *polycarbonates.*

PCT See *polycyclohexylenedimethylene terephthalate.*

PCTG See *glycol modified polycyclohexylenedimethylene terephthalate.*

PE copolymer See *polyethylene copolymer.*

PEEK See *polyetheretherketone.*

PEI See *polyetherimides.*

PEK See *polyetherketone.*

percent light transmittance See *transmittance.*

perfluoroalkoxy resins Thermoplastic polymers of perfluoroalkoxyethylenes having good creep, heat, and chemical resistance and processibility but low compressive and tensile strengths. Processed by molding, extrusion, rotational molding, and powder coating. Used in films, coatings, pipes, containers, and chemical apparatus linings. Also called PFA.

PES See *polyethersulfone.*

PET See *polyethylene terephthalate.*

PETG See *polycyclohexylenedimethylene ethylene terephthalate.*

PFA See *perfluoroalkoxy resins.*

phase transition See *phase transition properties.*

phase transition point The temperature at which a phase transition occurs in a physical system such as material. **Note:** An example of phase transition is glass transition. Also called phase transition temperature, transition point, transition temperature.

phase transition properties Properties of physical systems such as materials associated with their transition from one phase to another, e.g., from liquid to solid phase. Also called phase transition.

phase transition temperature See *phase transition point.*

phenolic resins Thermoset polymers of phenols with excess or deficiency of aldehydes, mainly formaldehyde, to give resole or novolak resins, respectively. Heat-cured resins have good dielectric properties, hardness, thermal stability, rigidity, and compressive strength but poor chemical resistance and dark color. Processed by coating, potting, compression, transfer, or injection molding and extrusion. Used in coatings, adhesives, potting compounds, handles, electrical devices, and auto parts.

photo bleaching See *photochemical bleaching.*

photochemical bleaching Complete loss of color of the material as a result of photodegradation of colored substances present in its surface layer. Also called photo bleaching.

photochemical degradation Degradation as a result of light-induced reactions such as photolysis. Also called photodegradation.

photodegradation See *photochemical degradation.*

physical sterilants See *physical sterilization agents.*

physical sterilization agents A physical agent, such as heat, used to kill all microorganisms, including spores, on or in an object. Also called physical sterilants. See also *gamma radiation, electron beam radiation, ultraviolet radiation.*

PI See *polyimides.*

plastics See *polymers.*

PMMA See *polymethyl methacrylate.*

PMP See *polymethylpentene.*

polyacrylates See *acrylic resins.*

polyallomer Crystalline thermoplastic block copolymers of ethylene, propylene, and other olefins. Have good impact strength and flex life and low density.

polyamide thermoplastic elastomers Copolymers containing soft polyether and hard polyamide blocks having good chemical, abrasion, and heat resistance, impact strength, and tensile properties. Processed by extrusion and injection and blow molding. Used in sporting goods, auto parts, and electrical devices. Also called polyamide TPE.

polyamide TPE See *polyamide thermoplastic elastomers.*

polyamides Thermoplastic aromatic or aliphatic polymers of dicarboxylic acids and diamines, of amino acids, or of lactams. Have good mechanical properties, chemical resistance, and antifriction properties. Processed by extrusion and molding. Used in fibers and molded parts. Also called PA.

polyaminobismaleimide resins Thermoset polymers of aromatic diamines and bismaleimides having good flow and thermochemical properties and flame and radiation resistance. Processed by casting and compression molding. Used in aircraft parts and electrical devices. Also called PABM.

polyarylamides Thermoplastic crystalline polymers of aromatic diamines and aromatic dicarboxylic anhydrides having good heat, fire, and chemical resistance, property retention at high temperatures, dielectric and mechanical properties, and stiffness but poor light resistance and processibility. Processed by solution casting, molding, and extrusion. Used in films, fibers, and molded parts.

polyarylsulfone Thermoplastic aromatic polyether-polysulfone having good heat, fire, and chemical resistance, impact strength, resistance to environmental stress cracking, dielectric properties, and rigidity. Processed by injection and compression molding and extrusion. Used in circuit boards, lamp housings, piping, and auto parts.

polybenzimidazoles Mainly polymers of 3,3',4,4'-tetraminonbiphenyl(diaminobenzidine) and diphenyl isophthalate. Have good heat, fire, and chemical resistance. Used as coatings and fibers in aerospace and other high-temperature applications. Also called PBI.

polybutylene terephthalate Thermoplastic polymer of dimethyl terephthalate and butanediol having good tensile strength, dielectric properties, and chemical and water resistance, but poor impact strength and heat resistance. Processed by injection and blow molding, extrusion, and thermoforming. Used in auto body parts, electrical devices, appliances, and housings. Also called PBT.

polycarbonate See *polycarbonates.*

polycarbonate polyester alloys High-performance thermoplastics processed by injection and blow molding. Used in auto parts.

polycarbonate resins See *polycarbonates.*

polycarbonates Polycarbonates are thermoplastics prepared by either phosgenation of dihydric aromatic alcohols such as bisphenol A or by transesterification of these alcohols with carbonates, e.g., diphenyl carbonate. Polycarbonates consist of chains with repeating carbonyldioxy groups and can be aliphatic or aromatic. They have very good mechanical properties, especially impact strength, low moisture absorption and good thermal and oxidative stability. They are self-extinguishing and some grades are transparent. Polycarbonates have relatively low chemical resistance and resistance to stress cracking. Processed by injection and blow molding, extrusion, thermoforming at relatively high processing temperatures. Used in telephone parts, dentures, business machine housings, safety equipment, nonstaining dinnerware, food packaging, etc. Also called polycarbonate, PC, polycarbonate resins.

polychlorotrifluoroethylene Thermoplastic polymer of chlorotrifluoroethylene having good transparency, barrier properties, tensile strength, and creep resistance, modest dielectric properties and solvent resistance, and poor processibility. Processed by extrusion, injection and compression molding, and coating. Used in chemical apparatus, low-temperature seals, films, and internal lubricants. Also called CTFE.

polycyclohexylenedimethylene ethylene terephthalate Thermoplastic polymer of cyclohexylenedimethylenediol, ethylene glycol, and terephthalic acid. Has good clarity, stiffness, hardness, and low-temperature toughness. Processed by injection and blow molding and extrusion. Used in containers for cosmetics and foods, packaging film, medical devices, machine guards, and toys. Also called PETG.

polycyclohexylenedimethylene terephthalate Thermoplastic polymer of cyclohexylenedimethylenediol and terephthalic acid having good heat resistance. Processed by molding and extrusion. Also called PCT.

polydispersity See *molecular weight distribution.*

polyester resins See *polyesters.*

polyester thermoplastic elastomers Copolymers containing soft polyether and hard polyester blocks having good dielectric strength, chemical and creep resistance, dynamic performance, appearance, and retention of properties in a wide temperature range but poor light resistance. Processed by injection, blow, and rotational molding, extrusion casting, and film blowing. Used in electrical insulation, medical products, auto parts, and business equipment. Also called polyester TPE.

polyester TPE See *polyester thermoplastic elastomers.*

polyesters A broad class of polymers usually made by condensation of a diol with dicarboxylic acid or anhydride. Polyesters consist of chains with repeating carbonyloxy group and can be aliphatic or aromatic. There are thermosetting polyesters, such as alkyd resins and unsaturated polyesters, and thermoplastic polyesters such as PET. The properties, processing methods and applications of polyesters vary widely. Also called polyester resins.

polyetheretherketone Semi-crystalline thermoplastic aromatic polymer having good chemical, heat, fire, and radiation resistance, toughness, rigidity, bearing strength, and processibility. Processed by injection molding, spinning, cold forming, and extrusion. Used in fibers, films, auto engine parts, aerospace composites, and electrical insulation. Also called PEEK.

polyetherimides Thermoplastic cyclized polymers of aromatic diether dianhydrides and aromatic diamine. Have good chemical, creep, and heat resistance and dielectric properties. Processed by extrusion, thermoforming, and compression, injection, and blow molding. Used in auto parts, jet engines, surgical instruments, industrial apparatus, food packaging, cookware, and computer disks. Also called PEI.

polyetherketone Thermoplastic having good heat and chemical resistance. thermal stability. Used in advanced composites, wire coating, filters, integrated circuit boards, and bearings. Also called PEK.

polyethersulfone Thermoplastic aromatic polymer having good heat and fire resistance, transparency, dielectric properties, dimensional stability, rigidity, and toughness, but poor solvent and stress cracking resistance, processibility, and weatherability. Processed by injection, blow, and compression molding and extrusion. Used in high temperature applications electrical devices, medical devices, housings, and aircraft and auto parts. Also called PES.

polyethylene copolymer Thermoplastics polymers of ethylene with other olefins such as propylene. Processed by molding and extrusion. Also called PE copolymer.

polyethylene terephthalate Thermoplastic polymer of ethylene glycol with terephthalic acid. Has good hardness, wear and chemical resistance, dimensional stability, and dielectric properties. High-crystallinity grades have good tensile strength and heat resistance. Processed by extrusion and injection and blow molding. Used in fibers, food packaging (films, bottles, trays), magnetic tapes, and photo films. Also called PET.

polyimides Thermoplastic aromatic cyclized polymers of trimellitic anhydride and aromatic diamine. Have good tensile strength, dimensional stability, dielectric and barrier properties, and creep, impact, heat, and fire resistance, but poor processibility. Processed by compression and injection molding, powder sintering, film casting, and solution coating. Thermoset uncyclized polymers are heat curable and have good processability. Processed by transfer and injection molding, lamination, and coating. Used in jet engines, compressors, sealing coatings, auto parts, and business machines. Also called PI.

polymers Polymers are high-molecular-weight organic or inorganic compounds the molecules of which comprise linear, branched, crosslinked or otherwise shaped chains of repeating molecular groups. Synthetic polymers are prepared by polymerization of one or more monomers. The monomers are low-molecular-weight substances with one or more reactive bonds or functional groups. Also called resins, plastics.

polymethyl methacrylate Thermoplastic polymer of methyl methacrylate having good transparency, weatherability, impact strength, and dielectric properties. Processed by compression and injection molding, casting, and extrusion. Used in lenses, sheets, airplane canopies, signs, and lighting fixtures. Also called PMMA.

polymethylpentene Thermoplastic polymer of 4-methyl-1-pentene having low density, good transparency, rigidity, dielectric and tensile properties, and heat and chemical resistance. Processed by injection and blow molding and extrusion. Used in laboratory ware, coated paper, light fixtures, auto parts, and electrical insulation. Also called PMP.

polyolefin resins See *polyolefins.*

polyolefins Polyolefins are a broad class of hydrocarbon-chain elastomers or thermoplastics usually prepared by addition (co)polymerization of alkenes such as ethylene. There are branched and linear polyolefins and some are chemically or physically modified. Unmodified polyolefins have relatively low thermal stability and a nonporous, nonpolar surface with poor adhesive properties. Processed by extrusion, injection molding, blow molding and rotational molding. Polyolefins are used more and have more applications than any other polymers. Also called olefinic resins, olefin resins, polyolefin resins.

polyphenylene ether nylon alloys Thermoplastics having improved heat and chemical resistance and toughness. Processed by molding and extrusion. Used in auto body parts.

polyphenylene sulfide High-performance engineering thermoplastic having good chemical, water, fire, and radiation resistance, dimensional stability, and dielectric properties, but decreased impact strength and poor processibility. Processed by injection, compression, and transfer molding and extrusion. Used in hydraulic components, bearings, electronic parts, appliances, and auto parts. Also called PPS.

polyphenylene sulfide sulfone Thermoplastic having good heat, fire, creep, and chemical resistance and dielectric properties. Processed by injection molding. Used in electrical devices. Also called PPSS.

polyphthalamide Thermoplastic polymer of aromatic diamine and phthalic anhydride. Has good heat, chemical, and fire resistance, impact strength, retention of properties at high temperatures, dielectric properties, and stiffness, but decreased light resistance and poor processibility. Processed by solution casting, molding, and extrusion. Used in films, fibers, and molded parts. Also called PPA.

polypropylene Thermoplastic polymer of propylene having low density and good flexibility and resistance to chemicals, abrasion, moisture, and stress cracking, but decreased dimensional stability, mechanical strength, and light, fire, and heat resistance. Processed by injection molding, spinning, and extrusion. Used in fibers and films for adhesive tapes and packaging. Also called PP.

polystyrene Polystyrenes are thermoplastics produced by polymerization of styrene with or without modification (e.g., by copolymerization or blending) to make impact resistant or expandable grades. They have good rigidity, high dimensional stability, low moisture absorption, optical clarity, high gloss and good dielectric properties. Unmodified polystyrenes have poor impact strength and resistance to solvents, heat and UV radiation. Processed by injection molding, extrusion, compression molding, and foam molding. Used widely in medical devices, housewares, food packaging, electronics and foam insulation. Also called polystyrenes, PS, polystyrol.

polystyrenes See *polystyrene.*

polystyrol See *polystyrene.*

polysulfones Thermoplastics, often aromatic and with ether linkages, having good heat, fire, and creep resistance, dielectric properties, transparency, but poor weatherability, processibility, and stress cracking resistance. Processed by injection, compression, and blow molding and extrusion. Used in appliances, electronic devices, auto parts, and electric insulators. Also called PSO.

polytetrafluoroethylene Thermoplastic polymer of tetrafluoroethylene having good dielectric properties, chemical, heat, abrasion, and fire resistance, antiadhesive properties, impact strength, and weatherability, but decreased strength, processibility, barrier properties, and creep resistance. Processed by sinter molding and powder coating. Used in nonstick coatings, chemical apparatus, electrical devices, bearings, and containers. Also called PTFE.

polyurethane resins See *polyurethanes.*

polyurethanes Polyurethanes (PUs) are a broad class of polymers consisting of chains with a repeating urethane group, prepared by condensation of polyisocyanates with polyols, e.g., polyester or polyether diols. PUs may be thermoplastic or thermosetting, elastomeric or rigid, cellular or solid, and offer a wide range of properties depending on composition and molecular structure. Many PUs have high abrasion resistance, good retention of properties at low temperatures and good foamability. Some have poor heat resistance, weatherability and resistance to solvents. PUs are flammable and can release toxic substances. Thermoplastic PUs are not crosslinked and are processed by injection molding and extrusion. Thermosetting PUs can be cured at relatively low temperatures and give foams with good heat insulating properties. They are processed by reaction injection molding, rigid and flexible foam methods, casting and coating. PUs are used in load bearing rollers and wheels, acoustic clamping materials, sporting goods, seals and gaskets, heat insulation, potting and encapsulation. Also called PUR, PU, urethane polymers, urethane resins, urethanes, polyurethane resins.

polyvinyl chloride Thermoplastic polymer of vinyl chloride, available in rigid and flexible forms. Has good dimensional stability, fire resistance, and weatherability, but decreased heat and solvent resistance and high density. Processed by injection and blow molding, calendering, extrusion, and powder coating. Used in films, fabric coatings, wire insulation, toys, bottles, and pipes. Also called PVC.

polyvinyl fluoride Crystalline thermoplastic polymer of vinyl fluoride having good toughness, flexibility, weatherability, and low-temperature and abrasion resistance. Processed by film techniques. Used in packaging, glazing, and electrical devices. Also called PVF.

polyvinylidene chloride Stereoregular thermoplastic polymer of vinylidene chloride having good abrasion and chemical resistance and barrier properties. Processed by molding and extrusion. Used in food packaging films, bag liners, pipes, upholstery, fibers, and coatings. Also called PVDC.

polyvinylidene fluoride Thermoplastic polymer of vinylidene fluoride having good strength, processibility, wear, fire, solvent, and creep resistance, and weatherability, but decreased dielectric properties and heat resistance. Processed by extrusion, injection and transfer molding, and powder coating. Used in electrical insulation, pipes, chemical apparatus, coatings, films, containers, and fibers. Also called PVDF.

post evacuation pressure Negative pressure (vacuum) in an apparatus after evacuation (removal) of gases, e.g., pressure in a sterilizer after evacuation of gaseous sterilant. See also *sterilization pressure*.

PP See *polypropylene*.

PPA See *polyphthalamide*.

ppm A unit for measuring small concentrations of material or substance as the number of its parts (arbitrary quantity) per million parts of medium consisting of another material or substance.

PPS See *polyphenylene sulfide*.

PPSS See *polyphenylene sulfide sulfone*.

pre-vacuum conditioning Process of bringing the material or apparatus to a certain condition, e.g., moisture content or temperature, prior to evacuation (creating vacuum) of the apparatus. Also called pre-vacuum conditioning cycle. See also *sterilization*.

pre-vacuum conditioning cycle See *pre-vacuum conditioning*.

pressure Stress exerted equally in all directions. See also *processing pressure*.

process characteristics See *processing parameters*.

process conditions See *processing parameters*.

process media See *processing agents*.

process parameters See *processing parameters*.

process pressure See *processing pressure*.

process rate See *processing rate*.

process speed See *processing rate*.

process time See *processing time*.

process velocity See *processing rate*.

processing additives See *processing agents*.

processing agents Agents or media used in the manufacture, preparation and treatment of a material or article to improve its processing or properties. The agents often become a part of the material. Also called process media, processing aids, processing additives.

processing aids See *processing agents*.

processing defects Structural and other defects in material or article caused inadvertently during manufacturing, preparation and treatment processes by using wrong tooling, process parameters, ingredients, part design, etc. Usually preventable. Also called processing flaw, defects, flaw. See also *cracking*.

processing flaw See *processing defects*.

processing methods Method names and designations for material or article manufacturing, preparation and treatment processes. **Note:** Both common and standardized names are used. Also called processing procedures.

processing parameters Measurable parameters such as temperature prescribed or maintained during material or article manufacture, preparation and treatment processes. Also called process characteristics, process conditions, process parameters.

processing pressure Pressure maintained in an apparatus during material or article manufacture, preparation and treatment processes. Also called process pressure. See also *pressure*.

processing procedures See *processing methods*.

processing rate Speed of the process in manufacture, preparation and treatment of a material or article. It usually denotes the change in a process parameter per unit of time or the throughput speed of material in a unit of weight, volume, etc. per unit of time. Also called process speed, process velocity, process rate.

processing time See *processing time*.

processing time Time required for the completion of a process in the manufacture, preparation and treatment of a material or article. Also called processing time, cycle time, process time. See also *time*.

PS See *polystyrene*.

PSO See *polysulfones*.

PTFE See *polytetrafluoroethylene*.

PU See *polyurethanes*.

PUR See *polyurethanes*.

PVC See *polyvinyl chloride*.

PVDC See *polyvinylidene chloride*.

PVDF See *polyvinylidene fluoride*.

PVF See *polyvinyl fluoride*.

R

radiation dosage See *radiation dose*.

radiation dose Amount of ionizing radiation energy received or absorbed by the material during exposure. Also called radiation dosage, ionizing radiation dose, ionizing radiation dosage. See also *radiation dose units*.

radiation dose units The units for measuring the amount of ionizing radiation energy received or absorbed by the material during the exposure. **Note:** The radiation dose units are usually expressed in the units of radiation energy (e.g., joule) per one unit of mass (e.g., kilogram) of material such as gray, rad and rem. Other units include sievert and rontgen. See also *radiation dose.*

radiation resistant materials Materials that resist degradation on long- and medium-term or repeated exposure to ionizing radiation, e.g., steel grades designed for nuclear reactors. Radiation damage to materials includes swelling, radiolysis, blistering, changes in electrical and mechanical properties, etc. There are different mechanisms of radiation damage but most can be linked to free-radical reactions. The resistance of materials to radiation can be improved by stabilizing them with agents that can neutralize free radicals, such as dimethyl sulfoxide, carbohydrates and various reducing agents. Also called radiation stabilized material.

radiation stabilized material See *radiation resistant materials.*

radioactive isotopes See *radioisotopes.*

radioactive materials Materials that have at least one ingredient emitting ionizing radiation, e.g., due to radioactive decay as in unstable isotopes.

radioisotopes Unstable isotopes that emit ionizing radiation due to radioactive decay. Also called radioactive isotopes, unstable isotopes. See also *ionizing radiation.*

rate of residual ethylene oxide dissipation The rate of dissipation of residual ethylene oxide trapped in the material that had been exposed to it, e.g., after ethylene oxide evacuation in sterilization. See also *ethylene oxide, sterilization rate.*

reaction injection molding system Liquid compositions, mostly polyurethane-based, of thermosetting resins, prepolymers, monomers, or their mixtures. Have good processibility, dimensional stability, and flexibility. Processed by foam molding with in-mold curing at high temperatures. Used in auto parts and office furniture. Also called RIM.

residence time See *dwell time.*

resins See *polymers.*

resorcinol modified phenolic resins Thermosetting polymers of phenol, formaldehyde, and resorcinol having good heat and creep resistance and dimensional stability.

rigid thermoplastic polyurethanes Rigid thermoplastic polyurethanes are not chemically crosslinked. They have high abrasion resistance, good retention of properties at low temperatures, but poor heat resistance, weatherability and resistance to solvents. Rigid thermoplastic polyurethanes are flammable and can release toxic substances. Processed by injection molding and extrusion. Also called rigid thermoplastic urethanes, nonelastomeric thermoplastic polyurethanes.

rigid thermoplastic urethanes See *rigid thermoplastic polyurethanes.*

RIM See *reaction injection molding system.*

S

SAN See *styrene acrylonitrile copolymer.*

SAN copolymer See *styrene acrylonitrile copolymer.*

SAN resin See *styrene acrylonitrile copolymer.*

shelf life Time during which a physical system, such as a material, retains its storage stability under specified conditions. Also called storage life.

silicone There are rigid thermoplastic and liquid silicones and silicone rubbers consisting of alternating silicone and oxygen atom chains with organic pendant groups, prepared by hydrolytic polymcondensation of chlorosilanes, followed by crosslinking. Silicone rubbers have good adhesion, flexibility, dielectric properties, weatherability, barrier properties, and heat and fire resistance, but decreased strength. Rigid silicones have good flexibility, weatherability, soil repelling properties, dimensional stability, but poor solvent resistance. Processed by coating, casting, and injection compression, and transfer molding. Used in coatings, electronic devises, diaphragms, medical products, adhesives, and sealants. Also called siloxane.

siloxane See *silicone.*

silver streaking See *silver streaks.*

silver streaks Scars or surface defects on injection moldings caused by the high velocity injection of a stream of molten material into the mold ahead of the normally advancing material front and its premature solidification. Also similar appearance defects resulting from exposure or stress. Also called silver streaking, splay marks.

SMA See *styrene maleic anhydride copolymer.*

SMA PTB alloy See *styrene maleic anhydride copolymer PBT alloy.*

softening point Temperature at which the material changes from rigid to soft or exhibits a sudden and substantial decrease in hardness. Also called softening temperature, softening range.

softening range See *softening point.*

softening temperature See *softening point.*

splay marks See *silver streaks.*

stability The ability of a physical system, such as a material, to resist a change or degradation under exposure to outside forces, including mechanical force, heat and weather. See also *degradation.*

starch modified low density polyethylene Biodegradable thermoplastic starch-grafted low-density polyethylene.

starch modified polypropylene Biodegradable thermoplastic starch-grafted polypropylene.

starch modified polyurethane Biodegradable thermoplastic starch-grafted polyurethane.

steam sterilization See *autoclave sterilization.*

sterilants See *sterilization agents.*

sterilization Process of killing of all microorganisms, including spores, on or in an object by chemical or physical means. **Note:** In microbiology and bioengineering.

sterilization agents A chemical or physical agent used to kill all microorganisms, including spores, on or in an object. Also called sterilants.

sterilization cycle time See *sterilization time.*

sterilization pressure Pressure maintained in a sterilizer such as an autoclave during sterilization cycle. See also *post evacuation pressure*.

sterilization rate The rate of killing of all microorganisms, including spores, on or in an object by chemical or physical means. See also *rate of residual ethylene oxide dissipation*.

sterilization time Time required for sterilization cycle to complete. **Note:** Includes sterilization only and not time required for pre- and post-processing (conditioning, evacuation, etc.) times Also called sterilization cycle time. See also *dwell time*.

storage life See *shelf life*.

storage stability The resistance of a physical system, such as a material, to decomposition, deterioration of properties or any type of degradation in storage under specified conditions.

strain The per unit change, due to force, in the size or shape of a body referred to its original size or shape. **Note:** Strain is nondimensional but is often expressed in unit of length per unit of length or percent.

stress cracking Appearance of external and/or internal cracks in the material as a result of stress that is lower than its short-term strength.

styrene acrylonitrile copolymer SAN resins are thermoplastic copolymers of about 70% styrene and 30% acrylonitrile with higher strength, rigidity and chemical resistance than polystyrene. Characterized by transparency, high heat deflection properties, excellent gloss, hardness and dimensional stability. Have low continuous service temperature and impact strength. Processed by injection molding, extrusion, injection-blow molding and compression molding. Used in appliances, housewares, instrument lenses for automobiles, medical devices, and electronics. Also called styrene-acrylonitrile copolymer, SAN, SAN resin, SAN copolymer.

styrene butadiene block copolymer Thermoplastic amorphous block polymer of butadiene and styrene having good impact strength, rigidity, gloss, compatibility with other styrenic resins, water resistance, and processibility. Used in food and display containers, toys, and shrink wrap.

styrene butadiene copolymer Thermoplastic polymers of butadiene and >50% styrene having good transparency, toughness, and processibility. Processed by extrusion, injection and blow molding, and thermoforming. Used in film wraps, disposable packaging, medical devices, toys, display racks, and office supplies.

styrene maleic anhydride copolymer Thermoplastic copolymer of styrene with maleic anhydride having good thermal stability and adhesion, but decreased chemical and light resistance. Processed by injection and foam molding and extrusion. Used in auto parts, appliances, door panels, pumps, and business machines. Also called SMA.

styrene maleic anhydride copolymer PBT alloy Thermoplastic alloy of styrene maleic anhydride copolymer and polybutylene terephthalate having improved dimensional stability and tensile strength. Processed by injection molding. Also called SMA PTB alloy.

styrene plastics See *styrenic resins*.

styrene polymers See *styrenic resins*.

styrene resins See *styrenic resins*.

styrene-acrylonitrile copolymer See *styrene acrylonitrile copolymer*.

styrenic resins Styrenic resins are thermoplastics prepared by free-radical polymerization of styrene alone or with other unsaturated monomers. The properties of styrenic resins vary widely with molecular structure, attaining the high performance level of engineering plastics. Processed by blow and injection molding, extrusion, thermoforming, film techniques and structural foam molding. Used heavily for the manufacture of automotive parts, household goods, packaging, films, tools, containers and pipes. Also called styrene resins, styrene polymers, styrene plastics.

styrenic thermoplastic elastomers Linear or branched copolymers containing polystyrene end blocks and elastomer (e.g., isoprene rubber) middle blocks. Have a wide range of hardnesses, tensile strength, and elongation, and good low-temperature flexibility, dielectric properties, and hydrolytic stability. Processed by injection and blow molding and extrusion. Used in coatings, sealants, impact modifiers, shoe soles, medical devices, tubing, electrical insulation, and auto parts. Also called TES.

T

temperature Property which determines the direction of heat flow between objects. **Note:** The heat flows from the object with higher temperature to that with lower.

tensile elongation See *elongation*.

tensile heat distortion temperature See *heat deflection temperature*.

tensile properties Properties describing the reaction of physical systems to tensile stress and strain. See also *tensile property tests*.

tensile property tests Names and designations of the methods for tensile testing of materials. Also called tensile tests. See also *tensile properties*.

tensile strain The relative length deformation exhibited by a specimen in tension. See also *elongation*.

tensile strength The maximum tensile stress that a specimen can sustain in a test carried to failure. **Note:** The maximum stress can be measured at or after the failure or reached before the fracture, depending on the viscoelastic behavior of the material. Also called tensile ultimate strength, ultimate tensile strength, UTS, tensile strength at break, ultimate tensile stress. See also *ASTM D638*.

tensile strength at break See *tensile strength*.

tensile stress The stress is perpendicular and directed to the opposite plane on which the forces act.

tensile tests See *tensile property tests*.

tensile ultimate strength See *tensile strength*.

terephthalate polyester Thermoset unsaturated polymer of terephthalic anhydride.

TES See *styrenic thermoplastic elastomers*.

test methods Names and designations of material test methods. Also called testing methods.

test variables Terms related to the testing of materials such as test method names.

tetrafluoroethylene propylene copolymer Thermosetting elastomeric polymer of tetrafluoroethylene and propylene having good chemical and heat resistance and flexibility. Used in auto parts.

tetrahydro-1,4-oxazine See *morpholine.*

thermal properties Properties related to the effects of heat on physical systems such as materials and heat transport. The effects of heat include the effects on structure, geometry, performance, aging, stress-strain behavior, etc.

thermal stability The resistance of a physical system, such as a material, to decomposition, deterioration of properties or any type of degradation in storage under specified conditions.

thermodynamic properties A quantity that is either an attribute of the entire system or is a function of position, which is continuous and does not vary rapidly over microscopic distances, except possibility for abrupt changes at boundaries between phases of the system. Also called macroscopic properties.

thermoplastic polyesters A class of polyesters that can be repeatedly made soft and pliable on heating and hard (flexible or rigid) on subsequent cooling.

thermoplastic polyurethanes A class of polyurethanes including rigid and elastomeric polymers that can be repeatedly made soft and pliable on heating and hard (flexible or rigid) on subsequent cooling. Also called thermoplastic urethanes, TPUR, TPU.

thermoplastic urethanes See *thermoplastic polyurethanes.*

three-membered heterocyclic compounds A class of heterocyclic compounds containing rings that consist of three atoms.

three-membered heterocyclic oxygen compounds A class of heterocyclic compounds containing rings that consist of three atoms, one or two of which is an oxygen.

time One of basic dimensions of the universe designating the duration and order of events at a given place. See also *processing time.*

tint See *color.*

toughness Property of a material indicating its ability to absorb energy by plastic deformation rather than crack or fracture.

TPO See *olefinic thermoplastic elastomers.*

TPU See *urethane thermoplastic elastomers.*

TPU See *thermoplastic polyurethanes.*

TPUR See *thermoplastic polyurethanes.*

transition point See *phase transition point.*

transition temperature See *phase transition point.*

transmittance The ratio of the light intensity transmitted by a body to the incident light intensity. Also called percent light transmittance, light transmission, luminous transmittance, optical transmittance, transmittancy, transparency, transparence.

transmittancy See *transmittance.*

transparence See *transmittance.*

transparency See *transmittance.*

U

UHMWPE See *ultrahigh molecular weight polyethylene.*

ultimate elongation See *elongation.*

ultimate tensile strength See *tensile strength.*

ultimate tensile stress See *tensile strength.*

ultrahigh molecular weight polyethylene Thermoplastic linear polymer of ethylene with molecular weight in the millions. Has good wear and chemical resistance, toughness, and antifriction properties, but poor processibility. Processed by compression molding and ram extrusion. Used in bearings, gears, and sliding surfaces. Also called UHMWPE.

ultraviolet light See *ultraviolet radiation.*

ultraviolet radiation Electromagnetic radiation in the wavelength range 13-400 nm below the short-wavelength limit of the visible light. **Note:** UV light comprises a significant portion of the natural sun light. Also called ultraviolet light, UV light, UV radiation. See also *physical sterilization agents.*

units See *units of measurement.*

units of measurement Systematic and non-systematic units for measuring physical quantities, including metric and US pound-inch systems. Also called units.

unstable isotopes See *radioisotopes.*

urea resins Thermosetting polymers of formaldehyde and urea having good clarity, colorability, scratch, fire, and solvent resistance, rigidity, dielectric properties, and tensile strength, but decreased impact strength and chemical, heat, and moisture resistance. Must be filled for molding. Processed by compression and injection molding, impregnation, and coating. Used in cosmetic containers, housings, tableware, electrical insulators, countertop laminates, adhesives, and coatings.

urethane polymers See *polyurethanes.*

urethane resins See *polyurethanes.*

urethane thermoplastic elastomers Block polyether or polyester polyurethanes containing soft and hard segments. Have good tensile strength, elongation, adhesion, and a broad hardness and service temperature ranges, but decreased moisture resistance and processibility. Processed by extrusion, injection molding, film blowing, and coating. Used in tubing, packaging film, adhesives, medical devices, conveyor belts, auto parts, and cable jackets. Also called TPU.

urethanes See *polyurethanes.*

UTS See *tensile strength.*

UV light See *ultraviolet radiation.*

UV radiation See *ultraviolet radiation.*

V

Vicat softening point The temperature at which a flat-ended needle of prescribed geometry will penetrate a thermoplastic specimen to a certain depth under a specified load using a uniform rate of temperature rise. **Note:** Vicat softening point is determined according to ASTM D1525 test for thermoplastics such as polyethylene which have no definite melting point. Also called Vicat softening temperature.

Vicat softening temperature See *Vicat softening point.*

vinyl ester resins Thermosetting acrylated epoxy resins containing styrene reactive diluent. Cured by catalyzed polymerization of vinyl groups and crosslinking of hydroxy groups at room or elevated temperatures. Have good chemical, solvent, and heat resistance, toughness, and flexibility, but shrink during cure. Processed by filament winding, transfer molding, pultrusion, coating, and lamination. Used in structural composites, coatings, sheet molding compounds, and chemical apparatus.

vinyl resins Thermoplastics polymers of vinyl compounds such as vinyl chloride or vinyl acetate. Have good weatherability, barrier properties, and flexibility, but decreased solvent and heat resistance. Processed by molding, extrusion, and coating. Used in films and packaging.

vinyl thermoplastic elastomers Vinyl resin alloys having good fire and aging resistance, flexibility, dielectric properties, and toughness. Processed by extrusion. Used in cable jackets and wire insulation.

vinylidene fluoride hexafluoropropylene copolymer Thermoplastic polymer of vinylidene fluoride and hexafluoropropylene having good antistick, dielectric, and antifriction properties and chemical and heat resistance, but decreased mechanical strength and creep resistance and poor processibility. Processed by molding, extrusion, and coating. Used in chemical apparatus, containers, films, and coatings.

vinylidene fluoride hexafluoropropylene tetrafluoroethylene terpolymer Thermosetting elastomeric polymer of vinylidene fluoride, hexafluoropropylene, and tetrafluoroethylene having good chemical and heat resistance and flexibility. Used in auto parts.

W

weight The gravitational force with which the earth attracts a body.

Y

yellowing Developing of yellow color in the material due to chemical or physical changes.

yellowness index A measure of the tendency of materials such as plastics to become yellow as a result of: long-term exposure to light, irradiation, etc.

(a) name: melt flow rate

(b) name: 100% modulus

(c) name: 100% modulus; test method: ASTM D 638-82; specimen type: injection molded slab; specimen thickness: 2mm

(d) name: 100% modulus; test method: ASTM D412, Die C

(e) name: 300% modulus

(f) name: crystalline melting temperature; specimen mass: 10.5 mg; test apparatus: Perkin-Elmer DSC-7 calorimeter

(g) name: deflection at failure; test method: flex to failure test; test note: extension of ASTM D790-86, customized test

(h) name: deflection at failure; test method: flex to failure test; test note: extension of ASTM D790-86, customized test; value note: extrapolated

(i) name: deflection at peak load; test method: flex to failure test; test note: extension of ASTM D790-86, customized test

(j) name: deflection temperature

(k) name: deflection temperature; load: 1.82 MPa

(l) name: deflection temperature; test method: ASTM D648; load: 1.82 MPa

(m) name: deflection temperature; test method: ASTM D648; specimen thickness: 3.2 mm; conditioning: unannealed

(n) name: Δa color change; test apparatus: MacBeth Color-Eye; note: standard white color plate behind specimen

(o) name: Δa color change; test method: ASTM D2244; test apparatus: Macveth 1500 series colorimeter; specimen thickness: 3.2 mm

(p) name: Δa color; test method: ASTM D2244; type: CDM; specimen thickness: 2 mm

(q) name: Δa color; type: CDM; specimen thickness: 3.2 mm

(r) name: Δa color; type: CIELAB; specimen thickness: 3.8 mm

(s) name: Δb color; test apparatus: MacBeth Color-Eye; note: standard white color plate behind specimen

(t) name: Δb color; test method: ASTM D2244; test apparatus: Macbeth 1500 series colorimeter; specimen thickness: 3.2 mm

(u) name: Δb color; test method: ASTM D2244; type: CDM; specimen thickness: 2 mm

(v) name: Δb color; test method: CIE

(w) name: Δb color; type: CDM; specimen thickness: 1.6 mm

(x) name: Δb color; type: CDM; specimen thickness: 3.2 mm

(y) name: Δb color; type: CIELAB; specimen thickness: 3.8 mm

(z) name: Δb color; test apparatus: Hunter colorimeter

(aa) name: ΔE color change; test method: CIE

(ab) name: ΔL color change; test apparatus: MacBeth Color-Eye; note: standard white color plate behind specimen

(ac) name: ΔL color change; test method: ASTM D2244; test apparatus: Macveth 1500 series colorimeter; specimen thickness: 3.2 mm

(ad) name: ΔL color; type: CIELAB; specimen thickness: 3.8 mm

(ae) name: Elmendorf tear resistance

(af) name: Elmendorf tear resistance (machine direction)

(ag) name: Elmendorf tear resistance (transverse direction)

(ah) name: Elmendorf tear resistance; specimen thickness: 0.254 mm

(ai) name: elongation

(aj) name: elongation (machine direction); test method: ASTM D882; specimen thickness: 0.127 mm

(ak) name: elongation @ break

(al) name: elongation @ break (machine direction)

(am) name: elongation @ break (machine direction); specimen thickness: 0.015 mm

(an) name: elongation @ break (machine direction); specimen thickness: 0.02 mm

(ao) name: elongation @ break (transverse direction); specimen thickness: 0.015 mm

(ap) name: elongation @ break (transverse direction); specimen thickness: 0.02 mm

(aq) name: elongation @ break; strain rate: 50.8 mm/min.

(ar) name: elongation @ break; test method: ASTM D412, Die C

(as) name: elongation @ break; test method: ASTM D638

(at) name: elongation @ break; test method: ASTM D638-82; specimen type: injection molded slab; specimen thickness: 2 mm

(au) name: elongation @ break; test method: ASTM D638-84; specimen type: ASTM Type I tensile bar; specimen size: 165 mm x 3.2 mm x 12.7 mm; standard deviation: 11%

(av) name: elongation @ break; test method: ASTM D638-84; specimen type: ASTM Type I tensile bar; specimen size: 165 mm x 3.2 mm x 12.7 mm; standard deviation: 12%

(aw) name: elongation @ break; test method: ASTM D638-84; specimen type: ASTM Type I tensile bar; specimen size: 165 mm x 3.2 mm x 12.7 mm; standard deviation: 26%

(ax) name: elongation @ break; test method: ASTM D638-84; specimen type: ASTM Type I tensile bar; specimen size: 165 mm x 3.2 mm x 12.7 mm; standard deviation: 4%

(ay) name: elongation @ break; test method: ASTM D638-84; specimen type: ASTM Type I tensile bar; specimen size: 165 mm x 3.2 mm x 12.7 mm; standard deviation: 7%

(az) name: elongation @ break; test method: ASTM D638-84; specimen type: ASTM Type I tensile bar; specimen size: 165 mm x 3.2 mm x 12.7 mm; standard deviation: 70%

(ba) name: elongation @ break; test method: ASTM D638-87b; specimen type: injection molded; strain rate: 50.8 cm/min

(bb) name: elongation @ break; test method: ASTM D638; strain rate: 50 mm/min

(bc) name: elongation @ break; test method: ASTM D638; temperature: 23°C; RH: 50%

(bd) name: elongation @ yield

(be) name: elongation @ yield; test method: ASTM D638

(bf) name: elongation @ yield; test method: ASTM D638; strain rate: 50 mm/min

(bg) name: elongation @ yield; test method: ASTM D638; temperature: 23°C; RH: 50%

(bh) name: elongation; temperature: 28°C

(bi) name: elongation; temperature: 93°C

(bj) name: elongation; test method: ASTM D1708; specimen type: film; specimen thickness: 0.25 mm

(bk) name: elongation; test method: ASTM D638-84; specimen type: ASTM Type I tensile bar; specimen size: 165 mm x 3.2 mm x 12.7 mm; standard deviation: 1%

(bl) name: elongation; test method: ASTM D638-84; specimen type: ASTM Type I tensile bar; specimen size: 165 mm x 3.2 mm x 12.7 mm; standard deviation: 10%

(bm) name: elongation; test method: ASTM D638-84; specimen type: ASTM Type I tensile bar; specimen size: 165 mm x 3.2 mm x 12.7 mm; standard deviation: 2%

(bn) name: elongation; test method: ASTM D638-84; specimen type: ASTM Type I tensile bar; specimen size: 165 mm x 3.2 mm x 12.7 mm; standard deviation: 5%

(bo) name: elongation; test method: ASTM D638; specimen type: ASTM type 1 tensile bar; specimen size: 165 mm x 12.7 mm x 3.2 mm

(bp) name: elongation; test method: ASTM D638; specimen type: ASTM type 1 tensile bar; specimen size: 165 mm x 12.7 mm x 3.2 mm; standard deviation: 12%

(bq) name: elongation; test method: ASTM D638; specimen type: ASTM type 1 tensile bar; specimen size: 165 mm x 12.7 mm x 3.2 mm; standard deviation: 13%

(br) name: elongation; test method: ASTM D638; specimen type: ASTM type 1 tensile bar; specimen size: 165 mm x 12.7 mm x 3.2 mm; standard deviation: 19%

(bs) name: elongation; test method: ASTM D638; specimen type: ASTM type 1 tensile bar; specimen size: 165 mm x 12.7 mm x 3.2 mm; standard deviation: 32%

(bt) name: elongation; test method: ASTM D638; specimen type: ASTM type 1 tensile bar; specimen size: 165 mm x 12.7 mm x 3.2 mm; standard deviation: 33%

(bu) name: elongation; test method: ASTM D638; specimen type: ASTM type 1 tensile bar; specimen size: 165 mm x 12.7 mm x 3.2 mm; standard deviation: 4%

(bv) name: elongation; test method: ASTM D638; specimen type: ASTM type 1 tensile bar; specimen size: 165 mm x 12.7 mm x 3.2 mm; standard deviation: 9%

(bw) name: energy to peak load; test method: flex to failure test; test note: extension of ASTM D790-86, customized test

(bx) name: energy to peak load; test method: flex to failure test; test note: extension of ASTM D790-86, customized test; value note: extrapolated

(by) name: failure mode; test method: flex to failure test; test note: extension of ASTM D790-86, customized test

(bz) name: flexural modulus

(ca) name: flexural modulus; offset: 1%; test method: ASTM D790-86; specimen type: injection molded

(cb) name: flexural modulus; test method: ASTM D790

(cc) name: flexural modulus; test method: ASTM D790; specimen type: ASTM type 1 tensile bar; specimen size: 165 mm x 12.7 mm x 3.2 mm; standard deviation: 10%

(cd) name: flexural modulus; test method: ASTM D790; specimen type: ASTM type 1 tensile bar; specimen size: 165 mm x 12.7 mm x 3.2 mm; standard deviation: 11%

(ce) name: flexural modulus; test method: ASTM D790; specimen type: ASTM type 1 tensile bar; specimen size: 165 mm x 12.7 mm x 3.2 mm; standard deviation: 2%

(cf) name: flexural modulus; test method: ASTM D790; specimen type: ASTM type 1 tensile bar; specimen size: 165 mm x 12.7 mm x 3.2 mm; standard deviation: 3%

(cg) name: flexural modulus; test method: ASTM D790; specimen type: ASTM type 1 tensile bar; specimen size: 165 mm x 12.7 mm x 3.2 mm; standard deviation: 4%

(ch) name: flexural modulus; test method: ASTM D790; specimen type: ASTM type 1 tensile bar; specimen size: 165 mm x 12.7 mm x 3.2 mm; standard deviation: 5%

(ci) name: flexural modulus; test method: ASTM D790; strain rate: 1.27 mm/min

(cj) name: flexural modulus; test method: ASTM D790; temperature: 23°C; RH: 50%

(ck) name: flexural strength

(cl) name: flexural strength; test method: ASTM D790

(cm) name: flexural strength; test method: ASTM D790; specimen type: ASTM type 1 tensile bar; specimen size: 165 mm x 12.7 mm x3.2 mm; standard deviation: 1%

(cn) name: flexural strength; test method: ASTM D790; specimen type: ASTM type 1 tensile bar; specimen size: 165 mm x 12.7 mm x3.2 mm; standard deviation: 2%

(co) name: flexural strength; test method: ASTM D790; specimen type: ASTM type 1 tensile bar; specimen size: 165 mm x 12.7 mm x3.2 mm; standard deviation: 3%

(cp) name: flexural strength; test method: ASTM D790; temperature: 23°C; RH: 50%

(cq) name: Gardner impact strength; temperature: 23°C

(cr) name: Gardner impact strength; test method: ASTM D3029-84; specimen type: injection molded; temperature: 23°C

(cs) name: glass transition temperature; specimen mass: 10.5 mg; test apparatus: Perkin-Elmer DSC-7 calorimeter

(ct) name: instrumented dart drop impact strength

(cu) name: instrumented dart impact (peak energy); specimen thickness: 3.2 mm; test method: ASTM D3763; standard deviation: 23%

(cv) name: instrumented dart impact (peak energy); specimen thickness: 3.2 mm; test method: ASTM D3763; standard deviation: 25%

(cw) name: instrumented dart impact (peak energy); specimen thickness: 3.2 mm; test method: ASTM D3763; standard deviation: 29%

(cx) name: instrumented dart impact (peak energy); specimen thickness: 3.2 mm; test method: ASTM D3763; standard deviation: 40%

(cy) name: instrumented dart impact (peak energy); specimen thickness: 3.2 mm; test method: ASTM D3763; standard deviation: 44%

(cz) name: instrumented dart impact (peak energy); specimen thickness: 3.2 mm; test method: ASTM D3763; standard deviation: 5%

(da) name: instrumented dart impact (peak energy); specimen thickness: 3.2 mm; test method: ASTM D3763; standard deviation: 50%

(db) name: instrumented dart impact (peak energy); specimen thickness: 3.2 mm; test method: ASTM D3763; standard deviation: 55%

(dc) name: instrumented dart impact (peak energy); specimen thickness: 3.2 mm; test method: ASTM D3763; standard deviation: 56%

(dd) name: instrumented dart impact (peak energy); specimen thickness: 3.2 mm; test method: ASTM D3763; standard deviation: 6%

(de) name: instrumented dart impact (peak energy); specimen thickness: 3.2 mm; test method: ASTM D3763; standard deviation: 60%

(df) name: instrumented dart impact (peak energy); specimen thickness: 3.2 mm; test method: ASTM D3763; standard deviation: 63%

(dg) name: instrumented dart impact (peak energy); test method: ASTM D3763; specimen thickness: 3.2 mm; standard deviation: 13%

(dh) name: instrumented dart impact (peak energy); test method: ASTM D3763; specimen thickness: 3.2 mm; standard deviation: 14%

(di) name: instrumented dart impact (peak energy); test method: ASTM D3763; specimen thickness: 3.2 mm; standard deviation: 15%

(dj) name: instrumented dart impact (peak energy); test method: ASTM D3763; specimen thickness: 3.2 mm; standard deviation: 19%

(dk) name: instrumented dart impact (peak energy); test method: ASTM D3763; specimen thickness: 3.2 mm; standard deviation: 2%

(dl) name: instrumented dart impact (peak energy); test method: ASTM D3763; specimen thickness: 3.2 mm; standard deviation: 26%

(dm) name: instrumented dart impact (peak energy); test method: ASTM D3763; specimen thickness: 3.2 mm; standard deviation: 29%

(dn) name: instrumented dart impact (peak energy); test method: ASTM D3763; specimen thickness: 3.2 mm; standard deviation: 3%

(do) name: instrumented dart impact (peak energy); test method: ASTM D3763; specimen thickness: 3.2 mm; standard deviation: 32%

(dp) name: instrumented dart impact (peak energy); test method: ASTM D3763; specimen thickness: 3.2 mm; standard deviation: 36%

(dq) name: instrumented dart impact (total energy)

(dr) name: instrumented dart impact (total energy); specimen thickness: 3.2 mm; test method: ASTM D3763; standard deviation: 15%

(ds) name: instrumented dart impact (total energy); specimen thickness: 3.2 mm; test method: ASTM D3763; standard deviation: 20%

(dt) name: instrumented dart impact (total energy); specimen thickness: 3.2 mm; test method: ASTM D3763; standard deviation: 27%

(du) name: instrumented dart impact (total energy); specimen thickness: 3.2 mm; test method: ASTM D3763; standard deviation: 28%

(dv) name: instrumented dart impact (total energy); specimen thickness: 3.2 mm; test method: ASTM D3763; standard deviation: 3%

(dw) name: instrumented dart impact (total energy); specimen thickness: 3.2 mm; test method: ASTM D3763; standard deviation: 31%

(dx) name: instrumented dart impact (total energy); specimen thickness: 3.2 mm; test method: ASTM D3763; standard deviation: 39%

(dy) name: instrumented dart impact (total energy); specimen thickness: 3.2 mm; test method: ASTM D3763; standard deviation: 40%

(dz) name: instrumented dart impact (total energy); specimen thickness: 3.2 mm; test method: ASTM D3763; standard deviation: 46%

(ea) name: instrumented dart impact (total energy); specimen thickness: 3.2 mm; test method: ASTM D3763; standard deviation: 5%

(eb) name: instrumented dart impact (total energy); specimen thickness: 3.2 mm; test method: ASTM D3763; standard deviation: 50%

(ec) name: instrumented dart impact (total energy); specimen thickness: 3.2 mm; test method: ASTM D3763; standard deviation: 57%

(ed) name: instrumented dart impact (total energy); specimen thickness: 3.2 mm; test method: ASTM D3763; standard deviation: 66%

(ee) name: instrumented dart impact (total energy); specimen thickness: 3.2 mm; test method: ASTM D3763; standard deviation: 70%

(ef) name: instrumented dart impact (total energy); specimen thickness: 3.2 mm; test method: ASTM D3763; standard deviation: 79%

(eg) name: instrumented dart impact (total energy); specimen thickness: 3.2 mm; test method: ASTM D3763; standard deviation: 84%

(eh) name: instrumented dart impact (total energy); test method: ASTM D3763; specimen thickness: 3.2 mm; standard deviation: 11%

(ei) name: instrumented dart impact (total energy); test method: ASTM D3763; specimen thickness: 3.2 mm; standard deviation: 12%

(ej) name: instrumented dart impact (total energy); test method: ASTM D3763; specimen thickness: 3.2 mm; standard deviation: 14%

(ek) name: instrumented dart impact (total energy); test method: ASTM D3763; specimen thickness: 3.2 mm; standard deviation: 2%

(el) name: instrumented dart impact (total energy); test method: ASTM D3763; specimen thickness: 3.2 mm; standard deviation: 22%

(em) name: instrumented dart impact (total energy); test method: ASTM D3763; specimen thickness: 3.2 mm; standard deviation: 26%

(en) name: instrumented dart impact (total energy); test method: ASTM D3763; specimen thickness: 3.2 mm; standard deviation: 30%

(eo) name: instrumented dart impact (total energy); test method: ASTM D3763; specimen thickness: 3.2 mm; standard deviation: 40%

(ep) name: instrumented dart impact (total energy); test method: ASTM D3763; specimen thickness: 3.2 mm; standard deviation: 5%

(eq) name: instrumented dart impact energy (peak); specimen thickness: 3.2 mm; test method: ASTM D3763; standard deviation: 12%

(er) name: instrumented dart impact energy (peak); specimen thickness: 3.2 mm; test method: ASTM D3763; standard deviation: 26%

(es) name: instrumented dart impact energy (peak); specimen thickness: 3.2 mm; test method: ASTM D3763; standard deviation: 27%

(et) name: instrumented dart impact energy (peak); specimen thickness: 3.2 mm; test method: ASTM D3763; standard deviation: 39%

(eu) name: instrumented dart impact energy (peak); specimen thickness: 3.2 mm; test method: ASTM D3763; standard deviation: 44%

(ev) name: instrumented dart impact energy (peak); specimen thickness: 3.2 mm; test method: ASTM D3763; standard deviation: 48%

(ew) name: instrumented dart impact energy (peak); specimen thickness: 3.2 mm; test method: ASTM D3763; standard deviation: 55%

(ex) name: instrumented dart impact energy (peak); specimen thickness: 3.2 mm; test method: ASTM D3763; standard deviation: 75%

(ey) name: instrumented dart impact energy (peak); specimen thickness: 3.2 mm; test method: ASTM D3763; standard deviation: 78%

(ez) name: instrumented dart impact; specimen thickness: 0.254 mm; temperature: 23°C; dart size: 12.7 mm; height: 660 mm

(fa) name: instrumented dart impact; test method: ASTM D3763-86; specimen thickness: 3.2 mm; standard deviation: 18%

(fb) name: instrumented dart impact; test method: ASTM D3763-86; specimen thickness: 3.2 mm; standard deviation: 22%

(fc) name: instrumented dart impact; test method: ASTM D3763-86; specimen thickness: 3.2 mm; standard deviation: 25%

(fd) name: instrumented dart impact; test method: ASTM D3763-86; specimen thickness: 3.2 mm; standard deviation: 29%

(fe) name: instrumented dart impact; test method: ASTM D3763-86; specimen thickness: 3.2 mm; standard deviation: 40%

(ff) name: instrumented dart impact; test method: ASTM D3763-86; specimen thickness: 3.2 mm; standard deviation: 5%

(fg) name: instrumented dart impact; test method: ASTM D3763-86; specimen thickness: 3.2 mm; standard deviation: 52%

(fh) name: instrumented dart impact; test method: ASTM D3763-86; specimen thickness: 3.2 mm; standard deviation: 6%

(fi) name: instrumented dart impact; test method: ASTM D3763-86; specimen thickness: 3.2 mm; standard deviation: 70%

(fj) name: melt flow rate; test method: ASTM D1237; temperature: 23°C; load 2.16 kg

(fk) name: melt flow rate; test method: ASTM D1238

(fl) name: modulus (machine direction); test method: ASTM D882; specimen thickness: 0.127 mm

(fm) name: notched Izod impact strength

(fn) name: notched Izod impact strength; specimen thickness: 3.2 mm

(fo) name: notched Izod impact strength; test method: ASTM D256

(fp) name: notched Izod impact strength; test method: ASTM D256-84; specimen thickness: 3.2 mm

(fq) name: notched Izod impact strength; test method: ASTM D256-84; specimen thickness: 3.2mm

(fr) name: notched Izod impact strength; test method: ASTM D256-84; specimen thickness: 3.2mm; standard deviation: 11%

(fs) name: notched Izod impact strength; test method: ASTM D256-84; specimen thickness: 3.2mm; standard deviation: 23%

(ft) name: notched Izod impact strength; test method: ASTM D256-84; specimen thickness: 3.2mm; standard deviation: 3%

(fu) name: notched Izod impact strength; test method: ASTM D256-84; specimen thickness: 3.2mm; standard deviation: 35%

(fv) name: notched Izod impact strength; test method: ASTM D256-84; specimen thickness: 3.2mm; standard deviation: 5%

(fw) name: notched Izod impact strength; test method: ASTM D256-84; specimen thickness: 3.2mm; standard deviation: 6%

(fx) name: notched Izod impact strength; test method: ASTM D256-84; specimen thickness: 3.2mm; standard deviation: 8%

(fy) name: notched Izod impact strength; test method: ASTM D256-88A; specimen type: injection molded; temperature: 23°C

(fz) name: notched Izod impact strength; test method: ASTM D256; specimen thickness: 3.2 mm; standard deviation: 1%

(ga) name: notched Izod impact strength; test method: ASTM D256; specimen thickness: 3.2 mm; standard deviation: 11%

(gb) name: notched Izod impact strength; test method: ASTM D256; specimen thickness: 3.2 mm; standard deviation: 2%

(gc) name: notched Izod impact strength; test method: ASTM D256; specimen thickness: 3.2 mm; standard deviation: 28%

(gd) name: notched Izod impact strength; test method: ASTM D256; specimen thickness: 3.2 mm; standard deviation: 3%

(ge) name: notched Izod impact strength; test method: ASTM D256; specimen thickness: 3.2 mm; standard deviation: 34%

(gf) name: notched Izod impact strength; test method: ASTM D256; specimen thickness: 3.2 mm; standard deviation: 5%

(gg) name: notched Izod impact strength; test method: ASTM D256; specimen thickness: 3.2 mm; standard deviation: 6%

(gh) name: notched Izod impact strength; test method: ASTM D256; specimen thickness: 3.2 mm; standard deviation: 8%

(gi) name: notched Izod impact strength; test method: ASTM D256; temperature: 23°C

(gj) name: peak load at failure; test method: flex to failure test; test note: extension of ASTM D790-86, customized test

(gk) name: peak load at failure; test method: flex to failure test; test note: extension of ASTM D790-86, customized test; value note: extrapolated

(gl) name: Rockwell hardness

(gm) name: Rockwell hardness; test method: ASTM D785

(gn) name: Shore hardness

(go) name: Shore hardness; test method: ASTM D2240

(gp) name: tear strength (machine direction); test method: ASTM D1004; specimen thickness: 0.127 mm

(gq) name: tensile impact strength; test method: ASTM D1822

(gr) name: tensile impact; test method: ASTM -D1822

(gs) name: tensile modulus

(gt) name: tensile modulus; specimen thickness: 0.254 mm

(gu) name: tensile modulus; test method: ASTM D638-84; specimen type: ASTM Type I tensile bar; specimen size: 165 mm x 3.2 mm x 12.7 mm; standard deviation: 1%

(gv) name: tensile modulus; test method: ASTM D638-84; specimen type: ASTM Type I tensile bar; specimen size: 165 mm x 3.2 mm x 12.7 mm; standard deviation: 2%

(gw) name: tensile modulus; test method: ASTM D638-84; specimen type: ASTM Type I tensile bar; specimen size: 165 mm x 3.2 mm x 12.7 mm; standard deviation: 5%

(gx) name: tensile modulus; test method: ASTM D638; specimen type: ASTM type 1 tensile bar; specimen size: 165 mm x 12.7 mm x 3.2 mm; standard deviation: 10%

(gy) name: tensile modulus; test method: ASTM D638; specimen type: ASTM type 1 tensile bar; specimen size: 165 mm x 12.7 mm x 3.2 mm; standard deviation: 12%

(gz) name: tensile modulus; test method: ASTM D638; specimen type: ASTM type 1 tensile bar; specimen size: 165 mm x 12.7 mm x 3.2 mm; standard deviation: 4%

(ha) name: tensile modulus; test method: ASTM D638; specimen type: ASTM type 1 tensile bar; specimen size: 165 mm x 12.7 mm x 3.2 mm; standard deviation: 5%

(hb) name: tensile modulus; test method: ASTM D638; specimen type: ASTM type 1 tensile bar; specimen size: 165 mm x 12.7 mm x 3.2 mm; standard deviation: 6%

(hc) name: tensile modulus; test method: ASTM D638; specimen type: ASTM type 1 tensile bar; specimen size: 165 mm x 12.7 mm x 3.2 mm; standard deviation: 8%

(hd) name: tensile modulus; test method: ASTM D882

(he) name: tensile strength

(hf) name: tensile strength (machine direction)

(hg) name: tensile strength (machine direction); specimen size: 0.102 mm x 6.4 mm x 101.6 mm; crosshead speed: 50.8 mm/min.

(hh) name: tensile strength (machine direction); specimen size: 0.127 mm x 6.4 mm x 101.6 mm; crosshead speed: 50.8 mm/min.

(hi) name: tensile strength (transverse direction)

(hj) name: tensile strength (transverse direction); specimen size: 0.102 mm x 6.4 mm x 101.6 mm; crosshead speed: 50.8 mm/min.

(hk) name: tensile strength (transverse direction); specimen size: 0.127 mm x 6.4 mm x 101.6 mm; crosshead speed: 50.8 mm/min.

(hl) name: tensile strength (transverse direction); specimen size: 0.02 mm

(hm) name: tensile strength @ break

(hn) name: tensile strength @ break (machine direction); specimen thickness: 0.015 mm

(ho) name: tensile strength @ break (machine direction); specimen thickness: 0.02 mm

(hp) name: tensile strength @ break (machine direction); test method: ASTM D882; specimen thickness: 0.127 mm

(hq) name: tensile strength @ break (transverse direction); specimen thickness: 0.015 mm

(hr) name: tensile strength @ break (transverse direction); specimen thickness: 0.02 mm

(hs) name: tensile strength @ break; test method: ASTM D638

(ht) name: tensile strength @ break; test method: ASTM D638-84; specimen type: ASTM Type I tensile bar; specimen size: 165 mm x 3.2 mm x 12.7 mm; standard deviation: 1%

(hu) name: tensile strength @ break; test method: ASTM D638-84; specimen type: ASTM Type I tensile bar; specimen size: 165 mm x 3.2 mm x 12.7 mm; standard deviation: 3%

(hv) name: tensile strength @ break; test method: ASTM D638-84; specimen type: ASTM Type I tensile bar; specimen size: 165 mm x 3.2 mm x 12.7 mm; standard deviation: 4%

(hw) name: tensile strength @ break; test method: ASTM D638-84; specimen type: ASTM Type I tensile bar; specimen size: 165 mm x 3.2 mm x 12.7 mm; standard deviation: 5%

(hx) name: tensile strength @ break; test method: ASTM D638-84; specimen type: ASTM Type I tensile bar; specimen size: 165 mm x 3.2 mm x 12.7 mm; standard deviation: 6%

(hy) name: tensile strength @ break; test method: ASTM D638-84; specimen type: ASTM Type I tensile bar; specimen size: 165 mm x 3.2 mm x 12.7 mm; standard deviation: 8%

(hz) name: tensile strength @ break; test method: ASTM D638-84; specimen type: ASTM Type I tensile bar; specimen size: 165 mm x 3.2 mm x 12.7 mm; standard deviation: 9%

(ia) name: tensile strength @ break; test method: ASTM D638; temperature: 23°C; RH: 50%

(ib) name: tensile strength @ yield

(ic) name: tensile strength @ yield (machine direction); specimen thickness: 0.015 mm

(id) name: tensile strength @ yield (machine direction); specimen thickness: 0.02 mm

(ie) name: tensile strength @ yield (machine direction); test method: ASTM D882; specimen thickness: 0.127 mm (if) name: tensile strength @ yield (transverse direction); specimen thickness: 0.015 mm

(ig) name: tensile strength @ yield (transverse direction); specimen thickness: 0.02 mm

(ih) name: tensile strength @ yield; test method: ASTM D638

(ii) name: tensile strength @ yield; test method: ASTM D638-84; specimen type: ASTM Type I tensile bar; specimen size: 165 mm x 3.2 mm x 12.7 mm

(ij) name: tensile strength @ yield; test method: ASTM D638-84; specimen type: ASTM Type I tensile bar; specimen size: 165 mm x 3.2 mm x 12.7 mm; standard deviation: 0.5%

(ik) name: tensile strength @ yield; test method: ASTM D638-84; specimen type: ASTM Type I tensile bar; specimen size: 165 mm x 3.2 mm x 12.7 mm; standard deviation: 0.9%

(il) name: tensile strength @ yield; test method: ASTM D638-84; specimen type: ASTM Type I tensile bar; specimen size: 165 mm x 3.2 mm x 12.7 mm; standard deviation: 1%

(im) name: tensile strength @ yield; test method: ASTM D638-84; specimen type: ASTM Type I tensile bar; specimen size: 165 mm x 3.2 mm x 12.7 mm; standard deviation: 11%

(in) name: tensile strength @ yield; test method: ASTM D638-84; specimen type: ASTM Type I tensile bar; specimen size: 165 mm x 3.2 mm x 12.7 mm; standard deviation: 2%

(io) name: tensile strength @ yield; test method: ASTM D638-84; specimen type: ASTM Type I tensile bar; specimen size: 165 mm x 3.2 mm x 12.7 mm; standard deviation: 5%

(ip) name: tensile strength @ yield; test method: ASTM D638-84; specimen type: ASTM Type I tensile bar; specimen size: 165 mm x 3.2 mm x 12.7 mm; standard deviation: 8%

(iq) name: tensile strength @ yield; test method: ASTM D638; specimen type: ASTM type 1 tensile bar; specimen size: 165 mm x 12.7 mm x 3.2 mm; standard deviation: 1%

(ir) name: tensile strength @ yield; test method: ASTM D638; specimen type: ASTM type 1 tensile bar; specimen size: 165 mm x 12.7 mm x 3.2 mm; standard deviation: 2%

(is) name: tensile strength @ yield; test method: ASTM D638; specimen type: ASTM type 1 tensile bar; specimen size: 165 mm x 12.7 mm x 3.2 mm; standard deviation: 3%

(it) name: tensile strength @ yield; test method: ASTM D638; strain rate: 50 mm/min

(iu) name: tensile strength @ yield; test method: ASTM D638; temperature: 23°C; RH: 50%

(iv) name: tensile strength; specimen thickness: 0.254 mm

(iw) name: tensile strength; temperature: 23°C

(ix) name: tensile strength; temperature: 28°C

(iy) name: tensile strength; temperature: 93°C

(iz) name: tensile strength; test method: ASTM D1708; specimen type: film; specimen thickness: 0.25 mm

(ja) name: tensile strength; test method: ASTM D412, Die C

(jb) name: tensile strength; test method: ASTM D638

(jc) name: tensile strength; test method: ASTM D638-82; specimen type: injection molded slab; specimen thickness: 2 mm

(jd) name: ultimate elongation

(je) name: ultimate elongation; specimen type: extruded strip; gauge length: 9.5 mm; specimen thickness: 0.51 - 0.76 mm; number of specimens: 2; standard deviation: 11

(jf) name: ultimate elongation; specimen type: extruded strip; gauge length: 9.5 mm; specimen thickness: 0.51 - 0.76 mm; number of specimens: 3; standard deviation: 1

(jg) name: ultimate elongation; specimen type: extruded strip; gauge length: 9.5 mm; specimen thickness: 0.51 - 0.76 mm; number of specimens: 3; standard deviation: 11

(jh) name: ultimate elongation; specimen type: extruded strip; gauge length: 9.5 mm; specimen thickness: 0.51 - 0.76 mm; number of specimens: 3; standard deviation: 16

(ji) name: ultimate elongation; specimen type: extruded strip; gauge length: 9.5 mm; specimen thickness: 0.51 - 0.76 mm; number of specimens: 3; standard deviation: 20

(jj) name: ultimate elongation; specimen type: extruded strip; gauge length: 9.5 mm; specimen thickness: 0.51 - 0.76 mm; number of specimens: 3; standard deviation: 32

(jk) name: ultimate elongation; specimen type: extruded strip; gauge length: 9.5 mm; specimen thickness: 0.51 - 0.76 mm; number of specimens: 3; standard deviation: 42

(jl) name: ultimate elongation; specimen type: extruded strip; gauge length: 9.5 mm; specimen thickness: 0.51 - 0.76 mm; number of specimens: 3; standard deviation: 54

(jm) name: ultimate elongation; specimen type: extruded strip; gauge length: 9.5 mm; specimen thickness: 0.51 - 0.76 mm; number of specimens: 3; standard deviation: 56

(jn) name: ultimate elongation; specimen type: extruded strip; gauge length: 9.5 mm; specimen thickness: 0.51 - 0.76 mm; number of specimens: 3; standard deviation: 58

(jo) name: ultimate elongation; specimen type: extruded strip; gauge length: 9.5 mm; specimen thickness: 0.51 - 0.76 mm; number of specimens: 3; standard deviation: 60

(jp) name: ultimate elongation; specimen type: extruded strip; gauge length: 9.5 mm; specimen thickness: 0.51 - 0.76 mm; number of specimens: 4; standard deviation: 16

(jq) name: ultimate elongation; specimen type: extruded strip; gauge length: 9.5 mm; specimen thickness: 0.51 - 0.76 mm; number of specimens: 4; standard deviation: 37

(jr) name: ultimate elongation; specimen type: extruded strip; gauge length: 9.5 mm; specimen thickness: 0.51 - 0.76 mm; number of specimens: 4; standard deviation: 40

(js) name: ultimate elongation; specimen type: extruded strip; gauge length: 9.5 mm; specimen thickness: 0.51 - 0.76 mm; number of specimens: 4; standard deviation: 64

(jt) name: ultimate elongation; test direction: perpendicular to flow; specimen type: injection molded; gauge length: 9.5 mm; specimen thickness: 3.2 mm; number of specimens: 4; standard deviation: 2

(ju) name: ultimate elongation; test direction: perpendicular to flow; specimen type: injection molded; gauge length: 9.5 mm; specimen thickness: 3.2 mm; number of specimens: 4; standard deviation: 38

(jv) name: ultimate elongation; test direction: perpendicular to flow; specimen type: injection molded; gauge length: 9.5 mm; specimen thickness: 3.2 mm; number of specimens: 4; standard deviation: 68

(jw) name: ultimate tensile strength

(jx) name: ultimate tensile strength; specimen type: extruded strip; gauge length: 9.5 mm; specimen thickness: 0.51-0.76 mm; number of specimens: 3; standard deviation: 11

(jy) name: ultimate tensile strength; specimen type: extruded strip; gauge length: 9.5 mm; specimen thickness: 0.51-0.76 mm; number of specimens: 3; standard deviation: 116

(jz) name: ultimate tensile strength; specimen type: extruded strip; gauge length: 9.5 mm; specimen thickness: 0.51-0.76 mm; number of specimens: 3; standard deviation: 18

(ka) name: ultimate tensile strength; specimen type: extruded strip; gauge length: 9.5 mm; specimen thickness: 0.51-0.76 mm; number of specimens: 3; standard deviation: 242

(kb) name: ultimate tensile strength; specimen type: extruded strip; gauge length: 9.5 mm; specimen thickness: 0.51-0.76 mm; number of specimens: 3; standard deviation: 30

(kc) name: ultimate tensile strength; specimen type: extruded strip; gauge length: 9.5 mm; specimen thickness: 0.51-0.76 mm; number of specimens: 3; standard deviation: 46

(kd) name: ultimate tensile strength; specimen type: extruded strip; gauge length: 9.5 mm; specimen thickness: 0.51-0.76 mm; number of specimens: 3; standard deviation: 54

(ke) name: ultimate tensile strength; specimen type: extruded strip; gauge length: 9.5 mm; specimen thickness: 0.51-0.76 mm; number of specimens: 3; standard deviation: 65

(kf) name: ultimate tensile strength; specimen type: extruded strip; gauge length: 9.5 mm; specimen thickness: 0.51-0.76 mm; number of specimens: 3; standard deviation: 70

(kg) name: ultimate tensile strength; specimen type: extruded strip; gauge length: 9.5 mm; specimen thickness: 0.51-0.76 mm; number of specimens: 3; standard deviation: 73

(kh) name: ultimate tensile strength; specimen type: extruded strip; gauge length: 9.5 mm; specimen thickness: 0.51-0.76 mm; number of specimens: 3; standard deviation: 99

(ki) name: ultimate tensile strength; specimen type: extruded strip; gauge length: 9.5 mm; specimen thickness: 0.51-0.76 mm; number of specimens: 4; standard deviation: 20

(kj) name: ultimate tensile strength; specimen type: extruded strip; gauge length: 9.5 mm; specimen thickness: 0.51-0.76 mm; number of specimens: 4; standard deviation: 40

(kk) name: ultimate tensile strength; specimen type: extruded strip; gauge length: 9.5 mm; specimen thickness: 0.51-0.76 mm; number of specimens: 4; standard deviation: 56

(kl) name: ultimate tensile strength; specimen type: extruded strip; gauge length: 9.5 mm; specimen thickness: 0.51-0.76 mm; number of specimens: 4; standard deviation: 65

(km) name: ultimate tensile strength; test direction: perpendicular to flow; specimen type: injection molded; gauge length: 9.5 mm; specimen thickness: 3.2 mm; number of specimens: 4; standard deviation: 31

(kn) name: ultimate tensile strength; test direction: perpendicular to flow; specimen type: injection molded; gauge length: 9.5 mm; specimen thickness: 3.2 mm; number of specimens: 4; standard deviation: 58

(ko) name: ultimate tensile strength; test direction: perpendicular to flow; specimen type: injection molded; gauge length: 9.5 mm; specimen thickness: 3.2 mm; number of specimens: 4; standard deviation: 99

(kp) name: ultimate tensile strength; test method: ASTM D882

(kq) name: unnotched Izod impact strength

(kr) name: unnotched Izod impact strength; test method: ASTM D256; temperature: 23°C

(ks) name: unnotched Izod impact strength; test method: ASTM D4812; temperature: 23°C

(kt) name: Vicat softening point

(ku) name: Vicat softening point; test method: ASTM D1525

(ks) name: Vicat softening temperature

(kt) name: yellowness index

(ku) name: yellowness index; specimen thickness: 3.2 mm

(kv) name: yellowness index; test method: ASTM D1925

(kw) name: yellowness index; test method: ASTM D1925; specimen thickness: 3.2 mm

(kx) name: yellowness index; test method: ASTM D1925; test apperatus: Gardner colorimeter; specimen size: 101.6 mm x 101.6 mm x 3.2 mm

(1) Hermanson, Nancy J., Steffens, John F., *The Physical and Visual Property Changes in Thermoplastic Resins After Exposure to High Energy Sterilization – Gamma Versus Electron Beam,* ANTEC 1993, conference proceedings – Society of Plastics Engineers, 1993.

(2) *Kynar Polyvinylidene Fluoride,* supplier technical report (PL705–Rev4–1–91) – Atochem North America, Inc., 1991.

(3) Sturdevant, Marianne F., *Sterilization Compatibility of Rigid Thermoplastic Materials,* supplier technical report (301–1548) – Dow Chemical Company, 1988.

(4) Goulder, G. K., Seymour, R. W., *Polyesters And Copolyesters For Use In Medical Applications,* supplier technical report – Eastman Performace Plastics.

(5) Sturdevant, Marianne F., *The Long-term Effects of Ethylene Oxide and Gamma Radiation Sterilization on the Properties of Rigid Thermoplastic Materials,* ANTEC 1990, conference proceedings – Society of Plastics Engineers, 1990.

(6) Hermanson, Nancy J., *Effects Of Alternate Carriers Of Ethylene Oxide Sterilant On Thermoplastics,* supplier technical report (301–02018) – Dow Chemical Company.

(7) Seller, David A., Robinson, Donald N., *Potential Applications For Kynar Flex PVDF In The Nuclear Industry,* supplier technical report – Elf Atochem North America.

(8) *Pellethane Polyurethane Elastomers – 2363 Series For Medical Applications,* supplier marketing literature (306–180–792XSMG) – Dow Chemical Company, 1992.

(9) *Medical Grades Of Santoprene Thermoplastic Rubber,* supplier technical report (TCD00793) – Advanced Elastomer Systems, 1993.

(10) *Santoprene Thermoplastic Rubber For Food And Beverage Contact,* supplier technical report (TCD00693) – Advanced Elastomer Systems, 1993.

(11) *Attane Ultra Low Density Ethylene–Octene Copolymers: Performance Plus Compared To LLDPE And EVA Resins In Flexible Packaging,* supplier marketing literature (305–1596–790) – Dow Chemical Company, 1989.

(12) *Trefsin 3281 Food/Medical Grade Thermoplastic Elastomer,* supplier technical report – Advanced Elastomer Systems, 1991.

(13) *Applied Innovations In Silicone Tubing,* supplier marketing literature (51–769C–91) – Dow Corning Medical.

(14) *Silicone Tubing Meets Biocompatibility and Sterilization Demands in Devices,* Biomaterials News, supplier newsletter (Form # 51–849; Volume 5, Issue 4) – Dow Corning Medical, 1991.

(15) *Udel Polysulfone Design Engineering Handbook,* supplier design guide (F–47178) – Amoco Performance Products, Inc., 1988.

(16) *The Radiation Response of Udel Polysulfone,* supplier technical report (Number: 101) – Amoco Performance Products, Inc.

(17) *Engineering Plastics For The Medical And Health Care Fields,* supplier marketing literature (G–F–50022) – Amoco Performance Products, Inc.

(18) *Radel Polyarylsulfone...A Family Of Thermoplastic Materials Engineered For High Performance In Harsh Environments,* supplier marketing literature (F–49896) – Amoco Performance Products, Inc.

(19) Atakent, A. I., *Gamma Radiation Study,* supplier technical report – Cyro Industries, 1990.

(20) *Torlon Engineering Polymers / Design Manual,* supplier design guide (F–49893) – Amoco Performance Products.

(21) *Plastics Technology Center Report,* competitor's technical report (Form 17479) – Atohaas, 1992.

(22) *Plexiglas Acrylic Resin From AtoHaas Clearly The Best.,* supplier marketing literature (PL–1700b) – Atohaas, 1993.

(23) Portnoy, R. C., *The Response Of Various Polyethylenes To High Energy Radiation,* ANTEC 1994, conference proceedings – Society of Plastics Engineers, 1994.

(24) Hu, C. B., Ma, M. T., Mcintyre, J., Nguyen, D., Myers, K.E., *The Effect Of Steam And Gamma Sterilization On The Thermal And Mechanical Properties Of Polyester Film,* ANTEC 1994, conference proceedings – Society of Plastics Engineers, 1994.

(25) *Lupolen, Lucalen Product Line, Properties, Processing,* supplier design guide (B 581 e/(8127) 10.91) – BASF Aktiengesellschaft, 1991.

(26) *Polystyrol Product Line, Properties, Processing,* supplier design guide (B 564 e/2.93) – BASF Aktiengesellschaft, 1993.

(27) *Ultrapek Product Line, Properties, Processing,* supplier design guide (B 607 e/10.92) – BASF Aktiengesellschaft, 1992.

(28) *Ultrason E, Ultrason S Product Line, Properties, Processing,* supplier design guide (B 602 e/10.92) – BASF Aktiengesellschaft, 1992.

(29) *Styrolux Product Line, Properties, Processing,* supplier design guide (B 583 e/(950) 12.91) – BASF Aktiengesellschaft, 1992.

(30) *Luran Product Line, Properties, Processing,* supplier design guide (B 565 e/10.83) – BASF Aktiengesellschaft, 1983.

(31) Johnson, James A., supplier written correspondence – BASF Corporation, 1994.

(32) Brookman, R. S., *A New PVC Based Polymer For Medical Applications,* ANTEC 1991, conference proceedings – Society of Plastics Engineers, 1991.

(33) Portnoy, R. C., Cross, V. R., *Method For Evaluating The Gamma Radiation Tolerance Of Polypropylene For Medical Device Applications,* ANTEC 1991, conference proceedings – Society of Plastics Engineers, 1991.

(34) Rakus, J., Brantley, T., Szabo, E., *Environmental Concerns With Vinyl Medical Devices,* ANTEC 1991, conference proceedings – Society of Plastics Engineers, 1991.

(35) *CyRex 200 Acrylic–Polycarbonate Alloy,* supplier marketing literature (1767–194–5BP) – Cyro Industries, 1994.

(36) *Vespel Parts And Radiation,* supplier technical report (E–73910) – Du Pont Company, 1989.

(37) *Exploring Vespel Territory – The Properties of Du Pont Vespel Parts,* supplier design guide (E–26800) – Du Pont Company.

(38) *Hytrel Radiation Resistance,* supplier technical report (E–68116) – Du Pont Company, 1984.

(39) *Handbook Of Properties For Teflon PFA,* supplier design guide (E–96679) – Du Pont Company, 1987.

(40) *The Effects Of Cobalt 60 Radiation Upon The Physical Properties Of Selected C–Flex Elastomers,* Technical Information

Document, supplier technical report – Consolidated Polymer Technologies, Inc.

(41) *Autoclaving/Steam & Gas Sterilization,* Technical Information Document, supplier technical report – Consolidated Polymer Technologies, Inc.

(42) Johnson, D., Draaijer, L. M., Schipper, G. P., *Sterilization Of "Cariflex" TR–1000 Series And "Kraton" 2000 Compounds,* supplier technical report – Shell Development Co., 1983.

(43) *Gamma Irradiaiton Studies On Polysulfone And Polycarbonate,* supplier technical report (project no.: 214M01; file no.: 7845–R) – Amoco Performance Products, Inc., 1982.

(44) *Radiation Resistance Of Viton,* supplier technical report (E–37758; VT–515.1) – Du Pont Company.

(45) *Sterile Packaging Of Tyvek,* supplier marketing literature (H–19226) – Du Pont Company.

(46) Paulino, J. C., Baccaro, L. E., *Effects Of Sterilization Procedures On Physical Properties of Engineering Thermoplastics,* supplier technical report – General Electric Company.

(47) *Lexan 8040 Film For Medical Packaging – Comparison Data,* supplier marketing literature (SP–2013 (7/88) RTB) – General Electric Company, 1988.

(48) *Lexan GR Resin Gamma Resistant Product Information Book,* supplier marketing literature (NBM–110) – General Electric Company.

(49) *GE Plastics Medical Applications,* supplier marketing literature (NBM–100 (1/89) RTB) – General Electric Company, 1989.

(50) *Ultem Resin: Advanced Technology For Reusable Medical Devices,* supplier marketing literature (ULT–314A 1/91) RTB) – General Electric Company, 1991.

(51) *Ultem Design Guide,* supplier design guide (ULT–201G (6/90) RTB) – General Electric Company, 1990.

(52) Bonifant, Benjamin C., *Designing With Amorphous Thermoplastics,* Medical Device & Diagnostic Industry, trade journal – Cannon Communications, Inc., 1988.

(53) *Noryl Extrusion Resins,* supplier design guide (CDX–265) – General Electric Company.

(54) *Properties Of Tenite Polypropylene P5M4R–034 Before And After Gamma Sterilization,* supplier technical report – Eastman Performance Plastics, 1992.

(55) *Tenite Polypropylene And Tenite Polyallomer Materials For Radiation Sterilization,* supplier technical report (MB–93) – Eastman Plastics, 1985.

(56) *Effect Of Sterilization By E–Beam On Selected Properties Of Ektar DN003 Copolyester,* supplier technical report (PPE–DN3–EBS) – Eastman Performance Plastics, 1992.

(57) *Effect Of Sterilization By E–Beam On Selected Properties Of Ektar MB DA003 Polymer,* supplier technical report (PMB–DA3–EBS) – Eastman Performance Plastics, 1992.

(58) *Ektar Polymers For Healthcare Products,* supplier marketing literature (P/MD–11) – Eastman Performance Plastics, 1991.

(59) *Ektar Plastics Used In Health Care Products And Packaging,* supplier marketing literature (PG 40) – Eastman Performance Plastics, 1990.

(60) *Ecdel Elastomers,* supplier design guide (MB–100A) – Eastman Plastics, 1990.

(61) *Laboratory Report – Gamma Sterilization Of Grilamid TR55,* supplier technical report – EMS–American Grilon Inc., 1984.

(62) *Technical Information Evoprene G,* supplier technical report (RDS 028/9240) – Evode Plastics.

(63) *Kuraray Eval Resin,* supplier design guide (5–2,000–507) – KurarayCo., Ltd.

(64) *Effects Of Radiation On Eval Resins,* supplier design guide (Technical Bulletin No. 140) – Eval Company Of America.

(65) *BF Goodrich Geon RX Medical Grade Molding Compounds,* supplier marketing literature (CIM–014F) – BFGoodrich Company, 1988.

(66) *BF Goodrich Geon Rigid Vinyl Gamma Sterilizable Compounds,* supplier marketing literature (RX–020) – BFGoodrich Company, 1990.

(67) *Extremely Clear, Radiation Tolerant Polypropylene Sheet For Medical Device Packaging,* supplier technical report – Exxon Chemical Company.

(68) *Design Handbook For Du Pont Engineering Plastics – Module II,* supplier design guide (E–42267) – Du Pont Engineering Polymers.

(69) *Pro–fax PF–511 High–Flow, Radiation–Resistant Polypropylene,* supplier marketing literature – Himont, 1989.

(70) *Vectra Polymer Materials,* supplier design guide (B 121 BR E 9102/014) – Hoechst, 1991.

(71) *Physical Properties Of Unfilled And Filled Polytetrafluoroethylene,* supplier design guide (Technical Service Note F12/13) – ICI PLC, 1981.

(72) *Perspex Acrylic Polymers,* supplier marketing literature (MG1 1/93) – ICI Acrylics, 1993.

(73) *Victrex Polymers For Medical Applications,* supplier technical report – ICI Advanced Materials.

(74) *Effects Of Gamma Ray Sterilization On Acrycal MP Impact Modified G Series Resins,* supplier technical report – Continental Polymers Inc.

(75) *Victrex PEK Properties And Processing,* supplier design guide (VP2/ October 1987) – ICI Advanced Materials, 1987.

(76) *Victrex PES Data For Design Unreinforced Grades,* supplier design guide – ICI Advanced Materials, 1987.

(77) *Victrex PEEK,* supplier design guide (VK2/0586) – ICI Advanced Materials, 1986.

(78) *Calibre Engineering Thermoplastics Basic Design Manual,* supplier design guide (301–1040–1288) – Dow Chemical Company, 1988.

(79) *Mechanical Properties Of Victrex Peek After Exposure To Steam And Hot Condensate,* supplier technical report – Victrex USA.

(80) *Ultros PETG Copolyester For Film And Sheet For Medical Device Packaging,* supplier marketing literature – Lustro Plastics Co.

(81) *Makrolon Polycarbonate For Medical Device Applications,* supplier design guide (55–A715(10)J/ 112–9/87) – Miles Inc., 1987.

(82) *Medical Milestones,* supplier marketing literature (KU–F–2019(10F)/ 112/6/92) – Miles Inc., 1992.

(83) *Product Information Makrolon Rx–2350,* supplier marketing literature (201 (1/89)) – Miles Inc., 1989.

(84) *Texin Thermoplastic Polyurethane For Medical Applications,* supplier marketing literature (KU–E 1002(5)L/ 427–5M–5/92) – Miles Inc., 1992.

(85) *Texin Thermoplastic Urethane Technical Charts,* supplier technical report – Miles Inc., 1994.

(86) *Applications Of Styrene–Butadiene–Styrene (SBS) And Styrene–Ethylene–Butylene–Styrene (SEBS) In The Medical Industry,* supplier technical report – Shell Chemical Company.

(87) *Kraton Thermoplastic Rubber Medial Products,* supplier technical report (SC:1032–88) – Shell Chemical Company, 1988.

(88) *Effects Of Electron Beam And Gamma Radiation Sterilization Of Thermoplastics,* supplier technical report (7126A) – Monsanto Company, 1990.

(89) *Foraflon PVDF,* supplier design guide (694.E/07.87/20) – Atochem S. A., 1987.

(90) *Plexiglas Acrylic Molding Pellets For Medical Applications,* supplier marketing literature (PL–1519a) – Rohm and Haas, 1986.

(91) *Plexiglas Acrylic Resin Biomaterials,* supplier marketing literature (PL–1700) – Rohm and Haas, 1990.

(92) *Kinel Polyimide Compounds, Properties, Applications,* supplier design guide – Rhone Poulenc.

(93) *Ultramid Nylon Resins Product Line, Properties, Processing,* supplier design guide (B 568/1e/4.91) – BASF Corporation, 1991.

(94) *Hostalen Polymer Materials,* supplier design guide (HDKR 101 E 9050/022) – Hoechst AG.

(95) *Tecoflex Medical Grade Aliphatic Polyurethanes,* supplier technical report – Thermedics Inc.

(96) *Kel–F 81 PCTFE Engineering Manual,* supplier design guide (98–0211–5944–1 (120.5) DPI) – 3M Industrial Chemical Products Division, 1990.

(97) *Ube Ultra–High Heat–Resistant Polimide Film Upilex,* supplier marketing literature – Ube Industries, Ltd.

(98) *Radiation Resistant Polypropylene,* supplier technical report (EPP 7037 10/87 4M JP) – El Paso Products Company, 1987.

(99) *From Rexene Technology: Radiation Resistant Polypropylenes,* supplier technical report (EPP 7037 10/87 4M JP) – Rexene Products Company, 1987.

(100) *K–Resin A Clear Choice For Medical Packaging Devices,* supplier technical report (55–8801) – Phillips 66 Company, 1988.

(101) *Engineering Properties Of Marlex Resins,* supplier design guide (TSM–243) – Phillips 66 Company, 1983.

(102) *Ryton Polyphenylene Sulfide Resins Engineering Properties Guide,* supplier design guide (1065(a)–89 A 02) – Phillips 66 Company, 1989.

(103) *Performance Polyolefins For The Healthcare Market,* supplier marketing literature (6615/591) – Quantum Chemical Corporation, 1991.

(104) Lucas, Ben M., Paton, Samuel J., Thakker, M. T., *Radiation Resistant Polypropylene,* Medical Plastics '90 International Conference, Malmo, Sweden, conference proceedings – Rexene Products Company, 1990.

(105) Haines, Jim, Hauser, Debbie, *Zylar Clear Alloys For Medical Applications,* supplier technical report – Novacor Chemicals Inc.

(106) *The Effects Of Gamma And Ethylene Oxide Sterilization On Styrene Methyl Methacrylate Copolymers And A Toughened Terpolymer,* Medical '91, Ghent, Belgium, conference proceedings – Novacor Chemicals Inc., 1991.

(107) Stubstad, James A., Hemmerich, Karl J., *Preventing Plastic Part Failure After Radiation Sterilization*, ANTEC 1994, conference proceedings – Society of Plastics Engineers, 1994.

(108) Hermanson, Nancy J., Hosman, Scott L., *Hospital Sterilants And Their Effects On Thermoplastic Materials*, ANTEC 1994, conference proceedings - Society of Plastics Engineers, 1994.